D0078731

ACS SYMPOSIUM SERIES 367

Cross-Linked Polymers

Chemistry, Properties, and Applications

Ray A. Dickie, EDITOR
Ford Motor Company

S. S. Labana, EDITOR
Ford Motor Company

Ronald S. Bauer, EDITOR
Shell Development Company

Developed from a symposium sponsored
by the Division of Polymeric Materials:
Science and Engineering
at the 193rd Meeting
of the American Chemical Society,
Denver, Colorado
April 5-10, 1987

American Chemical Society, Washington, DC 1988

Library of Congress Cataloging-in-Publication Data

Cross-linked polymers: chemistry, properties, and applications
(ACS symposium series, ISSN 0097-6156; 367)

"Developed from a symposium sponsored by the Division of Polymeric Materials: Science and Engineering at the 193rd meeting of the American Chemical Society, Denver, Colorado, April 5–10, 1987."

Bibliography: p.

Includes index.

1. Polymers and polymerization—Congresses.

I. Dickie, R. A., 1940– . II. Labana, Santokh S., 1936– . III. Bauer, Ronald S., 1932– . IV. American Chemical Society. Division of Polymeric Materials: Science and Engineering. V. Series.

QD380.C76 1988 547.7 88-3327
ISBN 0-8412-1471-9

PRINTED IN THE UNITED STATES OF AMERICA

ACS Symposium Series

M. Joan Comstock, *Series Editor*

Foreword

The ACS SYMPOSIUM SERIES was founded in 1974 to provide a medium for publishing symposia quickly in book form. The format of the Series parallels that of the continuing ADVANCES IN CHEMISTRY SERIES except that, in order to save time, the papers are not typeset but are reproduced as they are submitted by the authors in camera-ready form. Papers are reviewed under the supervision of the Editors with the assistance of the Series Advisory Board and are selected to maintain the integrity of the symposia; however, verbatim reproductions of previously published papers are not accepted. Both reviews and reports of research are acceptable, because symposia may embrace both types of presentation.

Contents

DEFORMATION, FATIGUE, AND FRACTURE: MECHANICAL PROPERTIES OF CROSS-LINKED POLYMERS

HIGH-PERFORMANCE POLYMER NETWORKS

Preface

CROSS-LINKED POLYMERS ARE USED in a wide variety of aerospace, automotive, building construction, and consumer product applications. Not all paints, adhesives, composites, and elastomers are cross-linked, but cross-linking systems are often used in these applications when resistance to solvents, resistance to high temperatures, and high mechanical performance are required. These important properties can be traced directly to the three-dimensional interconnected molecular network that is characteristic of cross-linked systems.

The first section of this book deals with current topics in network theory directed toward explaining the relationship between molecular architecture and macroscopic physical properties. The closely related questions of network formation and degradation are also discussed in this section. Deformation, fatigue, and fracture are discussed in the second section. The third section includes recent advances in cross-linking chemistry; several chapters outline applications of new systems and detail the relationship between network structure and application properties.

The editors acknowledge the support of the Ford Motor Company and of Shell Development Company throughout the preparation of this volume. Financial aid and organizational support for the symposium from which this volume was developed have been provided by the Division of Polymeric Materials: Science and Engineering of the American Chemical Society. Finally, the editors express their sincere thanks to the authors and reviewers who have made this volume possible through their hard work and cooperation.

RAY A. DICKIE
S. S. LABANA
Ford Motor Company
Dearborn, MI 48121–2053

RONALD S. BAUER
Shell Development Company
Houston, TX 77251–1380

December 21, 1987

CROSS-LINKING, STRUCTURE, AND DEGRADATION: MOLECULAR ARCHITECTURE AND PROPERTIES

Chapter 1

Cross-Linking and Structure of Polymer Networks

Karel Dusek[1] and William J. MacKnight[2]

[1]Institute of Macromolecular Chemistry, Czechoslovak Academy of Sciences, Prague 6, 162 06, Czechoslovakia
[2]Polymer Science and Engineering Department, University of Massachusetts, Amherst, MA 01003

The article is an overview on network formation theories, their experimental verification and application to more complex systems of industrial importance. Network formation controlled by specific or overall diffusion is also discussed.

The development of an understanding of formation-structure-properties relations for polyfunctional systems undergoing crosslinking and branching is much more difficult than for bifunctional systems leading to linear polymers. The degree-of-polymerization distribution becoming very broad when the gel point is approached, the presence of an "infinite" molecule (the gel) beyond the gel point and the limited number of experimental methods for examining the gel structure are among the main reasons. Therefore, realistic network formation theories verified by experiments are of major importance in understanding and predicting these relations. Experiments on model low-functionality systems as well as polyfunctional systems serve as input information for the formulation of a correct and concrete theoretical approach, they also give values of initial parameters controlling the crosslinking reaction. A combination of organic and physical chemistry with branching theory appears to be necessary in order to understand the network formation and structure of real systems.

A knowledge of the factors controlling the crosslinking reaction is one of the necessary prerequisites for selection and application of the branching theory. The crosslinking reaction can be controlled by:

- Chemical equilibrium or chemical kinetics: The reactivity of functional groups is independent of the size and topology of the

0097-6156/88/0367-0002$07.25/0

macromolecules they are bound to. Their reactivity may depend, however, on the state of the other groups in the monomer units. (first shell substitution effect). The temperature dependence of the rate constant(s), k, obeys the Arrhenius law - log k ∝ 1/T.

- Specific diffusion control: The rate constants depend on the size and topology of the molecule the group is bound to; i.e., they depend on the translation diffusion coefficient of the species. Then, they depend also on the viscosity of the system. Specific diffusion control is characteristic of fast reactions like fluorescence quenching. In polymer formation, specific diffusion control is responsible for the acceleration of chain polymerization due to the retardation of the termination by recombination of two macroradicals (Trommsdorff effect). Step reactions are usually too slow to exhibit a dependence on translational diffusion; also, the temperature dependence of their rate constants is of the Arrhenius type.

- Overall diffusion control. The reaction rate is controlled by segmental mobility (segmental diffusion). Such rate control is characteristic for reactions occuring in the glass transition region. It is typical for many thermosets because the glass transition temperature during curing may increase by 10^2 K.

The network formation theories are based mainly on the assumption of the validity of the mass action law and Arrhenius dependence of the rate constants. However, diffusion control can be taken into account by some theories in which whole molecules appear as species developing in time.

The network formation theories can be grouped into two major categories:

(1) Graph-like models not directly associated with dimensionality of space. The effect of spatial correlations such as cyclization can be taken into account only as an approximation (usually a mean-field approximation).

(2) Simulation of network formation in n-dimensional space using lattice or off lattice computer simulation.

The network formation theories of the first category can be divided into two groups depending on the way they generate the molecules:

(1a) statistical theories in which branched and crosslinked structures are generated from monomer units or larger structural fragments,

(1b) kinetic (coagulation) theories in which the branching process is described by an (infinite) set of kinetic differential equations based on the mass-action law, or by differential equations of the Smoluchowski type by which also other than chemical kinetic controls can be approximated.

Statistical Theory

In the statistical theories (1-7), the molecules and structures are generated from monomer (or other) units occuring in different reaction states. The reaction state is characterized by the number and type of reacted functional groups and type of bonds by which they are bound to neighboring units. An example is shown in Figure 1 for the simple case of a trifunctional monomer with groups of equal type. The trifunctional units can exist in four reaction states with 0, 1, 2 and 3 reacted groups which is equal to the number of bonds by which they are bound to their neighbors. It is only this distribution of units which develops in time and the distribution of the molecules and the gel is obtained by a proper combination of reacted functionalities of these units.

The reacting system can be represented by graphs (trees) in which the nodes represent monomer units. In the theory of branching processes this collection of graphs (Figure 2) - a molecular forest - is transformed into another forest - the forest of rooted trees. This transformation is performed by taking every node with the same probability and by putting it in generation 0 (root). Then, the other monomer units of the same molecule appear in the first, second, ... etc. generations. This transplantation of threes has two main consequences: (1) all nodes (monomer units) of the system can be found in generation zero (the root) and all units with at least one reacted functionality can be found in generation 1; (2) an n-mer appears in the root n times which means that the probability of finding an n-mer among the rooted trees is given by its weight fraction.

The transformation into the rooted forest was performed because it is easy to generate the rooted forest from monomer units. In the root, there are all units and the corresponding trees are generated by stepping from generation zero to generation 1, etc. by cascade substitution. If only the first shell substitution effect is operative, the normalized distribution of units in generations $g \geq 1$ is the same. This cascade substitution yields in an implicit form the weight-fraction distribution which can be used as a source of numerous pieces of information: it can be transformed into the number fraction, z-fraction, etc. distributions in the form of generalizing functions (pgf) and they yield in a simpler way the corresponding averages. Also, a special weighting can be applied to various paths of these pgf's in order to get radii of gyration or structure factors.

Beyond the gel point, the bonds issuing from a monomer unit can have finite or infinite continuation. If the continuation is finite, the issuing subtree is also only finite; if the continuation is infinite, the unit is bound via this bond to the "infinite" gel. The classification of bonds with respect to whether they have finite or infinite continuation enables a relatively detailed statistical description of the gel structure. The probability of finite continuation of a bond is called the extinction probability. The extinction probability is obtained in a simple way from the distribution of units in generation $g>0$. This distribution is obtained from the distribution of units in the root $g=0$ (for more details see Ref. 6).

Kinetic Theory

The kinetic theory works with whole molecules and the "infinite" gel is represented by an "infinite" molecule. The reaction can occur between any pair of molecules by which a larger molecule is (Figure

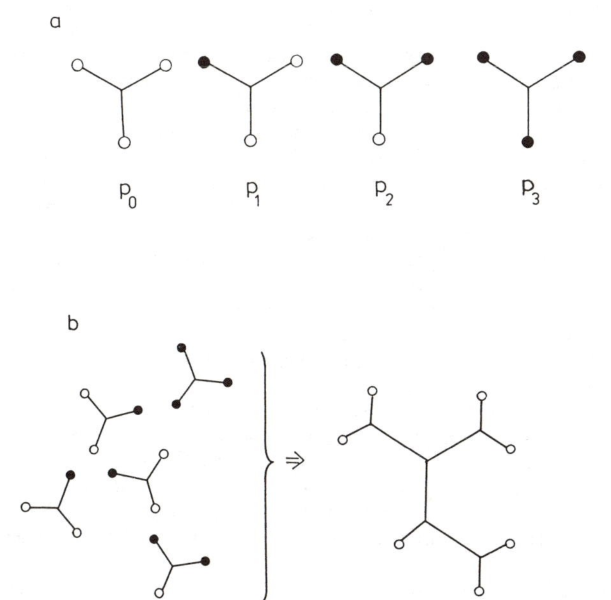

Figure 1. Statistical build-up of branched and crosslinked structure from units.
1a. Distribution of units of a trifunctional monomer with respect to the number of reacted functional groups: • reacted group, o unreacted group.
1b. Assamblage of a species (molecules) from monomer units.

3) formed (8-13). The reaction rate is proportional to the product of the numbers of unreacted functional groups in the respective reaction partners. The number of the corresponding kinetic differential equations is infinite. If groups of different type and/or reactivity are present, there exist several ways to build-up an oligomer of higher degree of polymerization. The infinite set of kinetic differential equations can almost always be transformed into a single partial differential equation for the generating function describing the whole distribution of molecules. The solution giving the whole degree-of-polymerization distribution was found possible either for the so-called random reaction (groups of equal and independent reactivity) or for step polyaddition (polycondensation) of a bifunctional monomer with dependent reactivity of functional groups. If the functionality is higher than two and the substitution effect is operative an analytical solution is not possible. However, the method of moments can be applied. The resulting equation is differentiated with respect to the variables of the generating function and a limited number of differential equations for moments is obtained which are then solved numerically. In this way, rather complex systems with respect to the reactivity of functional groups and reaction mechanisms can be treated.

The kinetic (coagulation) generation of structures has the advantage that it can deal with combined chemical and physical effects such as translational diffusion controlling coagulation, diffusion control of reaction determined by local fluctuations of reactive groups (12-14) or, possibly, size dependent effects on apparent reactivity of reactive groups such as cyclization and shielding. However, a number of these modifications are still to be investigated. The kinetic (coagulation) method has an important disadvantage: the gel is considered as the "largest" molecules and no information can be obtained about its interior. However, one can calculate the cycle rank, because it is determined by the number of a reacted functional groups per monomer unit and the number of units.

It is to be remarked that the process described by the infinite set of kinetic (coagulation) equations can be simulated by Monte-Carlo methods (15). The information on the number of molecules of the respective size is stored in the computer memory and weighting for selection of molecules is applied given by the number and reactivity of groups in the respective molecule.

Combination of Statistical and Kinetic Theory

The non-equivalence of the statistical and kinetic methods is given by the fact that the statistical generation is always a Markovian process yielding a Markovian distribution, e.g. in case of a bifunctional monomer the most probable or pseudo-most probable distributions. The kinetic generation is described by deterministic differential equations. Although the individual addition steps can be Markovian, the resulting distribution can be non-Markovian. An initiated step polyaddition can be taken as an example: the distribution is determined by the memory characterized by the relative rate of the initiation step (11).

To preserve the simplicity of the statistical generation from units without violation of the correlations resulting from the

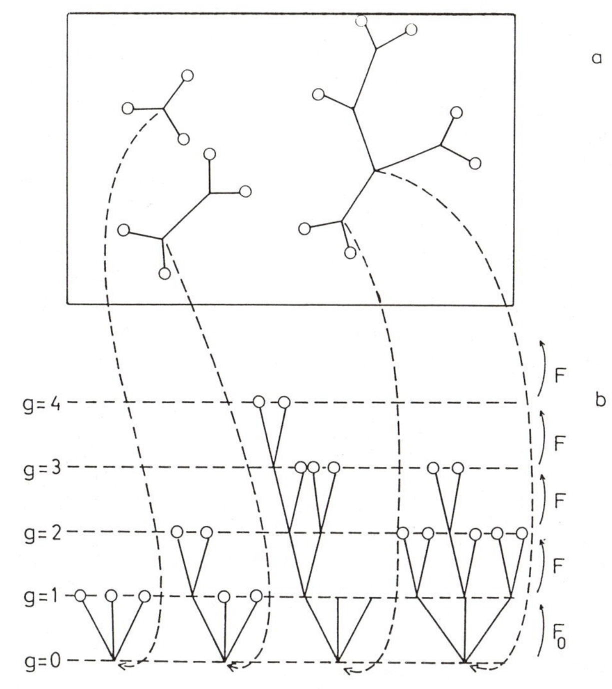

Figure 2. Transformation of a molecular forest into a forest of rooted trees and generation of this forest.

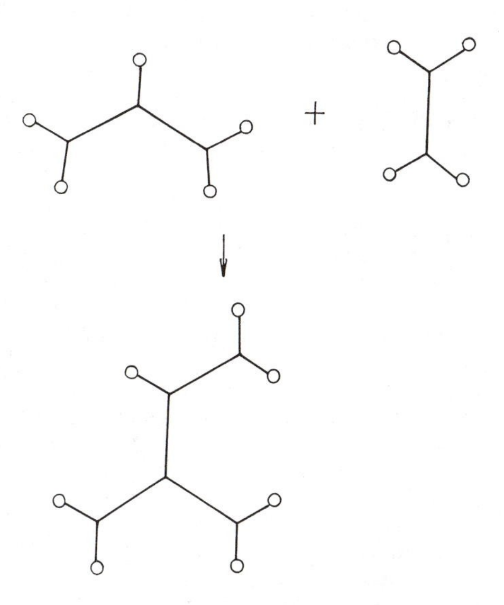

Figure 3. Generation of species using the kinetic method.

possible non-Markovian character of the distribution, the fact has been employed that connections between groups of independent reactivity can be cut and reformed again at random. It means that the structure generated from original polyfunctional units by the kinetic method will be identical with that generated in the following three steps (11):

(1) cutting the connections between groups of independent reactivity and labelling the points of cut. The distribution of units which results will be of lower functionality;

(2) applying the kinetic method to this new distribution. What results is a distribution of branched fragments (provided the reduction of functionality was sufficient to prevent gelation); these fragments still carry the labelled points of cut;

(3) recombining the labelled point of cut using the statistical method.

The application of this method is visualized using the example of a tricomponent system composed of a telechelic polymer with end groups of independent reactivity, a trifunctional crosslinking agent and a bifunctional chain extender where the latter two compounds exhibit a substitution effect on the reactivity of groups (Figure 4). The fragment method was recently applied to initiated crosslinking of a tetrafunctional monomer, to the treatment of postetherification in epoxy-amine curing, and to additional crosslinking by side reactions in the formation of polyurethanes.

Approximate Treatment of Cyclization

Intramolecular reactions always accompany intermolecular crosslinking. Their intensity depends on the structure of the constituent units and very much on the reaction mechanism. Thus, if the network is built up by step reactions from low functionality components, cyclization is relatively weak. On the contrary, chain vinyl-divinyl copolymerization yields highly cyclized products just in the beginning of the polymerization especially if the concentration of the polyvinyl monomer is higher. This case will be briefly commented on later in this article.

Weak cyclization can be treated as an approximation within the framework of the statistical branching theory. In the so-called spanning-tree approximation (cf. e.g. Refs. 16-18), the tree-like structure of the molecules is formally retained, but some of the reacted functionalities are classified as engaged in ring formation while the others are active in branching and structure growth. The ring-forming functionalities are considered as reacted from the point of view that they cannot react any more, but as unreacted from the point of view that they do not issue any bond to the next generation. The transformation of a molecule with cycles into a spanning tree and special labelling of the resulting functionality is visualized in Figure 5.

The main problem is to find the number of the ring forming functionalities and their distribution among the building units. The number of ring forming functionalities per unit is obtained by considering all probabilities of ring closures between an unreacted group of a unit in the root (remember that we have all units as roots) and all unreacted groups on units in all generations g, $g>0$. Allowance can be made for the case when cyclization can occur within

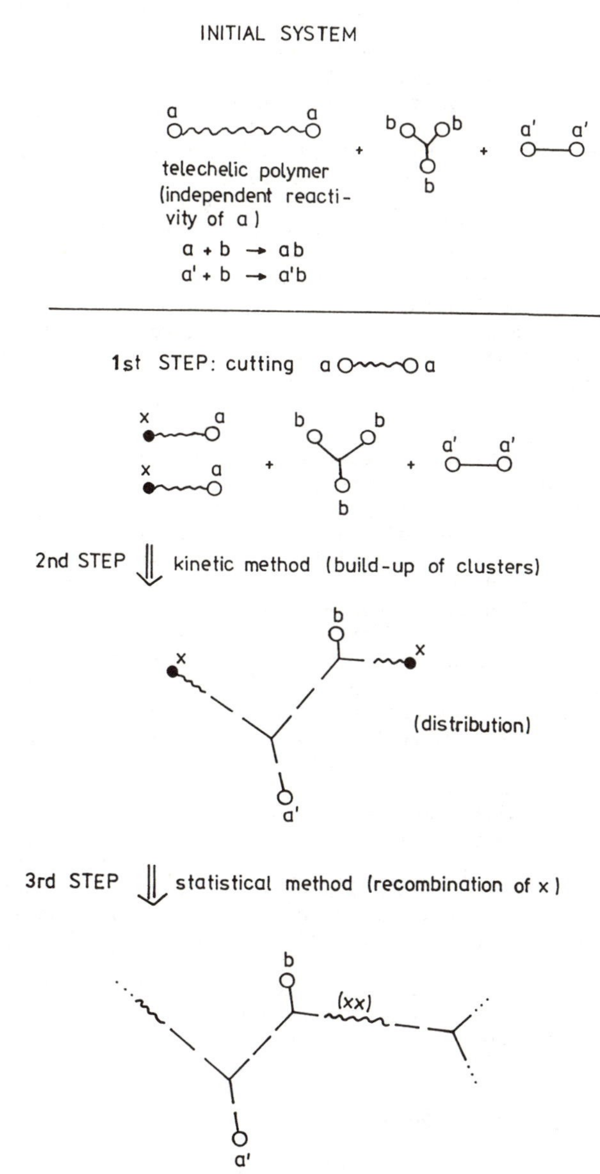

Figure 4. Example of combination of the statistical and kinetic methods for network build-up. (Reproduced with permission from Ref. 42. Copyright 1987 CRC Press).

one monomer unit, or when the smallest possible ring is composed of several monomer units. For obtaining the probability of ring closure between two groups at a given distance (counted in the number of monomer units), the Gaussian distribution of the square end-to-end distance of an equivalent chain is usually employed, but again allowance can be made for a non-Gaussian character of this distribution.

The disadvantage of this method is the following: although a considerable effort has been spent on counting rings of all possible size, the information concerning ring size distribution is lost in what results: the distribution of monomer units with respect to the number of groups reacted inter- and intramolecularly. In other-words, this procedure reduces the effect of rings to the effect of the first-shell substitution effect. Nevertheless, this approach gives results which are in relatively good agreement with experiments, if cyclization is weak. This approximation for the treatment of cyclization can be improved if the smallest ring is included into the distribution of building fragments and if its concentration is calculated exactly or measured experimentally. A number of experimental studies on cyclization especially at the gel point or in the pre-gel region should be mentioned. (24)

Cyclization in the post-gel stage is a special problem, because it is the property of the gel that it must contain rings (circuits) and the conformationally uncorrelated circuit closing was already implicitly considered by Flory (1). It can be seen that Monte Carlo simulation of the kinetic process gives the same results if intramolecular reactions are allowed and if their intensity is considered proportional to the products of the mean concentrations of the groups belonging to the reacting molecules (15). As the size of the system increases, the intramolecular contributions coming from molecules smaller than the largest one converge to zero (because their mean concentrations converge to zero) and only the concentration of groups in the largest molecule beyond the gel point remains finite, so that intramolecular reactions can occur. The majority of chains in these cycles is elastically active. However, there exist in the gel cyclic structures the chains of which are not active in the equilibrium retractive force. The calculation of the number of such chains is possible (18) (based again on the spanning tree approximation) and a recent experimental study has shown that this approach is a reasonable approximation.

The kinetic theory should be able to deal with the problem of cyclization in a more detailed way due to the special feature that it considers the growth of each molecule separately. The work is still to be done, however.

Computer Simulations in Space

Computer simulation in space takes into account spatial correlations of any range which result in intramolecular reaction. The lattice percolation was mostly used. It was based on random connections of lattice points of rigid lattice. The main interest was focused on the critical region at the gel point, i.e., on critical exponents and scaling laws between them. These exponents were found to differ from the so-called classical ones corresponding to Markovian systems irrespective of whether cyclization was approximated by the spanning-tree approach or not. (25) This was not surprising because fluctuations must become dominant very near the critical point and these also affect cyclization. The question was whether random percolation was suitable for predicting formation-structure relations. This theory has been found to be in considerable disagreement with experimental data with a number of real systems. The stiffness of the lattice and exclusion of any conformational rearrangements characteristic for usual polymeric systems were the main reasons. Much closer to reality is the off-lattice simulation by Eichinger et al. (19,20) where the variety of conformations of primary precursor chains can be taken into account. The increasing conversion is modelled by increasing diameter of the sphere of action which determines the reaction volume necessary for two groups to react and form a bond. The real situation may be different, however, particularly at higher conversions. The conformational rearrangements resulting from the cross-linking reaction are not taken into account either.

It seems that the simulation of diffusion controlled reactions of groups on polymer chains developed by Muthukumar et al. (21) that takes into account the bond formation by determined conformational rearrangement, can be adapted for the equilibrium situation, i.e. for systems controlled by pure chemical kinetics.

The rigid lattice simulations can be more realistic for fast reactions in which the propagation rate is comparable to or higher than the rate of the necessary conformational rearrangements. The free-radical chain polymerization of polyvinyl compounds is an example. The reaction is treated as an initiated reaction: the initiating species are distributed in the system at random and activated randomly. The bond formation starts at these points and only the activated growing end is able to add monomers. This closely simulates what happens in reality. This model (called the kinetic model) was developed by Manneville and deSeze and Herrmann (22) and generalized by Boots(23) who introduced continuous initiation by dispersed initiating centers which also corresponds to reality (initiation is slower than propagation). The simulation results in a rather inhomogeneous system of internally crosslinked clusters - a picture predicted also on the basis of experiments (see below).

Structural Parameters Supplied by Branching Theories

It may be of interest to list the most important structural parameters which can be calculated by applying the branching theory.

Pre-gel stage:

degree of polymerization distributions in an implicit form; in some cases it can be obtained in an explicit form;

- degree-of-polymerization or molecular weight averages
- compositional distribution averages in multicomponent system
- radii of gyration
- structure factors (scattering functions) used in static and dynamic light scattering

Gel point:

- critical conversion, critical molar ratios of functional groups

Post-gel stage:

- degree-of-polymerization or molecular weight distributions and averages of the sol, compositional distribution averages, radii of gyration, structure factors as for the system before the gel point,
- concentration of elastically active network chains or effective cycle rank important in equilibrium rubber elasticity, Figure 6
- amount of sol and gel (Figures 7 and 8)
- amount of material in size distribution of elastically active network chains and dangling chains important in viscoelastic behavior,
- intensity of interchain interactions (topological constraints) important in equilibrium elasticity as well as viscoelasticity,
- size distribution and averages of covalently bound clusters of chemically dissimilar units, e.g. a branched analogue of hard segment sequences in linear polyurethanes.

Newly developing experimental methods for looking into the structure of crosslinked systems may require additional information which very likely can be obtained by branching theories.

Experimental Verification of the Branching Theories

There exist a number of experimental methods for determination of structure sensitive parameters of a system undergoing branching and crosslinking. However, evaluation of some of the results requires application of a theoretical approach to the phenomenon the measurement is concerned with. Then, we may be testing two theories at once. The equilibrium elasticity is one example, since there exist alternative rubber elasticity theories. However, certain conclusions can always be made.

A very good agreement between theory and experiment has been achieved for the system diglycidyl ether of Bisphenol A - phenylglycidyl ether -4,4'-diaminodiphenylmethane (30). This indicates that, at least for networks formed by step reactions, the branching theories can describe the sol fraction well, even when cyclization is not negligible. However, again caution is advised against uncritical use of the sol fraction and branching theory in systems where side reactions are operative or possible. Not yet fully exploited is the possibility to analyze the sol fraction with respect to the molecular weight distribution and/or content of the monomers or species of the lowest molecular weights.

The measurement of the equilibrium modulus offers another possibility to compare the branching theory and experiment. However, in doing so one tests two theories: the network formation theory and the rubber elasticity theory and there are at present deeper uncertainties in the latter than in the former. Many attempts to analyze the validity of the rubber elasticity theories were in the past based on the assumption of ideality of networks prepared usually by endlinking. The ideal state can be approached but never reached experimentally and small deviations may have a considerable effect on the concentration of elastically active chains (EANC) and thus on the equilibrium modulus. The main issue of the rubber elasticity studies is to find which theory fits the experimental data best. This problem goes far beyond the network

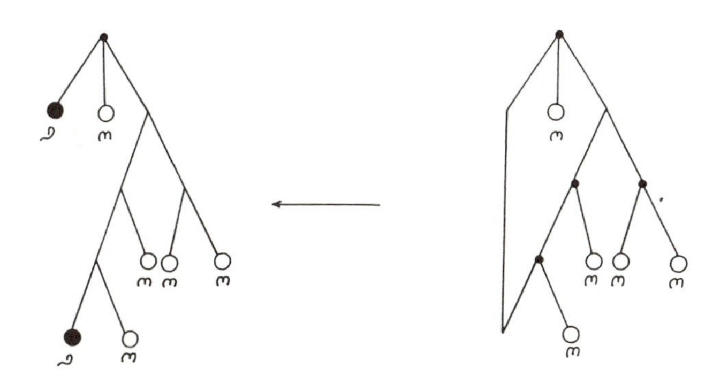

Figure 5. Transformation of a molecule with cycle into a spanning tree and labelling of the ring forming functionalities. (Reproduced with permission from Ref. 42. Copyright 1987 CRC Press).

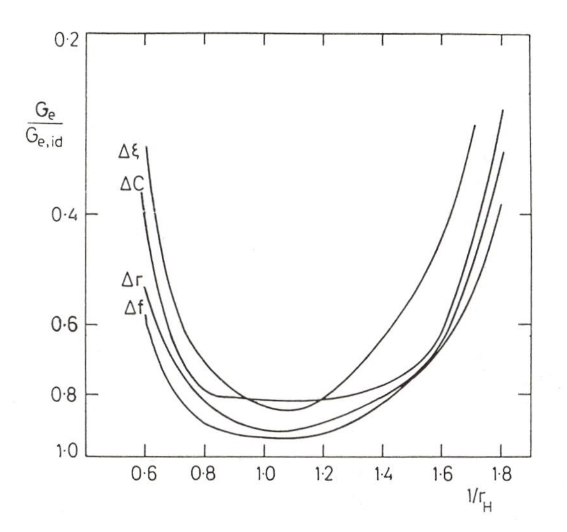

Figure 6. Expected change in the equilibrium modulus, G_e, with respect to its ideal value for a perfect network, $G_{e,id}$ produced by a 3% change in conversion, $\Delta\zeta$, functionality, Δf the molar ratio [OH]/[NCO], Δr, and cyclization, ΔC in dependence on conversion. The data refer to a system composed of a trifunctional telechelic polymer and difunctional coupling agent. (Reproduced with permission from Ref. 42. Copyright 1987 CRC Press.)

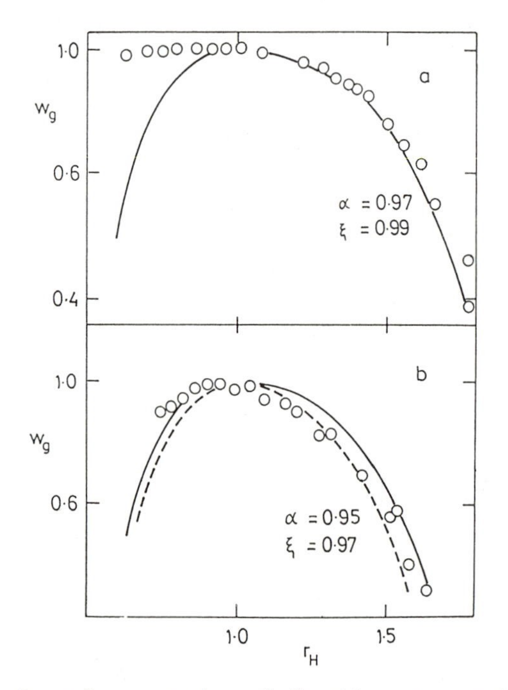

Figure 7. Dependence on the gel fraction, w_g, on the initial molar ratio of OH to NCO groups, r_H, for poly(oxypropylene) (POP) triol -4,4'-diphenyl-methane diisocyanate (HDI) networks. a) POP triol, M_n = 708 b) POP triol, M_n = 2630. The curves have been calculated for final conversion of the minority groups ζ indicated which corresponds to the value of conversion to intermolecular bonds α indicated. (Reproduced with permission from Ref. 42. Copyright 1987 CRC Press.)

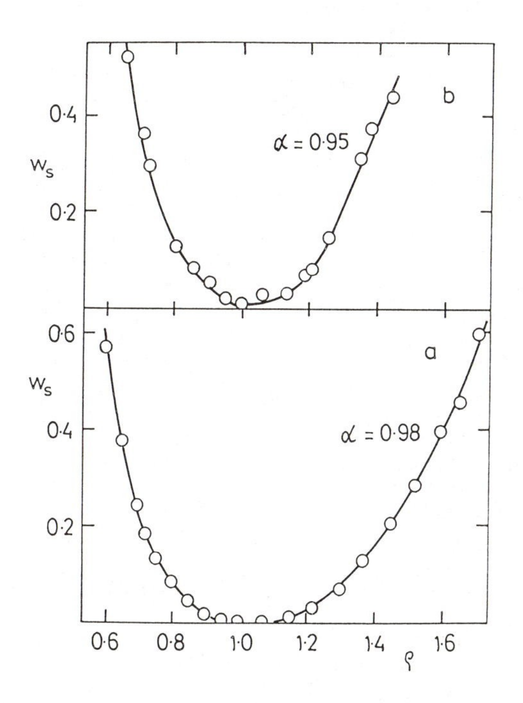

Figure 8. Sol fraction of the PPT-MDI networks with added monofunctional component (cyclohexanol or phenyl isocyanate keeping $r_H = 1$. for $\zeta<1$, $\zeta = [OH]_{PPT}/\sum[OH]$, for $\zeta>1$ $\zeta = \sum[NCO]/[NCO]_{MDI}$ (Reproduced with permission from Ref. 42. Copyright 1987 CRC Press).

formation theories and involves the study of deformations on the macro- and microscales. Certain conclusions can be made, however, about the compatibility of the results with a given theory. Thus, the investigation of network formation and equilibrium elasticity of polyetherurethane networks (28) has shown that the equilibrium modulus exceeds the limit of the Flory junction-fluctuation (phantom) theory and that it can be brought down to the phantom value neither by increasing tensile strain nor by swelling (Figure 9). In other words, there seem to exist in the network interchain topological constraints which are permanent and which belong to the type of "trapped enganglements" constraints. Similar conclusions were obtained for networks prepared from poly(oxypropylenediamines) and polyepoxides (30,31).

Diluent added during crosslinking has two main effects: it increases the population of elastically inactive cycles and it weakens the interchain constraints. Studies of poly(oxypropylene) triol-diisocyanate networks in the presence of diluent have shown that the effect of diluent on the equilibrium modulus is much stronger than would correspond to the effect of cycles (Figure 10) (32) which again corroborates the concept of permanent interchain constraints.

Equilibrium swelling has been often used for characterization of crosslinking density. The concentration dependence of the interaction parameter χ has been assumed either the same as in solution of a linear polymer of the same concentration or it has been calibrated by equilibrium elasticity measurements. The newer theories (Flory-Erman, tube, etc.) predict a variation of the intensity of the topological constraints with the degree of swelling which introduces another parameter into the swelling equation and makes the application of the swelling equation for a routine determination of the concentration of EANC's unsuitable. Two additional problems are to be mentioned: an analysis by Eichinger (36) has shown that either the additivity of the mixing and elasticity terms in the free energy equations does not apply or that the crosslinks exert in its vicinity an effect on conformation and elastic response of EANC's. Also the interaction of the crosslink with the diluent may be different than that of a chain segment. This has been demonstrated by the example of polyetherurethanes prepared from a poly(oxypropylene) diol and bulky triisocyanate. The swelling data could be reasonably explained only when the different interaction energies and interacting surfaces of crosslinks and chain segments were taken into account (37).

The existence of such constraints is likely. In a series of papers (cf. e.g. Ref. 33) it has been proved that cyclic molecules added as a diluent to a polyfunctional endlinking system become entrapped. It is not yet clear how much this entrapment contributes to the modulus. In a pure polyfunctional system, the situation is analogous in the sense that many cycles exist in the gel and that they are interpenetrated by newly formed network chains. Rigbi and Mark (34) even observed gelation and measured elastic properties of the gel formed in a bifunctional $A_2 + B_2$ system at high conversion. Gel formation was intertreped as due to ring formation and ring-ring interpenetration which resulted in catenane-like structures.

Newer rubber elasticity theories based on the tube model (35) consider special constraint release mechanisms which allow a physi-

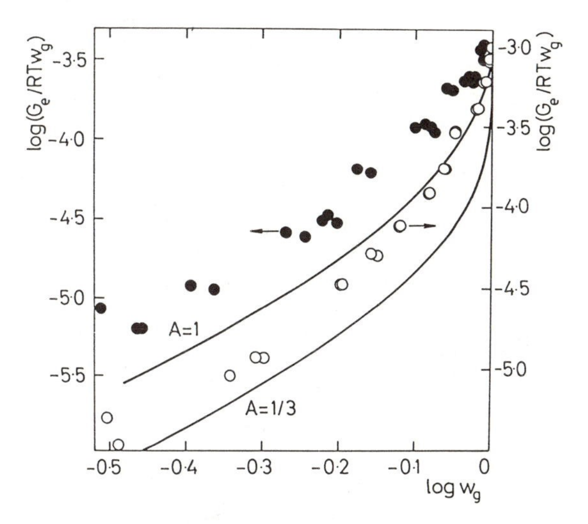

Figure 9. Reduced equilibrium modulus of polyurethane networks from POP triols and MDI in dependence on the sol fraction. • networks from POP triol M_n - 708, o networks from POP triol M_n = 2630. (-) calculated dependence using Flory junction fluctuation theory for the value of the front factor A indicated. (Reproduced from Ref. 57. Copyright 1982 American Chemical Society.)

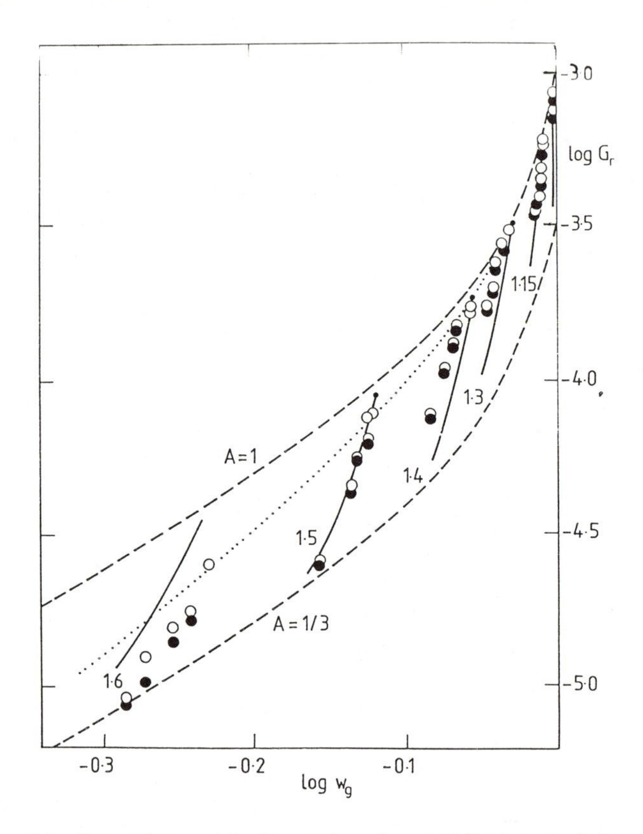

Figure 10. Dependence of the reduced equilibrium modulus of POP triol - MDI networks prepared in the presence of diluent. POP triol M_n= 708; stress-strain measurements in the presence of diluent (o) and after evaporation of the diluent (•). Flory theory for the values of the front factor A indicated, theoretical dependence including trapped interchain constraints Numbers at curves indicate the value of r_H. (.....) predicted dependence if only formation of elastically inactive cycles affected G_r. (Data from Ref. 32).

cal contribution to the force on top of the chemical contribution even in the limit of infinite strain (C_1 constant) or sufficiently high degree of swelling. Thus, for progress in the understanding of the rubber elasticity of crosslinked systems, the network formation theories offer valuable assistance.

Application of Branching Theories to More Complex Systems

The branching theories have been successfully applied to a number of systems of industrial importance which are usually rather complex. The examples of these applications are listed below:

- crosslinking and degradation (radiation curing) (38,39)
- more complex systems polyepoxide-polyamine, polyetherification (6)
- curing of epoxy resins with polycarboxylic acids or cyclic anhydrides (6)
- additional crosslinking by side reactions in polyurethanes (allophanate, urea, biuret, isocyanurate groups), (40,41)
- gelation and network structure in multicomponent systems where formation of chemically bound clusters (analogy of hard segments in linear polyurethanes) is possible, (42)
- melamine-formaldehyde-acrylic coatings, (43)
- multistage processes in which (branched) prepolymers are formed in several steps and then crosslinked. (44,45)

These applications will be commented on only briefly:

The theory of branching processes is able to treat the crosslinking-chain scission problems rigorously without the limitations imposed by the chain length and degree of degradation and crosslinking applied in the Saito-Inokuti theory based on integro-differential equations. It also can treat more complex systems.

In the polyepoxide-polyamine curing it is possible to take into account not only the substitution effect in the amino groups but also differences in reactivity. However, the curing of polyepoxides based on N,N diglycidylamines (e.g. diglycidylaniline and tetraglycidyldiaminodiphenylamethane) are still not understood quantitatively to the extent that a network formation model can be formulated. (46) Interdependence of reactivities (substitution effects) of amino and epoxy groups, strong cyclization, and existence of stereoisomers of different reactivity are among the main reasons. It is now well understood that polyetherification following the epoxy amine addition and started by the hydroxyl group formed in the epoxy-amine addition step should be considered as an initiated reaction. This means that the kinetic theory is to be applied instead of the statistical one. The procedure is also based on splitting and recombination of connections between groups of independent reactivity (6). If larger fragments than monomer units are used in network build-up, the exact solution obtained by the kinetic theory is approached relatively closely (47). However, the application to real polyfunctional systems requires further experimental work to show what the role of chain transfer is and whether the reactivity of all OH groups is the same.

The application of the branching theories to acid curing was rather successful: the discovery of the relative importance of the transesterification following the epoxy-carboxyl addition initiated the theoretical treatment of branching. The prediction was that gelation and network formation were possible although the number of bonds did not change as a result of transesterification. Also a special shape of the sol fraction dependence on transesterification conversion was predicted. The experiments were in agreement with these predictions. The study of the model reaction of anhydrides with epoxides in the presence of tertiary amines revealed that the reaction of the tertiary amine with epoxide was the initiating step. The tertiary amine remained chemically bound. The branching theory predicts that the gel point conversion should depend on the relative amount of the tertiary amine which was indeed found to be so. In polycarboxyl-polyepoxide systems, the gel point conversion is independent of the concentration of tertiary amine because it does not become chemically bound.

For multicomponent polyurethanes, gelation and network formation strongly depend on the relative reactivities of OH groups of the components, e.g. of a polydiol and low molecular weight triol. Also the size of the so-called hard clusters composed in this case of the triol and diisocyanate units depends considerably on reactivities of groups, functionality and ratios of reactants and on the crosslinking process (one-stage, two-stage). Because the internal structure of such networks can now be characterized, these systems represent an interesting object of study using combinations of a number of optical and mechanical methods. A change in temperature may bring about physical association of these clusters which may be, however, prevented by the rigidity of the network. If the polydiol (e.g. polybutadiene diol) chains are not compatible with the crosslinker, association and even phase separation already occurs in the early stages of the reaction which results in a specific morphology (48). The system is to be considered inhomogeneous and the branching theories based on the homogeneous distribution of reacting groups cannot be applied.

Recently, procedures for treating multistage processes have been developed (44,45). In a multistage process, branched prepolymers are formed first, they can be further modified by reaction with some other monomers and eventually crosslinked. this is a theoretically rather difficult problem especially if the substitution effect on reactivity of groups is operative even in a ring-free system. The problem has been solved by the branching theory: The strategy is based on the following operations: (a) track is being kept of all unreacted groups, so that they can be activated when necessary, (b) the reaction product in one stage becomes the initial component in the next stage, and (c) the weight-fraction distribution of branched prepolymers is generated in each step and is converted to the number fraction distribution. Use of the probability generating functions makes this transformation relatively easy. Systems based on multistage processes are important in a number of applications, especially in coatings.

Specific Features of Chain Crosslinking (Co)Polymerization

Chain crosslinking (co)polymerization involving a polyvinyl monomer represents a process where the application of the branching theory based on a (perturbed) tree-like model fails due to strong cyclization. Strong cyclization is characteristic for these systems not because of a special configuration of the monomers but because of the reaction mechanism. If one considers the situation in the beginning of the reactions when only very few molecules are dissolved in the monomers, then the probability of forming a crosslink via the reaction of a growing macroradical with a pendant double bond of another chain is very small because only very few such chains exist. The probability that an intramolecular crosslink will be formed is much higher because pendant double bonds on the same chain (growing macroradical) are available in the vicinity. (Figure 11) Therefore, the primary chain may become intramolecularly crosslinked several times before it becomes bound to another such species. Therefore, internally crosslinked compact microgel-like species are formed in the beginning of the reaction. However, exclusive formation of small rings within the divinyl molecule-cyclopolymerization-leads to ladder type polymers. There is ample experimental evidence for such special mechanisms and structure (49), e.g.:

- the fraction of doubly reacted divinyl molecules in the beginning of the reaction is already relatively high and it does not depend much on increasing conversion. At higher concentrations of the divinyl monomer, the fraction of these units becomes independent of the fraction of crosslinker in the feed. The increase of concentration of the divinyl monomer in the feed leads only to an increase of the number of pendant double bonds. It means that there exists a limit to the degree of intramolecular crosslinking in the species and eventually in the whole crosslinked system.
- the gel point conversion is shifted by a factor of 10^1-10^3 to higher values which demonstrates a huge wastage of crosslinks in the intramolecular reaction,
- the low conversion polymer has a compact spherical structure.

Therefore, concentrations of the polyvinyl monomer, the branching and network formation proceeds as shown in Figure 12. Microgel-like species are formed first which are highly internally crosslinked. The pendant double bonds in the interior are very immobile so that their reactivity is strongly diffusion controlled and they almost cannot react at all, even with the monomers. Only the more mobile pendant double bonds in the periphery of these species can enter into reactions with macroradicals and participate in interbinding of the species together.

Recently, Kloosterboer (50) accumulated a number of data on fast polymerization of diacrylates. He has also shown that the density of the system decreases with increasing crosslinking density. The initiated (kinetic) percolation model described the structure of the resulting system much better than any other model.

The free-radical crosslinking polymerization can be regarded as a special example of specific diffusion control, in which the tendency to microgel formation and decrease of apparent reactivity of internal double bonds depends on the size of the microgel which in turn depends on the molecular weight of the primary chain. Polymerization of diallyl monomers exhibits much less of these features (49) because the degree of polymerization of their primary chains is extremely low due to degradative chain transfer.

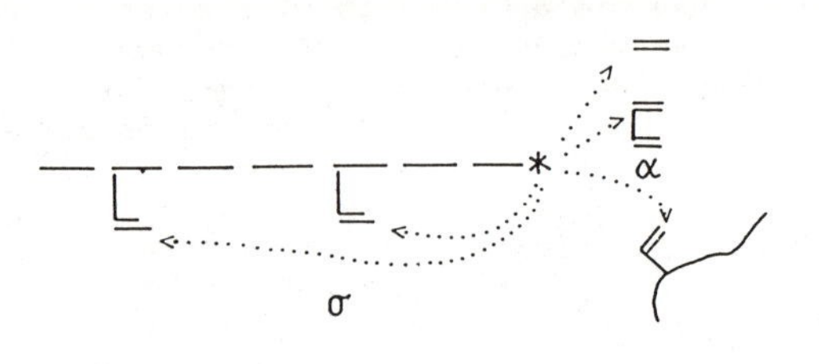

$$s = \sigma / (\sigma + \alpha)$$

Figure 11. Scheme for crosslinking chain copolymerization x free radical site, α inter-and σ intramolecular reaction.

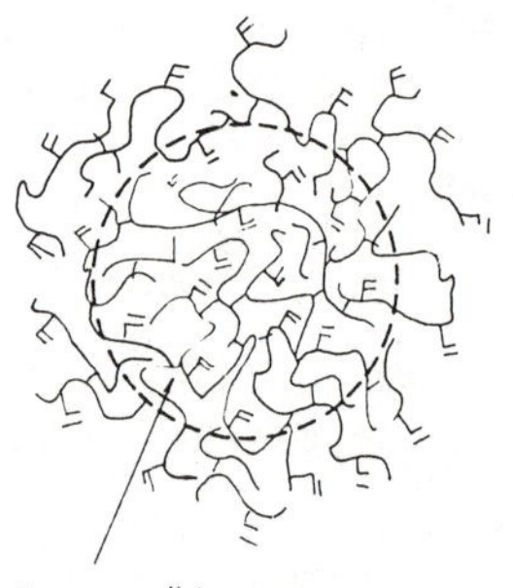

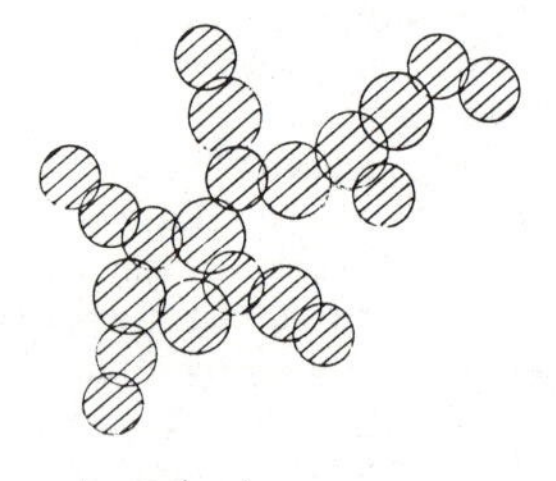

Figure 12. Network formation in chain crosslinking copolymerization (Reproduced with permission from Ref. 49. Copyright 1982 Applied Science Publishers).

Overall Diffusion Control in the Glass Transition Region

When the encounter probability of reactive groups and the rate of reaction becomes controlled by the segmental mobility (viscosity of the medium), overall diffusion control sets in. The overall diffusion control is typical of polymer systems in which, as a result of the chemical reaction, the system passes from the liquid (rubbery) state into the glassy state.

The reaction is postulated to proceed via an activated complex. In the simple case of the reaction of A and B groups, the activated complex -(AB)* is formed first and it eventually gives the product - the -AB- bond

$$\sim A + B\sim \underset{k_{-1}}{\overset{k_1}{\rightleftarrows}} \sim(AB)^*\sim \overset{k_2}{\rightarrow} \sim AB\sim \quad (1)$$

If k_1 and k_{-1} are much larger than k_2, the reaction is controlled by k_2. If however, k_1 and k_{-1} are larger than or comparable to k_2, the reaction rate becomes controlled by the translational diffusion determining the probability of collisions which is typical for specific diffusion control. The latter case is operative for fast reactions like fluorescence quenching or free-radical chain reactions. The acceleration of free-radical polymerization due to the diffusion-controlled termination by recombination of macroradicals (Trommsdorff effect) can serve as an example.

Step reactions (polycondensation, step polyadditions, etc) are relatively slow, so that their rates are controlled by the equilibrium given by k_1/k_{-1} . This ratio is not much affected by viscosity since the concentration of the activated complex is given by the residence time of the two groups in the "reaction volume". The two groups are hindered by increasing viscosity from associating (entering the reaction volume) and dissociating (going out of the reaction volume) by the same factor. However, the rate constant k_2 can itself become controlled by segmental diffusion.

There are numerous examples in the literature which demonstrate the onset of diffusion control during the reaction due to increasing glass transition temperature of the reacting system (cf. e.g. Ref. 52). Figure 13 shows an example of isothermal cure of an epoxy resin at different temperatures (53). If one assumes that the chemical kinetics (Arrhenius) control is operative one gets a good superposition of kinetic curves recorded at different temperatures for the region where $T_r > T_g$. When T_r approaches or becomes smaller than T_g, the kinetic curves start to diverge. The largest negative deviations of the reaction rate are observed for systems reacting at the lowest temperature, i.e. for systems where $T_r - T_g$ is the most negative. In some systems, the reaction rate becomes so slow that it stops before the reaction reaches completion.

The question arises when the reaction becomes diffusion controlled, i.e. when one can observe experimental deviations from the Arrhenius dependence, and when it becomes fully controlled by segmental diffusion.

The segmental mobility can be related to the T_r-T_g difference using a theory of the glass transition. For example, if the reaction rate is controlled by only segmental mobility in the glass transition region, the rate constant k_T should be (according to the Williams-Landel-Ferry) theory equal to

$$\log k_T - \log k_{Tg} = \frac{C_1(T_r - T_g)}{T_r - T_g + C_2} \quad (2)$$

where k_{Tg} is the value of k_T at $T_r = T_g$. Gordon considered k_{Tg} constant but even if it is not, log k_{Tg} can be neglected with respect to log k_T for T_r not much larger than (T_g - C_2). The C_2 constant is of the order of 50 K for the majority of polymeric materials so that k_T should reach the value 0 for $T_g - T_r = 50$. Because T_g depends on the conversion of the chemical reaction (α), k_T depends now on both temperature and conversion.

For mixed chemical kinetics and diffusion control, Rabinowich suggested the relation (51)

$$\frac{1}{k_T} = \frac{1}{k_c} + \frac{1}{k_d} \quad (3)$$

where k_c is the rate constant of a chemical kinetics controlled reaction given by ($k_c = A\exp(-E/RT_r)$) and k_d is the constant for diffusion controlled reaction determined by the T_r-T_g difference (for example through Eq.(2)).

Thus the rate of a crosslinking (curing) reaction can be written as

$$\frac{d\alpha}{dt} = k_T F(\alpha) \quad (4)$$

where α is the degree of conversion of a reactive group and $F(\alpha)$ is a function of α and initial composition and is given by the mechanism and order(s) of the reaction(s); $k_T = k_T(T)$ (Eq.(3)) for the Arrhenius regimen ($k_d >> k_c$) and $k_T = k_T(T,\alpha)$ for the mixed and the diffusion regimes (54,55).

Figure 14 shows the prediction of the variation of k_T when the Gibbs-DiMarzio theory is used for the description of the dependence of k_d on T_r-T_g (55). Characteristic is the relatively sharp downturn of log k_T in the vicinity of T_g of the reacting system which is also observed experimentally.

However, it is expected the reaction rates below T_g may be affected also by volume relaxation (physical aging) which was not taken into account and which will result in the dependence of k_T not only on T_r and α but also on time t. If we take the positive deviations of experiments in Figure 14 as a measure of the volume relaxation effect then the physical aging increases the apparent mobility although it leads also to a denser (and less mobile) state.

The understanding of the temperature and conversion dependence of the crosslinking kinetics is one of the prerequisites for understanding the changes in viscosity and viscoelastic properties as a function of reaction time and reaction temperature (56). Three main factors determine these relations: the reaction kinetics determined by temperature and conversion, the changes in structure determined primarily by conversion and the changes in T_g determined primarily also by conversion.

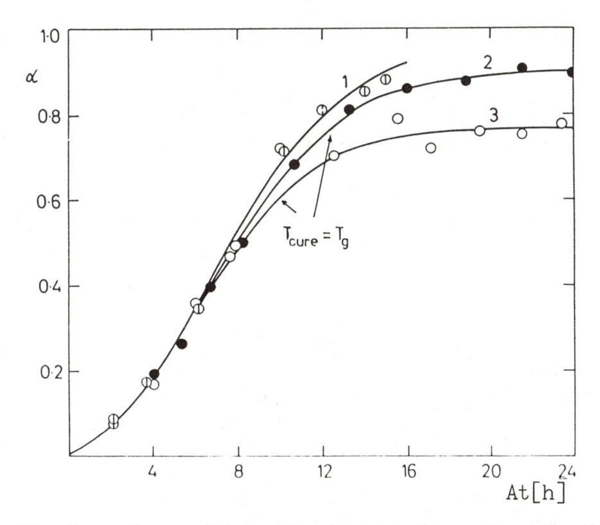

Figure 13. Superimposed kinetic curves (assuming the Arrhenius dependence of the rate constant on temperature) for the reaction of 4,4'-diamino-3,3'-dimethyldicyclohexylmethane and diglycidyl ether of Bisphenol A. Reaction temperature in °C: 1=100°, 2=64°, 3=40°. Divergence of curves indicates diffusion control. (Reproduced with permission from Ref. 6. Copyright 1986 Springer.)

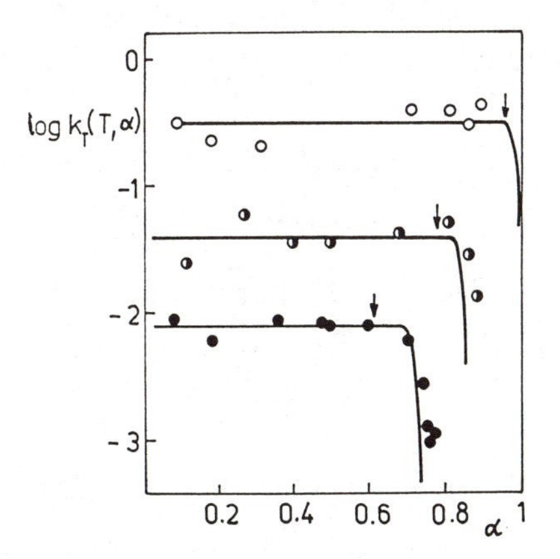

Figure 14. Data of Figure 14 processed according to Eqs. (3) and (4) calculated dependence. (Reproduced with permission from Ref. 55. Copyright 1987 W. de Gruyter.)

Conclusions

The branching theories in their present state can treat a number of complex branching reactions of industrial importance. It is to be stressed, however, that there does not exist any universal approach to all systems. The understanding of the reaction mechanism and kinetics is a necessary prerequisite for adaptation of the proper theory to give relations for structural parameters. Further progress in the network formation theory seems highly desirable particularly in the field of cyclization and diffusion control and in understanding the network structure-properties relations.

Literature Cited

1. Flory, P.J. Principles of Polymer Chemistry, Cornell Univ. Press, Ithaca, 1953.
2. Stockmayer, W.H. J. Chem. Phys. 1943, 11, 45.
3. Gordon, M. Proc. Roy. Soc. London 1962, A268, 240.
4. Gordon, M.; Malcolm, G.N. Proc. Roy. Soc. London 1966, A295, 29.
5. Dusek, K. Makromol. Chem., Suppl. 1979, 2, 35.
6. Dusek, K. Adv. Polym. Sci. 1986, 78, 1.
7. Burchard, W. Adv. Polym. Sci. 1982, 48, 1.
8. Kuchanov, S.I. Methods of Kinetic Calculations in Polymer Chemistry, Khimiya, Moscow, 1978.
9. Kuchanov, S.I.; Povolotskaya, E.S. Vysokomol. Soedin. 1982, A24, 2179.
10. Dusek, K. Polym. Bull. 1979, 1, 523.
11. Dusek, K.; Somvarsky, J. Polym. Bull. 1985, 13, 313.
12. Ziff, R.M. in Kinetics of Aggregation and Gelation, Family F. and Landau D.P., Editors, North-Holland, 1984, p. 191.
13. Leyvraz, F. in On Growth and Form, Stanley, H.E., and Ostrowski, N., Editors, M. Nijhoff Publ., Boston, 1986, p. 136.
14. Van Campen, N.G. Stochastic Processes in Physics and Chemistry, North-Holland, Amsterdam, 1981.
15. Mikes, J.; Dusek, K. Macromolecules 1982, 15, 93.
16. Gordon, M.; Scantlebury, G.R. Trans. Faraday Soc. 1964, 60, 604.
17. Gordon, M.; Ross-Murphy, S.B. Pure Appl. Chem. 1975, 43, 1.
18. Dusek, K.; Vojta V. Brit. Polym. J. 1977, 9, 164.
19. Leung, Y.K.; Eichinger, B.E. J. Chem. Phys. 1984, 80, 3877.
20. Leung, Y.K.; Eichinger, B.E. J. Chem. Phys. 1984, 80, 3885.
21. Balazs, A.C.; Anderson, C.; Muthukumar, M. Macromolecules, in press.
22. Hermann, H.J.; Stauffer, D.; Landau, D.P. J. Phys. 1983, A16, 1221.
23. Boots, H.M.J. in Integration of Polymer Science and Technology, Kleintjens, L.A.; Lemstra, P.J., Editors, Elsevier Appl. Sci. Publ., London, 1986, p. 204.
24. Stepto, R.F.T. in Developments in Polymerisation.3.,Haward, R.N., Editor, Applied Science Publ., London, 1982.
25. Stauffer, D.; Coniglio, A.; Adam, M. Adv. Polym. Sci. 1982, 44, 103.
26. Kajiwara, K.; Burchard, W.; Dusek, K., Kowalski, M.; Matejka, L.; Nerger, D.; Tuzar, Z., Z. Makromol. Chem. 1984, 185, 2025.

27. Konak, C.; Tuzar, Z.; Jakes, J.; Stepanek, P.; Dusek, K. Polym. Bull., in press.
28. Ilavsky, M.; Dusek, K. Polymer 1983, 24 981.
29. Ilavsky, M.; Dusek, K. Polym. Bull. 1982 8 359.
30. Dusek, K.; Ilavsky, M.; Lunak, S. Jr. in Crosslinked Epoxies, Sedlacek, B.; Kahovec, J., Editors. W. de Gruyter, Berlin 1987, p. 269.
31. Dusek, K.; Ilavsky, M.; Stokrova, S.; Matejka, L. in Crosslinked Epoxies, Sedlacek, B.; Kahovec, J. Editors. W. De Gruyter, Berlin, 1987, p. 279.
32. Ilavsky, M.; Dusek, K. Macromolecules 1986, 19, 2139.
33. Garrido, L., Mark, J.E.; Clarson, S.J.; Semlyen, J.A. Polym. Commun. 1985, 26, 55.
34. Rigbi, Z.; Mark, J.E. J. Polym. Sci., Polym. Phys. Ed. 1986, 24, 443.
35. Heinrich, G, Straube, E., Helmis, G. Adv. Polym. Sci. 1987, in press.
36. Eichinger, B.E., Polym. Mater. Sci. Eng. 1987, 56.
37. Petrovic, Z.; MacKnight, W.J.; Koningsveld, R.; Dusek, K. Macromolecules 1987, 20, 1688.
38. Demjanenko, M.; Dusek, K. Macromolecules 1980, 13, 571.
39. Dusek, K.; Demjanenko, M. Radiat. Phys. Chem. 1986, 28, 479.
40. Dusek, K., Ilavsky, M.; Matejka, L. Polym. Bull. 1984, 12, 33.
41. Dusek, K. Polym. Bull., 1987, 17, 481.
42. Dusek, K. in Telechelic Polymers, Coethals, J., Editor, CRC Press, Baton Rouge, FL, 1987, in press.
43. Bauer, D.R.; Dickie, R.A.; J. Polym. Sci., Polym. Phys. Ed. 1980, 18, 1977, 2015.
44. Dusek, K.; Scholtens, B.J.R.; Tiemersma-Thone, G.P.J.M., in press.
45. Tiemersma-Thone, C.P.J.M., Scholtens, B.J.R.; Dusek, K. Proc. First Internat. Conf. Ind. Appl. Mathematics, Paris-LaVillette, vanderBurgh, A.H.P.; Mattheij, R.M.M., Editors, 1987, p. 295.
46. Dusek, K.; Matejka, L. Polym. Mater. Sci. Eng. 1987, 56.
47. Riccardi, C.C.; Williams, R.; Dusek, K. Polym. Bull. 1987.
48. Bengston, B.; Feger, C. MacKnight, W.J.; Schneider, N.S. Polymer 1985, 26, 895.
49. Dusek, K. in Advances in Polymerisation.3, Haward, R.N., Editor, Applied Sci. Publ., London, 1982.
50. Kloosterboer, H. Adv. Polym. Sci. 1987, in press.
51. Rabinowitch, E. Trans. Faraday Soc. 1937, 33, 122.
52. Aronhime, M.T.; GIllham, J.K. Adv. Polym. Sci. 1986, 78, 83.
53. Lunak, S.; Vladyka, J.; Dusek, K. Polymer 1978, 19, 931.
54. Huguenin, F.G.A.E.; Klein, M.T. Ind. Eng. Chem., Chem. Prod. Res. Dev. 1985, 24, 166.
55. Havlicek, I.; Dusek, K. in Crosslinked Epoxies, Sedlacek, B.; Kahovec, J., Editors, W De Gruyter, Berlin, 1987, p. 425.
56. Apicella, A.; D'Amore, A.; Kenny, J.; Nicolais, L. ANTEC 86 1986, 557.
57. Dusek, K. Rubber Chem. Technol. 1982, 55, 1.

RECEIVED January 25, 1988

Chapter 2

Intramolecular Reaction

Effects on Network Formation and Properties

R. F. T. Stepto[1]

Department of Polymer Science and Technology, University of Manchester Institute of Science and Technology, Manchester, M60 1QD, England

It is shown that model, end-linked networks cannot be perfect networks. Simply from the mechanism of formation, post-gel intramolecular reaction must occur and some of this leads to the formation of inelastic loops. Data on the small-strain, shear moduli of trifunctional and tetrafunctional polyurethane networks from polyols of various molar masses, and the extents of reaction at gelation occurring during their formation are considered in more detail than hitherto. The networks, prepared in bulk and at various dilutions in solvent, show extents of reaction at gelation which indicate pre-gel intramolecular reaction and small-strain moduli which are lower than those expected for perfect network structures. From the systematic variations of moduli and gel points with dilution of preparation, it is deduced that the networks follow affine behaviour at small strains and that even in the limit of no pre-gel intramolecular reaction, the occurrence of post-gel intramolecular reaction means that network defects still occur. In addition, from the variation of defects with polyol molar mass it is demonstrated that defects will still persist in the limit of infinite molar mass. In this limit, theoretical arguments are used to define the minimal significant structures which must be considered for the definition of the properties and structures of real networks.

Networks formed from stoichiometric, end-linking polymerisation are often assumed to be perfect networks(1-4). However, such an assumption means that intramolecular reaction within gel molecules, which must occur for a network to be formed, never leads to inelastic chains. The assumption is unlikely to be true. The smallest loops which can occur must be elastically ineffective(5-9) and from chain

[1]Current address: Störklingasse 44, CH-4125 Riehen, Switzerland

0097-6156/88/0367-0028$06.00/0

statistics these are the most easily formed. Such loops have marked effects on the modulus.

The moduli of model polyurethane networks clearly show reductions below the values expected for perfect networks, with the reductions increasing with pre-gel intramolecular reaction(5-7). The reductions can be shown to be too large to come solely from pre-gel loop formation(9), some must occur post-gel. In addition, extrapolation to conditions of zero pre-gel intramolecular reaction, by increasing reactant concentrations, molar masses of reactants or chain stiffness, still leaves a residual proportion of inelastic chains due to gel-gel intramolecular reaction. It is basically a law-of-mass-action effect(9). The numbers of reactive groups on gel molecules are unlimited, intramolecular reaction occurs, and some of this gives inelastic chains. Only a small amount of such reaction has a marked effect on the modulus.

In the present paper, theoretical arguments and modulus measurements are used to deduce the significant gel structures which lead to inelastic loop formation and to quantify the network defects and reductions in modulus which may be expected, even in the limit of no pre-gel intramolecular reaction. In this limit all the existing theories and computer simulations of polymerisations including intramolecular reaction(8,10,11) predict that perfect networks are formed.

Experimental Data

The experimental data to be considered are shown in Figure 1. They refer to previously published data on hexamethylene diisocyanate(HDI) reacting with polyoxypropylene(POP) triols and tetrols in bulk and in nitrobenzene(5-7,12) that is, to $RA_2 + RB_f$ polymerisations. M_c is the molar mass of chains between elastically effective junction points. A/M_c has been determined directly from small-strain compression measurements on swollen and dry networks using the equations

$$\sigma = G(\Lambda - \Lambda^{-2}) \tag{1}$$

$$G = ART\rho\phi_2{}^{1/3}(V_u/V_F)^{2/3}/M_c \tag{2}$$

σ is the stress per unit unstrained area, G the shear modulus, Λ the deformation ratio, ρ the density of the dry network, ϕ_2 the volume fraction of polymer present in the network, V_F the volume at formation. A=1 for affine behaviour (expected) and 1-2/f for phantom behaviour(1,3). M_c^o is the molar mass for the perfect network, essentially the molar mass of a chain of ν bonds, the number which can form the smallest loop (5-7); see Figure 2. M_c/M_c^o is equal to the proportional reduction in modulus below the that expected for the perfect, dry network assuming affine behaviour. $p_{r,c}$ is the extent of intramolecular reaction at gelation in the polymerisations used to prepare the networks, and is defined by the equation

$$p_{r,c} = (\alpha_c)^{\frac{1}{2}} - (f-1)^{-\frac{1}{2}} \tag{3}$$

where α_c is the product of extents of reaction of A and B groups at gelation. All the reactions studied used stoichiometric amounts of A

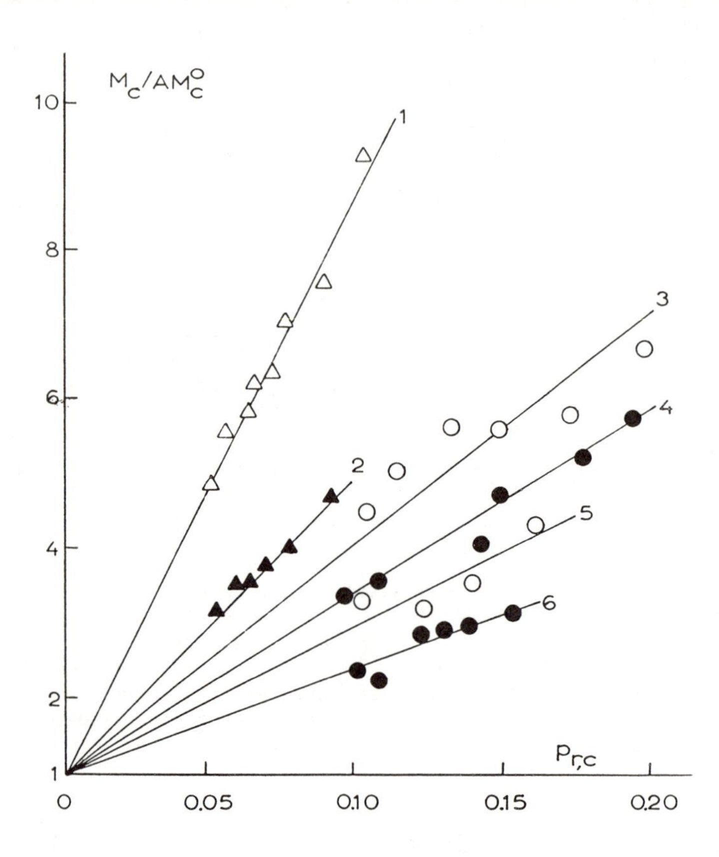

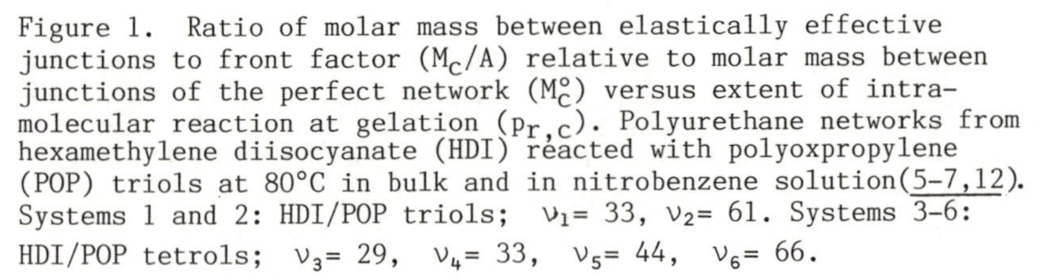

Figure 1. Ratio of molar mass between elastically effective junctions to front factor (M_c/A) relative to molar mass between junctions of the perfect network (M_c^o) versus extent of intramolecular reaction at gelation ($p_{r,c}$). Polyurethane networks from hexamethylene diisocyanate (HDI) reacted with polyoxpropylene (POP) triols at 80°C in bulk and in nitrobenzene solution(5-7,12). Systems 1 and 2: HDI/POP triols; ν_1= 33, ν_2= 61. Systems 3-6: HDI/POP tetrols; ν_3= 29, ν_4= 33, ν_5= 44, ν_6= 66.

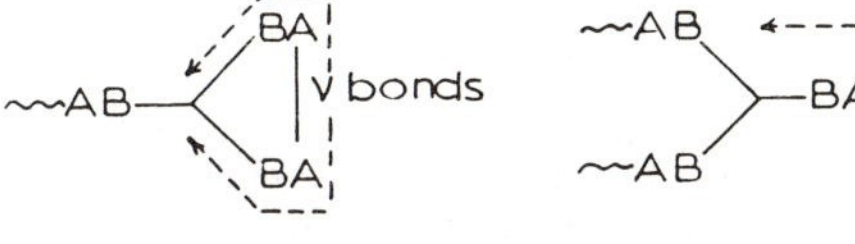

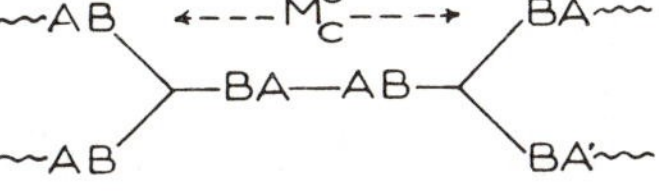

(a)

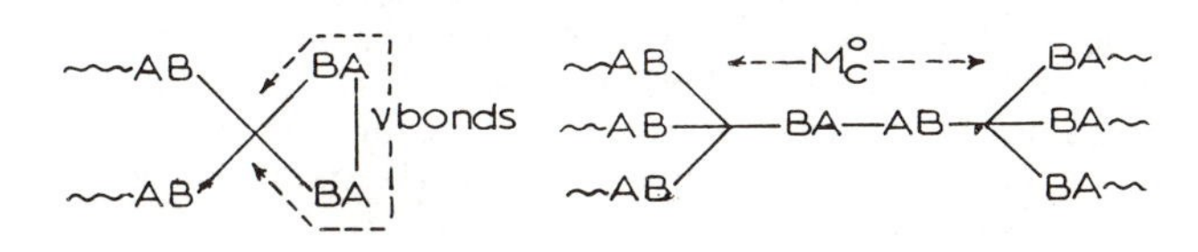

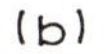

Figure 2. Illustrating the equivalence between the chain forming the smallest loop of ν bonds, and the chain between junction points in the perfect network (of molar mass M_c^o). (a) RA_2 + RB_3 and (b) RA_2 + RB_4 polymerisations and networks.

and B groups and $p_{r,c}$ was varied by carrying out polymerisations at various initial dilutions in nitrobenzene as solvent.

Curves 1 and 2, and 3 to 6 in Figure 1 refer, respectively, to HDI/POP triol and HDI/POP tetrol polymerisations with different values of ν. Marked reductions in modulus occur even for bulk reaction systems, which give the points at the lowest values of $p_{r,c}$ for the different systems. More inelastic chains are formed in trifunctional as compared with tetrafunctional networks for a given value of $p_{r,c}$ (cf. curves 1 and 2 with 3 to 6. In addition, for a given functionality, as ν decreases the proportion of inelastic loops increases. Similar results have been obtained for polyester-forming systems using POP triols and diacid chlorides(13).

Interpretation of Results

The lines drawn in Figure 1 have been extrapolated to $M_c/AM_c^o = 1$ at $p_{r,c} = 0$, predicting that perfect affine networks are formed by ideal gelling systems. The assumptions implicit in this prediction may be examined assuming for the present that intramolecular reaction leads only to the smallest loops. The reasoning may be generalised to include two-membered loops. Given equal opportunities for the formation of loops of all sizes, the smallest (one-membered) loops are from chain-statistical considerations the most readily formed(14,15). In this context, it should be remembered that loops are formed from the beginning of a reaction onwards and initially the smallest loops are formed exclusively.

The smallest loops, as they occur in the networks at complete reaction, are illustrated in Figure 3, together with the elastically active function points lost. For f=3, each smallest loop leads to the loss of two junction points and for f=4 only one junction point per smallest loop is lost. Notwithstanding that more complex ring structures will occur, the greater loss of junction points per smallest loop, and indeed per next smallest loop(16) for f=3 compared with f=4 networks is the basic reason why the former networks (curves 1 and 2 in Figure 1) show larger reductions in modulus per pre-gel loop than the latter networks (curves 3 to 6).

To a first approximation, which neglects changes in average chain structure, the loss in elastically active junction point concentration may be translated directly into loss in concentration of elastically active chains and increase in the value of M_c. For a perfect network in the dry state, the concentration of elastically active chains is given by the equations

$$\rho/M_c^o = n^o = (f/2)N_B \tag{4}$$

where N_B is the concentration of RB_f units (junction points) in the $RA_2 + RB_f$ polymerisation. Structures 3(a) and 3(b) show that for f=3 the actual concentration will be

$$\rho/M_c = n = (f/2)N_B(1 - 2fp_{r,e}) \tag{5}$$

where $p_{r,e}$ is the extent of intramolecular reaction at the end of the reaction, with $N_B.f.p_{r,e}$ the number of intramolecularly reacted B(orA) groups. Hence,

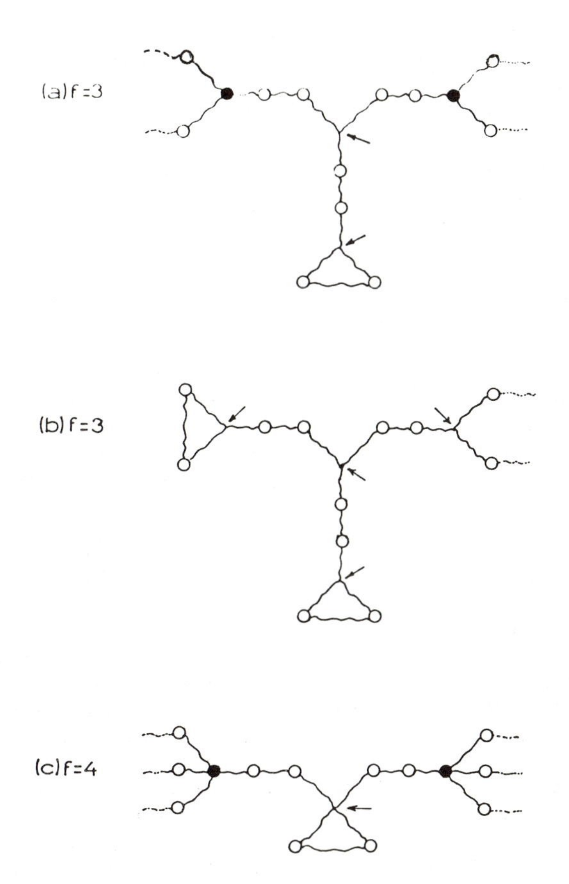

Figure 3. Smallest loop structures in the gel in, (a) and (b), RA_2 + RB_3 polymerisations and, (c), RA_2 + RB_4 polymerisations. Junction points lost indicated with arrows. ● - elastically active points. 0 - reacted pairs of groups (-AB-).

$$M_c/M_c^o = 1/(1 - 6p_{r,e}) \quad (6)$$

Equation 6 may be rearranged to allow evaluation of $p_{r,e}$ from the values of M_c/A deduced from the modulus measurements with

$$p_{r,e} = \frac{1}{6}(1 - \frac{1}{A}.(AM_c^o/M_c)) \quad (7)$$

Similar arguments for f=4 give

$$p_{r,e} = \frac{1}{4}(1 - \frac{1}{A}(AM_c^o/M_c)) \quad (8)$$

as only one junction point is lost per smallest loop. Although the detailed arguments leading to Equations 7 and 8 are based on smallest loops, the equations themselves are valid provided each inelastic loop causes the loss of two junction points when f=3 and one junction point when f=4, irrespective of the size of the loop.

To evaluate $p_{r,e}$ from modulus measurements, a value of A has to be assumed and used together with the experimental value of M_c/AM_c^o. Accordingly, Figures 4 and 5 show $p_{r,e}$ plotted versus $p_{r,c}$ for affine (A=1) and phantom (A = 1 - 2/f) chain behaviour using the results in Figure 1. One condition that should be obeyed in the plots is that $p_{r,e} > p_{r,c}$, as $p_{r,e}$ includes pre-gel and post-gel reaction. The lines $p_{r,e} = p_{r,c}$ are indicated in the figures and it is apparent that the inequality is satisfied only for affine behaviour (Figure 4), as expected for the small-strain measurements used. The condition $p_{r,e} > p_{r,c}$ need not be met if a significant number of complex ring structures are formed pre-gel which then become elastically active when incorporated in the gel. The present interpretation of the plots in Figures 4 and 5 makes the simplifying assumption that all pre-gel ring structures remain elastically inactive in the final network.

The points in Figure 4 do not show a tendency for $p_{r,e}$ to tend to zero as $p_{r,c}$ tends to zero. That is, even in the limit of a perfect gelling system, inelastic loops are formed post-gel. Extrapolation to $p_{r,c} = 0$ gives $p_{r,e}^o$, the extent of reaction leading to inelastic loops at complete reaction in the perfect gelling system. The values of $p_{r,e}^o$ range from about 9% to 18% for the system studied. As expected from considerations of pre-gel intramolecular reaction, the values of $p_{r,e}^o$ are smaller for f=3 compared with f=4 and they increase as ν decreases, there being less opportunity for intramolecular reaction at lower functionalities(14).

The derived values of $p_{r,e}^o$ may be reconverted to values of M_c/M_c^o for $p_{r,c} = 0$ using Equations 7 and 8. Figure 6 shows M_c/M_c^o versus $p_{r,c}$, as in Figure 1 but with A=1 and the values of M_c/M_c^o at $p_{r,c} = 0$ consistent with those of $p_{r,e}^o$ from Figure 4. The curves give just as satisfactory a fit to the data as the straight lines in Figure 1. Because it is not possible to have concentrations of reactive groups higher than those in bulk, points at lower values of $p_{r,c}$ than those shown cannot be obtained for the particular systems studied. Thus, there are uncertainties in the values of $p_{r,e}^o$ from Figure 4 and in the intercepts shown in Figure 6. However, merely the existence of

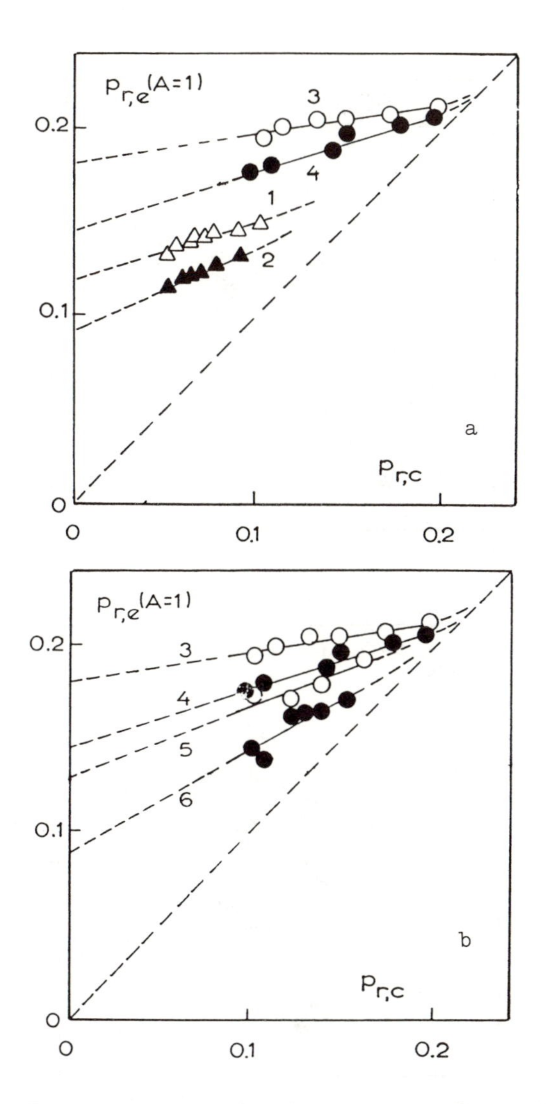

Figure 4. Extent of intramolecular reaction at complete reaction leading to inelastic loops ($p_{r,e}$) versus $p_{r,c}$. (a) Systems 1-4 of Figure 1; (b) systems 3-6 of Figure 1. Values of $p_{r,e}$ derived from modulus measurements assuming affine behaviour and each loop leading to two elastically active junction points lost for f=3 networks (1 and 2) and one junction point lost for f=4 networks (3-6). Intercepts at $p_{r,c}$ = 0 define values of $p_{r,c}$ for ideal gelling systems denoted $p^{o}_{r,e}$.

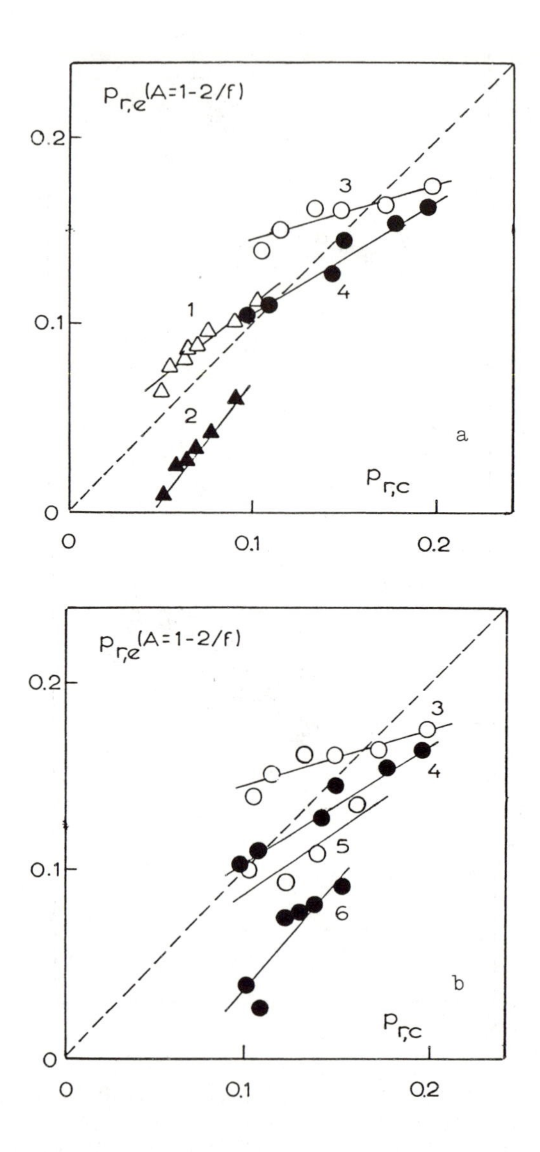

Figure 5. $p_{r,e}$ versus $p_{r,c}$. As Figure 4(a) and (b) but with values of $p_{r,e}$ derived assuming phantom chain behaviour.

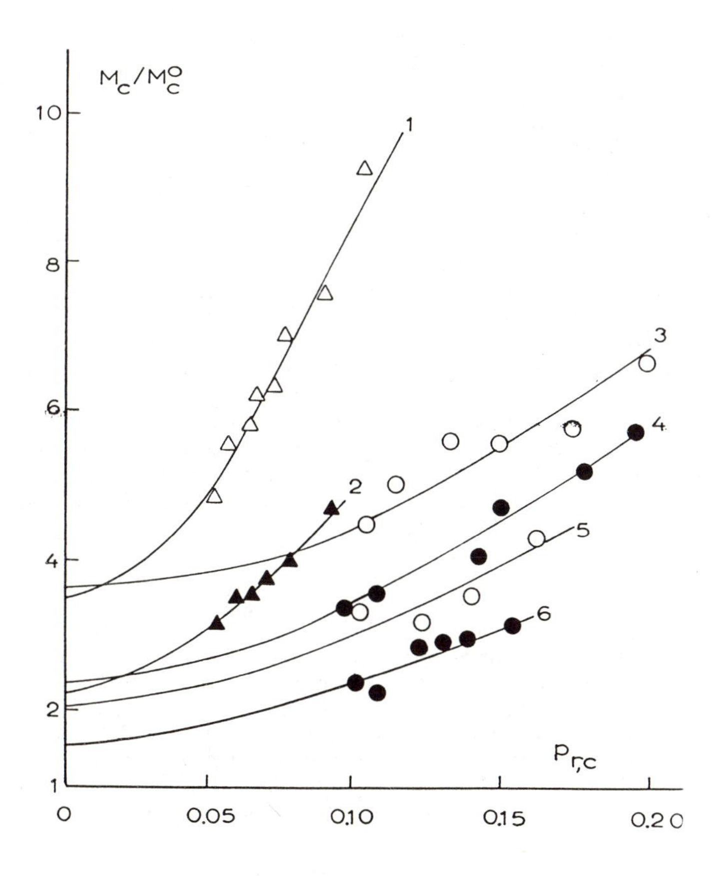

Figure 6. M_c/M_c° versus $p_{r,c}$ with curves drawn consistent with the values of $p_{r,e}^\circ$ from Figure 4.

non-zero intercepts in Figure 4 shows that post-gel intramolecular reaction indeed leads to network imperfections. It should be emphasised that, at their present stage of development, theories and calculations of polymerisation and network formation including intramolecular reaction(8,10,11) cannot treat this phenomenon. They all predict that $p_{r,e} = 0$ when $p_{r,c} = 0$.

The Universal Occurrence of Imperfections (Law-of-Mass-Action Effect)

The reaction systems discussed do not have particularly large values of ν. Hence, it may be argued that for triols and tetrols of higher molar mass, networks with $M_c = M_c^o$ would have been obtained, at least for bulk reaction systems. The decrease of $p_{r,e}^o$ for a given functionality as ν increases (Figure 4) indicates a trend in this direction. However, for $M_c = M_c^o$, $p_{r,e}^o$ must be equal to zero. That such a value is unlikely to be obtained, irrespective of ν, is apparent from Figure 4 and is emphasised by using the present value $p_{r,e}^o$ to predict a value for infinite ν. Thus, Figure 7 shows $p_{r,e}^o$ versus $(f-2)/(\nu b^2)^{3/2}$. The abscissa is proportional to the probability of pre-gel intramolecular reaction(14,15). The data used in Figure 7 are given in Table I, together with the extrapolated values ($p_{r,e}^{o,\infty}$). The values of b were obtained from analyses of the gel points of the various reaction systems(5-7,12).

Table I. Values of parameters characterising pre-gel intramolecular reaction (ν,b,$(f-2)/(\nu b^2)^{3/2}$) (5-7,12) and the extents of post-gel intramolecular reaction which, in the limit of ideal gelling systems, lead to inelastic loop formation at complete reaction ($p_{r,e}^o$). The values of $p_{r,e}^o$ define the indicated values of M_c/M_c^o and the reductions in shear moduli of the dry networks relative to those of the perfect networks ($G/G^o = M_c^o/M_c$). The values of $p_{r,e}^o$ in the limit of reactants of infinite molar mass ($\nu = \infty$) are denoted $p_{r,e}^{o,\infty}$ in the text

System	f	ν	M_c^o/g mol^{-1}	b/nm	$(f-2)/(\nu b^2)^{3/2}$ /nm^{-3}	$p_{r,e}^o$	M_c/M_c^o	G/G^o
1.HDI/LHT240	3	33	635	0.247	0.350	0.119	3.50	0.29
2.HDI/LHT112	3	61	1168	0.222	0.192	0.092	2.23	0.45
	3	∞	∞	-	0	0.060	1.57	0.64
3.HDI/OPPE1	4	29	500	0.240	0.926	0.181	3.62	0.28
4.HDI/OPPE2	4	33	586	0.237	0.792	0.145	2.38	0.42
5.HDI/OPPE3	4	44	789	0.234	0.534	0.129	2.07	0.48
6.HDI/OPPE4	4	66	1220	0.215	0.376	0.090	1.56	0.64
	4	∞	∞	-	0	0.030	1.14	0.88

The last two columns in Table I give the intercept values of M_c/M_c^o in Figure 6 and the corresponding values of shear moduli of the dry networks relative to those of the perfect networks. The values of G/G^o for $\nu = \infty$ are estimates of the <u>maximum</u> values obtainable for model, end-linking networks formed from reactants of high molar mass. It is clear that significant reductions in modulus remain.

The positive intercepts in Figure 7 show that post-gel(inelastic) loop formation is influenced by the same factors as pre-gel intramolecular reaction but is not determined solely by them. The important conclusion is that imperfections still occur in the limit of infinite reactant molar masses or very stiff chains ($\nu b^2 \to \infty$). They are a demonstration of a law-of-mass-action effect. Because they are intercepts in the limit $\nu b^2 \to \infty$, spatial correlations between reacting groups are absent and random reaction occurs. Intramolecular reaction occurs post-gel simply because of the unlimited number of groups per molecule in the gel fraction. The present values of $p^{o,\infty}_{r,e}$ (0.06 for f=3 and 0.03 for f=4 are derived from modulus measurements, assuming two junction points per lost per inelastic loop in f=3 networks and one junction point lost per loop in f=4 networks. The assumption is strictly correct if only the smallest loops are formed. However, as the intercepts relate to random reaction, classical polymerisation theory may be used tc investigate whether they can at least be consistent with defects arising only from simple loops.

Significant Structures for Inelastic Loop Formation under Conditions of Random Reaction

The condition of random reaction means that intramolecular reaction occurs only post-gel. This corresponds to the condition treated by Flory(17) whereby intramolecular reaction occurs randomly between groups on the gel. Accordingly, various types of pairs of reacting groups may be defined, namely, sol-sol, sol-gel and gel-gel. Only the last can lead to inelastic loops. Hence, the first task is to find the total amount of gel-gel reaction. This has been done for an RA_f polymerisation(16). The equations are simpler than those for an RA_2 + RB_f polymerisation and to within the accuracy required they may be used for an RA_2 + RB_f polymerisation after appropriate transformation.

The principle of the calculation is to derive an expression for the number of unreacted groups on sol molecules (Γ_s) by summing over finite species of all sizes and to find the number of unreacted groups on the gel (Γ_g) as the difference

$$\Gamma_g = \Gamma - \Gamma_s \tag{9}$$

where Γ is the total number of unreacted groups at the given extent of reaction (p). The number of unreacted groups on gel molecules cannot be found directly as the molecules have unlimited numbers of chain ends. The probability that a randomly chosen group at p is on the gel is then

$$\gamma_g = \Gamma_g/\Gamma \tag{10}$$

Given random reaction, the number of gel-gel pairs of groups reacting at p, i.e. the number of ring structures (loops) forming at p, is

$$\Gamma_{gg} = \Gamma\,\gamma_g.\,\gamma_g = fN_f(1-p)\gamma_g^2 \tag{11}$$

where $fN_f(1-p)$ is the expression for Γ, with fN_f the initial number of groups on the N_f monomers present intially. Integration of Γ_{gg} between the gel point and complete reaction, i.e.

$$\Gamma_{gg,e} = \int_{1/(f-1)}^{1} \Gamma_{gg}dp = fN_f \int_{1/(f-1)}^{1} (1-p)\gamma_g^2 dp \tag{12}$$

gives the total of number of loops formed by the end of the reaction. Equation 12 may be written more simply as

$$\Gamma_{gg,e} = fN_f.p_{r;e}^{o;\infty} = fN_f \int_{1/f-1}^{1} p_r dp \tag{13}$$

where $p_r = (1-p)\gamma_g^2$ is number of loops forming at p per initial number of groups and

$$p_{r;e}^{o;\infty} = \int_{1/(f-1)}^{1} p_r dp \tag{14}$$

is the total extent of loop formation between gel and complete reaction for an ideal gelling system of infinite molar mass. To within an accuracy of about 10%(16), $p_{r;e}^{o;\infty}$ may be compared with the $p_{r,e}^{o,\infty}$ for stoichiometric $RA_2 + RB_f$ polymerisations discussed earlier with reference to Table I.

Figure 8 shows p_r versus p for RA_3 and RA_4 polymerisations. The integrals under the curves, which have to be evaluated numerically, give $p_{r;e}^{o;\infty} = 0.0604$ for f=3 and 0.1306 for f=4. Comparison with the experimentally derived values in Table I gives the results summarised in Table II.

Table II. Comparison of calculated and experimentally derived values of the final extents of intramolecular reaction for $RA_2 + RB_f$ polymerisations, under conditions of the random reaction of functional groups ($p_{r,e}^{o,\infty}$)

f	$p_{r,e}^{o,\infty}$ calc.	$p_{r,e}^{o,\infty}$ expt.	expt./calc.	active junctions lost per loop	elastic chains lost per loop
3	0.0604	0.06	1	2	3
4	0.1306	0.03	0.23	0.23	0.46

The equality of the values for f=3 implies that random gel-gel reaction in fact leads to an average of two elastically active junctions or three elastic chains lost per pair of gel-gel groups reacted. For f=4, the experimentally derived value 0.03 was on the basis of one elastically active junction lost per pair of gel-gel groups reacted. Hence, the ratio of the experimentally derived and the calculated values, 0.03/0.13 = 0.23, is the average number of elastic-

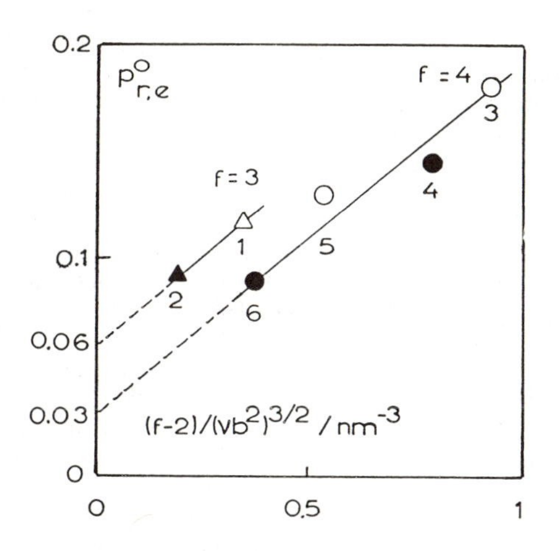

Figure 7. Dependence of post-gel intramolecular reaction leading to inelastic loops in the limit of ideal gelling systems ($p^{\circ}_{r,e}$) on the parameters affecting pre-gel intramolecular reaction ($(f-2)/(\nu b^2)^{3/2}$).

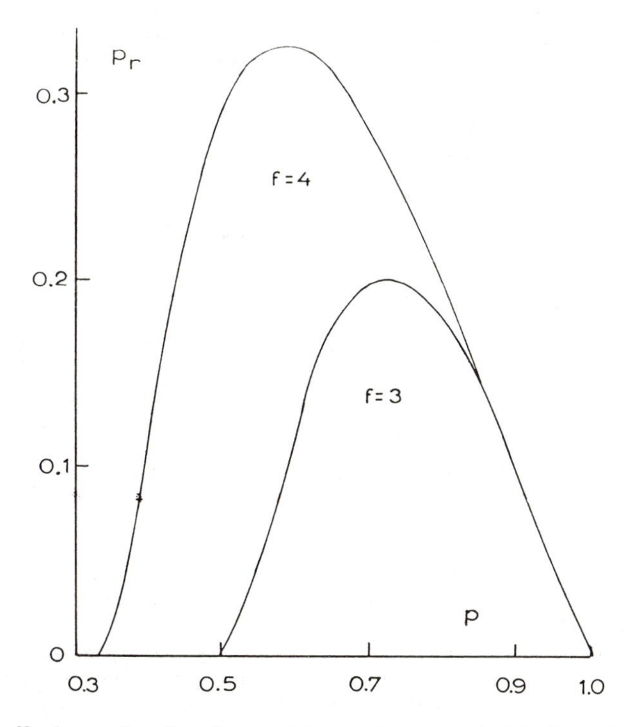

Figure 8. Number of gel-gel reacting pairs per initial number of groups (p_r) versus extent of reaction (p) in RA_3 and RA_4 polymerisations. The integrals under the curves give $p^{\circ;\infty}_{r;e} = 0.0604(RA_3)$ and $0.1306(RA_4)$.

ally active junctions actually lost per pair of gel-gel groups reacted. The agreement between the experimental and calculated values of $p_{r;e}^{o;\infty}$ for f=3 and the relatively large ratio of 0.23 for f=4 indicate that the sizes of loop which need to be considered for inelastic loop formation are indeed small.

To investigate the significance of the values of the ratios of experimentally derived and calculated values of $p_{r;e}^{o;\infty}$, let us proceed on the basis that all inelastic loops are the smallest loops that can form (i.e. one-membered loops, as in Figure 3). A more complete, but still approximate consideration including two-membered loops has been given elsewhere(16). The ratio of unity for f=3 means that one-membered loops are the only significant defects which occur regarding inelastic chains. The ratio of 0.23 for f=4 means that some larger loops occur in which chains are elastically active. It is now a question of defining the significant structures which need to be considered to give a value of 0.23 junctions lost per loop. Such structures will define the minimum sizes of structure which are needed to define the formation and topology of actual tetrafunctional networks.

It is impossible to specify in detail all the isomeric structures which occur as a network is forming. However, with regard to one-membered loop formation, two extreme types may be delineated, namely, linear and symmetric isomers. They are illustrated in Figure 9 for an RA_4 polymerisation. Linear isomers are able to form the smallest number of one-membered loops and symmetric isomers the largest number. In an RA_f polymerisation, for linear isomers of n units, the total number of pairs of unreacted ends for intramolecular reaction is

$$(f-2)((f-1)(n-\tfrac{1}{2}) + (f-2)(n-1)(n-2)/2 + (f-3)(n-1)/2),\ n > 1$$

and

$$\binom{f-1}{2},\ n = 1 \tag{15}$$

The number of pairs which can form one-membered loops is

$$\binom{f-1}{2} + (n-1)\binom{f-2}{2} \tag{16}$$

For symmetric isomers of m generations or $((f-1)^m - 1)/(f-2)$ units, the total number of pairs is

$$(f-1)^m((f-1)^m - 1)/2 \tag{17}$$

and the number for one-membered loop formation

$$\binom{f-1}{2}(f-1)^{m-1} \tag{18}$$

The preceding expressions have been evaluated for f=3 and f=4 and the fractions of one-membered loops formed for given values of n or m are shown in Table III.

As indicated previously, for f=3 only the simplest structure is relevant. For f=4, reactions within linear isomers of up to three units and reactions within symmetric isomers of up to four units

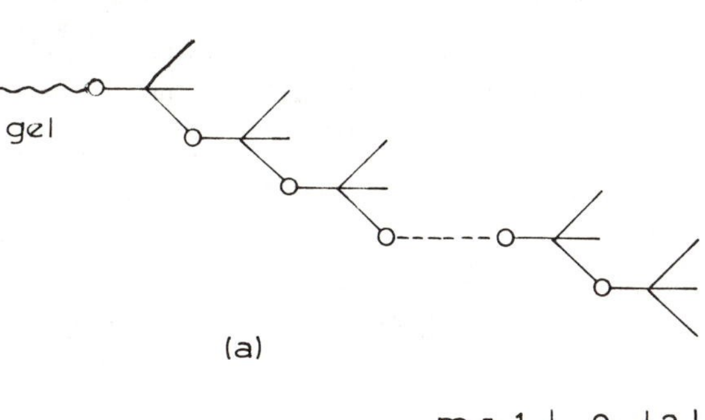

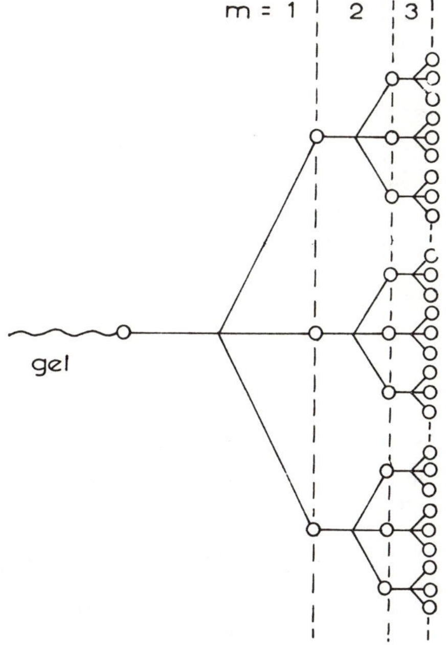

Figure 9. Isomeric structures in RA_4 gels, (a) linear and (b) symmetric. Each contains n units, divided, in the symmetric case, into m generations.

Table III. Total numers of loops and numbers and fractions of one-membered loops formed by random intramolecular reaction within linear and isomeric gel structures of different numbers of units(n) and generations (m). The fractions marked * agree with the experimentally deduced concentrations of inelastic junction points or chains on the basis of one-membered loops

m	n	total	1-membered	fraction 1-membered
f=3: linear isomers				
	1	1	1	1*
	2	3	1	0.33
f=3: symmetric isomers				
1	1	1	1	1*
2	3	6	2	0.33
f=4: linear isomers				
	1	3	3	1
	2	10	4	0.40
	3	21	5	0.24*
	4	36	6	0.17
f=4: symmetric isomers				
1	1	3	3	1
2	4	36	9	0.25*
3	13	351	27	0.08

have to be accounted for to give fractions of one-membered loops approximately equal to the experimentally deduced value of 0.23. The significant structures are illustrated in Figure 10. They are the minimum ones within which all the probabilities of intramolecular reaction have to be evaluated if network defects are to be accounted for and absolute values of moduli or concentrations of elastic chains in real networks predicted. Intramolecular reaction within these structures is significant even in the limit of reactants of infinite molar mass ($\nu \rightarrow \infty$). For actual systems, with finite values of ν, the numbers of inelastic loops actually forming will depend on the detailed chain statistics within the structutes. The structures should be considered in the sol and the gel fractions and changes in their concentrations with extent of reaction also included. The inclusion of two-membered loops(16) shows that for f=3, linear isomers up to n=2 and symmetric isomers up to n=1 probably have to be considered and for f=4, linear isomers up to n=7 and symmetric isomers up to n=4 (or m=2).

Conclusions

Intramolecular reaction always occurs in non-linear polymerisations and because of this end-linking polymerisations never lead to perfect networks. The amount of pre-gel intramolecular may be small

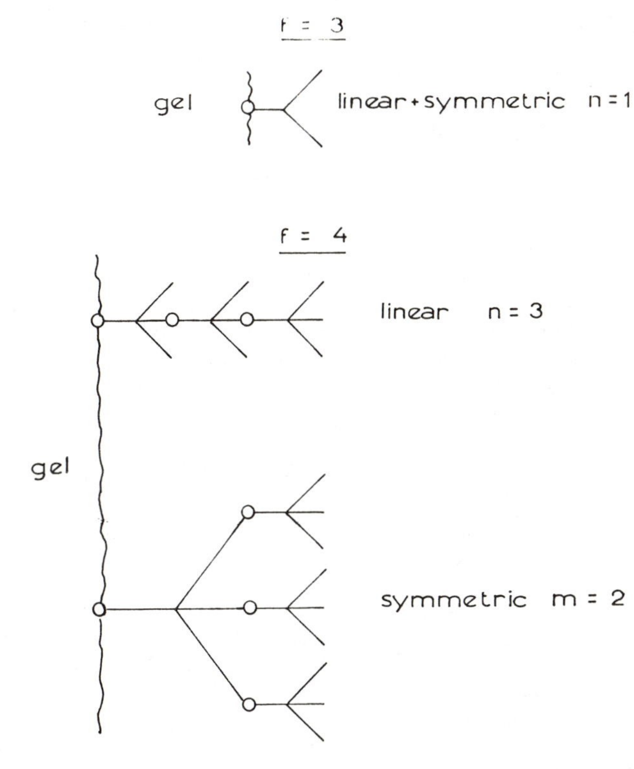

Figure 10. Significant structures for the formation of inelastic chains in trifunctional and tetrafunctional networks.

for reactants of high molar mass. However, significant post-gel intramolecular reaction always occurs, In the actual reaction systems discussed here, intramolecular reaction from both sources leads to large reductions in shear moduli below the values expected for perfect networks.

With end-linking polymerisations, the monitoring of network formation may easily be achieved through the determination of gel points. Such monitoring is important for the interpretation of networks structure and properties. For example, the systemic variation of the gel point by varying the dilution of given reactants is particularly useful. Reductions in modulus are found to depend on the same factors as pre-gel intramolecular reaction. They increase similarly with dilution and with decrease in reactant molar masses.However, they are less for networks of higher functionality, whereas intramolecular reaction is higher. This distinction is understandable, simple in terms of numbers of elastic chains lost per one-membered loop.

The absolute values of the reductions in moduli, or increases in M_c, can be interpreted in terms of small, inelastic loops. The small-strain, static moduli measured are consistent with affine chain behaviour, showing, on the basis of one-membered loops, that between about 10 and 20% of groups react to form inelastic loops by the end of a polymerisation. For the more concentrated systems a significant proportion of this comes from post-gel intramolecular reaction $(p_{r,e} - p_{r,c})$.

Extrapolation of $p_{r,e}$ to the limit of zero pre-gel intramolecular reaction for given reaction systems shows that post-gel intramolecular reaction always results in network defects, with significant increases in M_c above M_c^o. Such post-gel intramolecular reaction is characterised as $p_{r,e}^o$. The variation of $p_{r,e}^o$ with intramolecular-reaction parameters shows that even in the limit of infinite molar mass, i.e. no spatial correlation between reacting groups, inelastic loops will be formed. The formation may be considered as a law-of-mass-action effect, essentially the random reaction of functional groups. Intramolecular reaction under such conditions $(p_{r,e}^{o,\infty})$ must be post-gel and may be treated using classical polymerisation theory.

The case of RA_f polymerisations has been evaluated and the experimentally deduced values of $p_{r,e}^{o,\infty}$ interpreted in terms of one-membered loop formation within structures of defined size. The knowledge of such structures is important for theories of non-linear polymerisations and network formation. Theories have to simplify in some way the infinite numbers of structures which actually occur and it is important that those significant for intramolecular reaction are retained. In this context, it should be noted that present theories assume that ideal gelling systems lead to perfect networks. The approximation needs to be removed as it is inconsistent with the structure of the growing gel, a molecule with unlimited pairs of groups for intramolecular reaction.

Literature Cited

1. Mark, J. E. Adv. Polymer Sci. 1982, 44, 1.
2. Batsberg, W.; Kramer, O. In Elastomers and Rubber Elasticity; Mark, J. E.; Lal, J., Eds.; Amer. Chem. Soc. Symp. Series No. 193; American Chemical Society: Washington, DC, 1982; p.439.

3. Flory, P. J. Br. Polymer J. 1985, 17, 96.
4. Brotzman, R. W.; Flory, P.J. Polymer Preprints 1985, 26(2), 51.
5. Stanford, J. L.; St.pto, R. F. T. In Elastomers and Rubber Elasticity; Mark, J. E.; Lal, J., Eds.; Amer. Chem. Soc. Symp. Series No. 193; American Chemical Society: Washington, DC, 1982; p.377.
6. Stanford, J. L.; Stepto, R. F. T.; Still, R. H. In Reaction Injection Molding and Fast Polymerization Reactions; Kresta, J. E., Eds.; Plenum Publishing Corp.: New York, 1982; p.31.
7. Stanford, J. L.; Stepto, R. F. T.; Still, R. H. In Characterization of Highly Cross-linked Polymers; Labana, S. S.; Dickie, R. A., Eds.; Amer. Chem. Soc. Symp. Series No. 243; American Chemical Society: Washington, DC, 1984; p.1.
8. Lloyd, A.; Stepto, R. F. T. Br. Polymer J. 1985, 17, 190.
9. Stepto, R. F. T. Polymer Preprints 1985, 26(2), 46; In Advances in Elastomers and Rubber Elasticity; Lal, J.; Mark, J. E.,Eds.; Plenum Publishing Corp.: New York, 1986; p.329.
10. Dušek, K. Br. Polymer J. 1985, 17, 185.
11. Shy, L.Y.; Eichinger, B. E. Br. Polymer J. 1985, 17, 200.
12. Demirörs, M. Ph.D. Thesis, Univ. of Manchester, 1985.
13. Faṣina, A. B.; Stepto, R. F. T. Makromol. Chem. 1981, 182, 2479.
14. Ahmad, Z.; Stepto, R. F. T. Colloid and Polymer Sci. 1980, 258, 663.
15. Stepto, R. F. T. In Developments in Polymerisation - 3; Haward, R. N., Ed.; Applied Science Publishers Ltd.: Barking, 1982; Chapter 3.
16. Stepto, R. F. T. In Biological and Synthetic Polymer Networks; Kramer, O., Ed.; Elsevier Applied Science Publishers, Barking, 1987 in press
17. Flory, P. J. Principles of Polymer Chemistry; Cornell University Press: Ithaca, New York, 1953.

RECEIVED October 7, 1987

Chapter 3

Selective Quenching of Large-Scale Molecular Motions by Cross-Linking in the Strained State

Ole Kramer

Department of Chemistry, University of Copenhagen, Universitetsparken 5, DK-2100 Copenhagen, Denmark

It is shown without the need of a molecular theory of rubber elasticity or any assumption that chain entangling gives a large contribution to the equilibrium modulus. The contribution is equal to the rubber plateau modulus at fairly high degrees of cross-linking. High-vinyl polybutadiene was cross-linked in the strained state. This allows prevention of disentanglement by reptation and similar large scale molecular motions as well as experimental separation of the elastic contributions from cross-links and chain entangling. It is also argued that the molecular motions at the entrance to the rubber plateau must be local motions which involve chain segments smaller than the molecular weight between cross-links, only.

The large scale molecular motions which take place in the rubber plateau and terminal zones of an uncross-linked linear polymer give rise to stress relaxation and thereby energy dissipation. For narrow molecular weight distribution elastomers non-catastrophic rupture of the material is caused by the disentanglement processes which occur in the terminal zone, e.g., by the reptation process. In practical terms it means that the ´green strength´ of the elastomer is poor. The green strength can be improved by broadening the molecular weight distribution or alternatively by preventing the molecular motions responsible for rupture from occuring. Thus, understanding and controlling these molecular motions is of both fundamental and practical interest.

The stress relaxation properties of a high molecular weight polybutadiene with a narrow molecular weight distribution are shown in Figure 1. The behavior is shown in terms of the apparent rubber elasticity stress relaxation modulus for three differrent extension ratios and the experiment is carried on until rupture in all three cases. A very wide rubber plateau extending over nearly 6 decades in time is observed for the smallest extension ratio. However, the plateau is observed to become narrower with increasing extension

0097-6156/88/0367-0048$06.00/0

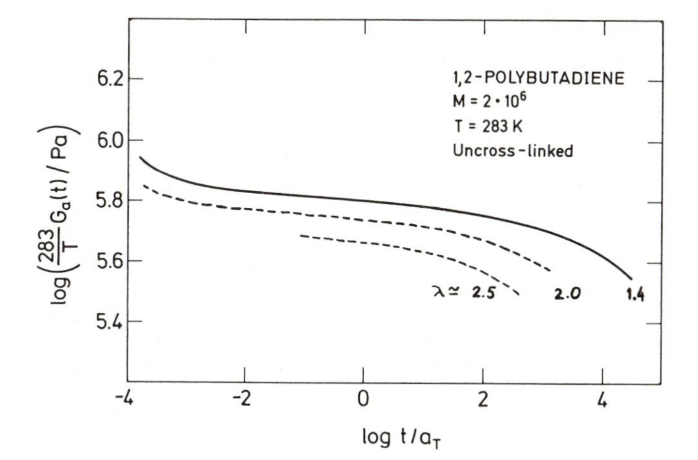

Figure 1. Stress relaxation curves for three different extension ratios. Uncross-linked high-vinyl polybutadiene with a weight average molecular weight of 2 million and a reference temperature of 283 K. G_a is the apparent rubber elasticity modulus calculated from classical affine theory. (Solid line is data from Ref. 1).

ratio. If it is true that reptation (2) is responsible for the decrease of modulus in the terminal zone, the observed behavior indicates either that the reptation process is strain dependent or that an additional relaxation process becomes important at large strains. The latter possibility has led to the proposal of a new type of molecular motion by Doi and Edwards (3).

Reptation. Gradual disentanglement by the reptation process is illustrated in Figure 2. The reptation mechanism was proposed by de Gennes in 1971 (2). The main idea is the following: Lateral diffusion of a linear chain meets high resistance since the other chains cannot be intersected. The lengthwise diffusion along the overall contour of the chain meets little resistance which means that it is a surprisingly fast process unless the molecular weight becomes exceedingly high. In the present context, it is important that reptation can be prevented by a single cross-link anywhere along the chain.

Contour Length Relaxation. Doi and Edwards have proposed an additional, faster relaxation mechanism for which they use the term ´contour length relaxation´ (5). As shown in Figure 3, contour length relaxation is a process by which a deformed linear (or star shaped) chain should retract towards the center of mass of the chain. Since the overall contour length increases upon deformation, the proposal by Doi and Edwards is that the deformed chain would want to resume the same chain density along the overall contour of the chain as that of the undeformed chain (5). The overall contour length is usually taken as the ´primitive path´. The primitive path is defined as the shortest path connecting the two ends of the chain with the same topology as the chain itself relative to the surrounding obstacles (7). The primitive path is therefore much smaller than the contour length along the backbone of the chain.

Contour length relaxation is a process which should play no role at small deformations. However, it should contribute substantially to the non-linear properties at large deformations. The overall contour length of the deformed chain should contract to the equilibrium length before reptation begins, thereby relaxing a substantial fraction of the stress (5). According to the Doi and Edwards model, the relaxation time for reptation is much longer than the relaxation time for the contour length relaxation process, especially in the case of very high molecular weight polymers. The ratio is 6N, where N is the number of steps of the primitive path (one step corresponds approximately to the distance between two consecutive entanglement points) (8). Contour length relaxation is a more difficult process to quench by cross-linking than the reptation process. Referring to Figure 3, it can be seen that a single cross-link exactly at the midpoint of the chain would have no effect whatsoever. A cross-link near each of the two chain ends would prevent the contour length relaxation process completely. Using random cross-linking, fairly high degrees of cross-linking are required to prevent the contour length relaxation process altogether.

Effect of Cross-linking

The effect of cross-linking in the unstrained state is shown in Figure 4 for a high-vinyl polybutadiene with a molecular weight of

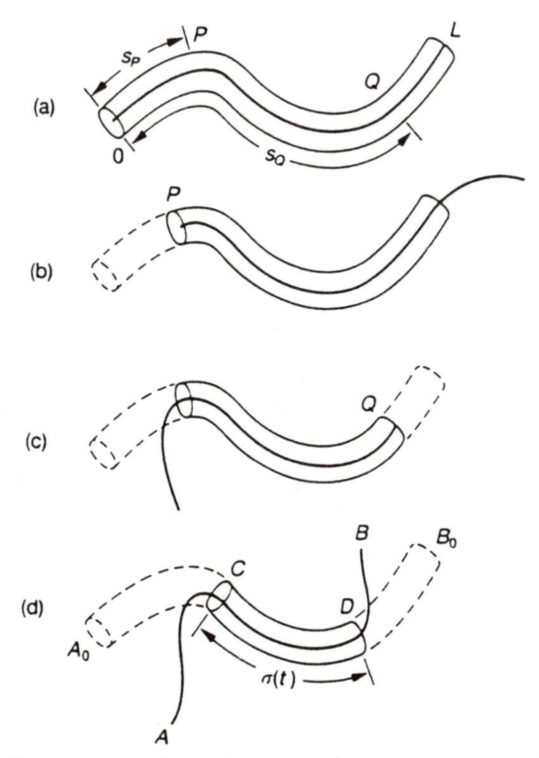

Figure 2. Four successive situations of a reptating chain. The solid line indicates the overall contour of the chain. Terminal stress relaxation should occur when the chain gradually disengages from its original surroundings. (a) The initial conformation of the primitive chain and the original tube. (b) and (c) As the chain moves right or left, some parts of the chain leave the original tube. The parts of the original tube which have become empty of the chain disappear (dotted line). (d) The conformation at a later time t. Reptation may be prevented by a single cross-link anywhere along the chain. (Reproduced with permission from Ref. 4. Copyright 1986 Oxford University Press.)

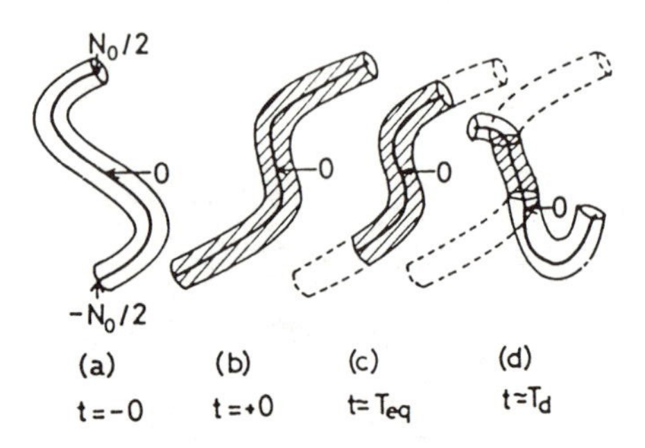

Figure 3. Contour length relaxation after a large step strain. (a) Before deformation, the conformation is in equilibrium. (b) Immediately after deformation, the chain and the tube are deformed affinely. (c) After time T_{eq}, the chain contracts inside the tube and recovers the equilibrium contour length. (d) After time T_d, the chain escapes from the deformed tube by reptation. The contour length relaxation process is more difficult to quench by cross-linking than the reptation process. (Reproduced with permission from Ref. 6. Copyright 1980 John Wiley & Sons.)

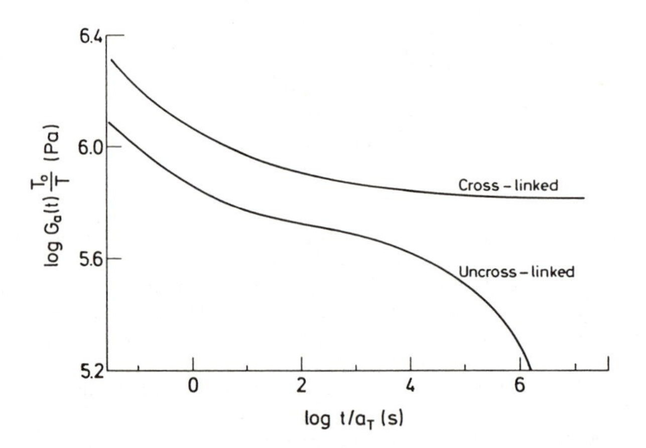

Figure 4. Stress relaxation curves for high-vinyl polybutadiene with a weight average molecular weight of 291,000 and a reference temperature of 263 K. The cross-linked sample has about 60 cross-links per chain. (Reproduced with permission from ref. 9. Copyright 1985 O. Kramer.)

291 000 which exhibits a much narrower plateau zone than that shown in Figure 1. It can be seen that the terminal zone has disappeared in the case of the cross-linked sample. It means that the molecular motions responsible for terminal relaxation have been quenched by the cross-links. Although the two curves will converge in the rubber-to-glass transition region, they seem to be superposable by a vertical shift at the entrance to the rubber plateau region. This means that the rates of relaxation in this region are insensitive to the presence of chemical cross-links. Unlike the processes responsible for terminal relaxation which could be prevented by cross-linking, the molecular motions responsible for relaxation at the entrance to the plateau zone must be rather local processes. They must involve chain segments smaller than the molecular weight between cross-links.

The degree of cross-linking for the cross-linked sample in Figure 4 is fairly high, of the order of 60 cross-linked points per chain. It means that the equilibrium modulus is higher than the plateau modulus of the uncross-linked polymer and that the equilibrium modulus is reached quickly. Elastic equilibrium is reached in less than two minutes at 323 K which is 68 K above the glass-transition temperature. At low degrees of cross-linking, the equilibrium modulus would be lower than the plateau modulus and the time to reach elastic equilibrium much longer.

The Contribution of Chain Entangling to the Equilibrium Modulus. An important question is whether chain entangling contributes to the equilibrium modulus of cross-linked elastomers. It is a problem which is of both practical and theoretical interest and which has become the source of much controversy. In his new theory (10,11), Flory specifically made the assumption that chain entangling makes no contribution to the modulus at elastic equilibrium. This assumption seems to be confirmed by a number of experiments which primarily are based on end-linked networks (12,13). Other workers believe that the cross-links trap the entangled structure thereby preventing disentanglement. Using the Langley method (14), Graessley and coworkers (15,16) found that the contribution from chain entangling depends on the degree of cross-linking and that the contribution approaches the rubber plateau modulus in highly cross-linked networks where the structure is trapped almost completely. Ferry and coworkers found the same result, using the two-network method (17,18).

Comparison of Figures 1 and 4 shows that an increase in the molecular weight from 291000 to 2 million delays terminal relaxation by more than two orders of magnitude. This is in agreement with the prediction for reptation which gives a dependence on molecular weight to the 3 to 3.4 power (19). The plateau modulus in Figure 1 is fairly constant until disentanglement sets in. It indicates that chain entangling would contribute strongly to the equilibrium modulus if disentanglement is prevented.

The importance of this observation may be reinforced by considering a hypothetical experiment, namely stress relaxation of a linear polymer with a molecular weight of 2 billion, i.e., 1000 times higher than that of the polymer in Figure 1. Although such a polymer may be difficult to make and handle experimentally, its stress relaxation properties can be predicted reliably by scaling (20). Relative to Figure 1, such an increase in molecular weight would move the terminal zone to the right by about 10 orders of magnitude, making the rubber plateau nearly entirely flat for about 15 orders of magnitude in time. It would be possible to move the terminal zone to shorter times by an increase in temperature. However, it would be impossible to perform stress relaxation experiments at high enough temperatures and at long enough times to reach the terminal zone. At least without causing thermal degradation of the polymer. Thus, the elastic effect of chain entangling would not relax below the rubber plateau value in the experimentally accessible time range since no disentanglement would occur.

The experimental stress relaxation results presented in Figures 1 and 4 combined with the conclusions from the hypothetical case clearly indicate the following: There is no reason to believe that the elastic contribution from chain entangling relaxes to zero at elastic equilibrium unless the highly entangled structure is allowed to disentangle. The simplest method to prevent disentanglement is to permanently trap the entangled structure by the introduction of chemical cross-links in the form of covalent bonds.

If we accept an elastic contribution from chain entangling in cross-linked networks, the problem is to find the relative magnitudes of the contributions from chain entangling and from cross-links. Since the two effects work in parallel in ordinary networks, it is necessary to know the concentration of effective cross-links and to have a molecular theory which correctly relates the modulus to the concentration of cross-links. The contribution from chain entangling is then found as the difference between the observed and the calculated modulus. This seems to be an almost hopeless task unless the network structure is very simple and the contribution from chain entangling is large.

The challenge is therefore to develop an experiment which allows an experimental separation of the contributions from chain entangling and cross-links. The Two-Network method developed by Ferry and coworkers (17,18) is such a method. Cross-linking of a linear polymer in the strained state creates a composite network in which the original network from chain entangling and the network created by cross-linking in the strained state have different reference states. We have simplified the Two-Network method by using such conditions that no molecular theory is needed (1,21).

Simplified Two-Network Method. As stated above, introduction of cross-links in the strained state means that the original highly entangled network and the cross-link network have different reference states (17,21). The reference state for the cross-link network is the state in which the cross-links are introduced. Relative to that state we should expect no contribution from the cross-links to the external stretching force. By conducting all force measurements relative to that state, all forces and changes in forces must come from the other network, the original entanglement network.

The simplified two-network experiment is performed in the following manner: A thin strip of the uncross-linked polymer is stretched by about 60% and maintained with constant length throughout the remainder of the experiment. The force is monitored at all times. After a predetermined relaxation period, the temperature is decreased to below the glass transition temperature to quench all overall conformational changes. The sample is cross-linked in the glassy state with high energy (10 MeV) electrons, thereby permanently trapping the remaining entangled structure. The temperature is then raised to a temperature of 8K above the glass transition temperature in order to allow the trapped free radicals to react.

In the next step, the temperature is raised to the stress relaxation temperature, allowing a direct comparison of the forces before, f, and after cross-linking, f_c. The sample is then allowed to reach elastic equilibrium. This is accomplished by first heating the sample to a higher temperature (323 K) and subsequently cooling it back to the stress relaxation temperature to measure the equilibrium force, f_e. The three forces f, f_c and f_e may now be directly compared at the same extension and temperature. Figure 5 shows the measured forces for a typical experiment for a high-vinyl polybutadiene with a molecular weight of 2 million. The stress relaxation temperature is 283 K in this case, i.e., 28 K above the glass transition temperature in order to get close to the terminal zone.

The results for three different degrees of cross-linking are shown in Table 1. The quantity $100(f-f_c)/f$ is a measure of the percentage change in the stretching force caused by cross-linking and the subsequent relaxation during heating to the stress relaxation temperature. The quantity $100(f-f_e)/f$ is a measure of the relative difference between the stress relaxation force and the equilibrium force. Table 1 shows very small values of the quantity $100(f-f_c)/f$ for all three degrees of cross-linking. This proves that the cross-links do not contribute to the stretching force in this experiment where the sample is held at constant length. They serve the purpose of trapping the entangled structure, only. The small values of the quantity $100(f-f_e)/f$ show that the equilibrium force is only slightly smaller than the stress relaxation force just before cross-links are introduced in all three cases. This demonstrates that the trapped entangled structure gives rise to a large equilibrium force, since the cross-links contribute nothing to the measured forces irrespective of degree of cross-linking.

Table 1. Results from cross-linking in the strained state with extension ratios of about 1.6.

Sample	7	9	8
Dose/kGy	50	100	200
$\frac{f-f_c}{f}$ 100, %	3.0	2.9	3.6
$\frac{f-f_e}{f}$ 100, %	8.9	7.9	5.4

SOURCE: Data are from ref. 1.

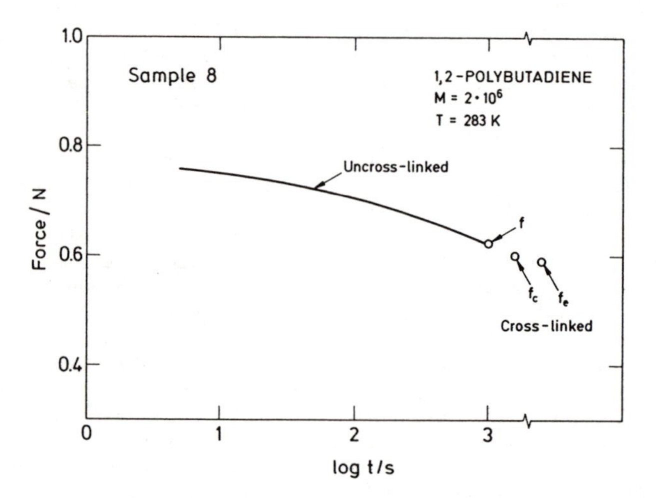

Figure 5. Stretching force at constant length and a temperature of 283 K for the same polymer as in Figure 1. The sample is uncross-linked for the first 1000 s and then rapidly cooled to the glassy state and cross-linked with high energy electrons. The force f_c is obtained by subsequent heating to the stress relaxation temperature of 283 K. The equilibrium force f_e at 283 K is obtained after heating to 323 K followed by cooling back to 283 K. (Reproduced with permission from Ref. 1. Copyright 1988 Elsevier Applied Science Publishers.)

Thus, the simplified Two-Network experiment shows by a direct comparison of forces at constant length that the trapped entangled structure of a well cross-linked elastomer contributes to the equilibrium modulus by an amount that is approximately equal to the rubber plateau modulus. The modulus contribution from the trapped entangled structure will be less for lower molecular weights and especially at low degrees of cross-linking (14).

The small values of the quantity $100(f-f_e)/f$ prove that chain scission is absent or small for this system. This is important since the interpretation of the experimental results would be difficult or even impossible in the case of substantial chain scission. It should also be stressed that the functionality of the cross-links need not be known, and that a distribution of cross-link functionalities as well as inhomogeneous cross-linking is unimportant. The reason being that the cross-links merely serve the purpose of trapping the entangled structure.

Finally, it should be pointed out that no molecular theory of rubber elasticity is required and that no assumptions were made in order to reach above conclusions.

The Simplified Two-Network Method and Contour Length Relaxation. An attempt was made to study the contour length relaxation process by the simplified Two-Network method. It is possible to make samples which have been cross-linked in the strained state before the contour length relaxation process takes place (Batsberg, W.; Graessley, W.W.; Kramer, O., unpublished data). However, all samples which were tested at large extension ratios broke before elastic equilibrium was reached. The first direct evidence of the contour length relaxation process has been obtained on high molecular weight poly(ethylethylene), using small angle neutron scattering on highly strained samples (22).

Conclusions

Cross-linking in the strained state can be used to quench the large scale molecular motions selectively. It is found that chain entangling contributes strongly to the equilibrium modulus if cross-links are introduced before disentanglement takes place. The contribution from chain entangling at elastic equilibrium is approximately equal to the rubber plateau modulus for elastomers of very high molecular weight and fairly high degrees of cross-linking. This result is obtained without the need of a molecular theory of rubber elasticity and without making any assumptions. Cross-linking in the strained state could not be used to study the contour length relaxation process due to sample rupture at large extensions. The molecular motions responsible for stress relaxation at the entrance to the rubber plateau zone are argued to be local processes which involve chain segments smaller than the molecular weight between cross-links, only.

Acknowledgment

Support from the Danish Council for Natural Science and the Danish Council for Science and Industrial Research under the FTU program are gratefully acknowledged.

Literature Cited

1. Batsberg, W.; Hvidt, S.; Kramer, O.; Fetters, L.J. In Biological and Synthetic Polymer Networks; Kramer, O., Ed.; Elsevier Applied Science Publishers: London, 1988; p xx
2. de Gennes, P.G. J. Chem. Phys. 1971, 55, 572.
3. Doi, M.; Edwards, S.F. J. Chem. Soc. Faraday Trans. 2 1978, 74, 1789, 1802 and 1818.
4. Doi, M.; Edwards, S.F. The Theory of Polymer Dynamics; Clarendon Press: Oxford, 1986, p 194.
5. Ibid., p 247.
6. Doi, M. J. Polym. Sci., Polym. Phys. Ed. 1980, 18, 1891.
7. Ref. 4, p 192.
8. Graessley, W.W. Adv. Polym. Sci. 1982, 47, 67.
9. Kramer, O. Brit. Polym. J. 1985, 17, 129.
10. Flory, P.J. Proc. Roy. Soc. London (A) 1976, 351, 351.
11. Flory, P.J. J. Chem. Phys. 1977, 66, 5720.
12. Mark, J.E. Makromol. Chem. 1979, Suppl. 2, 87.
13. Llorente, M.A.; Mark, J.E. Macromolecules 1980, 13, 681.
14. Langley, N.R. Macromolecules 1968, 1, 348.
15. Dossin, L.M.; Graessley, W.W. Macromolecules 1979, 12, 123.
16. Pearson, D.S.; Graessley, W.W. Macromolecules 1980, 13, 1001.
17. Kramer, O.; Carpenter, R.L.; Ty, V.; Ferry, J.D. Macromolecules 1974, 7, 79.
18. Ferry, J.D. Polymer 1979, 20, 1343.
19. Doi, M. J. Polym. Sci., Polym. Lett. Ed. 1981, 19, 265.
20. de Gennes, P.G. Scaling Concepts in Polymer Science; Cornell Univ. Press: Ithaka, 1979.
21. Batsberg, W.; Kramer, O. J. Chem. Phys. 1981, 74, 6507.
22. Mortensen, K.; Batsberg, W.; Kramer, O.; Fetters, L.J. In Biological and Synthetic Polymer Networks; Kramer, O., Ed.; Elsevier Applied Science Publishers: London, 1988; p. xx.

RECEIVED October 7, 1987

Chapter 4

Rubber Elasticity Modulus of Interpenetrating Heteropolymer Networks

Christos Tsenoglou

Department of Chemistry and Chemical Engineering, Stevens Institute of Technology, Castle Point, Hoboken, NJ 07030

This is a theoretical study on the structure and modulus of a composite polymeric network formed by two intermeshing co-continuous networks of different chemistry, which interact on a molecular level. The rigidity of this elastomer is assumed to increase with the number density of chemical crosslinks and trapped entanglements in the system. The latter quantity is estimated from the relative concentration of the individual components and their ability to entangle in the unmixed state. The equilibrium elasticity modulus is then calculated for both the cases of a simultaneous and sequential interpenetrating polymer network.

This is a theoretical study on the entanglement architecture and mechanical properties of an ideal two-component interpenetrating polymer network (IPN) composed of flexible chains (Fig. 1a). In this system molecular interaction between different polymer species is accomplished by the simultaneous or sequential polymerization of the polymeric precursors [1]. Chains which are thermodynamically incompatible are permanently interlocked in a composite network due to the presence of chemical crosslinks. The network structure is thus reinforced by chain entanglements trapped between permanent junctions [2,3]. It is evident that, entanglements between identical chains lie further apart in an IPN than in a one-component network (Fig. 1b) and entanglements associating heterogeneous polymers are formed in between homopolymer junctions. In the present study the density of the various interchain associations in the composite network is evaluated as a function of the properties of the pure network components. This information is used to estimate the equilibrium rubber elasticity modulus of the IPN.

Structural Characteristics of the Network

Let us assume that each of the two networks participates with a v_i volume fraction in the IPN (i = 1,2) and is composed of elastomeric

0097-6156/88/0367-0059$06.00/0

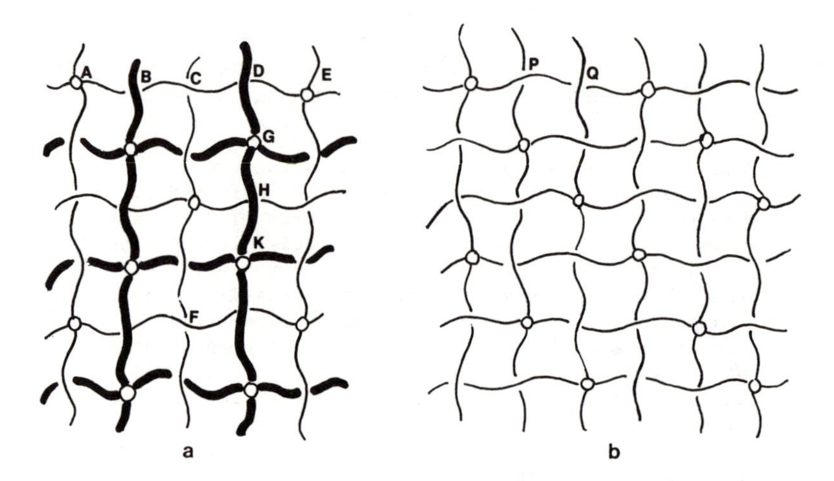

Fig. 1. (a) An idealized representation of an interpenetrating polymer network. Lines of unequal thickness signify different polymeric species, crosslinks are represented by circles and entanglements by line interceptions. The molecular weight of the AE step is equal to M_{c1}, of BD is M_{e12}, EF is M_{e11}, GK is M_{c2}, and DH is M_{e21}. Here $N_1 = 3$ and $N_2 = 1$. (b) A single-component elastomeric network. The molecular weight of the PQ step is equal to M_{e10} and $N_{10} = 2$.

strands of molecular weight M_{ci}. In the absence of any other component ($v_1 = 1 = 1-v_2$), M_{e10} is the molecular weight between trapped entanglements along a (1) strand and ρ_{10} the density of the pure system. The number of primitive steps (i.e., subsegments defined by two consecutive chain junctions) per strand is then equal to:

$$1 + N_{10} = 1 + M_{c1}/M_{e10} \tag{1}$$

where N_{10} is the number of steps caused by trapped entanglements. Similar quantities can be defined for the pure network (2), by replacing the index (1) with (2).

The equivalence between steps, valid in a single component network, does not exist in an interpenetrating polymer network due to the diversity of interacting species. If ij signifies a step lying on an i chain and confined by two entanglements with j polymers, then N_{ij} is the average number of such subsegments per strand and M_{eij} its average molecular weight (Fig. 1). The total number of steps per i strand is then given by:

$$1 + N_i = 1 + N_{i1} + N_{i2} \tag{2}$$

where:

$$N_{ij} = M_{ci}/M_{eij} \tag{3}$$

and N_i signifies the number of steps caused by trapped entanglements. In an ideally structured IPN, without any phase separation, it is reasonable to assume that the frequency of entanglements between chains of the same species decreases linearly with the fractional participation of this polymer in the system:

$$M_{e10}/M_{e11} = N_{11}/N_{10} = v_1 \tag{4a}$$

$$M_{e20}/M_{e22} = N_{22}/N_{20} = v_2 \tag{4b}$$

It is also assumed that the fraction of trapped entanglements along a chain of the one species caused by associations with chains of the other species is equal to the fractional participation of the steps of the second species in the total step population:

$$N_{12}/N_1 = \nu_2 N_2/(\nu_1 N_1 + \nu_2 N_2) \tag{5a}$$

$$N_{21}/N_2 = \nu_1 N_1/(\nu_1 N_1 + \nu_2 N_2) \tag{5b}$$

where ν_i signifies the number density of the i chains in the IPN:

$$\nu_i = \rho_{i0} v_i N_A/M_{ci} \tag{6}$$

and N_A is the Avogadro number. Physical symmetry requires that the number density of the 12 steps is equal to that of the 21 steps:

$$\nu_1 N_{12} = \nu_2 N_{21} \tag{7}$$

The fact that Equation 7 can easily be proven from Equation 5 shows the self-consistency of this assumption.

Combination of Equations 1-7 results in the following expressions for N_i and M_{eij}:

$$N_1 = N_{11} + N_{12} = M_{c1}(v_1/M_{e10} + 1/M_{e12}) \tag{8a}$$

$$N_2 = N_{21} + N_{22} = M_{c2}(1/M_{e21} + v_2/M_{e20}) \tag{8b}$$

where

$$M_{e12}\ v_2(\rho_{20}/\rho_{10})^{1/2} = M_{e21}\ v_1(\rho_{10}/\rho_{20})^{1/2} = (M_{e10}M_{e20})^{1/2} \tag{9}$$

These relationships are adequate to quantitatively define the architecture of the IPN.

Equilibrium Elasticity Modulus

Simultaneous IPN. According to the statistical theory of rubber elasticity, the elasticity modulus (E_e), a measure of the material rigidity, is proportional to the concentration of elastically active segments (ν_e) in the network [3,4]. For negligible perturbation of the strand length at rest due to crosslinking (a reasonable assumption for the case of a simultaneous IPN), the modulus is given by:

$$E_e = 3k_B T \nu_e \tag{10}$$

where k_B is the Boltzmann constant and T the absolute temperature. For an elastomer with no defects, and with the assumption that steps caused by trapped entanglements are as effective in storing elastic energy as steps due to crosslinks, ν_e is evaluated by the following relationship:

$$\nu_e = \nu_1(1 + N_{11} + N_{12}) + \nu_2(1 + N_{12} + N_{22}) \tag{11}$$

The presence of dangling segments in the IPN decreases the structural solidity of the system since steps associated with the tethered chains have a finite lifetime subject to the dynamics of the free end [5,6]. If ϕ_i is the fraction of strands in the (i) network that are crosslinked on both ends, then:

$$\nu_e = \phi_1\nu_1\ (1 + \phi_1 N_{11} + \phi_2 N_{12}) + \phi_2\nu_2\ (1 + \phi_1 N_{21} + \phi_2 N_{22}) \tag{12}$$

Combination of this last relationship with Equations 6 and 8 results in:

$$\frac{E_e}{3RT} = \phi_1 v_1\left(\frac{\rho_{10}}{M_{c1}}\right) + \phi_2 v_2\left(\frac{\rho_{20}}{M_{c2}}\right) + \left(\phi_1 v_1\left(\frac{\rho_{10}}{M_{e10}}\right)^{1/2} + \phi_2 v_2\left(\frac{\rho_{20}}{M_{e20}}\right)^{1/2}\right)^2 \tag{13}$$

where R is the ideal gas constant. In the absence of the second intermeshing networks ($v_2 = 0$) the theory reduces to an expression similar to the result by Mancke et al. [5] for a single component elastomer. On the other hand, in the case of one of the system components (e.g., 2) being an oligomer or a low-molecular weight solvent, the inability of the small molecules to form to entanglements is accounted for by setting $M_{e20} \to \infty$.

An alternative way to express the rigidity of the composite network is in terms of the rigidities of its individual components. Starting from Equation 13 it can easily be shown that:

$$E_e = E_{e1} + E_{e2} + 2\phi_1\phi_2 v_1 v_2 \sqrt{E_{N1}E_{N2}} \tag{14}$$

where E_{ei} is the equilibrium modulus of the network (i) alone, stripped of its intermeshing heteropolymer neighbor; i.e., a network with a $(1-\phi_i)$ fraction of defects, a polymer chain concentration equal to $(\rho_{i0}v_i)$ and a molecular weight between trapped entanglements equal to M_{ei0}/v_i [5]:

$$E_{ei} = RT\ \phi_i \rho_{i0} v_i (1/M_{ci} + \phi_i v_i / M_{ei0}) \tag{15}$$

and E_{Ni} is the rubbery plateau modulus of a pure, undiluted fluid (i):

$$E_{Ni} = 3RT\ \rho_{i0}/M_{ei0} \tag{16}$$

It is remarkable that the predicted modulus from Equation 14 is higher in value than the one predicted by a simple linear additivity law due to the reinforcing effect of the heteropolymer couplings.

Sequential IPN. The preceding analysis does not apply to the case of a sequential IPN. The formation of this system originates with the synthesis of the network (1). Then, network (1) is swollen with monomer (2) which is subsequently polymerized in situ to form a second network. Due to perturbed chain dimensions, the modulus of the first network is higher than the corresponding modulus in the unswollen state by a factor equal to $v_1^{-2/3}$ [6]:

$$E_e = 3k_BT\ (v_1^{-2/3}\ \nu_{e1} + \nu_{e2}) \tag{17}$$

where ν_{e1} and ν_{e2}, the number densities of elastically active steps in the two networks, are given by the first and second term on the right of Equation 12. Following the same sequence of thoughts which gave Equation 13 an expression for the equilibrium modulus of a sequential IPN is obtained:

$$\frac{E_e}{3RT} = \phi_1 v_1^{1/3}\left(\frac{\rho_{10}}{M_{c1}}\right) + \phi_2 v_2\left(\frac{\rho_{20}}{M_{c2}}\right) + \phi_1^2 v_1^{4/3}\left(\frac{\rho_{10}}{M_{e10}}\right) + \phi_2^2\ v_2^2\left(\frac{\rho_{20}}{M_{e20}}\right) + $$

$$+ \phi_1\phi_2\ (v_1 + v_1^{1/3})\ v_2\left(\frac{\rho_{10}\rho_{20}}{M_{e10}M_{e20}}\right)^{1/2} \tag{18}$$

Because v_1 is a fractional quantity, Equation 18 always predicts modulus values larger than the corresponding expression for a simultaneous IPN(Equation 13). For the special case of a network with no defects or trapped entanglements ($\phi_i \rightarrow 1$, $M_{ei} \rightarrow \infty$), an earlier result derived by Sperling is recovered (Ref. [1], Equation 4.8).

Discussion

Starting from simple statistical considerations concerning the coupling between similar or dissimilar chains, a model was constructed for the architecture of a class of IPN where interaction among phases occurs on a molecular level. It was shown(Equation 13) that the contributions to the modulus due to crosslinks are subject to linear additivity, while a square root additivity rule holds for the contributions due to entanglements.

Comparisons of the theory with experiment can not be presently made due to the lack of data on well characterized molecular IPN. Indications about its validity can, however, be deduced by examining its consistency at extreme cases of material behavior. The agreement at the one-component limit, for example, provided that the rubber is not very weak (ϕ not very small), has been successfully demonstrated by Ferry and coworkers [5]. A useful result is obtained at the version of the theory applicable to the fluid state (i.e., at the limit of zero crosslinking). From the last two terms of Equation 13, the following relationship can be derived for the plateau [7] and time dependent relaxation modulus of miscible polymer blends:

$$E_{Blend}^{\frac{1}{2}} = v_1 E_1^{\frac{1}{2}} + v_2 E_2^{\frac{1}{2}} \tag{19}$$

This simple mixing rule demonstrates satisfactory agreement with experimental evidence from experiments with binary fluid blends [7]. Furthermore, it is similar in form with the result from a continuum theory approach by Davis [8], applicable for IPN with dual phase continuity but which are not mixed on a molecular level. This last model involves an exponent equal to 1/5 instead of 1/2 and is quite successful in predicting the experimental evidence [1] from permanent networks.

The molecular theory presented in this paper is not antagonistic but complementary to the macroscopic continuum models of multicomponent polymeric materials [9]. This is due to the fact that material homogeneity is rarely the case in IPN, and that some phase separation does occur upon polymerization. As a result, a two-phase material is produced. Due to partial miscibility each of the two phases contains both polymeric species but at different compositions. For the modeling of this heterogeneous system it is suggested here to first use the molecular approach for the description of each of the homogeneous phases and then combine these two results with a suitable phenomenological model (depending on the prevailing morphology) for the description of the multiphase IPN as a whole.

Literature Cited

1. Sperling, L.H., Interpenetrating Polymer Networks and Related Materials, Plenum Press, New York, 1981, pp. 38,51,160.
2. Ferry, J.D., Viscoelastic Properties of Polymers, 3rd ed., Wiley, New York, 1980, p. 409.
3. Graessley, W.W., Adv. Polymer Sci., 16, 1 (1974).
4. Treloar, L.R.G., The Physics of Rubber Elasticity, Oxford, 1958, p. 75.
5. Mancke, R.G., Dickie, R.A., and Ferry, J.D., J. Polym. Sci., Part A-2, 6, 1783 (1968).
6. Flory, P.J., Principles of Polymer Chemistry, Cornell U. Press, 1953, pp. 492, 462.
7. Tsenoglou, C., Polymer Preprints, 28, (2), 198, (1987), and J. Polymer. Sci. Phys. Ed., submitted.
8. Davis, W.E.A., J. Phys. D., Appl. Phys., 4, 1176 (1972).
9. Dickie, R.A., in Polymer Blends, Vol. 1, (D.R. Paul and S. Newman Eds.), Academic Press (1978), p. 353.

RECEIVED October 7, 1987

Chapter 5

Star-Branched Nylon 6

Effects of Branching and Cross-Linking on Polymer Properties

Lon J. Mathias and Allison M. Sikes

Department of Polymer Science, University of Southern Mississippi, S.S. Box 10076, Hattiesburg, MS 39406–0076

Nylon 6 two-armed and three-armed systems prepared via anionic bulk polymerization had different characteristics than their linear counterparts. While little differences were found in IR and NMR spectra, changes in performance characteristics were discovered. Both two-armed and three-armed species were found to form crosslinked networks at high ratios of initiator to monomer. It was also found that the two-armed and three-armed species have slightly lower melt temperatures, but higher flow temperatures than their linear counterparts.

Branched polymeric materials have different properties than their linear counterparts. In the case of star-branched polymers (multiple branches radiating from a single site), enhanced engineering properties are possible from increased chain entanglements. The initial goal of this research was to create a material with enhanced performance properties via a star-branched network.

Nylon 6 is well known for its high performance characteristics. It is easily polymerized hydrolytically (the commercial process), anionically, and cationically. Nylon 6 is already well characterized making it ideal for star versus linear material studies.

The details of the anionic polymerization of nylon 6 have been extensively reviewed (1-8) and will only be discussed briefly as they affect the star-polymerization of nylon 6. Nylon 6 is polymerized anionically in a two-step process (Figure 1). The first step, creation of the activated species 3, is the slow step. The ε-caprolactam monomer 1 reacts in the presence of a strong base (such as sodium hydride) to form the caprolactam anion 2. This anion reacts with more caprolactam monomer to form 3. The reaction of this activated species with lactam anions occurs rapidly to form the nylon 6 polymer 4.

0097–6156/88/0367–0066$06.00/0

Polymerization is assumed to occur only through the imide linkages--crucial to the formation of a star-branched nylon 6 species. If an imide-containing species (such as N-acetylcaprolactam) is added to the reaction, the slow step of the reaction is by-passed, allowing polymerization to take place very rapidly. This should also be the case for a tri-imide for generating star nylon species.

A star-branched nylon 6 may be created by using a "star initiator"--a species with three or more imide linkages as activated sites. It was initially assumed that reaction would take place only at the imide sites and that cross-linking would not occur, although the bis-caprolactam species is known as a crosslinking agent. (9-11) Star species have been generated in the past (9,11-12), but have not been completely characterized or evaluated for performance behavior.

Experimental Section

Materials. Four initiating systems have been synthesized that are not available commercially: 1,3,5-trimesoyl tris-caprolactam (TTC); phthaloyl bis-caprolactam (1,2-PBC); isophthaloyl bis-caprolactam (1,3-PBC); terephthaloyl bis-caprolactam (1,4-PBC). These materials were all characterized with FTIR, NMR, and thermal analysis. To date, only the TTC and 1,3-PBC initiators have consistently polymerized caprolactam (the 1,2 and 1,4 species have not been tested at this point). The commercially available N-acetylcaprolactam was also used to produce a linear nylon.

Preparation of Linear and Star Nylon 6. The star-branched initiator, trimesoyl-tris-caprolactam (TTC) 6 was synthesized from the commercially available trimesoyl chloride 5 using the route shown in Figure 2. (6) Trimesoyl acid chloride in benzene was slowly added to a stirring solution of ε-caprolactam, pyridene, and benzene. After addition was complete, the solution was heated to 70° C for 30 minutes to assure conversion to the initiator species. Single and difunctional analogs were synthesized using the same reaction scheme for direct comparison of the star-branched and linear aromatic initiator systems.

Star branched nylons were formed in an *in situ* process using the TTC initiator (Figure 3) according to a known procedure (13). Varying amounts of ε-caprolactam monomer 1 were reacted with the TTC initiator 6 using sodium hydride as the catalyst. Linear nylon species were generated in a similar manner using the single and disubstituted initiators as well as N-acetylcaprolactam. Polymerizations were carried out under nitrogen and at either 150° C or 170° C. The nylon polymers formed very rapidly (usually within 3-20 minutes) with the use of the aromatic initiators and were highly colored, ranging from bright orange to dark brown. In contrast, nylon made with N-acetylcaprolactam was straw yellow in color and polymerization took place less rapidly.

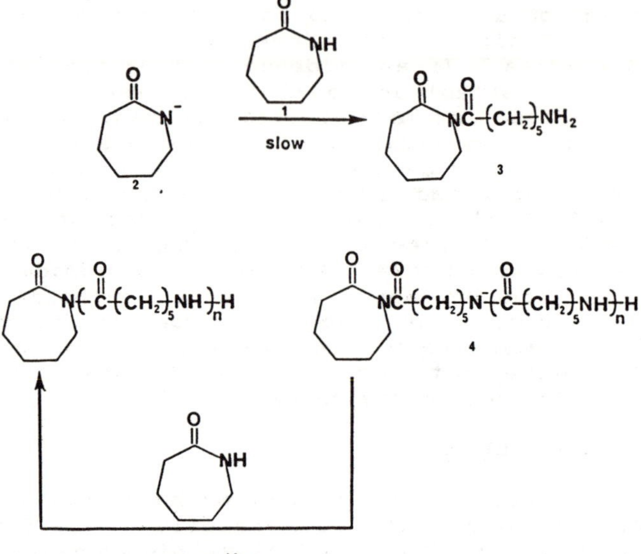

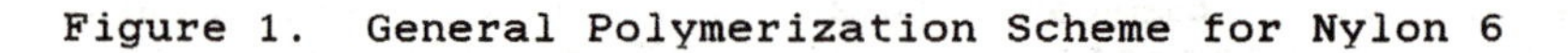
Figure 1. General Polymerization Scheme for Nylon 6

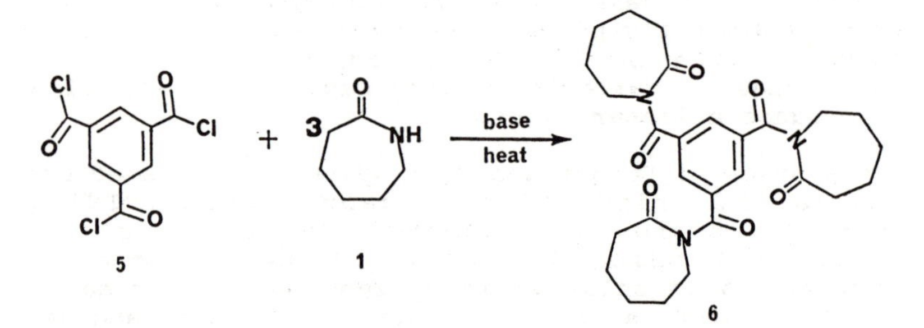

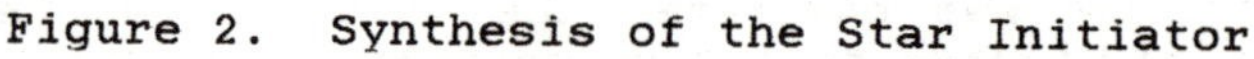
Figure 2. Synthesis of the Star Initiator

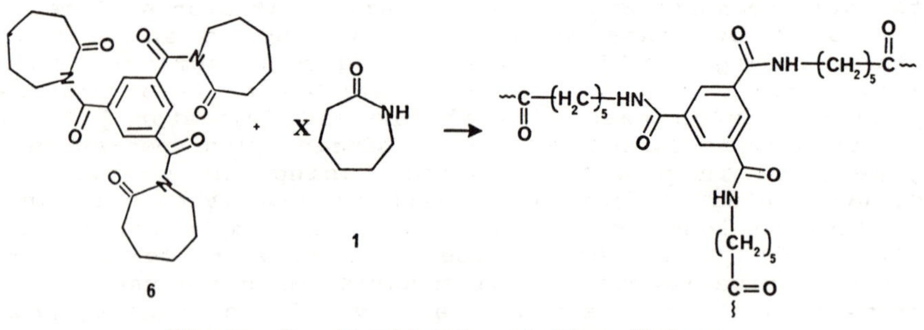

Figure 3. Synthesis of Star Nylons

The nylon samples were originally polymerized in dried test tubes under nitrogen. The caprolactam monomer and initiators were melted either in a flame or in a silicon oil bath. Polymerization was initiated by using 50% NaH as an oil dispersion. After polymerization, the sample "plugs" were removed from the tube by breaking the glass. These "plugs" were then melt-pressed between aluminum foil sheets to form a thick film. This method proved unsatisfactory for *in situ* studies. Later samples were polymerized in dried, untreated beakers in a silicon oil bath with nitrogen flowing over the top to control moisture absorption. NaH was added when the caprolactam and initiator system reached 150° C. Heating was continued until the system was completely hard and had shrunken away from the sides of the beaker. This block of material was easily removed from the beaker and was then cut into disks of approximately 1 mm thickness on a lathe. About 2 cm of the top of the block remained which usually contained some unpolymerized material. After removing most of this non-polymeric material by cutting and scraping with a razor blade, parts of the remaining sample were melt-pressed at 250° C. Both the melt-pressed and *in situ* disks were subjected to the same test procedures.

Polymer Characterization. Melt-pressed films of the nylon samples were examined using a Nicolett 5DX FTIR. The samples were pressed at 240° C - 260° C at 4,000 to 20,000 psi. Samples in 88% formic acid were precipitated into methanol and also examined as KBR pellets. All samples were scanned a minimum of 100 times.

NMR spectra were obtained using either a JEOL 90FXQ spectrometer or a Bruker MSL 200 with samples in both the solid and solution states. Solution spectra were obtained in 93.6% H_2SO_4 at ambient temperature.

Intrinsic viscosities were determined at 25° C using a Cannon-Ubbelohbe viscometer with 93.6% H_2SO_4 as the solvent.

DSC, TGA, and DMA were performed using a DuPont 9000 system. DSC runs were performed at scan rates of 10° C/minute and repeated at least one time. TGA runs were also carried out at 10° C/minute. DMA scans were at 5° C/minute with a 0.2 mm oscillation amplitude.

Tensile testing was performed on an Instron A1020C at elongation rates of 50%/min and 100%/min. A minimum of 7 samples were tested per material type. Both melt-pressed films and *in situ* disks were examined. Some samples were also conditioned in a humidity chamber before testing to insure that the samples contained the same amount of water which acts as plasticizer.

Results and Discussion

The chemical nature of the nylon materials was first determined with solution NMR using H_2SO_4 as the solvent.

The unusual properties of the star-branched system became evident when materials made with a low ratio of caprolactam to star initiator formed insoluble gels (2% polymer in H_2SO_4). Higher ratio star-branched polymers went into solution, although the aromatic initiating groups were not visible. Spectra of linear and star nylon 6 are shown in Figure 4. There is little difference between the top spectrum, a soluble star nylon, and the bottom spectrum, that of a linear nylon 6. Solid plugs of the materials, machined to fit the rotor of the solid-state NMR, have been examined briefly with only slight differences between linear nylon 6 and the star systems detected.

The thin films of the nylon samples that were melt-pressed for FTIR examination gave similar results. Linear samples pressed out easily to make thin films that were excellent for IR evaluation. Star branched materials would not press thin enough to give good spectra even at maximum press pressure. It was originally hoped that star and linear nylons would have different crystal structure forms (14); unfortunately, FTIR has shown little differences other than those that occur from differences in sample preparation (Figure 5).

The molecular weight determination of the star nylon samples has been made difficult because of the gelation problem. High monomer ratio star branched samples only go into solution upon heating (40-60° C). Samples of higher star initiator content would not dissolve even with extended heating, and retained their original shapes even when highly swollen with solvent. The two-arm nylons also displayed this tendency to a limited extent. Nylons made with N-acyl-caprolactams dissolved fairly rapidly with no evidence of the extensive gelation found in the two-arm and star species. Intrinsic viscosities for the single-armed samples ranged from 1.8-4.0 dl/g, 2.9-5.1 dl/g for the two-arm species, and 1.4-4.0 dl/g for the star samples that were soluble. The differences in the intrinsic viscosities between the star and linear samples may not be significant or may be due to smaller hydrodynamic volume of the star-branched species.

Thermal analysis has shown that the two-armed and star-branched materials have slightly lower melting points than the single-armed species (210-215° C versus 218-225° C for the linear species). TGA scans of the anionically produced materials show that the degradation temperatures are lower than that of commercial nylon 6 produced by hydrolytic polymerization (Figure 6). The degradation may be partly caused by residual NaOH (15) (the by-product from the use of the NaH initiator). Several star nylon and linear nylon samples were treated by stirring in $CHCl_3$ for 1 week in order to remove low molecular weight species and initiation by-products. TGA scans of these materials show great improvement in the degradation temperature (dashed curve, Figure 6). DMA analysis has shown that the star-branched systems still retain part of their

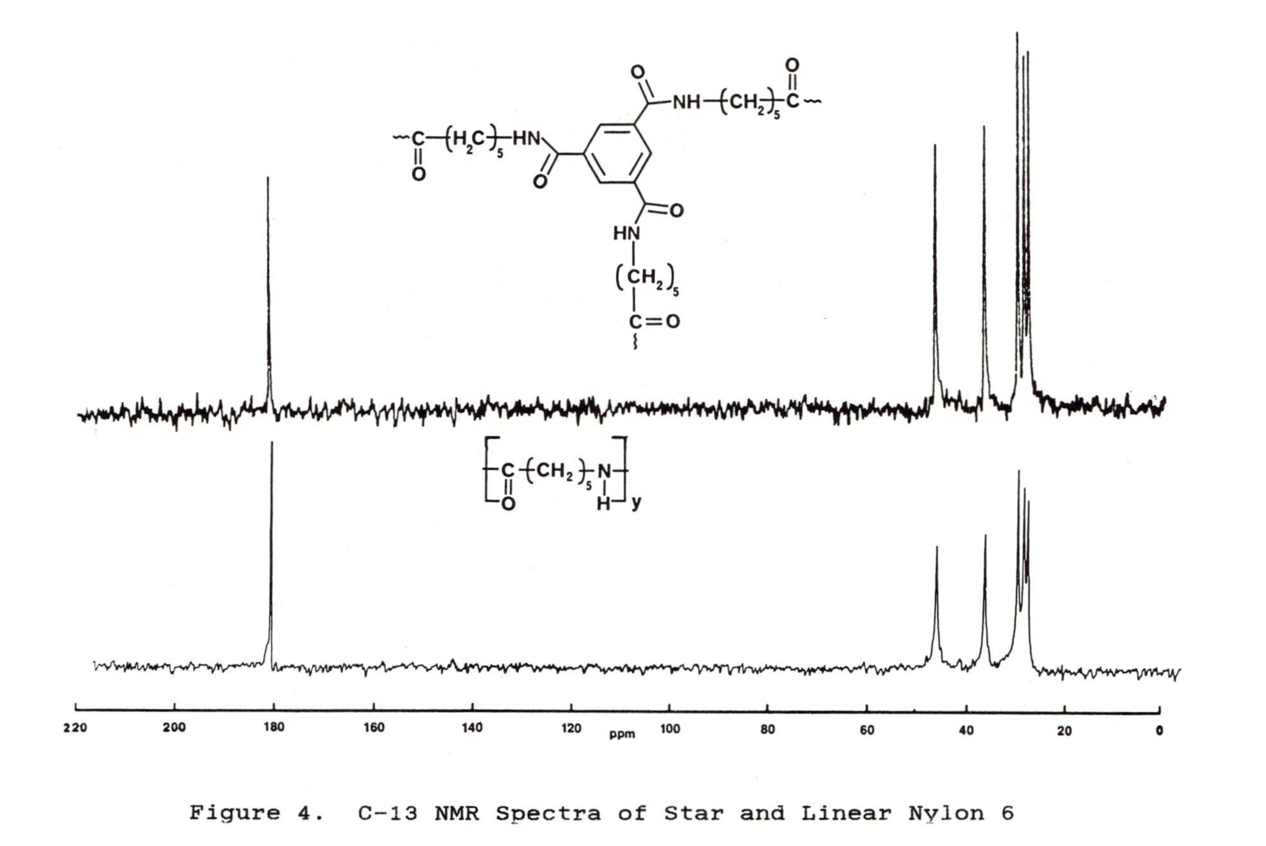

Figure 4. C-13 NMR Spectra of Star and Linear Nylon 6

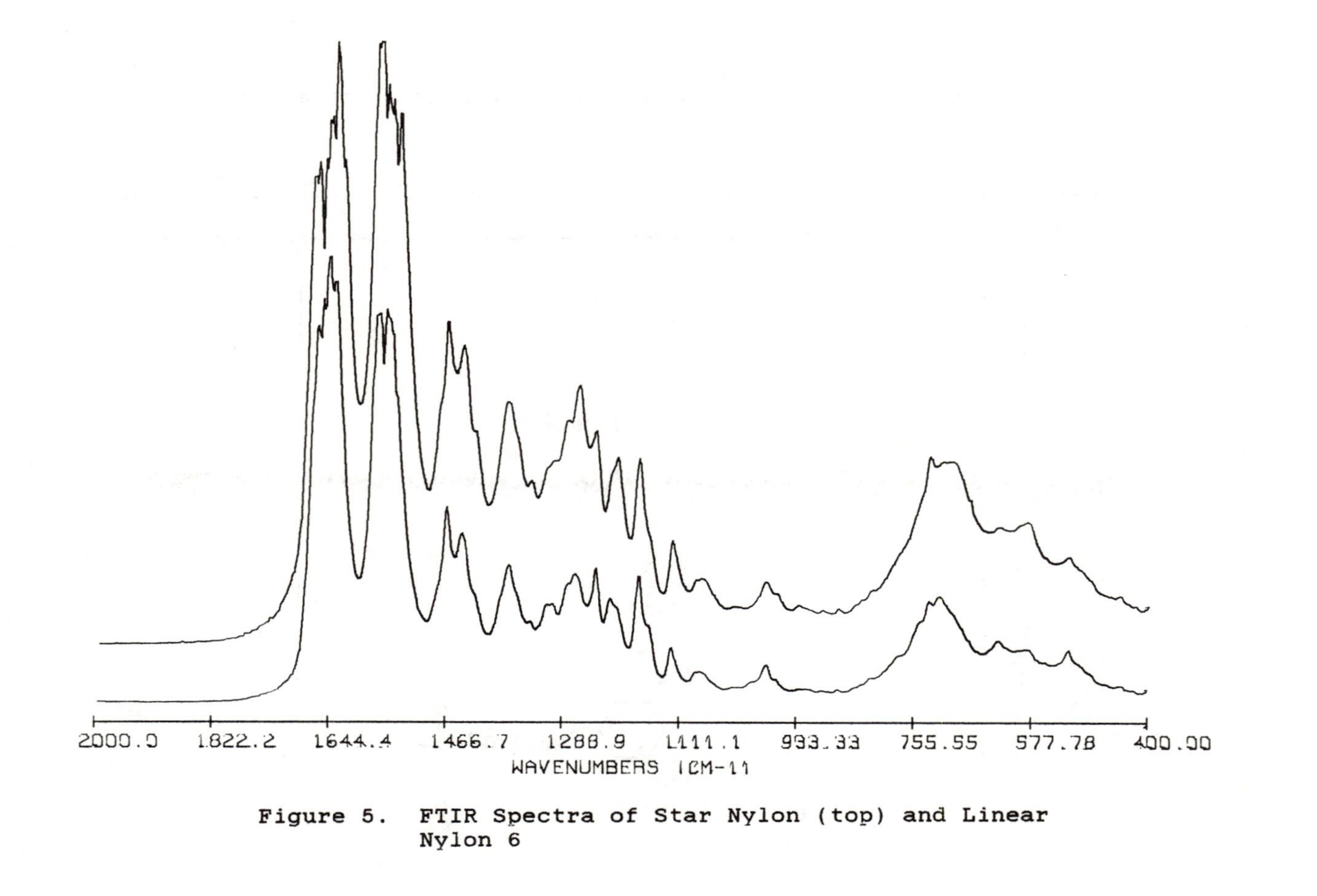

Figure 5. FTIR Spectra of Star Nylon (top) and Linear Nylon 6

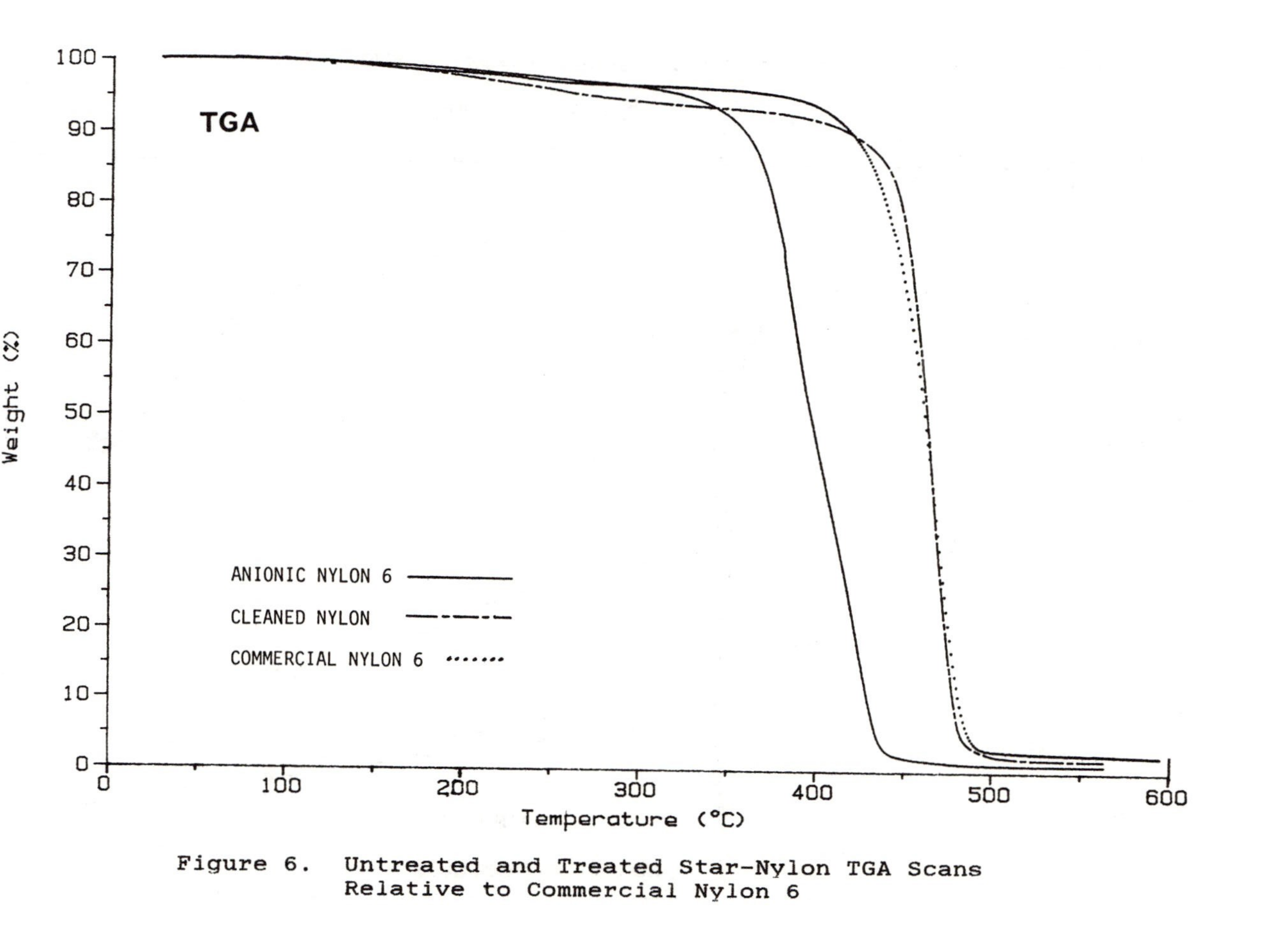

Figure 6. Untreated and Treated Star-Nylon TGA Scans Relative to Commercial Nylon 6

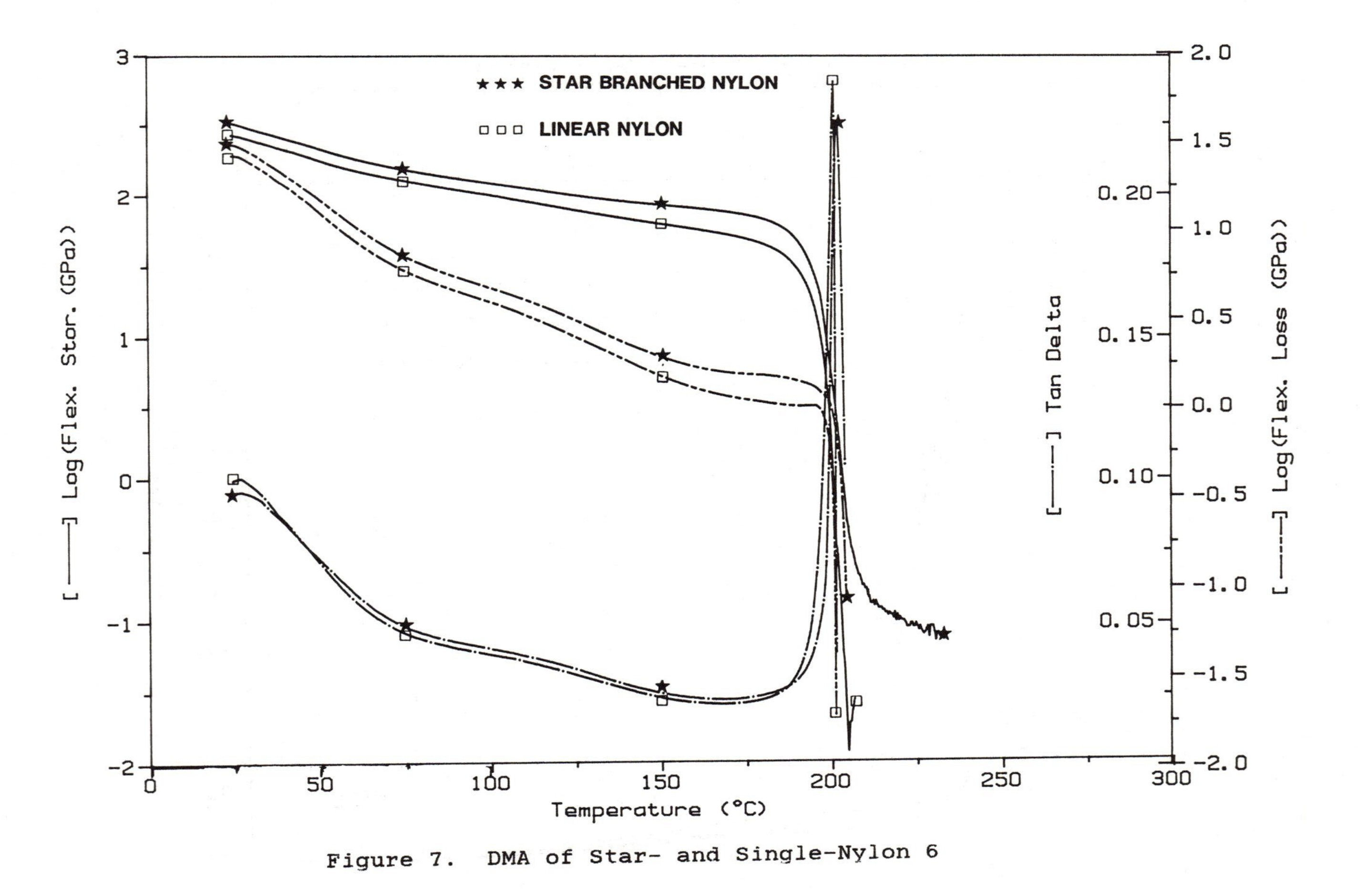

Figure 7. DMA of Star- and Single-Nylon 6

mechanical integrity even after melting which occurs slightly below the single-armed polymer. As shown in Figure 7, the latter nylon (curve marked with □) fails completely at a lower temperature than that of the star species (marked with *) indicating that no network formation is present to maintain properties. The two-armed curve falls somewhat between that of the star and the single-armed species.

Tensile testing of both disk and film samples has shown that differences in the nylon samples are dependent upon sample preparation. Samples conditioned in a desiccator have lower elongation and higher moduli than samples conditioned in a humidity chamber. Average values obtained from materials conditioned in atmosphere are given in Table 1. From the results given in this table, no clear trend is seen. Unexpectedly, the two-armed species has the highest elongation values.

Table 1. Instron Results (in situ samples)

Arms	Stress at Yield (MPa)	Tensile Strength (MPa)	Tensile Modulus (MPa)	%E
Single	160	140	2 200	80
Two	130	120	2 100	140
Three	150	120	2 200	80

Conclusions

The star-branched and two-armed systems appear to have enhanced entanglements or crosslinked networks that allow them to maintain their forms at temperatures above the linear nylon 6 melt temperature. Both the star-branched and two-armed species have melt temperatures slightly below that of single-armed nylon 6. The di-armed nylon appears to have properties similar to both the single-armed and the star species.

Literature Cited

1. Moore, J. A. and Pahls, T. J., Angew. Makromol. Chemie., 1976, 55, 141-154.

2. Greenley, R. Z., Stauffer, J. C. and Kurz, J. E., Macromolecules, 1969, 2, 561-567.

3. Scelia, R. P., Schonfeld, S. E. and Donaruma, L. G., J. Appl. Polym. Sci., 1964, 8, 1363-1369.

4. Scelia, R. P., Schonfeld, S. E. and Donaruma, L. G., J. Appl. Polym. Sci., 1967, 11, 1299-1313.

5. Reimschussel, H. K., J. Polym. Sci., Macromol. Rev., 1977, 12, 65.

6. Donaruma, L. G., Scelia, R. P. and Schondeld, S. E., J. Heterocyclic Chem., 1964, 1, 48-50.

7. Sekiguchi, H. and Coutin, B., J. Polym. Sci., Polym. Chem. Ed., 1973, 11, 1601-1614.

8. Mikheyev, V. N., Kharitonov, V. M. and Sokolova, N. P., Polym. Sci. USSR, 1980, 21, 2121-2128.

9. Kralicek, J., Kubanek, V. and Kondelikova, J., Collection Czechoslov. Chem. Commun., 1974, 39, 742.

10. Kondelikova, J., Tuzar, Z., Kralicek, J., Sandova, K., Strohalmova, M. and Kubanek, V., Angew, Makromol. Chemie, 1977, 64, 123-31.

11. Hendrick, R. M. and Gabbert, J. D., U. S. Patent 4 031 164, June 21, 1977.

12. Korshak, V. V., Frunze, T. M., Kurashev, V. V., Shleifman, R. B. and Donilevskaya, L. B., Vysokomolekul. Soedin., 1966, 8, 519-525.

13. Mathias, L. J., Vaidya, R. A. and Canterberry, J. B., J. Chem. Educ., 1984, 61, 805-807.

14. Parker, J. P., J. Appl. Polym. Sci., 1977, 21, 821-837.

15. Mukherjee, A. K. and Goel, D. K., J. Appl. Polym. Sci., 1978, 22, 361-368.

RECEIVED October 7, 1987

Chapter 6

Network Formation and Degradation in Urethane and Melamine–Formaldehyde Cross-Linked Coatings

David R. Bauer

Ford Motor Company, P.O. Box 2053, Dearborn, MI 48121

Isocyanate and melamine-formaldehyde resins are commonly used as crosslinkers in automotive coatings. Coatings based on these different crosslinkers have substantially different cure kinetics, network structure, and durability. Formation and degradation of crosslink structure in urethane and melamine crosslinked coatings are compared in this paper. Key differences in cure chemistry and kinetics which result differences in coating performance are identified. The chemistries of network structure degradation on exposure to UV light and water are discussed in terms of their effect on ultimate durability.

High solids coatings currently in use in the automotive industry as topcoats are typically thermoset coatings. They derive most of their physical properties from the formation of a highly crosslinked network structure. Two crosslinker chemistries are commonly employed. They are the reaction of isocyanate groups with polyols to form urethane crosslinks or the reaction of melamine-formaldehyde resins with polyols. The two classes of crosslinkers (isocyanates and melamines) yield coatings with significantly different process requirements and properties. For example, formulation of isocyanate coatings requires two components while melamine crosslinked coatings can be formulated with sufficient stability in a single component. Cure requirements can be significantly different and coating performance is strongly influenced by choice of crosslinker. For example, urethane coatings are generally more resistent to damage by industrial fallout and in stabilized formulations have better durability than similar melamine coatings. Use of isocyanates and melamine-formaldehyde resins as crosslinkers in coatings have been the subject of several recent reviews (1-4). The purpose of this paper is to review the specific differences in chemistry between the two coating systems responsible for the observed differences in properties. This review is based primarily on work in our laboratory on crosslinking chemistry, kinetics, network formation, and degradation in melamine-formaldehyde

0097–6156/88/0367–0077$06.00/0

and isocyanate crosslinked coatings (5-20). The paper is divided into three parts; first, the materials and techniques used are briefly described, second, differences in network formation and network structure are discussed, and third, the effect of ultraviolet light and water on network structure are compared particularly with regard to stabilization and ultimate durability.

Materials and Techniques

Polyols. Typical polyols used in automotive topcoats include acrylic copolymers and polyesters which have varied number of hydroxyl groups. Acrylic copolymers ranging in number average molecular weight from 1,000 to 10,000 and containing 15-40% by weight of a hydroxy functional comonomer such as hydroxyethyl acrylate have been studied. The acrylic copolymers were prepared by conventional free radical solution polymerization.

Melamine-formaldehyde Resins. Characterization of melamine-formaldehyde resins has been reviewed by Christensen (21,22). For the high solids coatings of interest here, there are basically two classes of melamine-formaldehyde resins: fully alkylated and partially alkylated. Typical structures are shown in Figure 1. The resins differ primarily in their level of amine and methylol functionality. The classifications are to some extent arbitrary since the range of structures in melamine resins is more or less continuous. As will be discussed below, the distinction between full and partial alkylation is important since the mechanism of cure depends on whether or not an alkoxy group is adjacent to an amine group or to another alkoxy group. The level of methylol functionality is also important since this group undergoes self-condensation. Melamine crosslinked coatings are generally formulated with an excess of alkoxy to hydroxy. The excess is largest for fully alkylated melamines and can be as high as 3:1.

Isocyanate Crosslinkers. A wide variety of both aromatic and aliphatic isocyanate crosslinkers are used in coatings (4). Aliphatic isocyanates are used when external durability is required. The isocyanate crosslinker studied in this work is the biuret of hexamethylene diisocyanate (Figure 1). Although resins based on triisocyanurates have been claimed to be superior in durability (23), biuret based triisocyanates are more commonly used. Urethane coatings are generally formulated with a ratio of isocyanate to hydroxy of around 1:1.

Fourier Transform Infrared Spectroscopy. Typical infrared spectra of cured and degraded melamine and urethane coatings are shown in Figures 2 and 3 (12,14). In both coatings the polyol is an acrylic resin. Spectra were obtained in transmission on KRS-5 salt plates. Spectra can also be obtained from KBr pellets of pulverized coating, by reflection from polished substrates, by attenuated total reflection, and by photoacoustic spectroscopy. The latter two techniques probe surface chemistry and are particularly useful in photodegradation studies. A description of the advantages and disadvantages of the different sampling techniques has been given (24,25). The specific changes used to follow cure and degradation will be discussed in the sections below.

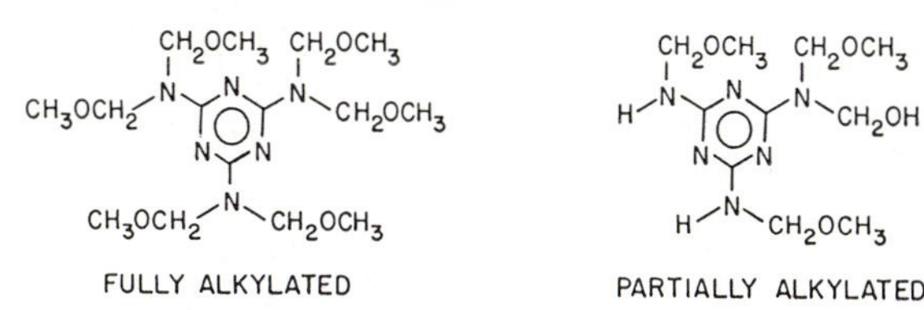

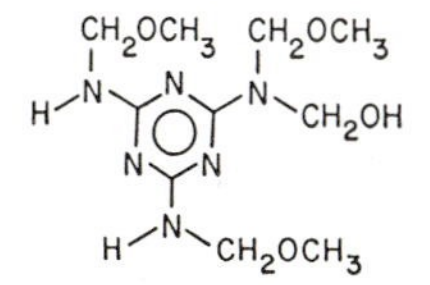

ISOCYANATE:

$$OCN-(CH_2)_6-N\begin{cases} \overset{O}{\overset{\|}{C}}-NH(CH_2)_6-NCO \\ \underset{O}{\underset{\|}{C}}-NH(CH_2)_6-NCO \end{cases}$$

BIURET OF HEXAMETHYLENE DIISOCYANATE

Figure 1. Typical melamine–formaldehyde and isocyanate cross-linkers.

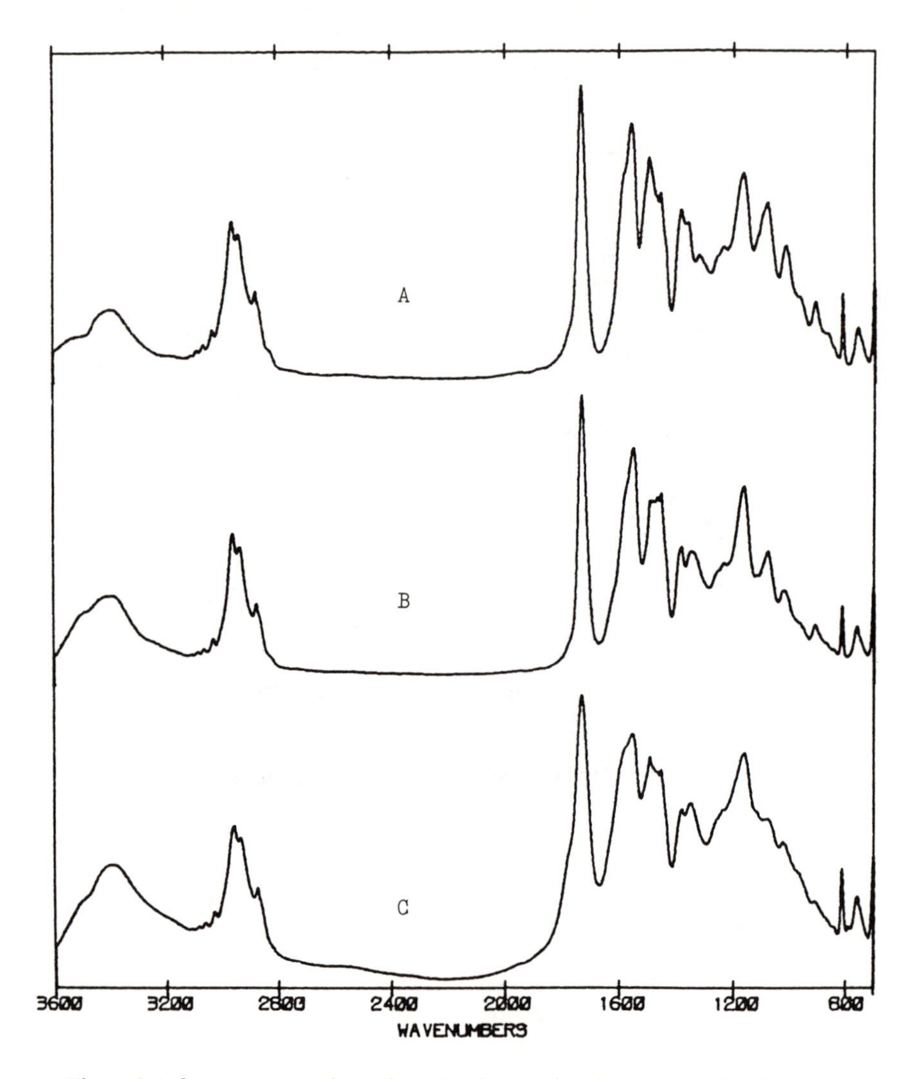

Figure 2. Infrared spectra of acrylic–melamine coating. Spectrum A denotes cured coating; B denotes coating exposed to condensing humidity; and C denotes coating exposed in QUV weathering chamber. (Reproduced with permission from ref. 12. Copyright 1984 John Wiley.)

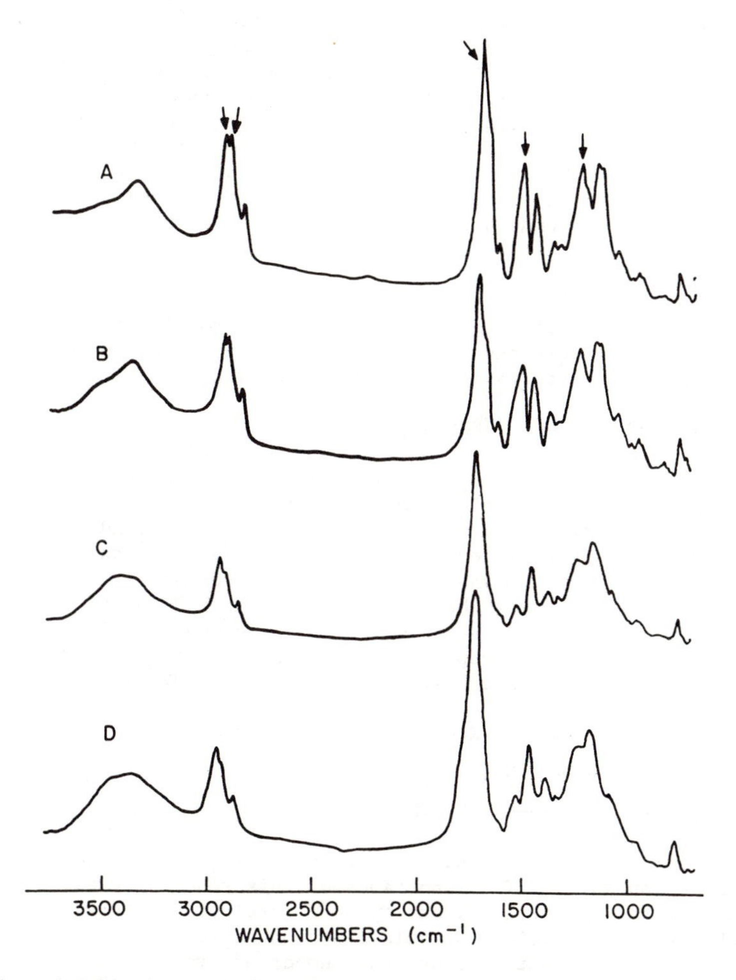

Figure 3. Infrared spectra of acrylic–urethane coating. Spectrum A denotes cured coating; B denotes coating exposed to condensing humidity; C denotes coating exposed in QUV weathering chamber; and D denotes coating exposed in a "dry" QUV. (Reproduced from ref. 14. Copyright 1986 American Chemical Society.)

Magic Angle C-13 Nuclear Magnetic Resonance. Crosslinked coatings are insoluble and thus cannot be studied using conventional NMR techniques. Within the past 12 years, modifications of the solution NMR experiment have been developed (including dipolar decoupling, magic angle sample spinning, and cross polarization) which yield high resolution NMR spectra from solid polymers (26-28). Typical cross polarization spectra of melamine and urethane coatings are shown in Figure 4. These spectra provide similar structural information to the FTIR spectra. The principal advantage of solids NMR in studying crosslinked polymers is it ability to provide dynamic information as well as structural information. Measurements of various relaxation times can provide information about the mobility of specific groups leading to a better understanding of the crosslink structure.

Network Formation

Differences in Crosslink Chemistry. The main crosslinking reaction in isocyanate-polyol coatings is the reaction of the isocyanate group with hydroxy groups to form a urethane crosslink.

$$\text{-NCO + ROH} \longrightarrow \text{-NHC(=O)OR} \quad \text{(U1)}$$

Reaction U1 can be followed by monitoring the disappearance of the strong NCO band at 2280 cm^{-1}. This reaction has been shown to obey second order kinetics with an activation energy of around 8-9 Kcal/mole (14). Uncatalyzed bake temperatures are generally greater than 100 C for these isocyanates. Catalysts such as tertiary amines, tin and zinc catalysts can be used to lower the bake temperature (4). The only significant side reaction in urethane coatings is the reaction of isocyanate groups with water in the coating. This leads to the formation of urea crosslinks via reactions U2-U4,

$$\text{-NCO + } H_2O \longrightarrow \text{-NHCOOH} \quad \text{(U2)}$$

$$\text{-NHCOOH} \longrightarrow \text{-}NH_2 + CO_2 \quad \text{(U3)}$$

$$\text{-}NH_2 \text{ + -NCO} \longrightarrow \text{-NHCONH-} \quad \text{(U4)}$$

The rate limiting step in this side reaction is the reaction of isocyanate and water. The importance of this reaction depends on the polarity of resin, the level and type of catalyst, and the humidity in the bake oven. The effect of this reaction is to reduce the number of urethane crosslinks but increase the number of urea crosslinks. The total crosslink density will likely decrease since two isocyanate groups are necessary to form one urea link while only one isocyanate group is required to form a urethane link. If excess isocyanate groups are present, this reaction will consume the excess isocyanate and increase the crosslink density. This reaction can occur not only in the bake oven but also on aging of the coating if unreacted isocyanate groups are present after cure.

In contrast to the fairly simple crosslinking chemistry in urethane coatings, crosslinking in melamine-formaldehyde coatings can be quite complex. The following crosslinking reactions can be written for melamine coatings (29):

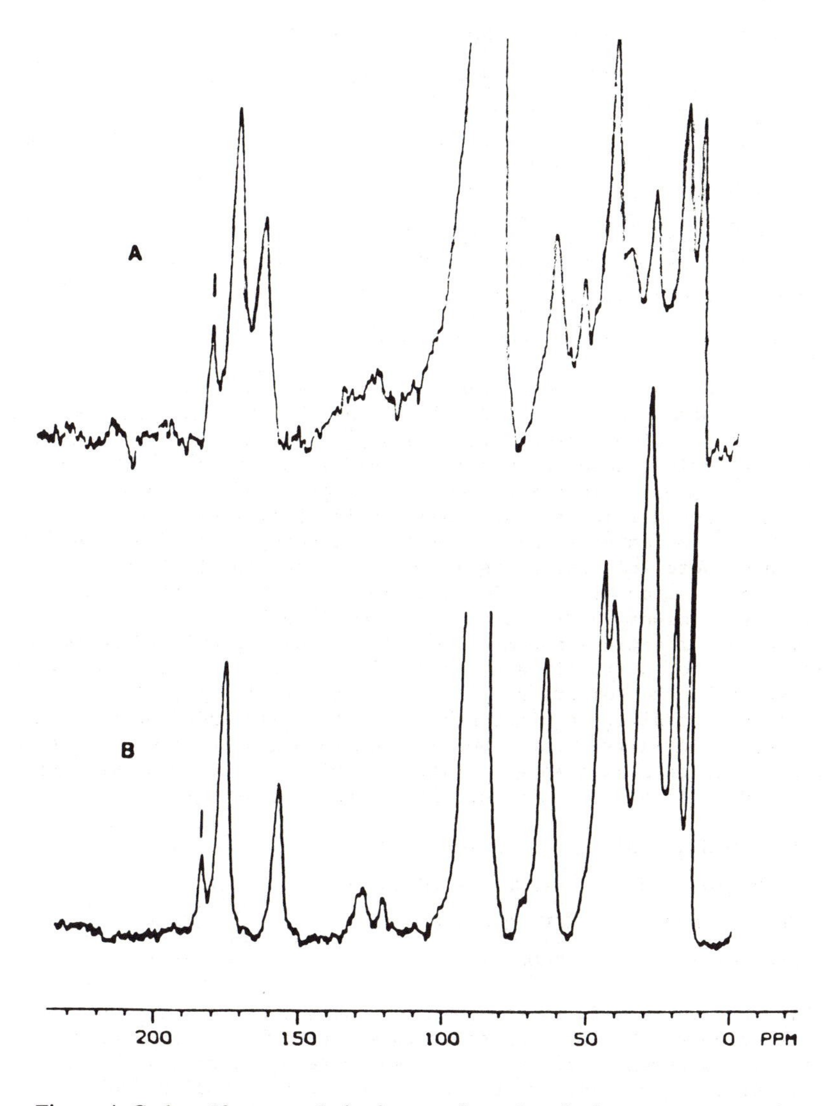

Figure 4. Carbon-13 cross-polarization, magic-angle spinning NMR spectra from an acrylic–melamine (A) and acrylic–urethane (B) coating. (Reproduced with permission from ref. 28. Copyright 1986 Elsevier Sequoia.)

$$-NCH_2OCH_3 + ROH \longrightarrow -NCH_2OR + CH_3OH \quad (M1)$$

$$2\ -NCH_2OCH_3 + H_2O \longrightarrow -NCH_2N- + H_2C{=}O + 2CH_3OH \quad (M2)$$

$$-NCH_2OCH_3 + -NH \longrightarrow -NCH_2N- + CH_3OH \quad (M3)$$

$$2\ -NCH_2OCH_3 \longrightarrow -NCH_2N- + CH_3OCH_2OCH_3 \quad (M4)$$

$$-NCH_2OCH_3 + -NCH_2OH \longrightarrow -NCH_2OCH_2N- + CH_3OH \quad (M5)$$

$$-NCH_2OCH_3 + H_2O \longrightarrow -NCH_2OH + CH_3OH \quad (M6)$$

$$-NCH_2OH \longrightarrow -NH + H_2C{=}O \quad (M7)$$

$$-NCH_2OH + -NH \longrightarrow -NCH_2N- + H_2O \quad (M8)$$

$$2\ -NCH_2OH \longrightarrow -NCH_2N- + H_2C{=}O + H_2O \quad (M9)$$

$$-NCH_2OH + ROH \longrightarrow -NCH_2OR + H_2O \quad (M10)$$

$$2\ -NCH_2OH \longrightarrow -NCH_2OCH_2N- + H_2O \quad (M11)$$

These reactions form polymer melamine crosslinks (M1 and M10), melamine-melamine crosslinks (M2, M3, M4, M5, M8, M9, and M11) or interconvert functional groups (M6 and M7). The importance of the different reactions depends on the catalyst level and type, the bake conditions, and most importantly on the structure of the melamine resin. Reaction M11 occurs only under basic conditions (used in the preparation of melamine-formaldehyde crosslinkers) and can be ignored in coatings where acid catalysts are used. Reaction M10 is slow compared to reaction M1 (5). The reactions involving water probably make at most a minor contribution under normal bake conditions. The most important reactions appear to be M1 for fully alkylated melamines and M1 and M9 for partially alkylated melamines (4-6). Reaction M4 occurs at high bake temperatures and catalyst levels (3). The mechanism and catalyst requirements of reaction M1 depend on melamine structure. Fully alkylated melamines require high bake temperatures or strong acid catalysts (e.g., PTSA) to achieve cure. The mechanism as proposed by Blank for this reaction is as follows (29):

$$-NCH_2\text{-}O\text{-}CH_3 + H^+ \rightleftarrows -NCH_2\text{-}O(H^+)\text{-}CH_3 \quad \text{fast}$$

$$-NCH2\text{-}O(H^+)\text{-}CH_3 \rightleftarrows -NCH_2^+ + CH_3OH \quad \text{slow}$$

$$-NCH_2^+ + ROH \rightleftarrows -NCH_2\text{-}O(H^+)\text{-}R \quad \text{fast}$$

$$-NCH_2\text{-}O(H^+)\text{-}R \rightleftarrows -NCH_2\text{-}O\text{-}R + H^+ \quad \text{fast}$$

Formally the reaction is first order in alkoxy functionality, however, the rate of conversion at long times is affected by the rate of the back reactions which is controlled by the rate of evaporation of methanol from the coating so that the kinetics are complex (30). Empirical models of the kinetics of this reaction have been derived (7,31). The activation energy for this reaction is around 12.5 Kcal/mole (7). Since there are generally excess methoxy groups, the side reactions that occur (e.g., reaction M4) result in an increase in crosslink density. The evolution of methanol, formaldehyde, and water during cure will contribute to solvent popping and may affect the final appearance particularly of metallic colors. Reaction U1 does not result in volatile emissions.

A mechanism for reaction M1 in partially alkylated melamines has been proposed by Blank (29):

$$-N(H)CH_2\text{-}O\text{-}CH_3 + HA \rightleftarrows -N(H)CH_2\text{-}O(H^+)\text{-}CH_3 + A^- \qquad \text{fast}$$

$$-N(H)CH_2\text{-}O(H^+)\text{-}CH_3 \rightleftarrows -N{=}CH_2 + CH_3OH + HA \qquad \text{slow}$$

$$-N{=}CH_2 + ROH \rightleftarrows -N(H)CH_2\text{-}O\text{-}R \qquad \text{fast}$$

This mechanism requires the presence of an amine hydrogen and involves general acid catalysis rather than specific acid catalysis. Not all of the alkoxy groups on a partially alkylated melamine will be adjacent to an NH (or NCH_2OH) group. Thus cure in these melamines will be by a mixture of general and specific acid catalysis. Since strong acids are generally not added to formulations containing partially alkylated melamines, the alkoxy groups adjacent to an NH group will react preferentially to those next to another alkoxy group. This may account for the observation that reaction M1 does not go to completion in partially alkylated melamines even when there is an excess of alkoxy groups (5,6). The extent of reaction M9 has also been measured in these melamines (5,6). These reactions can be followed by measuring the intensity of the acrylic OH band at 3520 cm^{-1}, the melamine OH band at 3460 cm^{-1} and the methoxy band at 915 cm^{-1} in the infrared (Figure 2).

Differences in Network Structure. Network formation depends on the kinetics of the various crosslinking reactions and on the number of functional groups on the polymer and crosslinker (32). Polymers and crosslinkers with low functionality are less efficient at building network structure than those with high functionality. Miller and Macosko (32) have derived a network structure theory which has been adapted to calculate "elastically effective" crosslink densities (4-6,8,9). This parameter has been found to correlate well with physical measures of cure (6,8). There is a range of crosslink densities for which acceptable physical properties are obtained. The range of bake conditions which yield crosslink densities within this range define a cure window (8,9).

The calculation requires measurements of the extent of reaction and of the distribution of functionality on the polymer and crosslinker. The functionality of the isocyanate crosslinker is formally three although it must be recognized that all of these materials have a distribution of functionality. Fully alkylated melamines have a functionality of 6. Coatings are formulated with an excess of alkoxy to hydroxy in order to achieve rapid cure (7). Thus, not all of the melamine alkoxy groups will react (ignoring side reactions) and the effective functionality will be less than six.

The differences in network formation in urethane and melamine coatings are mostly due to differences in kinetics. Urethane and melamine cure kinetics are different in two fundamental ways: the order of the reaction and the activation energy of the reaction. The activation energy of the isocyanate reaction is lower than that of the melamine reactions. This means that isocyanate crosslinking will be less sensitive to variations in temperature than melamine crosslinking. Thus the cure window for the urethane coating will be wider in temperature than in the melamine coating. As shown in Figure 5 the effect can be large. The cure window of the high solids

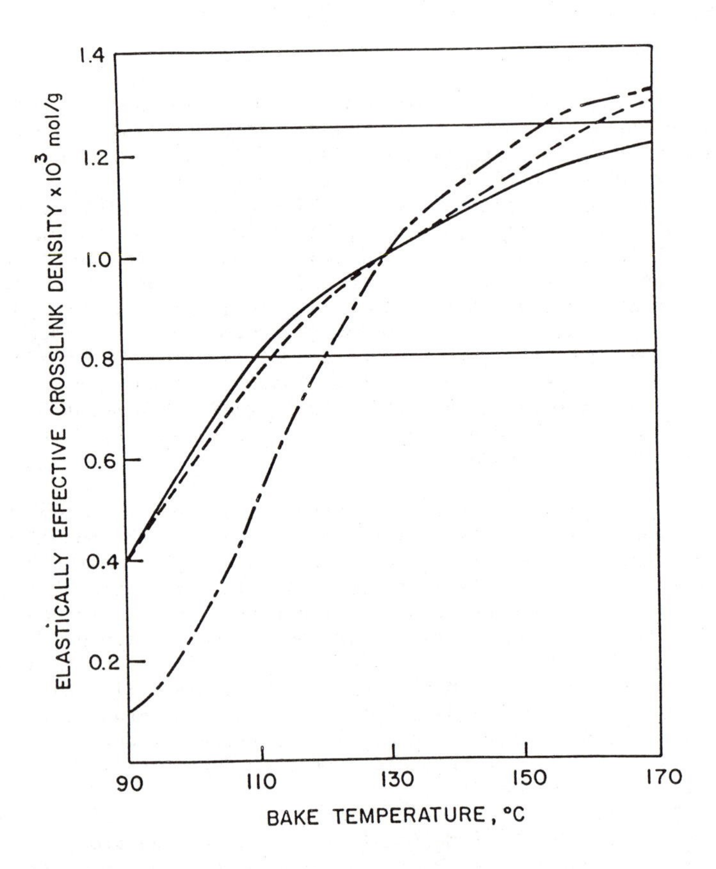

Figure 5. Elastically effective cross-link density versus bake temperature for a high solids acrylic–urethane (—), a high solids acrylic–melamine (— - —), and a low solid acrylic–melamine (- - -). The horizontal lines define the cure window. (Reproduced from ref. 14. Copyright 1986 American Chemical Society.

urethane coating is much larger than that for a melamine coating using the same polyol. The difference in order of reaction is also important. As shown in Figure 6, the extent of reaction at high conversion (where physical properties are achieved) is less sensitive to bake time for urethane crosslinking than for melamine crosslinking. The more rapid reaction of the urethane coating at short times and the lower activation energy precludes the use of these crosslinkers in single component formulations in automotive applications. Most applications involve metering the two components in the spray gun allowing for the possibility of going "off-ratio". Fortunately as shown in Figure 7 the network structure is not terribly sensitive for variations in the isocyanate to hydroxy ratio (14).

Another key difference between urethane and melamine network structures is the flexibility of the crosslink formed. Crosslink flexibility can be estimated from NMR measurements of the mobility of different groups. The mobility of the melamine triazine ring can be used to compare the flexibilities of polyol-melamine ether linkages with melamine-melamine $-N-CH_2-N-$ linkages. The mobility of the acrylic chain can be used to compare the flexibilities of urethane and melamine crosslinks in acrylic coatings. The results (12-14) suggest that acrylic-melamine crosslinks are significantly more flexible than melamine-melamine crosslinks and that acrylic-urethane crosslinks are more flexible than acrylic-melamine crosslinks. This is one reason why urethanes are preferred for coating flexible substrates. The reason for the increased flexibility may be due either to inherently higher flexibility of the urethane link or to the flexibility of the hexamethylene units in the isocyanate biuret crosslinker relative to the rigid triazine ring of the melamine crosslinker.

Network Degradation

Hydrolysis. In principle, both urethane and melamine polyol crosslinks are susceptible to hydrolysis,

$$-NHCO_2-R + H_2O \longrightarrow -NH_2 + CO_2 + ROH \qquad (U5)$$

$$-N-CH_2-O-R + H_2O \longrightarrow -N-CH_2OH + ROH \qquad (M12)$$

The methylol groups in reaction M12 can either deformylate (reaction M7) to amine or self condense (reaction M9) to yield melamine-melamine crosslinks.

Hydrolysis of urethane linkages is catalyzed most effectively by base rather than acid (33). In uncatalyzed formulations the measured rate of hydrolysis is slow (14). Less than 10% of the crosslinks are hydrolyzed after 2000 hours of exposure to condensing humidity at 50 C (Figure 3). The measured activation energy for urethane hydrolysis is around 20 kcal/mole (34). Using these data it can be concluded that for uncatalyzed formulations hydrolysis is negligible during normal service life.

In contrast, the rate of hydrolysis of acrylic-melamine crosslinks can be significant. Hydrolysis in acrylic-melamine crosslinks can be followed by measuring the rate of generation of polymer hydroxy groups (OH stretch region of IR) or the disappearance of residual melamine methoxy groups (Figure 2). It has been found that the rate of hydrolysis is very sensitive to the concentration of acid catalyst in the coating and the nature of the melamine

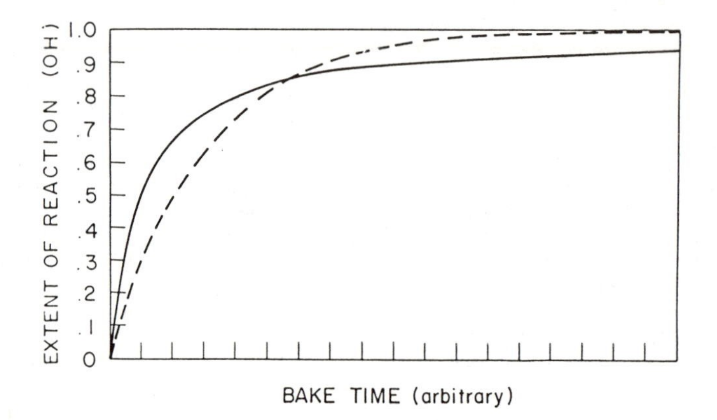

Figure 6. Extent of hydroxy reaction versus bake time for reaction with isocyanate (—) and a fully alkylated melamine (— - —). The bake temperatures were adjusted to achieve 85% conversion at the same time.

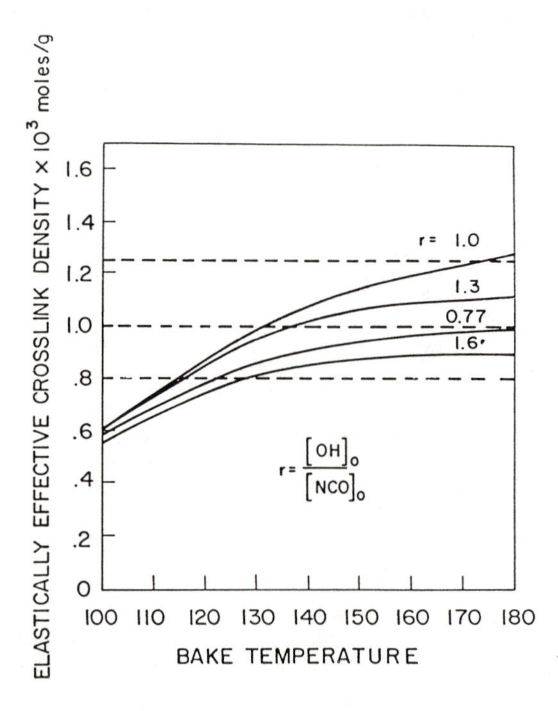

Figure 7. Elastically effective cross-link density versus bake temperature for an acrylic–urethane coating as a function of polyol to isocyanate ratio.

crosslinker (10). Hydrolysis of fully alkylated melamines is fast if strong acid catalysts are present but very slow if they are absent (or leached out of the coating). Hydrolysis of partially alkylated melamines is catalyzed by acrylic acid incorporated into the polymer backbone to promote cure in these coatings. After 1000 hours exposure to condensing humidity at 50 C, about 80% of the acrylic-melamine crosslinks are hydrolyzed in partially alkylated melamine coatings. Hydrolysis of the residual methoxy groups also occurs but at a much slower rate (10). The activation energy for hydrolysis is roughly 22 Kcal/mole. From this activation energy it can be estimated that a significant fraction of the acrylic-melamine crosslinks can be expected to be hydrolyzed during the service life of the coating. In contrast, hydrolysis in fully alkylated melamines in negligible after the strong acid catalysts have been leached from the coating. Formation of melamine-melamine crosslinks compensates for the loss of acrylic-melamine crosslinks in partially alkylated melamine coatings. The net crosslink density stays roughly constant. Coating rigidity increases on hydrolysis since melamine-melamine linkages are less flexible than polymer-melamine linkages (13). Formation of melamine-melamine linkages in fully alkylated melamines requires extensive hydrolysis. Thus, coatings with fully alkylated melamines maintain their flexibility longer than partially alkylated melamines (13). The lack of sensitivity to acid catalyzed hydrolysis of urethane coatings relative to melamine coatings may, in part, explain the superior resistance of urethane coating to industrial fallout and acid rain.

Photodegradation. Both urethane and melamine crosslinked coatings undergo photooxidation on exposure to ultraviolet (UV) light. The rate of photooxidation (as measured by increase in carbonyl intensity in the infrared spectrum) is similar in urethane and melamine coatings and appears to depend primarily on the nature of the polyol (17). One difference between the two coating types, however, is that while photooxidation in urethane coatings is independent of humidity, the rate of photooxidation in melamine crosslinked coatings increases with increasing humidity (18). In addition to carbonyl formation, photooxidative attack results in scission and/or crosslinking on the polyol. In the case of acrylic copolymers, the presence of styrene comonomer results in significant crosslink formation on exposure to UV light (12). The net change in crosslink structure on exposure is the sum of attack on the polyol and specific attack on the urethane and melamine crosslinks in the coating as discussed below.

In the case of the urethane coating, urethane crosslinks are rapidly broken on exposure to UV light as evidence by the loss of amide II and amide IV bands in the infrared spectrum, Figure 3. The rate of scission is independent of the humidity of exposure confirming that the reaction is not a hydrolysis reaction. The rate of scission is greatly slowed (by factors as great as 20) by the addition of hindered amine light stabilizers (HALS) (17). HALS function by inhibiting free radical propagation suggesting that urethane crosslink scission is a free radical driven process. After the crosslinks are broken, the crosslinker can be lost from the coating (14,17). The extent of loss depends on the humidity of exposure and the nature of the polyol.

Polymer-melamine crosslinks are also broken during exposure to UV light. Although some crosslink scission occurs in the dark by direct hydrolysis, the rate of scission (as measured by the rate of disappearance of residual methoxy functionality) increases dramatically with increasing UV light intensity (11). Scission occurs with UV light in the absence of humidity but at a much reduced rate. At high UV light intensities, the rate of crosslink scission is more or less independent of melamine type (11). Melamine-melamine crosslink formation accompanies polymer-melamine crosslink scission but only under humid exposure conditions. The addition of HALS reduces the rate of acrylic-melamine crosslink scission though not by as much as it reduces the rate of polymer-urethane crosslink scission. The mechanism for polymer-melamine crosslink scission in the presence of UV light is clearly different from that in the presence of just water. A mechanism involving free radical attack on the crosslink resulting in the $-N=CH_2$ intermediate has been proposed (15). In the presence of water, this group hydrolyses to form a methylol group which results in melamine-melamine crosslink formation.

The influence of humidity on chemical degradation during exposure to UV light is just one way that urethane and melamine coatings differ. Another difference between urethane and melamine coatings involves their response to hindered amine light stabilizers. HALS are not only significantly more effective in urethane coatings than in melamine coatings but they are also more persistent (17). These differences are reflected in differences in the kinetics of HALS consumption and the formation and decay of HALS based nitroxide (19,20). The key chemical difference between melamine and urethane coatings that can account for most of the observed effects is that melamine coatings release formaldehyde during degradation (primarily from reaction M9 following polymer-melamine crosslink scission). Formaldehyde emission has been observed on exposure to water and UV light (15). The rate of formaldehyde emission increases with increasing humidity and light intensity. Not all of the formaldehyde produced in the coating escapes. As much as 75% of the formaldehyde reacts in the coating. Formaldehyde is easily oxidized to performic acid which is a strong oxidant. This can account for the increase in oxidation rate of the melamine coating with humidity. Peracids can also have a strong effect on HALS and nitroxide stabilization chemistry.

It is clear from the above that substantial changes in network structure occur on degradation. It is less clear how these changes affect coating physical properties particularly coating appearance. The most important measures of appearance are gloss and distinctness of image. In pigmented coatings gloss decreases on exposure as the polymer coating erodes away and exposes pigment particles. Intuitively, it seems likely that a decrease in crosslink density on exposure would promote coating erosion. In clearcoats, on the other hand, the most serious appearance problem is crack formation. Cracking can be caused by excessive crosslinking on degradation. For example, styrene containing acrylic-urethanes crack before non-styrene containing acrylic-urethanes (35). This can be attributed to the formation of rigid crosslinks between acrylic chains at the styrene groups as discussed above. Other changes may also lead to crack formation. For example, clearcoats made from partially alkylated melamines which rapidly form rigid melamine-melamine crosslinks on exposure can be expected to crack before clearcoats made from fully alkylated melamines. Urethanes which do not undergo the melamine self-condensation reaction ought to be even more resistant to cracking. The effect on durability of changes in crosslink structure is clearly an area that requires further work.

Conclusion

Differences in crosslink formation and degradation between coatings using isocyanate and melamine-formaldehyde crosslinkers have been compared and their effects on physical properties discussed. The reaction which forms urethane crosslinks is second order while the reaction which forms polymer melamine crosslinks is pseudo first order. The activation energy for urethane crosslink formation is substantially lower than that for melamine crosslink formation. These differences in kinetics result in a wider cure window for the urethane coating at the expense of requiring two component formulation due to lack of shelf stability. Melamine coatings are much more sensitive to acid catalyzed hydrolysis than are urethane coatings. This may account for the improved resistance of urethane coatings to attack by acids. Both urethane and melamine crosslinks are broken on exposure to ultraviolet light. The scission appears to involve free radical attack on the crosslink. In the melamine case, melamine-melamine crosslinks are formed and formaldehyde is emitted into the coating. Formaldehyde based chemistry appears to have a profound effect on both the degradation chemistry of these coatings and also on the performance of hindered amine light stabilizers. These stabilizers are found to be much more effective and more persistent in urethane coatings than in melamine coatings.

Literature Cited

1. Santer, J. O. Prog. Org. Coat., 1984, 12, 309.
2. Nakamichi, T. Prog. Org. Coat., 1986, 14, 23.
3. Bauer, D. R. Prog. Org. Coat., 1986, 14, 193.
4. Potter, T. A.; Schmelzer, H. G.; Baker, R. D. Prog. Org. Coat., 1984, 12, 321.
5. Bauer, D. R.; Dickie, R. A. J. Polym. Sci., Polym. Phys., 1980, 18, 1997.
6. Bauer, D. R.; Budde, G. F. Ind. Eng. Chem., Prod. Res. Dev., 1981, 20, 674.
7. Bauer, D. R.; Budde, G. F. J. Appl. Polym. Sci., 1983, 28, 253.
8. Bauer, D. R.; Dickie, R. A. J. Coat. Technol., 1982, 54 (no. 685), 57.
9. Bauer, D. R.; Dickie, R. A. In Computer Application in the Polymer Laboratory; Provder, T., Ed.; American Chemical Society Symposium Series No. 313: Washington, DC, 1986; p. 256.
10. Bauer, D. R. J. Appl. Polym. Sci., 1982, 27, 3651.
11. Bauer, D. R.; Briggs, L. M. In Characterization of Highly Crosslinked Polymers; Labana, S. S.; Dickie, R. A., Eds.; American Chemical Society Symposium Series No. 243: Washington, DC, 1984; p. 271.
12. Bauer, D. R.; Dickie, R. A.; Koenig, J. L. J. Polym. Sci., Polym. Phys., 1984, 22, 2009.
13. Bauer, D. R.; Dickie, R. A.; Koenig, J. L. Ind. Eng. Chem., Prod. Res. Dev., 1985, 24, 121.

14. Bauer, D. R.; Dickie, R. A.; Koenig, J. L. Ind. Eng. Chem., Prod. Res. Dev., 1986, 25, 289.
15. Gerlock, J. L.; Dean, M. J.; Korniski, T. J.; Bauer, D. R. Ind. Eng. Chem., Prod. Res. Dev., 1986, 25, 449.
16. Gerlock, J. L.; Bauer, D. R.; Briggs, L. M.; Dickie, R. A. J. Coat. Technol., 1985, 57 (no. 722), 37.
17. Gerlock, J. L.; Dean, M. J.; Bauer, D. R. Ind. Eng. Chem., in press.
18. Gerlock, J. L.; Bauer, D. R.; Briggs, L. M. In Polymer Stabilization and Degradation; Klemchuk, P., Ed.; American Chemical Society Symposium Series No. 280: Washington, DC, 1985; p. 119.
19. Gerlock, J. L.; Bauer, D. R.; Briggs, L. M. Polym. Deg. Stab., 1986, 14, 53.
20. Gerlock, J. L.; Riley, T.; Bauer, D. R. Polym. Deg. Stab., 1986, 14, 73.
21. Christensen, G. Prog. Org. Coat., 1977, 5, 255.
22. Christensen, G. Prog. Org. Coat., 1980, 8, 211.
23. Shindo, M. Polym. Mat. Sci. Eng. Proc., 1983, 49, 169.
24. Carter III, R. O.; Bauer, D. R. Polym. Mat. Sci. Eng. Proc., 1987, 57, in press.
25. Briggs, L. M.; Carter III, R. O.; Bauer, D. R. Ind. Eng. Chem., 1987, 26, 667.
26. Scheafer, J.; Stejskal, E. O.; Buchdahl, R. Macromolecules, 1975, 8, 291.
27. Havens J. R.; Koenig, J. L. Appl. Spectrosc., 1983, 37, 226.
28. Bauer, D. R. Prog. Org. Coat., 1986, 14, 45.
29. Blank, W. J. J. Coat. Technol., 1979, 51, 61.
30. Meijer, E. W. J. Polym. Sci., Polym. Chem., 1986, 24, 2199.
31. Lazzara, M. G. J. Coat. Techno., 1984, 56, 19.
32. Miller, D. R.; Macosko, C. W. Macromolecules, 1976, 9, 206.
33. Widmaier, J. M.; Balmer, J. P.; Meyer, G. C. Polym. Mat. Sci. Eng. Proc., 1987, 56, 96.
34. Gerlock, J. L.; Braslaw, J.; Mahoney, L. R.; Ferris, F. C. J. Polym. Sci., Polym. Chem., 1980, 18, 541.
35. Boch, M.; Uerdingen, W. Org. Coat. Plast. Chem., 1980, 43, 59.

RECEIVED October 7, 1987

Chapter 7

Effect of Ionizing Radiation on an Epoxy Structural Adhesive

Thomas W. Wilson[1], Raymond E. Fornes, Richard D. Gilbert, and Jasper D. Memory

Fiber and Polymer Science Program, North Carolina State University, Raleigh, NC 27695

The epoxy resin formed by tetraglycidyl 4,4'-diamino diphenyl methane and 4,4'-diamino diphenyl sulfone was characterized by dynamic mechanical analysis. Epoxy specimens were exposed to varying dose levels of ionizing radiation (0.5 MeV electrons) up to 10,000 Mrads to assess their endurance in long-term space applications. Ionizing radiation has a limited effect on the mechanical properties of the epoxy. The most notable difference was a decrease of approximately 40°C in Tg after an absorbed dose of 10,000 Mrads. Sorption/desorption studies revealed that plasticization by degradation products was responsible for a portion of the decrease in Tg.

Graphite fiber reinforced composites are being utilized in an increasing number of structural applications. One such use for these materials would be in space. Many polymeric matrix systems do not maintain their integrity in such an environment (vacuum, ionizing radiation, temperature extremes). The epoxy resin TGDDM (tetraglycidyl 4,4'-diamino diphenyl methane) cured with DDS (4,4'-diamino diphenyl sulfone) appears to be ideal for structural use in long-term space applications (1-3).

Dynamic mechanical testing (4,5) was employed as a probe to investigate the influence of ionizing radiation on the mechanical properties of an epoxy since it can provide several types of information (e.g. E^*, E', E'' and $\tan\delta$).

Experimental

Epoxy film specimens of two different weight/weight ratios of TGDDM/DDS were evaluated (73/27 and 80/20). The films were cast between teflon sheets using a spacer. The cure cycle was 1 hr at

[1]Current address: Dental Research Center, University of North Carolina, Chapel Hill, NC 27514

0097-6156/88/0367-0093$06.00/0

150°C and 5 hrs at 177°C under vacuum. Specimens were sealed in evacuated aluminum foil bags and irradiated with 0.5 MeV electrons at six different dose levels, viz. 1000, 2000, 3000, 4000, 5000 and 10,000 Mrads.

Autovibron (Imass) testing was conducted according to recommended procedures (Autovibron manual). Thin film were tested at a frequency of 11 Hz in a tensile mode. Specimens were scanned from -120°C to +320°C at 2.5°C/min.

Results and Discussion

In Figure 1, the elastic modulus (E') and loss tangent (tanδ) for 73/27 and 80/20 TGDDM/DDS epoxy are shown. For the 73/27 ratio, E' decreases monotonically up to 155°C. Between 155°C and 240°C, E' first decreases and then increases. The changes in E' indicate additional curing reactions (6). That is, during fabrication the system vitrifies at the cure temperature before all the available functional groups have reacted. Once the cure temperature is exceeded, further reactions occur. Due to further crosslinking E' increases between 200°C and 240°C. Above 240°C, all functional groups have apparently reacted. The modulus declines rapidly to a rubbery state above 280°C.

The tanδ spectrum for the 73/27 TGDDM/DDS has three distinct transitions. The broad, low intensity peak ca. -60°C is the γ-transition. In epoxy systems the γ-transition arises from a crankshaft rotational motion of the glycidyl portion of the molecule after it has been reacted (6). Up to five different relaxation mechanisms have been measured in this region (7), but only one is resolved by the present technique. The next damping peak occurs at 200°C, as a result of the additional curing reactions mentioned previously. The curing peak disappears upon subsequent testing (6). The loss peak associated with Tg is ca. 280°C. For the present work, Tg will be taken as the maximum value of tanδ (8).

The additional curing reactions in the 80/20 TGDDM/DDS should be more evident than in the 73/27 TGDDM/DDS due to the greater excess of epoxide groups in the former. The curing reactions are more pronounced as evidenced by a larger decrease in E' and a curing peak of greater magnitude and breadth than in the 73/27 specimen.

The effects of radiation on E' and tanδ for 73/27 TGDDM/DDS epoxy are illustrated in Figure 2. At 1000 Mrads, the curing reactions appear almost complete since the curing peak ca. 210°C has diminished. There is no evidence of additional curing reactions at higher doses. This observation agrees with work by Netravali et al. (9) who reported that radiation induces additional cure. An analogous response occurs in the 80/20 specimens (Figure 3).

The obvious effect of increasing radiation dosages is a decrease in Tg (Table I). The Tg of a control is 40°C higher than the Tg of specimens irradiated to 10,000 Mrads. Also, the α-transition region broadens with increasing dose.

Property changes are evident in the rubbery plateau region at 40°C above Tg (ie. E'(Tg+40)). E'(Tg+40) decreases 20% as a function of dose up to 5000 Mrads. However between 5000 and 10,000 Mrads, E'(Tg+40) increases and is only 6% below the control value.

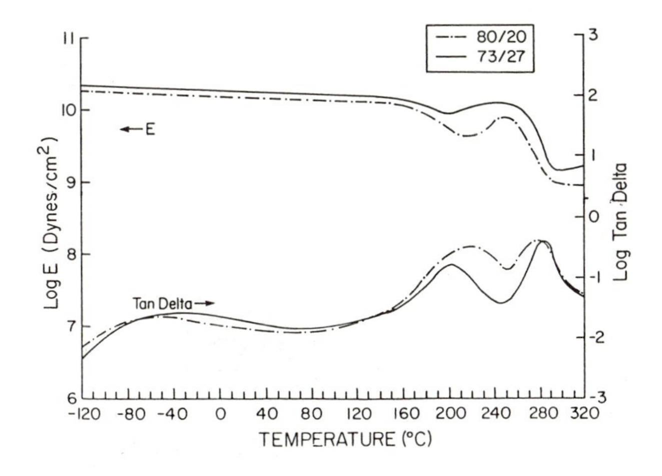

Figure 1. Dynamic mechanical spectra of 73/27 and 80/20 TGDDM/DDS epoxy.

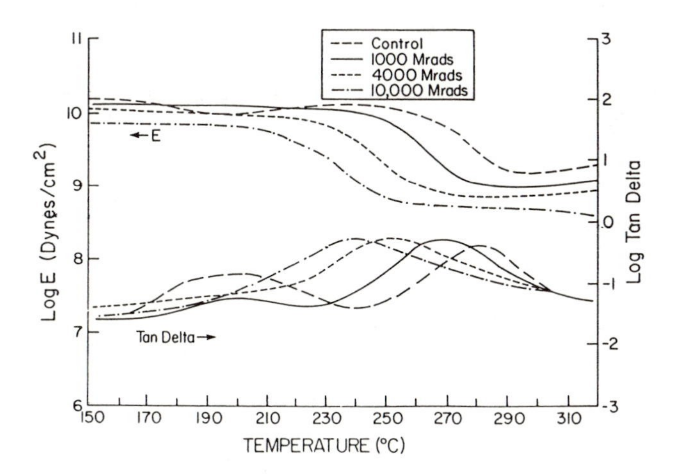

Figure 2. Dynamic mechanical spectra of irradiated 73/27 TGDDM/DDS epoxy.

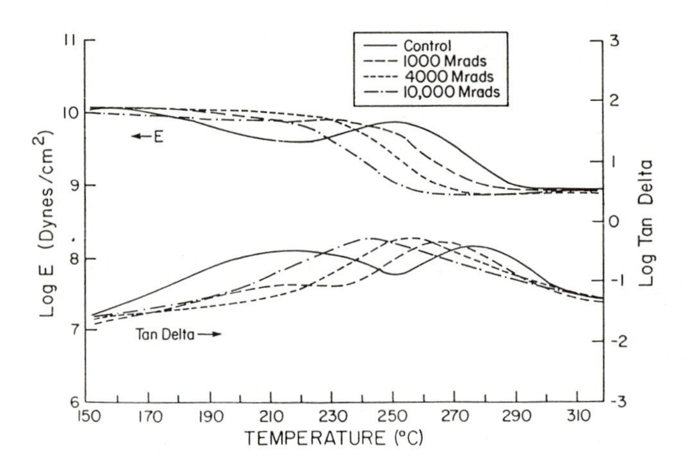

Figure 3. Dynamic mechanical spectra of irradiated 80/20 TGDDM/DDS epoxy.

Table I. Glass transition temperature as a function of dose for 73/27 and 80/20 TGDDM/DDS epoxy

Dose (Mrads)	Tg (°C)	
	73/27 TGDDM/DDS	80/20 TGDDM/DDS
control	283.3 ± 1.3[1]	275.4 ± 2.9
1000	271.8 ± 2.6	261.4 ± 3.9
2000	262.3 ± 4.5	257.6 ± 2.7
3000	260.7 ± 2.6	256.7 ± 3.7
4000	253.3 ± 3.9	251.7 ± 4.5
5000	252.6 ± 2.9	250.6 ± 1.8
10,000	237.7 ± 3.7	237.7 ± 3.4

[1] expressed as mean Tg ± the range for a 95% confidence interval.

The α-transition intensity increases with dose (tanδ=0.40 for the control; tanδ=0.54 at 10,000 Mrads). This increase indicates a "loosening" of the network structure.

The data suggest that some radiation induced degradation has occurred. The 40°C decrease in Tg and the broadened transition region reveal that a wider range of crosslink densities are contributing to the α-transition. Increases in transition intensity suggest a greater molecular weight between crosslinks (Mc). Another possibility is plasticization from degradation products or from an increase in the number of free chain ends. Plasticization of network polymers has been shown to cause decreases in Tg, increases in transition intensity and decreases in E'(Tg+40) (4,5,10-12).

The changes in room temperature elastic modulus (E'(rt)) suggest chain scission and crosslinking. The 73/27 specimens display a decrease of 13% in E'(rt) up to a dose of 3000 Mrads. Above 3000 Mrads, E'(rt) increases and is 5% greater than the control at 10,000 Mrads. The decrease in E'(rt) is presumed to occur due to a relaxation of internal stresses. After fabrication, internal stresses appear since some bonds are not free to relax. Relaxation arises from radiation induced chain scission of the more highly strained bonds in the system. The increase in E'(rt) between 3000 and 10,000 Mrads is due to additional crosslinking. Analogous behavior has been shown for elastomers exposed to ionizing radiation (13).

In the 80/20 specimens, the changes in E'(rt) follow a different pattern. There is little change in E'(rt) up to a dose of 2000 Mrads. Between 2000 and 4000 Mrads, E'(rt) decreases 13%. At 10,000 Mrads, E'(rt) is almost 9% greater than a control. Previous DSC studies have shown that during the first 2000 Mrads, there is substantial additional cure (14). Competition between chain scission of strained bonds and further curing reactions could "balance out" the changes in E'(rt). Above 2000 Mrads chain scission causes a decrease in E'(rt). Additional crosslinking contributes to the 9% increase in E'(rt) at 10,000 Mrads.

The temperature of the γ-transition does not change, but it decreases in magnitude (control=0.024; 10,000 Mrads=0.013). The decrease in magnitude is unexpected since further reaction in epoxy systems has been shown to increase γ-peak intensity (6). Plasticizers, ie. degradation products, are known to "freeze out" the molecular motions which constitute this transition. The discrepancy between further cure due to ionizing radiation and a lower transition intensity may be due to degradation products interfering with the molecular motions of the transition.

Thus, radiation causes degradation, chain scission and further crosslinking. To distinguish between chain scission, crosslinking and degradation products, sorption/desorption studies were conducted. Selected specimens of 73/27 and 80/20 TGDDM/DDS were allowed to absorb acetonitrile to equilibrium. The acetonitrile was desorbed, and the samples were tested.

After a sorption/desorption cycle on a control, the Tg is 2°C below the mean Tg of the untreated specimens. The α-peak intensity is unchanged. The sorption/desorption cycle does not appear to alter the mechanical properties or final network structure. The Tg at the 10,000 Mrad dose level increases approximately 10°C for the 80/20 and 73/27 samples after a sorption/desorption cycle (73/27 Tg=248°C, 80/20 Tg=250°C). The magnitude of the tanδ peak decreases about 20%, indicating that degradation products which plasticize the network had been removed.

The weight loss after the sorption/desorption cycle is 1.3% for the specimens irradiated to 10,000 Mrads. The low weight loss indicates limited degradation. The glass transition temperature should have returned to the value of the unirradiated epoxy (280°C) if degradation products are the only species plasticizing the network.

It is possible that either Mc has increased by degradation of the network structure or the resin is internally plasticized by free chain ends. If Mc has increased, then the modulus in the rubbery plateau region for irradiated specimens should be less than that of a control. As discussed above, E'(Tg+40) decreases up to a dose of 5000 Mrads. Between 5000 and 10,000 Mrads, E'(Tg+40) increases but remains 6% below the control. For the 73/27 and 80/20 samples (10,000 Mrads) which have been sorbed/desorbed, E'(Tg+40) is 18.5% greater than the control.

The Tg's of the irradiated specimens should be >280°C. To account for the experimental observations, the following explanation is suggested. As a sample is irradiated, chain scission and crosslinking occur. Chain scission will create a number of free chain ends. The α-transition will shift to lower temperature and intensify due to internal plasticization and an increase in free volume caused by the greater number of chain ends (4). The rubbery plateau region is not affected by an increase in the number of chain ends if Mn (number average molecular weight) is very large (5). Therefore, free chain ends plasticize the network, and Tg decreases, even though the system has been crosslinked further.

Conclusions

The most noticeable property change is a decrease in the glass transition temperature of the epoxy resin as a function of absorbed dose. The decrease in Tg is due to plasticization by degradation products and free chain ends from chain scission.

The epoxy resins have a number of unreacted functional groups. Ionizing radiation causes these groups to react.

Chain scission is the predominant process at lower dose levels (<5000 Mrads), yielding a decrease in elastic modulus at ambient temperature. Additional crosslinking at high doses (>5000 Mrads) results in an increase in elastic modulus at ambient temperature and in the rubbery region above Tg.

Literature Cited

1. Wolf, K. W. Ph.D. Thesis, North Carolina State University, Raleigh, 1982.
2. Sykes, G. F; Milkovich, J. M.; Herakovich C. T. Polym. Mats.: Sci. and Eng. 1985, 52, 598.
3. Naranong, N. Masters Thesis, North Carolina State University, Raleigh, 1980.
4. Nielsen, L. E. Mechanical Properties of Polymers and Composites; Marcel Dekker: New York, 1974; Vol. 1.
5. Murayama, T. Dynamic Mechanical Analysis of Polymeric Materials, Elsevier: New York, 1978.
6. Keenan, J. D; Seferis, J. C.; Quinlivan, J. T. J. Appl. Polym. Sci. 1979, 24, 2375.
7. Pangrle, S.; Chen, A.; Wu, C. C.; Geil, P. H. Bull. Am. Phys. Soc. 1985, 30(3), 437.
8. Roller, M. B. J. Coatings Tech. 1982, 54, 33.
9. Netravali, A. N.; Fornes, R. E.; Gilbert R. D.; Memory, J. D. J. Appl. Polym. Sci. 1984, 29, 311.
10. Murayama, T.; Bell, J. P. J. Polym. Sci.: Part A-2 1970, 8, 437.
11. Morgan, R. J.; O'Neal, J. E. Polym.-Plast. Technol. Eng. 1978, 10, 49.
12. McKague, E. L., Jr.; Reynolds, J. D.; Halkias, J. E. J. Appl. Polym. Sci. 1978, 22, 1643.
13. Traeger, R. K.; Castonguay, T. T. J. Appl. Polym. Sci. 1966, 10, 535.
14. Wilson, T. W. Ph.D. Thesis, North Carolina State University, Raleigh, 1986.

RECEIVED October 7, 1987

Chapter 8

Frequency-Dependent Dielectric Analysis

Monitoring the Chemistry and Rheology of Thermosets During Cure

D. Kranbuehl, S. Delos, M. Hoff, L. Weller, P. Haverty, and J. Seeley

Department of Chemistry, College of William and Mary, Williamsburg, VA 23185

Previous published reports have demonstrated how the frequency dependence of $\epsilon^*(\omega)$ can be used to qualitatively monitor the viscosity of a curing resin. The key is in using the frequency dependence in the Hz to MHz range to separate and determine parameters governing ionic and dipolar mobility. This paper reports on the use of the frequency dependence to determine ionic and dipolar diffusion processes. The quantitative relationship of the ionic and dipolar mobility parameters to the viscosity and degree of cure during the cure reaction of TGDDM epoxy is discussed. The temperature and degree of cure dependence of the ionic and dipolar mobility is analyzed in terms of an Arrhenius and WLF dependence. As an example of an application, the in-situ on-line measurement capability of the technique to measure the cure processing parameters in a thick laminate during cure in an autoclave is reported. The results support the use of the WLF equation for analyzing the advancement of a curing reaction in TGDDM epoxies.

Frequency dependent complex impedance measurements made over many decades of frequency provide a sensitive and convenient means for monitoring the cure process in thermosets and thermoplastics [1-4]. They are of particular importance for quality control monitoring of cure in complex resin systems because the measurement of dielectric relaxation is one of only a few instrumental techniques available for studying molecular properties in both the liquid and solid states. Furthermore, it is one of the few experimental techniques available for studying the polymerization process of going from a monomeric liquid of varying viscosity to a crosslinked, insoluble, high temperature solid.

In the past, impedance or dielectric studies have been examined as an experimental technique to monitor the flow properties, effects of composition, and the advancement of a reaction during cure [1]. Until a paper by Zukas et al [2], little emphasis had been placed on the frequency dependence except to note the shift in position and magnitude of impedance maxima and minima. Furthermore, most measurements on curing systems reported results in terms of

0097-6156/88/0367-0100$06.00/0

extensive parameters, such as conductance G, capacitance C or their ratio dissipation D [1-4]. Frequency dependent impedance measurements have long been used to provide information on a molecular level about the mobility of charged species and dipolar groups in liquids and solids [5]. Over the recent years, our laboratory and others [3,4,6-11] have focused on using the frequency dependence of the impedance to determine and separate the ionic and dipolar contributions to the impedance. Further, measurements are made in terms of the intensive geometry independent parameter, the complex permittivity ϵ^*. The angular frequency dependence, ω, and magnitude of $\epsilon^*(\omega)$ are determined by the time scale of the ionic groups' and the polar groups' mobility as well as by the number and charge of the ionic/polar species. The emphasis of this work is on continuous measurement of the frequency dependence of ϵ^* to measure both the ions and the dipolar groups' changes in mobility and then to relate this changing mobility to cure processing properties such as viscosity and degree of cure (see Figure 1). Thus, rather than focusing on extensive electrical impedance properties such as C, G, or D, the ionic and dipolar mobility parameters are used as molecular probes of the cure reaction. These molecular probes are used to sense in-situ the viscosity and degree of cure as a function of time.

A major long-range objective of this research is to develop on-line instrumentation using commercially available instruments, novel sensor techniques, and a molecular understanding of the frequency dependence of the impedance for quantitative nondestructive material evaluation and closed loop "smart" cure cycle control. The key to achieving this goal is to relate the chemistry of the cure cycle process to the dielectric properties of the polymer system by correlating the time, temperature, and frequency dependent impedance measurements with chemical and rheological measurements. Measurement of the wide variation in magnitude of the complex permittivity with both frequency and state of cure, coupled with other characterization work, have been shown to have the potential to determine: resin quality, composition and age; cure cycle window boundaries; onset of flow and point of maximum flow; extent of and completion of reaction; evolution of volatiles; T_g; crosslinking and molecular weight buildup [3,4,6-11].

In previous published reports we demonstrated how the frequency dependence of $\epsilon^*(\omega)$ can be used to qualitatively monitor the viscosity of a curing resin [6,7]. The key is in using the frequency dependence in the Hz to MHz range to separate and determine parameters governing ionic and dipolar mobility. This paper focuses on the use of the frequency dependence to determine ionic and dipolar diffusion processes. The quantitative relationship of the ionic and dipolar mobility parameters to the viscosity and degree of cure during the cure reaction of a TGDDM epoxy is discussed. The temperature and degree of cure dependence of the ionic and dipolar mobility is analyzed in terms of an Arrhenius and WLF dependence. As an example of an application, the in-situ on-line measurement capability of the technique to measure the cure processing parameters in a thick laminate during cure in an autoclave is reported.

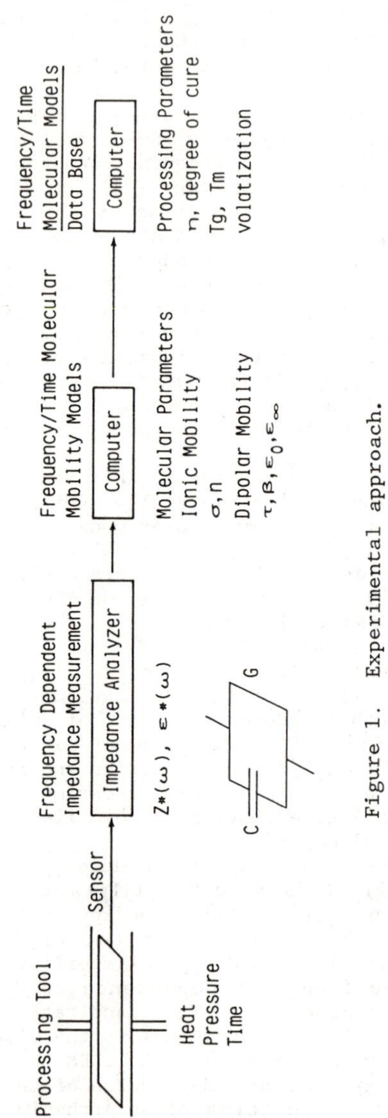

Figure 1. Experimental approach.

Experimental

Measurements were made with a now commercially available patented Dek Dyne permittivity sensor. The sensor, developed in our laboratory, consists of a fine array of two comb electrodes mounted on a thin inert substrate. The active surface area is 2 cm^2 and the thickness is 3 mm. The surface area, thickness and substrate can be varied to withstand the resin's reaction conditions. In many cases, a polyimide film is adequate for cure reactions below 200°C. At higher temperatures polyimides begin to loose some structural integrity and a higher temperature substrate such as Al_2O_3 is more reliable. The sensor was designed for use with conventional, commercially available bridges such as Hewlett Packard, GenRad or Tetrahedron.

We used a Hewlett-Packard 4192A LF Impedance Analyzer with the sensor. Continuous measurements of both the real and imaginary components of the complex permittivity, $\epsilon^* = \epsilon' - i\epsilon''$, were made over a range of 10^{-2} to 10^6 in magnitude. Since the sensor we used was inert, constructed from noble metals and a high temperature ceramic and does not contain any solid state circuitry, it is capable of being used in high temperature cure reactions as well.

The Impedance Analyzer was controlled by a 9836 Hewlett-Packard computer which also controlled the time-temperature of the press. Measurements at frequencies from 5 to 5 x 10^6 Hz were taken at regular intervals during the cure cycle and converted to the complex permittivity. Further details of the experimental procedure has been given elsewhere [10].

Viscosity measurements were made using a Rheometrics System IV-dynamic mechanical spectrometer.

Glass transition and degree of conversion measurements were made using a Perkin-Elmer DSC-7 differential scanning calorimeter.

The tetraglycidyl 4, 4' diaminodiphenyl methane epoxy (TGDDM) resins used were Hercules 3501-6 (catalyzed resin) and 3502 (uncatalyzed resin). The crosslinking reagent is 4, 4' diaminodiphenyl sulfone and the catalyst in 3501-6 is $BF_3:NH_2R$ [12]. These were supplied by Hercules through NASA-Langley Research Center and stored in a freezer until used.

Theory

Measurements of capacitance C and conductance G were used to calculate the complex permittivity $\epsilon^* = \epsilon' - i\epsilon''$

$$\epsilon' = \frac{C \text{ material}}{C_o} \qquad\qquad \epsilon'' = \frac{G \text{ material}}{C_o 2\pi f} \tag{1}$$

where C_o is the effective air replaceable capacitance and f is the frequency. Both the real and the imaginary parts of ϵ^* have a dipolar and an ionic component [5].

$$\epsilon' = \epsilon'_d + \epsilon'_i \tag{2}$$

$$\epsilon'' = \epsilon''_d + \epsilon''_i$$

The dipolar component arises from diffusion of bound charge or molecular dipole moments. The frequency dependence of the polar component may be represented by the Cole-Davidson function:

$$\epsilon^*_d = \epsilon_\infty + \frac{\epsilon_o - \epsilon_\infty}{(1+i\omega\tau)^\beta} \tag{3}$$

where ϵ_o and ϵ_∞ are the limiting low and high frequency values of ϵ, τ is a characteristic relaxation time and β is a parameter which measures the distribution in relaxation times. The dipolar term is generally the major component of the dielectric signal at high frequencies and in highly viscous media.

The ionic component, ϵ_i^*, often dominates ϵ^* at low frequencies, low viscosities and/or higher temperatures. The presence of mobile ions gives rise to localized layers of charge near the electrodes. Since these space charge layers are separated by very small molecular distances on the order of $\mathring{A}$, the corresponding space charge capacitance can become extremely large, with ϵ' on the order of 10^6. Johnson and Cole, while studying formic acid, derived empirical equations for the ionic contribution to ϵ^* [13]. In their equations, ϵ'_i is frequency dependent due to these space charge ionic effects and has the form

$$\epsilon'_i = C_o Z_o \sin\frac{(n\pi)}{2}\omega^{-(n+1)}\left(\frac{\sigma}{8.85\text{x}10^{-14}}\right)^2 \tag{4}$$

where $Z^* = Z_o(i\omega)^{-n}$ is the electrode impedance induced by the ions and n is an empirical parameter between 0 and 1 [5,13,14]. The imaginary part of the ionic component of the permittivity has the form

$$\epsilon''_i = \frac{\sigma}{8.85\text{x}10^{-14}\omega} - C_o Z_o \cos\frac{(n\pi)}{2}\omega^{-(n+1)}\left(\frac{\sigma}{8.85\text{x}10^{-14}}\right)^2 \tag{5}$$

where σ is the conductivity ($\text{ohm}^{-1}\ \text{cm}^{-1}$), an intensive variable, in contrast to conductance G(ohm^{-1}) which is dependent upon cell and sample size. The first term in Eq. 5 is due to the conductance of ions translating through the medium. The second term is due to electrode polarization effects. The second term, due to electrode polarization, makes dielectric measurements increasingly difficult to interpret and use as the frequency of measurement becomes lower.

Electrode polarization, represented by the second term in equation (5), in general is a significant and difficult to account for factor at frequencies below 10 Hz and/or for high values of σ usually associated with a highly fluid resin state. The frequency dependence ϵ^* due to dipolar mobility is generally observed at frequencies in the KHz and MHz regions. For this reason an analysis of the frequency dependence of ϵ^*, equations 3 and 5, in the Hz to

MHz range is, in general, optimum for measuring both the ionic mobility parameter σ and the dipolar mobility parameter τ.

The magnitude of the ionic mobility σ and the rotational mobility of the dipole τ depends on the extent of the reaction and the physical state of the material (5). As such, σ and τ determined from the frequency dependence of $\epsilon^*(\omega)$, provide two molecular probes for monitoring the reaction advancement and the viscosity during cure.

Results and Discussion

The variation in the magnitude of ϵ'' with frequency and with time is well represented by Figure 2 for the 165°C isothermal run of the uncatalyzed resin's reaction. Note that ϵ'' changes by over 10^6 in magnitude during the course of the cure reaction and that the measurement sensitivity of 10^{-2} on the non-log plot Figure 3 can be used to monitor the long time, final stages of cure which continues for hours. As shown previously (6,8), a plot of ϵ''* frequency or conductance, Figure 4, is a particularly informative representation of the cure because as seen from equations 1-5, the over-lap of $\epsilon''(\omega)$ for differing frequencies indicates that ionic diffusion is the dominant physical process affecting the loss (first term of eq. 5). Similarly the peaks in $\epsilon''*\omega$ for individual frequencies indicate dipolar or bound charge diffusion processes are contributing to ϵ'' (Equation 3).

The frequency dependence of the loss ϵ'' is used first to determine σ by determining from a computer analysis or a plot of $\epsilon''*\omega$ (Figure 4), the frequency region where $\epsilon''*\omega$ is a constant. Over this frequency region the value of σ is determined from the $1/\omega$ dependence of ϵ'', eq. 5. The ionic contribution ϵ''_i is substrated from ϵ'' measured to determine the dipolar component ϵ''_d. The time at which a peak occurs in the dipolar portion, ϵ''_d, for a particular frequency, ω, is used to determine the time of occurrence of the corresponding mean relaxation time $\tau = 1/\omega$.

In a parallel experiment, the extent of the reaction α is measured using the partial heat to a particular time divided by the total heat of the isotherm plus the residual heat of a subsequent 10°/min ramp. Figures 5 and 6 show the observed relationship of ln σ and ln τ to α. As expected for a similar degree of advancement α, the ionic mobility σ increases with temperature. Similarly for the same value of α, the dipolar mobility increases. An increase in dipolar mobility corresponds to a shorter relaxation time. Thus τ decreases as temperature increases. Somewhat unexpected, both ln σ and ln τ exhibit a nearly linear dependence on α. Curvature in the ln σ and ln τ versus α plot is most pronounced for small values of α and at the highest temperature. There is no evidence of a break in the ln σ or ln τ dependence on α which would indicate gel.

The rate of change of log σ with time and α approaches 0 as the advancement of the reaction approaches completion. The point at which $\frac{d\sigma}{dt}$ and $\frac{d\sigma}{d\alpha}$ decreases sharply, appears to indicate T_g. The values of α at this point, .72 at 165°, .80 at 180°, and .88 at 195° are in good agreement with our α_{T_g} values determined by DSC scans.

The correlation during cure of -log σ with log η (viscosity) measured at 10 radians/sec and the ability to use frequency dependent $\epsilon^*(\omega)$ measurements to determine σ, thereby accurately

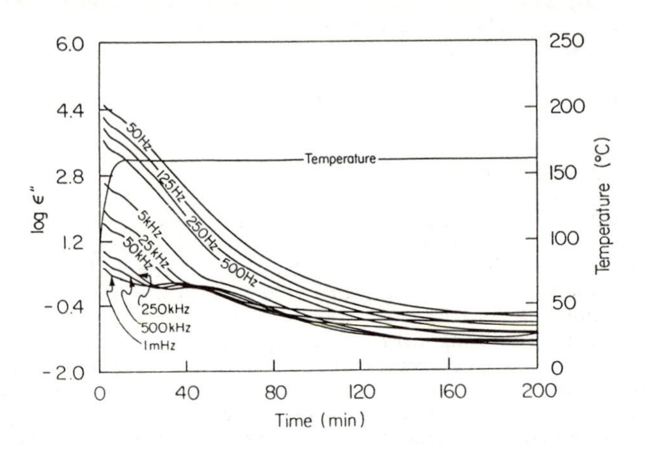

Figure 2. Log ε" vs. time for the uncatalyzed epoxy during a 165C isothermal cure.

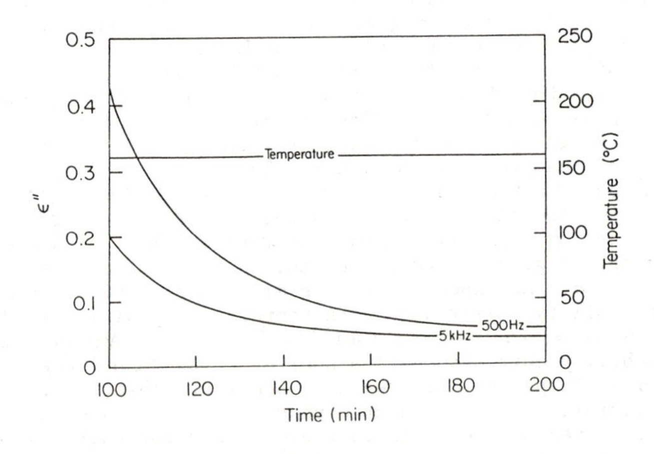

Figure 3. ε" vs. time for the uncatalyzed epoxy during a 165C isothermal cure showing the long time sensitivity at 500 Hz and 5kHz.

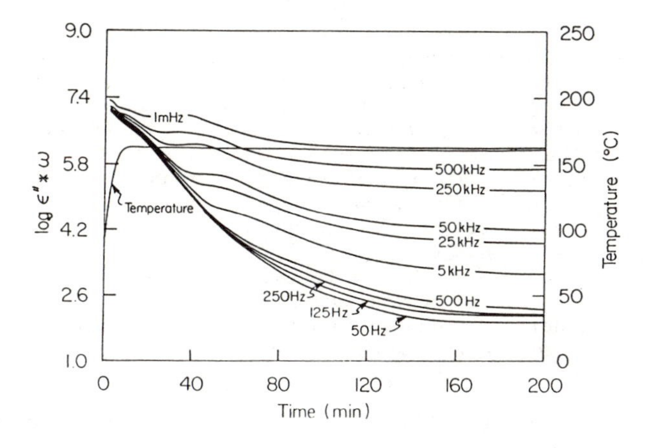

Figure 4. Log(ϵ"*ω) vs. time for the uncatalyzed epoxy during a 165C isothermal cure.

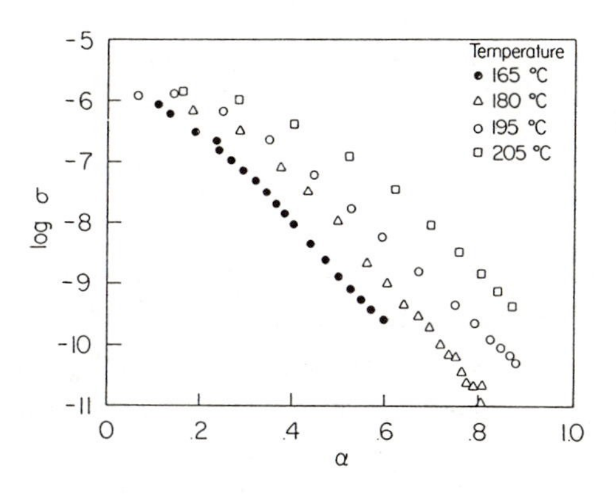

Figure 5. Log σ (ionic diffusion) vs. α (degree of cure) for the uncatalyzed epoxy during isothermal cure.

detecting points of maximum flow during cure, was first shown by us several years ago [6,7]. The qualitative correlation of $-\log \sigma$ and $\log \eta$ is clearly demonstrated in Figure 7. The quantitative relationship of $\log \sigma$ and $\log \eta$ is seen for two isothermal curves in Figure 8. To a first approximation one might look for a plot of $\log (\sigma)$ vs $\log (\eta)$ to be linear. Figure 8 shows there is a break in the σ-η dependence as the resin approaches a viscosity of 10^3 poise, a value often associated with gel. The break in the curve at this point undoubtably is due to the fact that the ionic and dipolar diffusion processes reflect a molecular viscosity [5], while the viscosity measured by the rheometer reflects a macroscopic resistance to flow. At gel the macroscopic viscosity begins to rise rapidly with small changes in α. In contrast, the α dependence of the molecular viscosity or the rate of change in the ionic diffusion remains unchanged as seen in Figure 5.

The relation of $\log \tau$ to $\log \eta$ is shown in Figure 9. The relationship is indeed approximately linear at 165°. This apparent linear dependence is due in part to the smaller range of η and the fact that the relaxation times all occur before η reaches 10^3 poise, that is before gel, the region of curvature in the $\log \sigma$ versus $\log \eta$ plot Figure 8.

The temperature dependence at constant α of σ and τ has been fit to an Arrhenius dependence

$$\ln (\sigma, \tau) = \frac{EA}{RT} \tag{6}$$

and a modified WLF dependence

$$\log \left(\frac{\sigma_o}{\sigma} , \frac{\tau}{\tau_o} \right) = \frac{C_1 (T - T_o)}{T - T_\infty} \tag{7}$$

where T_o is equal to an arbitrary reference temperature. Equation (7) is reported by Ferry (15) to provide a more objective fit than the standard WLF equation

$$\log \left(\frac{\sigma_o}{\sigma} , \frac{\tau}{\tau_o} \right) = \frac{C_1 (T - T)}{C_2 + T - T_o} \tag{8}$$

T_∞ is a fixed temperature approximately 50° below the glass transition temperature and at which $\log \left(\frac{\sigma_o}{\sigma} , \frac{\tau}{\tau_o} \right)$ becomes infinite. The WLF fit was determined by finding a value of T_∞ such that a plot of $\log \left(\frac{\sigma_o}{\sigma} , \frac{\tau}{\tau_o} \right)$ versus $(T - T_o) / (T - T_\infty)$ is linear and passes through the origin. Then C_1^o can be determined from the slope and C_2^o from the relation $T - T_\infty = C_2^o$. The results for $\alpha = .4$ and $T_o = 165°$ are

Ionic (σ)

$E_A = 1.2$ kJ/mole $\quad C_1^o = 7.7 \quad T_\infty = 20^oC \quad C_2^o = 145$

Dipolar (τ)

$E_A = 3.3$ kJ/mole $\quad C_1^o = 10.5 \quad T_\infty = 20^oC \quad C_2^o = 145$

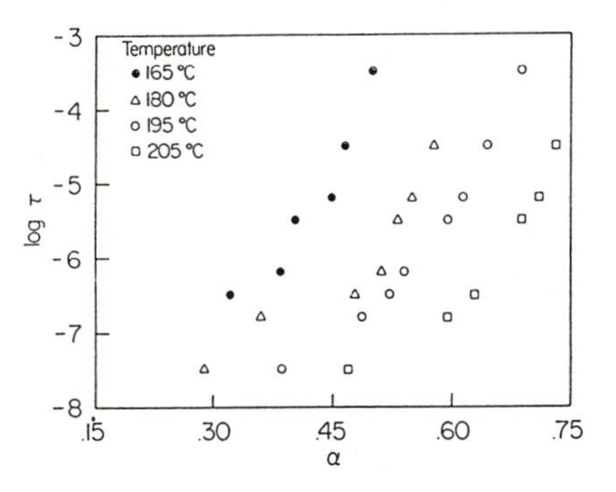

Figure 6. Log τ (relaxation time) vs. α (degree of cure) for the uncatalyzed epoxy during isothermal cure.

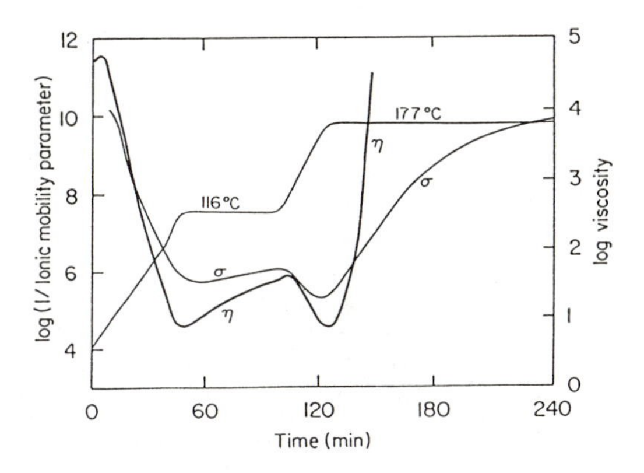

Figure 7. Simultaneous measurement during cure of the ionic mobility and viscosity for the catalyzed epoxy.

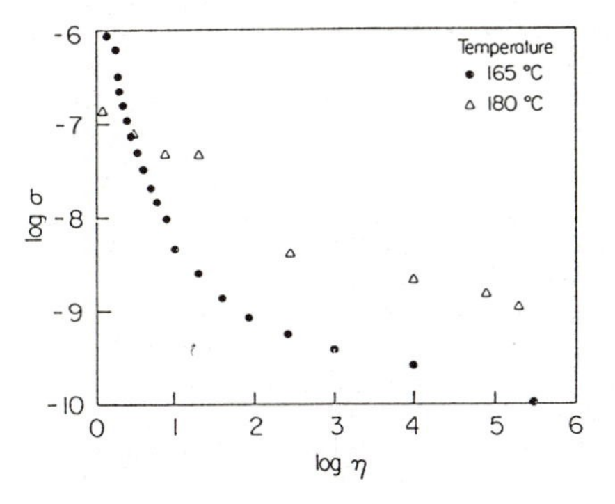

Figure 8. Log σ vs. log η for the uncatalyzed epoxy during isothermal cure.

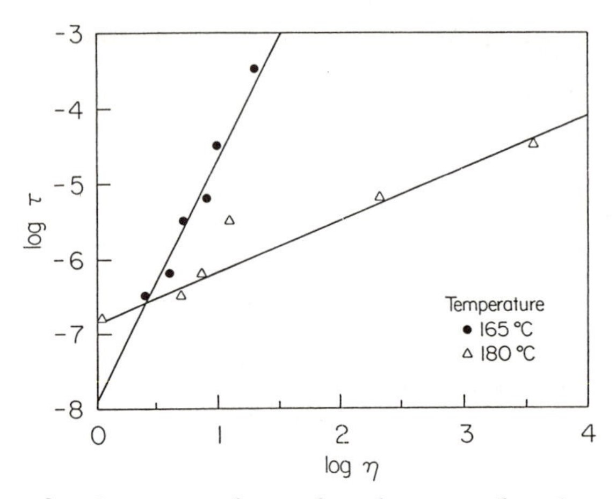

Figure 9. Log τ vs. log η for the uncatalyzed epoxy during isothermal cure.

The T_∞ value of 20° suggests an approximate Tg = 70° for α = .4, a value which is within the error of an experimentally determined Tg for a 165° sample quenched at α = .4.

The values of E_A and C_1 governing ionic mobility are less than those for dipolar mobility. This is consistent with the dipolar groups being larger than the ionic groups and requiring more energy as well as a larger volume to relax. C_1 is inversely proportional to the fractional free volume. Values of E_A and C_1 at other values of α show an increase during cure. The increase in E_A and C_1 is consistent with an increase in the energy and a decrease in available free volume governing ionic and dipolar diffusion with the buildup of the crosslink density. The value of T_∞ would also be expected to increase with α. The temperature range and the accuracy of the 4 values of σ and τ were insufficient to show the expected poor fit of the Arrhenius dependence nor was their accuracy sufficient to warrant fitting to independent values of T_∞ for σ and τ.

Figures 10 and 11 show a manufacturing application in which a resin's flow properties are measured in-situ at a particular point in a thick laminate during cure in an autoclave. The sensor was inserted on the tool surface and in the center of a thick 192 TGDDM graphite epoxy laminate. Figure 10 shows the noise free raw data taken by the center sensor during cure in the 8 x 4 foot production size autoclave. Using the procedures described, the ionic mobility was measured at both the tool surface and the center of the thick laminate. In Figure 11, the sensor values of σ show a 10 to 20 minute time lag in the point of maximum flow on the surface versus the laminate's center. Measurements of σ versus η, as shown in

Figures 7 and 8, allow Figure 11 to be quantitatively interpreted in terms of viscosity as well.

Conclusion

Frequency dependent dielectric measurements using impedence bridges in the Hz to MHz region are ideal for separating and measuring the ionic (σ) and dipolar (τ) mobility. The value of σ and τ can be used to quantitatively measure the viscosity, degree of cure and T_g. The value during cure of the Arrhenius and WLF constants reflect the buildup of the crosslink network. The results support the use of the WLF equation for analyzing the advancement of a curing reaction in TGDDM epoxies, a result which is in agreement with similar studies conducted on DGEBA epoxies.

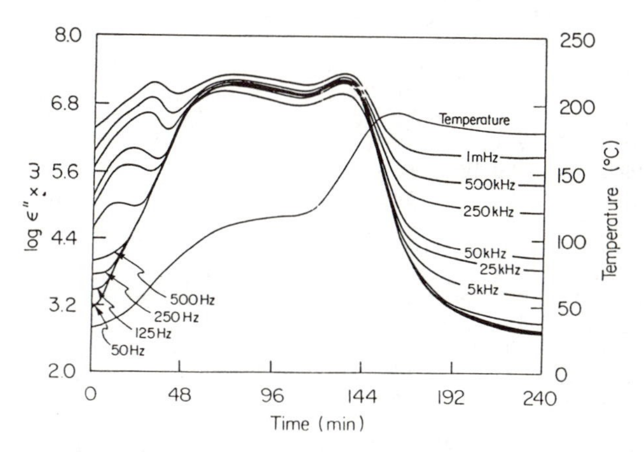

Figure 10. Raw data from the sensor in the middle of a 1" thick 3501-6 graphite epoxy laminate during cure in a production size autoclave.

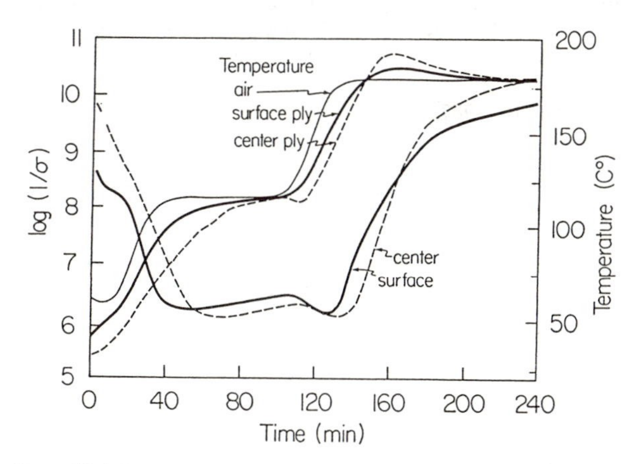

Figure 11. Log ($1/\sigma$) vs. time for a thick graphite laminate (192 ply) catalyzed epoxy monitored during cure in the autoclave.

Acknowledgments

Questions regarding the sensor and the instrumentation should be directed to D. Kranbuehl.

The work was made possible in part through the support of the National Aeronautics and Space Administration - Langley Research Center Research Grant No. NAG 1-237.

Literature Cited

1. May C., Chemorheology of Thermosetting Resins, ACS Sym. Ser. 227, Am. Chem. Soc. (1983).
2. Zukas W., MacKnight W., and Schneider N., Chemorheology of Thermosetting Resins, ACS Sym. SER. 227, 223, Am. Chem. Soc. (1983).
3. Kranbuehl, D., Developments in Reinforced Plastics -5, 181-204 Elsevier Appl. Science Publishers Ltd. (1986).
4. Senturia, S. and Sheppard, N., Adv. Poly. Sci. 80, 1 (1986).
5. Hill, N., Vaughan, W., Price, A., and Davis, M., Dielectric Properties and Molecular Behavior, Van Nostrand, London (1969).
6. Kranbuehl, D. E., Delos, S. E., Yi, E. C., Hoff, M. S. and Whitham, M. E., ACS Polym. Mater. Sci. and Eng. 53, 191 (1985).
7. Kranbuehl, D. E., Delos, S. E., Jue, P. K., Jarvie, T. P., and Williams, S. A., Nat'l SAMPE Symp. Ser. 29, 1251 (1984).
8. Kranbuehl, D., Delos, S., Hoff, M., and Weller, L., Nat'l SAMPE Symp. Ser. 31, 1087 (1986).
9. Kranbuehl, D. E., Delos, S. E., and Jue, P. K., National SAMPE Symp. Ser., 28, 608 (1983); SAMPE Journal, 19 (4), 18, (1983).
10. Kranbuehl, D. E., Delos, S. E., Jue, P. K., Polymer, 27, 11 (1986).
11. Kranbuehl, D., Delos, S., Hoff, M., and Weller, L. ACS Polym. Mater. Sci. and Eng. 54, 535 (1986).
12. Morgan, R. and Mones E., J. Appl. Polymr Sci. 33, 999 (1987).
13. Johnson, J. and Cole, R., J. Am. Chem. Soc. 73, 4536 (1951).
14. MacDonald, J. Ross, Trans., Faraday Soc. 66 (4), 943 (1970).
15. Ferry, J. D., Viscoelastic Properties of Polymers, John Wiley and Sons, Inc. (1961), Ch. 11., 2nd ed.

RECEIVED December 16, 1987

Chapter 9

Performance Characteristics of the Fluorescence Optrode Cure Sensor

R. L. Levy and S. D. Schwab

McDonnell Douglas Research Laboratories, P.O. Box 516, St. Louis, MO 63166

The autoclave curing of carbon-epoxy laminates is the most critical and costly stage in the fabrication of composites. A reliable, low-cost cure sensor suitable for use in manufacturing is needed to implement real-time control of the process. A second-generation FOCS which measures fluorescence intensity and wavelength with a fiber-optic spectrofluorometer and a tool-mounted optrode were used to monitor the curing of Hercules 3501-6 laminates. Changes of fluorescence intensity during cure follow changes in resin viscosity up to the gel-point, after which it increases in proportion to the degree of cure (DOC). The fluorescence wavelength as a function of cure-time produces a highly characteristic signal profile which is reproducible and independent of fluctuations in fluorescence intensity. Both FOCS signals detect changes in DOC during late stages of cure. Factors affecting the performance of the second-generation FOCS are discussed.

The autoclave curing of carbon-epoxy laminates is the most critical and costly stage in the fabrication of composite structures. Extensive efforts are currently in progress (1) to transform the autoclave curing process from a "skilled craft" to a science-based operation, resulting in lower cost and higher quality composites. Computer modeling (1,2), and optimization and real-time control of the process (1) are envisioned as vehicles towards this end (1-4). However, implementation of the advances anticipated in these projects depends on availability of a reliable, low-cost cure sensor which is suitable for use in the manufacturing environment. To be useful in manufacturing, a candidate sensor must be reliable, economic, and simple to operate. Toward these goals, a novel cure sensor called "Fluorescence Optrode Cure Sensor" (FOCS) was developed at MDRL in 1983-84 (5,6). This cure sensor, based on the combination of fiber-optic fluorometry and viscosity/degree-of-cure dependence of the resin fluorescence (5), has a potential for acceptance in the manufacturing environment. A simple tool-mounted "optrode" (5) was used to suc-

0097-6156/88/0367-0113$06.00/0

cessfully demonstrate the capability of the first-generation FOCS to monitor changes in fluorescence intensity during laminate cure which follow the changes in the resin viscosity/degree-of-cure. Observation of substantial shifts in the wavelength of the fluorescence emission maximum during cure provided the impetus to develop a second-generation FOCS capable of monitoring the changes in both fluorescence intensity and wavelength during cure (6). The second-generation FOCS hardware, software, and the optrode-laminate interface have been undergoing continuous "evolutionary" development and modifications. This paper describes the performance characteristics of the second-generation FOCS which were observed during the relatively early stages of development, i.e., September 1985-September 1986. Papers dealing with subsequent work on the FOCS are now in preparation (7).

Experiment

The second-generation FOCS is shown in Figure 1. It consists of a He-Cd laser excitation source (Omnichrome model 139), a polychromator (Instruments SA model HR-320), an optical multichannel analyzer (either PAR-OMA2 or PAR-OMA3), and a coupler interface of the type described by Hirshfeld et al. (8) which couples the excitation light (441.6 nm) into the optical fiber (Quartz Products QSF 1000) and

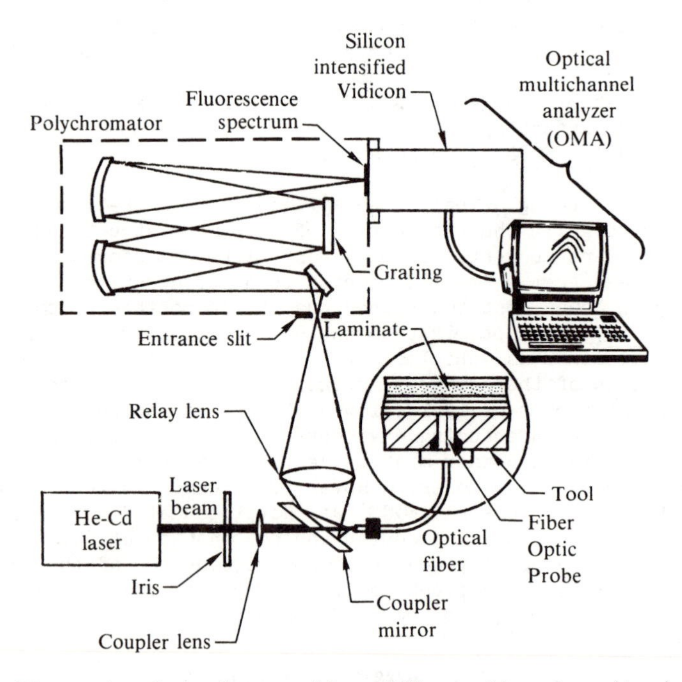

Figure 1. Second-generation FOCS capable of monitoring fluorescence intensity and wavelength.

transmits the collected fluorescence to the polychromator entrance slit. Data aquisition is controlled by a DEC model PRO 380 computer which triggers the OMA to output a signal-averaged spectrum every 13-20 seconds. Each spectrum is then transferred to the DEC PRO 380 computer for curve fitting via the IEEE 488 port. Cure temperature is recorded by a thermocouple (Doric model 400A) connected to the DEC PRO 380's real-time interface board.

Laminate curing with the second-generation FOCS was monitored with a tool-mounted optrode arrangement shown schematically in Figure 2. Six-ply laminates made from Hercules AS-4/3501-6 carbon/epoxy prepreg tape were stacked on a microscope slide. A silicone rubber dam enclosed all of the laminate except for a narrow channel for allowing resin flow. The glass slide with the laminate was then placed in a special clamping tool which permitted the optrode fluorescence excitation and emission to be coupled to the laminate as shown in Figure 2. Experiments were also performed with the optrode in direct contact with the laminate by elimination of the glass plate.

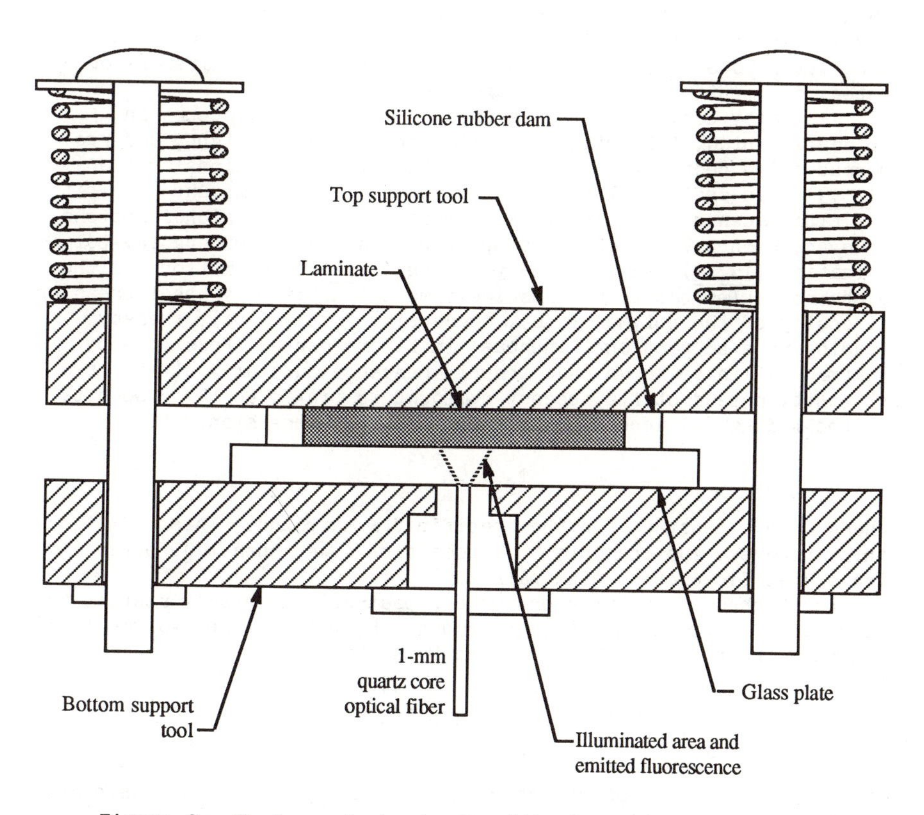

Figure 2. Tool-mounted optrode with glass-plate interface.

Results and Discussion

The mechanical properties of carbon-epoxy composites are strongly dependent on the chemorheological events taking place during the autoclave cure stage (2-4). The overall variation of the resin viscosity during the pre-gel portion of the cure cycle, i.e., the viscosity-time profile, along with the autoclave pressure, determine the extent of resin flow that can take place and thus strongly affect the ultimate mechanical properties and dimensions of the composites. Therefore, monitoring and control of the resin viscosity are of great importance (1-4).

Exploiting viscosity/degree-of-cure dependent fluorescence of the epoxy resin in conjunction with fiber optic fluorometry or spectrofluorometry produces a novel cure sensor. The inherent characteristics of this sensor offer the promise of meeting the demanding criteria for acceptance in the composite manufacturing factory. At present, cure sensors based on dielectrometry (9), microdielectrometry (10) and ultrasonic attenuation (11) which have been in evolutionary development for over 10 years are not routinely used in manufacturing. The second-generation FOCS, which is based on fiber-optic spectrofluorometry, on the VDF behavior of the epoxy resin, and on rapid processing of spectral data, simultaneously monitors the changes in resin fluorescence intensity (I_f) and the wavelength of maximum emission (λ_{max}) during cure (6). In contrast, the first-generation FOCS (5), which is based on a fiber-optic filter fluorometer, can monitor only the changes in I_f.

Viscosity-dependent fluorescence (VDF) typically occurs in molecules which, following absorption of excitation light, undergo nonradiative decay by intramolecular twisting or torsional motions (12-19). In ordinary low-viscosity solvents, VDF compounds exhibit low fluorescence due to fast deactivation by such torsional motions. However, when a VDF probe is dispersed in a monomer undergoing polymerization (15,16), such as epoxy curing (20,21), the motions of the probe molecules become progressively more inhibited by the increasing viscosity of the polymer (i.e., decreasing free-volume), thereby leading to a proportional increase in fluorescence. Forster and Hoffman (13) have shown that the fluorescence quantum yield Φ_f of a viscosity-dependent fluorescent compound is linked to the viscosity of the medium (η) according to the following expression:

$$\Phi_f = C\eta^{2/3}$$

where C is a constant for the specific VDF compound. This relationship, however, applies only to viscosities lower than 1000 poise and for a specific probe-medium pair.

For higher viscosities or for the post-gel and glassy states, the Φ_f of a VDF compound has been linked to the polymer free-volume (V_f) according to the expression derived by Loutfy (19)

$$\Phi_f = \left(\frac{K_r}{K^o_{nr}}\right) \exp\left(\beta\frac{V_o}{V_f}\right)$$

where K_r is the rate of radiative decay and K^o_{nr} is the intrinsic rate of molecular relaxation of the probe molecule, V_o is the occupied

(Van der Waals) volume of the probe molecule, and β is a constant for the particular probe.

Our experiments on the use of VDF probes for monitoring epoxy cure kinetics (20-22) lead to the observation of VDF behavior in tetraglycidyldiaminodiphenyl-methane (TGDDM), the main monomer constituent of the Ciba-Geigy MY720 resin used for fabrication of carbon-epoxy composites, as well as in the MY720 itself. Thus, the changes in the resin viscosity and DOC can be monitored without addition of a probe (20-22). Since LC-purified TGDDM also showed VDF behavior, it was initially assumed that TGDDM itself fluoresces. Subsequent experiments, however, indicated that one or more resin impurities are responsible for the observed VDF behavior. Work on separation and identification of impurities or synthesis by-products in the LC-purified TGDDM and MY720 resins is currently under way in our laboratory.

Typical FOCS I_f and λ_{max} signal profiles recorded during curing of Hercules 3501-6 laminate are shown in Figures 3 and 4 along with the temperature-time profile of the standard cure cycle used in our experiments. Analysis of the FOCS intensity and wavelength signal

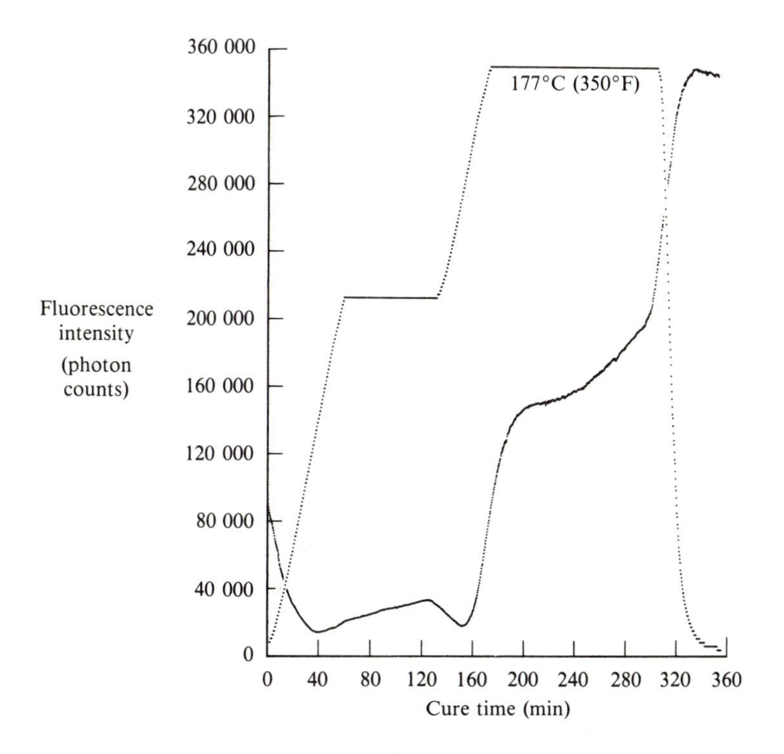

Figure 3. FOCS fluorescence intensity signal during curing of Hercules 3501-6 laminate.

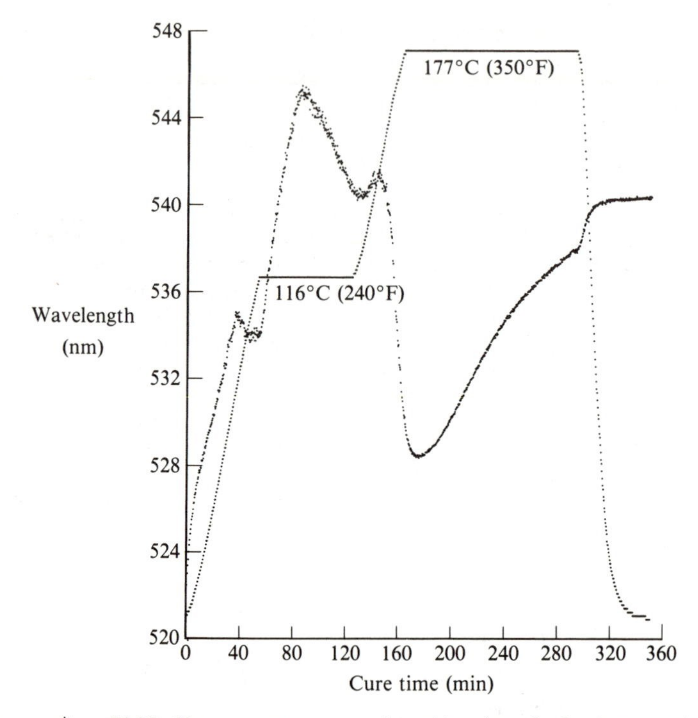

Figure 4. FOCS fluorescence wavelength signal during curing of Hercules 3501-6 laminate.

profiles recorded during curing of approximately 40 individual laminate specimens made of Hercules 3501-6 prepreg tape permitted the following observations and generalizations:

1. Changes in the fluorescence intensity during cure follow changes in the resin viscosity up to the gel-point as shown in Figure 3. Fluorescence intensity continues to increase beyond the gel-point at a rate that is proportional to the rate of cure. In fact, under standard cure-cycle conditions, the profile of fluorescence intensity as a function of cure-time signal follows the pattern of the profile of viscosity as a function of cure-time obtained under the same conditions with a Rheometrics dynamic mechanical spectrometer (1). Furthermore, the dynamic mechanical spectrometer ceases to provide the viscosity/degree-of-cure information early in the cure cycle (i.e., when the cure state approaches the gel-point), whereas FOCS signals continue to follow curing to completion.
2. Both FOCS signals (fluorescence intensity and wavelength maximum) successfully monitor cure-induced changes during the late stages of cure. It is particularly encouraging that the rate of signal change remains high during the late stages of cure, indicating sensitivity of the FOCS to small differences in the cure state. Lack of sensitivity to the minor changes encountered during the

late stages of cure is considered a common limitation of other cure sensors.

3. The run-to-run reproducibility of the profile shape of the FOCS fluorescence-intensity signal is good; however, the reproducibility of the absolute intensity values is unsatisfactory. The run-to-run variations in the fluorescence intensities are caused by the differences in resin thickness at the small area "viewed" by the optrode. In addition, substantial resin flow takes place during cure, causing the resin thickness to vary as a function of cure time. However, since this variation in resin thickness might be reproducible from run-to-run (if other cure parameters remain unchanged), it may be possible to develop a suitable signal-normalization procedure. Such normalization would minimize the effect of the inherent variation in resin thickness.
4. Changes in the fluorescence maximum wavelength λ_{max} as a function of temperature and cure produce a highly characteristic signal profile which is reproducible and also reveals the main chemorheological events, such as minimum viscosity and the gel-point with either reversal of the direction of signal change or with a distinct change of slope. The molecular origin of the changes in λ_{max} is complex in this case because both the dielectric environment of the fluorophores and the mobility of the dipoles surrounding the fluorophores change during cure. Therefore, the models employed for the analysis of the dependence of λ_{max} upon the dielectric properties of the solvent or the medium surrounding the fluorophores (23) cannot be applied to this case.
5. Changes in the FOCS fluorescence wavelength signal during the early stages of cure appear to be strongly dependent on the "freshness" of the laminate and the total time the laminate has been kept at room temperature prior to initiation of the cure cycle. This desirable feature of the FOCS could be exploited as an early indicator of the condition of the laminate at the onset of curing.

Fiber-optic fluorometry. Fiber-optic waveguides provide excellent means for delivering excitation energy to fluorescing media in remote, hostile, or inaccessible environments (8) such as reactors and plant streams, and for guiding the emitted fluorescence to a detector or a spectrometer, thus facilitating in-situ monitoring of the resin fluorescence (5,6). In our case, the autoclave represents a hostile environment (350°F and 100 psi) which is inaccessible for optical measurements by conventional methods.

Tool-mounted optrode (TMO). The cure-monitoring experiments described here were conducted with a "tool-mounted" optrode (TMO) arrangement (5,6) (Figure 2) which is ideally suited for the manufacturing environment where minimum interference with the laminate lay-up work is desirable. The use of a tool-mounted optrode is as simple as the use of tool-mounted thermocouples currently in wide use. Indeed, the TMO provides viscosity/degree-of-cure information on the cure state of the surface layer only. However, knowledge of the cure state of the surface layer permits determination of the cure states in the bulk based on the available models (1,2).

The optrode-laminate interface. The exact geometrical arrangement in which a tool-mounted optrode is in contact with the layer of resin covering the carbon fibers and the dimensions of the layer "viewed" by the optrode strongly affect the level of the FOCS intensity signal at a given cure time.

Since the thickness of the resin layer covering the carbon fibers is not uniform and varies with cure time, the initial magnitude of the FOCS intensity signal varies from run to run and has to be corrected or normalized to the initial signal level corresponding to a given thickness of the resin film. The introduction of a glass plate between the fiber tip and the laminate circumvents this problem because a larger laminate area is then viewed by the optrode and most of the signal comes from fluorescence originating from a thinner layer of resin. The use of a glass plate in the optrode-laminate interface is one acceptable configuration but many other arrangements are possible and some of these are under investigation (7).

Inner-filter effects. The absorption of the fluorescence excitation and emission by the specimen is referred to as the "inner-filter" effect; this effect has been treated in the literature (24-27). The inner-filter effect reduces the signal levels and distorts the emission spectrum and the intensity-concentration relationship. The effect is more pronounced in right-angle fluorescence measurements (27) than in the "front face" configuration in which the fluorescence is viewed from the same side as the excitation beam.

The magnitude of the inner-filter effect in our case depends strongly on the thickness of the resin layer viewed by the optrode and becomes pronounced only during the late stages of cure when the absorbance of the resin at 441.6 nm increases. A particularly pronounced impact of the inner-filter effect on the intensity signal profile of the FOCS is observed when the optrode monitors curing of the neat Hercules 3501-6 resin. The mathematical corrections for inner-filter effects assume, of course, a constant absorbance of the specimen, whereas in our case the absorbance of the resin at the excitation-light wavelength is a function of the degree-of-cure and therefore does not lend itself to the existing treatments.

Acknowledgment

The authors are thankful to C. Andrew, P. Zuker, K. Keply and K. Thiele for their skillful assistance in computer interfacing and software development, and to Dr. D. P. Ames for helpful discussions.

Initial phases of this work were performed under the MDC Independent Research and Development program. The research is currently supported by AFML contract #F33615-85-C-5024.

Literature Cited

1. Mallow, A. R.; Muncaster, F. R.; Campbell, F. C. Proc. Am. Soc. for Composites, 1st Tech. Conf., Technomic Publ. 1986, p. 171.
2. Loos, A. C.; Springer, G. S. J. Composite Mater. 17, 135 (1983).
3. Tajima, Y. A. Polym. Composites 3, 168 (1982).
4. Apicella, A. Developments in Reinforced Plastics; Pritchard, G., Ed.; vol. 5, p. 151 (1986).

5. Levy, R. L. Polym. Mater. Sci. Eng. 54, 321 (1986).
6. Levy, R. L. Review of Prog. in Quant. NDE, 3-8 August 1986, LaJolla, California.
7. Schwab, S. D.; Levy, R. L. in preparation.
8. Hirschfeld, T.; Deaton, T.; Milanovich, F.; Klainer, S. Optical Eng. 22, 527 (1983).
9. Krambuehl, D. E.; Delos, S. E.; Jue, P. K. SAMPE Journal 19(4), 18 (1983).
10. Senturia, S. D.; Sheppard, N. F.; Lee, H. L.; Day, D. R. J. Adhesion 15, 69 (1982).
11. Lindrose, A. M. Exp. Mech. 18, 227 (1978).
12. Oster, G.; Nishijima, Y. J. Am. Chem. Soc. 78, 1581 (1956).
13. Forster, Th.; Hoffmann, G. Z Physik, Chem. (n.F.) 75, 63 (1971).
14. Sarafy, S.; Muszkat, K. A. J. Am. Chem. Soc. 93, 4119 (1971).
15. Loutfy, R. O. Macromolecules 14, 270 (1981).
16. Loutfy, R. O. J. Polym. Sci., Polym. Phys. Ed. 20, 825 (1982).
17. Jaraudias, J. J. Photochem. 12, 35 (1980).
18. Tredwell, C. J.; Osborne, A. D. J. Chem. Soc. Faraday Trans. 1627 (1980).
19. Loutfy, R. O. Pure & Appl. Chem. 58, 1239 (1986).
20. Levy, R. L.; Ames, D. P. Proc. Org. Coat. Appl. Polym. Sci. 48 (1), 116 (1983)
21. Levy, R. L.; Ames, D. P. Adhesive Chemistry-Developments and Trends; Lee, L., Ed.; Plenum Press, 1984, p. 245.
22. Levy, R. L. Polym. Mater. Sci. Eng. 50, 125 (1984).
23. MacGregor R. B.; Weber, G. Ann,NY, Acad. Sci. 366, 140 (1981).
24. Mode, V. Alan; Sisson, D. H. Anal. Chem. 46, 200 (1974).
25. Holland, J. F.; Teets, R. E.; Kelley, P. M.; Timnick, A. Anal. Chem. 49, 706 (1977).
26. Novak, A. Coll. Czech. Chem. Commun. 43, 2869 (1978).
27. Hirschfeld, T. Spectrochim. Acta. 34A, 693 (1978).

RECEIVED February 16, 1988

DEFORMATION, FATIGUE, AND FRACTURE: MECHANICAL PROPERTIES OF CROSS-LINKED POLYMERS

Chapter 10

Deformation Kinetics of Cross-Linked Polymers

T. S. Chow

Xerox Corporation, Webster Research Center, Webster, NY 14580

A unified approach to the glass transition, viscoelastic response and yield behavior of crosslinking systems is presented by extending our statistical mechanical theory of physical aging. We have (1) explained the transition of a WLF dependence to an Arrhenius temperature dependence of the relaxation time in the vicinity of T_g, (2) derived the empirical Nielson equation for T_g, and (3) determined the Chasset and Thirion exponent (m) as a function of cross-link density instead of as a constant reported by others. In addition, the effect of crosslinks on yield stress is analyzed and compared with other kinetic effects -- physical aging and strain rate.

The time and temperature dependent properties of crosslinked polymers including epoxy resins (1-3) and rubber networks (4-7) have been studied in the past. Crosslinking has a strong effect on the glass transition temperature (T_g), on viscoelastic response, and on plastic deformation. Although experimental observations and empirical expressions have been made and proposed, respectively, progress has been slow in understanding the nonequilibrium mechanisms responsible for the time dependent behavior.

The purpose of this paper is to establish the fundamental links between the glass transition, viscoelastic relaxation, and yield stress by investigating the relaxation processes in polymers. The relationship between temperature and relaxation time scale is represented by a shift factor (a). At temperature T

0097-6156/88/0367-0124$06.00/0

> T_g, the stress relaxation data can be described by the WLF temperature dependence (8). However, an Arrhenius type of dependence is usually observed for viscoelastic response in the glassy region. The phenomenon is generally true for amorphous polymers as well as crosslinking systems. A typical transition from a WLF dependence to an Arrhenius temperature dependence for an epoxy resin (1) is shown in Figure 1. The change in the relaxation mechanism for deformation near T_g will be explained by our physical aging theory (9-12) recently developed for amorphous polymers. The same basic approach will be extended here to discuss the effect of crosslinking as a chemical aging process. The relaxation time will then be used to determine the above mentioned physical properties of crosslinked polymers.

RELAXATION TIME

On the basis of the idea of continuous conversion of the number of holes (free volumes) and the number of phonons in a polymer lattice, we have introduced (9) the physical picture of quantized hole energy states ε_i with i = 1, 2, . . . L. The problem is to determine the distribution of the ensemble characterized by a set of hole numbers $\{n_i\}$ with $\Sigma_i n_i = n$. The ratio of $n_i/N = f_i$ is the ith contribution to the free volume fraction ($f = \Sigma_i f_i$). Minimizing the excess Gibbs free energy due to hole introduction with respect to n_i, the equilibrium distribution of the free volume fraction is obtained (9)

$$\bar{f}(T) = \bar{f}_r \exp\left[-\frac{\bar{\varepsilon}}{R}\left(\frac{1}{T} - \frac{1}{T_r}\right)\right] \tag{1}$$

where $\bar{\varepsilon} = \Sigma_i \varepsilon_i \bar{f}_i/\bar{f}$ is the mean hole energy, R is the gas constant and the subscript r refers to the condition at $T = T_r$ which is a fixed quantity near T_g.

The nonequilibrium glassy state, $\delta(t) = f(t) - \bar{f}$, is determined by solving the kinetic equations which describe the local motion of holes in response to molecular fluctuations during vitrification and physical aging. The solution is (11)

$$\delta(T, t_e) = -\frac{\bar{\varepsilon}}{R}\int_0^{t_e} \frac{q\bar{f}}{T^2}\,\phi(t_e - t')\,dt' \tag{2}$$

where t_e is the physical aging time and q is the cooling (< 0) rate. The relaxation function $\phi(t)$ has been derived (10) as the probability of the holes having not reached their equilibrium states for a quenched and annealed glass, and has the form

$$\phi(t) = \exp[-(t/\tau)^{\beta}] \tag{3}$$

where β defines the shape of the hole energy spectrum. The relaxation time τ in Equation 3 is treated as a function of temperature, nonequilibrium glassy state (δ), crosslink density and applied stresses instead of as an experimental constant in the Kohlrausch-Williams-Watts function. The macroscopic (global) relaxation time τ is related to that of the local state (λ) by $\tau = \tau_r \lambda^{1/\beta} = \tau_r a$ which results in (11)

$$\ln a = \ln \lambda/\beta = \frac{1}{\beta}\left(\frac{1}{\bar{f}+\delta} - \frac{1}{\bar{f}_r}\right) \tag{4}$$

The above equation suggests that the Doolittle equation (13) has to be modified to include the nonequilibrium contribution. When $\delta = 0$, Equation 4 can be written in the form of the WLF equation which is known to be valid for $T > T_g$. In order to acquire a deep insight into the change in the deformation mechanism shown in Figure 1, we look at Equation 4 in the vicinity of T_g which is approximated by

$$\ln a = -\left[\bar{\alpha}_r (T - T_r) + \delta\right]/\beta\bar{f}_r^{\,2} \tag{5}$$

where $\bar{\alpha}_r = \bar{\varepsilon}\bar{f}_r/RT_r^2$. The prediction of Equations 2, 3 and 5 is shown in Figure 2 as the solid curve by using the input parameters (11): $\bar{\varepsilon} = 2.51$ kcal/mol, $\beta = 0.48$, $\bar{f}_r = 0.0336$ and $\tau_r = 25$ min for poly(vinyl acetate) (PVAc) with $T_r = 308$K. These parameters have been chosen to describe the volume relaxation and the equation of state for PVAc. The circles in Figure 2 are experimental data from the shear creep measurement (14) with a (297K) = 1. We have shown (12) that the cooling rate has little effect on the calculated slope: $\partial \log a/\partial T$ in the glassy state ($T < T_g - 10$K). Following Equation 5, we obtain the activation energy

$$E_a \simeq -RT_r^{\,2}\,\partial \ln a(T,\delta)/\partial T = (1-\mu)\bar{\varepsilon}/\beta\bar{f}_r \tag{6}$$

where

$$\mu = -\bar{\alpha}_r^{\,-1}\,\partial\delta(T,\delta)/\partial T \tag{7}$$

It reaches a constant value of 0.8 for $T < T_g$ -10K and approaches zero for $T > T_g$. The change of E_a in Figure 2 from $\bar{\varepsilon}/\beta\bar{f}_r = 155.6$ to 30.5 kcal/mol $\simeq (1 - \mu)\bar{\varepsilon}/\beta\bar{f}_r$ as PVAc is cooled

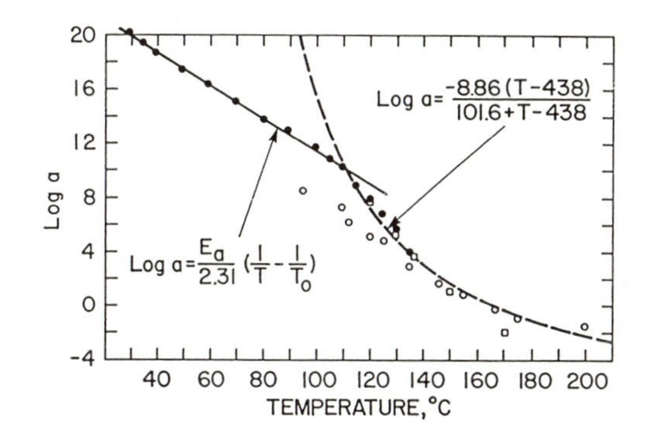

Figure 1. The transition of a WLF dependence to an Arrhenius temperature dependence for the global shift factor of an epoxy resin[1].

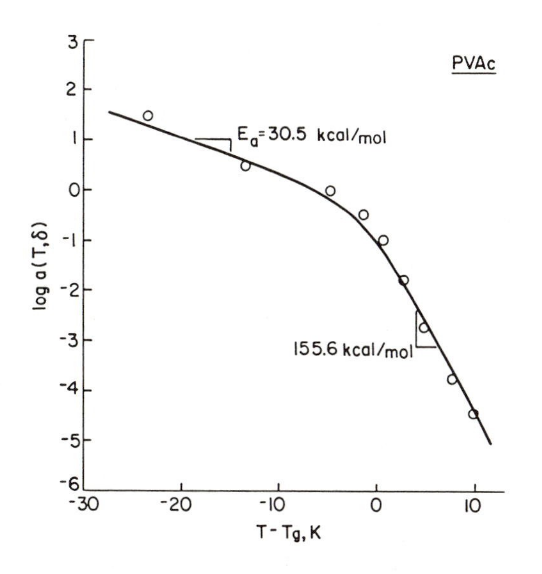

Figure 2. Comparison of calculated (curve) and measured (circles)[14] global shift factor in the vicinity of T_g.

through the glass transition region can now be interpreted in terms of μ. Struik (15) has introduced the same constant μ in the glassy state to characterize the physical aging rate observed in his isothermal creep experiments. Recently, we have derived (12) the Struik equation for the shift factor at longer aging times from Equation 5

$$a(T, t_e) \sim t_e^{\mu} \tag{8}$$

The exponent μ is no longer an empirical constant but can be calculated directly from the same molecular parameters ($\bar{\varepsilon}$, β, $\bar{f}_r$, τ_r) mentioned earlier for PVAc using Equations 2, 5 and 7. The value of μ for the epoxy resin in Figure 1 is lower than that of amorphous polymers in the glassy state.

In the case of crosslinked polymers, the global relaxation time τ has to be generalized by including a shift factor for the crosslink density (a_v)

$$\tau\ (T, \delta, v) = \tau_r\ a(T, \delta) a_v \tag{9}$$

where $v = \rho/M_c$ is the number of moles of network chains per unit volume, M_c is the number of average molecular weight between crosslinked junctions, and ρ is the density. Equation 9 will be discussed and used to determine the physical properties from the rubbery to glassy states of crosslinked polymers in the rest of the paper.

GLASS TRANSITION.

The glass transition temperature provides a good indication of the segmental mobility of polymer chains. The formation of crosslinking slows down the molecular motion of chain molecules and the global relaxation time has to increase. If the slowing down process is similar to that of the physical aging in the form of Equation 8, a_v takes the form $a_v \sim t_c^{\eta}$ where t_c is the curing time and η characterizes the double-logarithmic crosslinking rate. Again, this power law is valid not at $t = 0$ but for longer curing times when the system has already passed the initial phase of crosslinking. Figure 3 shows the log-log plot of crosslink density versus curing time of nature rubber (4-6) which reveals $v \sim t_c^{0.55}$. Defining $\gamma = \eta/0.55$, we obtain

$$a_v = (v/v_0)^{\gamma} \qquad \text{with } \gamma > 0 \tag{10}$$

where v_0 is the reference crosslink density. In analyzing the effects of crosslink density on the physical properties of already crosslinked systems, τ_r and T_r in Equations 9 and 5, respectively, refer to the crosslinked polymer with crosslink density v_o. We choose T_r to be near to T_{go}, the glass transition temperature at the reference crosslink density. In contrast to those reported in the literature (6-7), Equation 10 reveals that polymers relax slower rather than faster at higher crosslink densities.

The glass transition is a nonequilibrum phenomenon. T_g depends on the global relaxation time and is determined by a nonequilibrium criterion (11-12)

$$d\tau/dT = -1/|q| \text{ at } T = T_g \quad (11)$$

for amorphous materials vitrified under cooling rate $q = -|q|$. The effect of crosslinking on T_g can be evaluated from Equations 9-11 which gives

$$T_g - T_{go} = (\gamma/\theta) \ln (v/v_o)$$

$$\simeq (\gamma/\theta)(v - v_o)/v_o, \qquad (v - v_o)/v_o \ll 1 \quad (12)$$

where $\theta = \bar{\varepsilon}/\beta\bar{f}_r RT_r^2$. The approximate expression of Equation 12 assumes the form of Nielson's empirical equation (2) which has provided a good description of experimental data for a wide range of crosslinked systems (16).

VISCOELASTIC RESPONSE.

The shear relaxation modulus can in general be written as an integral over the relaxation time spectrum H. At the same time Equation 3 can also be used. Thus, we have

$$\frac{G(t) - G_\infty}{G_o - G_\infty} = \int_{-\infty}^{\infty} H(u) \exp(-t/u) \, d \ln u$$

$$= \exp\left\{ -\left[\frac{t}{\tau(T, \delta, v)} \right]^\beta \right\} \quad (13)$$

where G_o and G_∞ are unrelaxed and relaxed moduli, respectively and t is the load time. We have taken the Laplace inversion of Equation 13 and determined (17) the normalized relaxation spectrum $H(\tilde{\tau}, \beta)$ as a function of the nondimensional relaxation time, $\tilde{\tau} = u/\tau$, the ratio of local to global relaxation times, and β. When Equations 3 and 5 are used simultaneously in analyzing experimental data, we have found that $\beta \simeq 1/2$ for most amorphous polymers which will also be assumed for lightly crosslinking systems.

It has been reported (4-6) that elastomers undergo very long-term relaxation processes in stress relaxation and creep experiments. The long time behavior of shear modulus can be represented by (18)

$$G(t) \sim t^{-m} \tag{14}$$

Rather than treating m as the usual empirical exponent, we shall acquire a deeper insight into the double logarithmic relaxation rate. Using Equations 10 and 13, we obtain

$$m = -\frac{d \log a}{d \log t}$$

$$= \beta(1-\mu)(t/\tau_o)^{\beta} a_v^{-\beta}$$

$$\equiv m_o a_v^{-\beta} = m_o (v/v_o)^{-\gamma\beta} \tag{15}$$

When the relaxation time $\tau_o = \tau_r a$ is long, m_o is a slow varying function of time and may be approximated by a constant. In addition, the relaxation rate is known to be quite small and, therefore, can be related to the loss tangent by the equation (19)

$$m = \frac{2}{\pi} \tan\Delta, \quad m\pi/2 << 1 \tag{16}$$

The parameters

$$\gamma = 3.64$$

$$m_o = 0.065 \text{ at } v_o = 0.46 \times 10^{-4} \text{ mol/cm}^3 \tag{17}$$

are obtained by fitting the tan Δ data[4] near the maxima for the lightly crosslinked rubbers. A plot of log tan Δ versus log v is shown in Figure 4. The circles represent experimental values and the solid line is obtained from Equations 15 and 16. With the positive γ and the crosslink density dependent m, Equation 15, γ obtained in Equation 17 differs completely from the reported (6-7) $\gamma = -15$. We shall also see later that γ deduced here from the viscoelastic data is consistent with γ for the yield behavior.

Differentiating Equation 14 with respect to the limit $\ell n\, t$, we obtain the approximate relaxation spectrum

$$H(u) \simeq -\frac{dG}{d\,\ell n\, t}\bigg|_{t=u} = m\left(\frac{u}{\tau_o a_v}\right)^{-m} \tag{18}$$

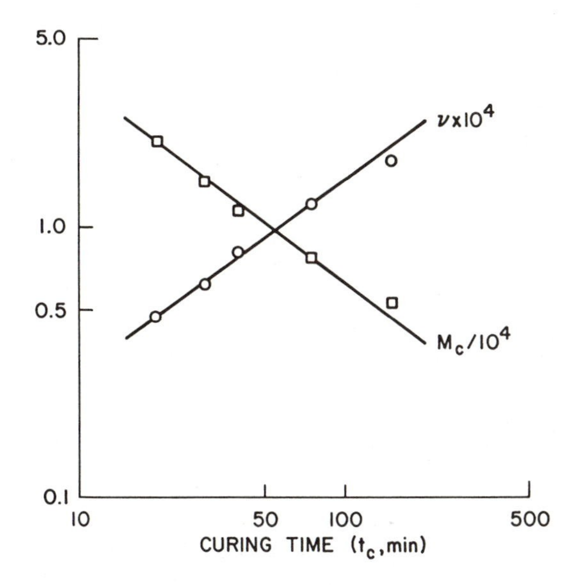

Figure 3. Cross-link density versus curing temperature. Straight lines are drawn through data points of nature rubber (NR)[4].

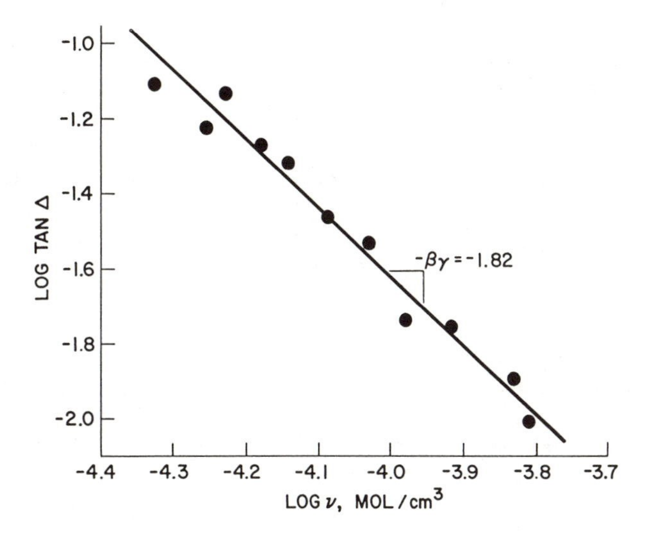

Figure 4. Determination of γ from the log Δ versus log v data of NR[4].

Substitution of Equation 18 into the integral representation of Equation 13 gives

$$\frac{G(t) - G_\infty}{G_\infty} \simeq m \int_{-\infty}^{\infty} \left(\frac{u}{\tau_o a_v} \right)^{-m} \exp\left(- \frac{t}{u} \right) d\,\ell n\, u$$

$$= \Gamma(m+1) \left(\frac{t}{\tau_o a_v} \right)^{-m}$$

Since the Gamma function Γ (m + 1) $\simeq$ 1 for m << 1 (see Equations 15 and 17), the above equation becomes

$$G(t) = G_\infty \left[1 + \left(\frac{t}{\tau_o a_v} \right)^{-m} \right], \text{ large } t \tag{19}$$

which has the form of the Chasset and Thirion empirical equation except that it has the crosslink density dependent m. The equilibrium modulus is a function of the crosslink density in accordance with the theory of rubber elasticity and is given by (4)

$$G_\infty = 0.71 \, v \, RT \tag{20}$$

Differing from the previous studies (5-7) where the parameters G_∞, m, $\tau_o a_v$ have been treated as constants, we find that they depend on cross-link density which is consistent with the measurements of Dickie and Ferry (4). Figure 5 shows the dependence of the viscoelastic relaxation on cross-link density. The solid curves are calculated from Equations 17, 19 and 20 by using a value of τ_o = 2.5 x 10^2 hrs at T = 25°C. Figure 5 resembles the corresponding figure in ref. 5.

YIELD STRESS

The relaxation phenomenon which has been discussed so far is within the linear viscoelastic range. Under large deformation, the global relaxation time has to include the contribution from the external work Δw done on the lattice site and takes the form (20)

$$\tau(T, \delta, v) \exp(-\Delta w / 2\beta RT) \tag{21}$$

The yield behavior is also a kinetic phenomenon and has been treated as the nonlinear cooperative deformation (21). Consider a tensile yield stress σ_y acting on N polymer lattice sites which include n holes. The external work done on each lattice cell is $-\sigma_y\Omega_{11}$ where Ω_{11} is the tensile activation volume. By using a mean field average, the work done by σ_y on a hole cell during the yielding of polymers is (12)

$$\Delta w = -\sigma_y \Omega_{11} \frac{N}{n} \simeq -\sigma_y \Omega_{11}/\bar{f}_r \tag{22}$$

The yield occurs when the product of the applied strain rate (e) and the global relaxation time reaches the order of unity, i.e., $\dot{e}\tau \sim 1$. Thus, we obtain

$$\sigma_y - \sigma_{yo} = (\beta RT/\Omega)[\ell n\ (\dot{e}/\dot{e}_o) + \mu\ \ell n\ (t_e/t_{eo}) + \gamma\ \ell n\ (v/v_o)] \tag{23}$$

where $\Omega = \Omega_{11}/2\bar{f}_r$. In addition to the contribution from strain rate and annealing time (at longer times), Equation 23 also includes the effect of crosslinks on yield stress. A comparison of Equation 23 with experimental data (3) on the dependence of σ_y on crosslinks of epoxy resins is shown in Figure 6. The slope, $\partial\sigma_y/\partial \log v$, gives 2.303 $(\beta RT/\Omega)\gamma$. The parameter $\beta RT/\Omega$ is usually determined from the measurement of $\partial\sigma_y/\partial \log e$ which has a value ranging from 10 to 20 Kg/cm^2 at the room temperature (21). Figure 6 gives $\gamma \sim 3.3$ to 6.5 for epoxy resins which is close to the value of γ obtained for crosslinked rubbers (see Equation 17). Since γ is greater than $\mu < 1$, the effect of crosslinks is more readily seen than either the effect of physical aging or strain rate on the yield stress.

CONCLUSION

The relaxation process defines the relationships between the physical properties of polymers and their structure. The transition of a WLF dependence to an Arrhenius temperature dependence of the relaxation time in the vicinity of T_g is related to the physical aging rate μ which can be calculated from the nonequilibrium glassy state in terms of the same set of molecular parameters. By extending our theory of physical aging, the effect of crosslinking is discussed as a chemical aging process. The approach provides the fundamental links between the glass transition, viscoelastic response and yield stress for crosslinking systems. The empirical Nielson equation for the glass transition temperature as a function of the crosslink density is derived from a non-equilibrium criterion in terms of the global relaxation time.

In contrast to the existing analyses, the present theory predicts that (1) the relaxation time scale gets longer rather

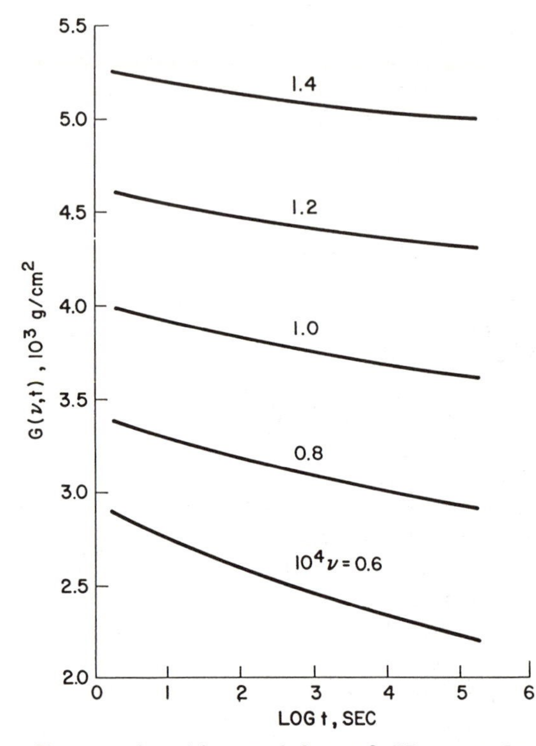

Figure 5. Shear relaxation modulus of NR as a function of cross-link density at 25°C.

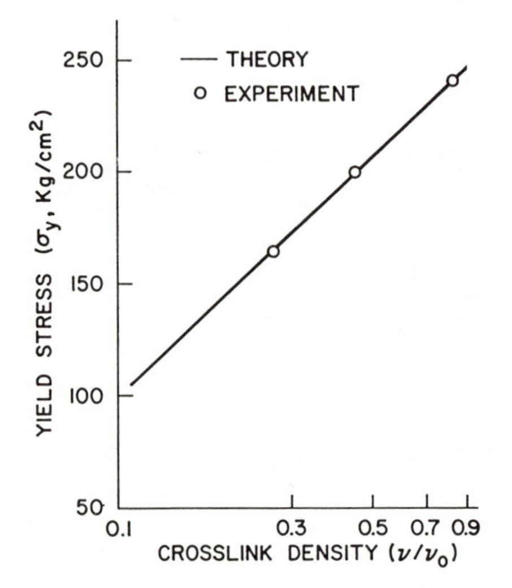

Figure 6. The effect of cross-link density on the yield stress of an epoxy resin. Circles are experimental data[3].

than shorter at higher crosslink densitities and behaves like $a_v \sim v^\gamma$ with $\gamma \sim 3.5 - 6.5$ for epoxy resins as well as crosslinked rubbers, and (2) the Chasset and Thirion exponent is not a constant but has the form $m \sim v^{-\beta\gamma}$ which is consistent with Dickie and Ferry's data. The effect of cross-links on yield stress is analyzed and compared with other kinetic effects -- physical aging and strain rate. Finally, a self-consistent theory is emerged not only for the viscoelastic response but also for T_g and yield behavior. The theory compares well with published experimental data.

REFERENCES

1. Halpin, J.C. In Compoisite Materials Workshop; Tsai, S.W.; Halpin, J.C.; Pagano, N.J., ed.; Technomic. Stamford, Conn., 1968; p. 87.
2. Nielson, L.E. J. Macromol. Sci. - Revs. Macromol. Chem., 1969, C3, 69.
3. Lohse, F.; Schmid, R.; Batzer, H.; Fisch, W. Brit. Polym. J., 1969, 1, 110.
4. Dickie, R.A.; Ferry, J.D. J. Phys. Chem., 1966, 70, 2594.
5. Chesset, R.; Thirion, P. In Physics of Non-Crystalline Solids; Prins, J.A., ed.; North-Holland. Amsterdam, 1965, p. 345.
6. Plazek, D.J. J. Polym. Sci., 1966, A-2, 4, 745.
7. Curro, J.G.; Pincus, P. Macromol., 1983, 16, 559.
8. Williams, M.L.; Landel, R.F.; Ferry, J.D. J. Am. Chem. Soc., 1955, 77, 3701.
9. Chow, T.S. J. Chem. Phys., 1983, 79, 4602; Macromol., 1984, 17, 2336.
10. Chow, T.S. J. Noncryst. Solids, 1985, 75, 209.
11. Chow, T.S. J Rheology, 1986, 30, 729.
12. Chow, T.S. J. Polym. Sci., 1987, 25B, 137.
13. Doolittle, A.K. J. Appl. Phys., 1951, 22, 1471.
14. Knauss, W.G.; Kenner, V.H. J. Appl. Phys., 1980, 51, 5131.
15. Struik, L.C.E. Physical Aging in Amorphous Polymers and Other Materials; Elsevier: Amsterdam 1978.
16. Reid, C.G.; Greenberg, A.R. Proc. ACS Polym. Mater. Sci. Eng., 1987, 56, 764.
17. Chow, T.S. Mat. Res. Soc. Symp. Proc., 1987, 79, 345.
18. Schapery, R.A. Polym. Eng. Sci., 1969, 9, 295.
19. Ferry, J.D. Viscoelastic Properties of Polymers; 3rd Ed., Wiley: NY, 1980; p. 91.
20. Chow, T.S. Polym. Eng. Sci., 1984, 24, 915, 1079.
21. Ward, I.M. Mechanical Properties of Solid Polymers; 2nd Ed., Wiley: NY, 1983; p. 377.

RECEIVED December 16, 1987

Chapter 11

Plastic Deformation in Epoxy Resins

S. M. Lee

Composite Materials Department, Ciba-Geigy Corporation, Fountain Valley, CA 92708

The plastic deformation of several amine and anhydride cured diglycidyl ether of bisphenol A (DGEBA) epoxy resins has been investigated. The yield stresses of the materials were measured in compression over a temperature range below the glass transition temperature T_g. The experimental results are reasonably intereperted by the Argon theory in which a thermally activated molecular deformation mechanism dictates the phenomenological yield behavior. The molecular parameters estimated from the theory are compared with the resin chemical structures to demonstrate the important structural features controlling the plasticity of epoxy resins. The similarity in deformation mechanism below T_g between thermoplastics and thermosets is evident. The role of crosslinks in the yield behavior of thermosets is also discussed.

Epoxy resins, even in highly cross-linked form, can undergo plastic deformation when fracture is suppressed, for example, under compression loading. The variety of epoxy resins offers a wide range of molecular structures that exhibit different yield behavior at the macroscopic level. The study of plastic deformation in different epoxy resins can help understand the structure/property relationship of plasticity in thermoset resins.

Attempts have been made in the past to explain the plasticity of glassy polymers based on different theories (a review of which can be found in (1)). Most of the earlier approaches are based on models non-specific about the molecular deformation process. The theory by Argon

0097-6156/88/0367-0136$06.00/0

(1-4), however, gives a unique description of the specific molecular response during plastic deformation. In the theory, the molecular chain segment undergoes a thermally activated rotation to result in macroscopic plastic flow. The model does not involve the free volume concept implicitly or explicitly included in many other models. The direct treatment of micro-mechanism by Argon allows the determination of molecular parameters governing the plasticity of glassy polymers.

The Argon theory has successfully interpreted the yield behavior of a large number of amorphous thermoplastic polymers (3,4). For thermosets, Yamani and Young (5) applied the theory to explain the plastic deformation of a diglycidyl ether of bisphenol A (DGEBA) epoxy resin cured with various amount of triethylene tetramine (TETA). They found that the theory gave a reasonable description for the resins below the glass transition temperatures T_g.

This paper rerports an investigation of the yield behavior of several amine and anhydride cured DGEBA resin systems. The Argon theory is used to assess the controlling molecular parameters from the experimental results. Such parameters are then compared with the known chemical structures of the resins. The mechanisms of plastic flow in thermoset polymers such as epoxies is demonstrated.

Experimental Details

Materials Description. Three CIBA-GEIGY epoxy/hardener systems were studied: Araldite 6010/906, Araldite 6010/HY 917 and Araldite 6010/972 with stoichiometries 100/80, 100/80 and 100/27, respectively. Araldite 6010 was a DGEBA epoxy resin. The hardeners 906, HY 917 and 972 were, respectively, methyl nadic anhydride (MNA), methyltetrahydro phthalic anhydride (MTPHA) and methylene dianiline (MDA). These systems were investigated previously for the matrix controlled fracture in composites (6-8). The curing cycles used can be found in (6). The ideal chemical structures of the systems are shown in Table I. Neat resins were thoroughly degassed and cast into 1.27 cm thick plates for preparation of test specimens.

Compression Test. Compression tests similar to that described in (5) were conducted for yield stress σ_y and modulus E measurement. Rectangular neat resin specimens (1.27 cm x 1.27 cm x 2.54 cm) cut from the cast resin plates were tested under compression, as shown in Figure 1, in an universal testing machine at a loading rate of 0.05 cm/min. For each resin system studied, tests were conducted at several temperature levels between -60 and 60 degree C. All specimens were instrumented with strain gages for

TABLE I. Chemical structures of the systems studied

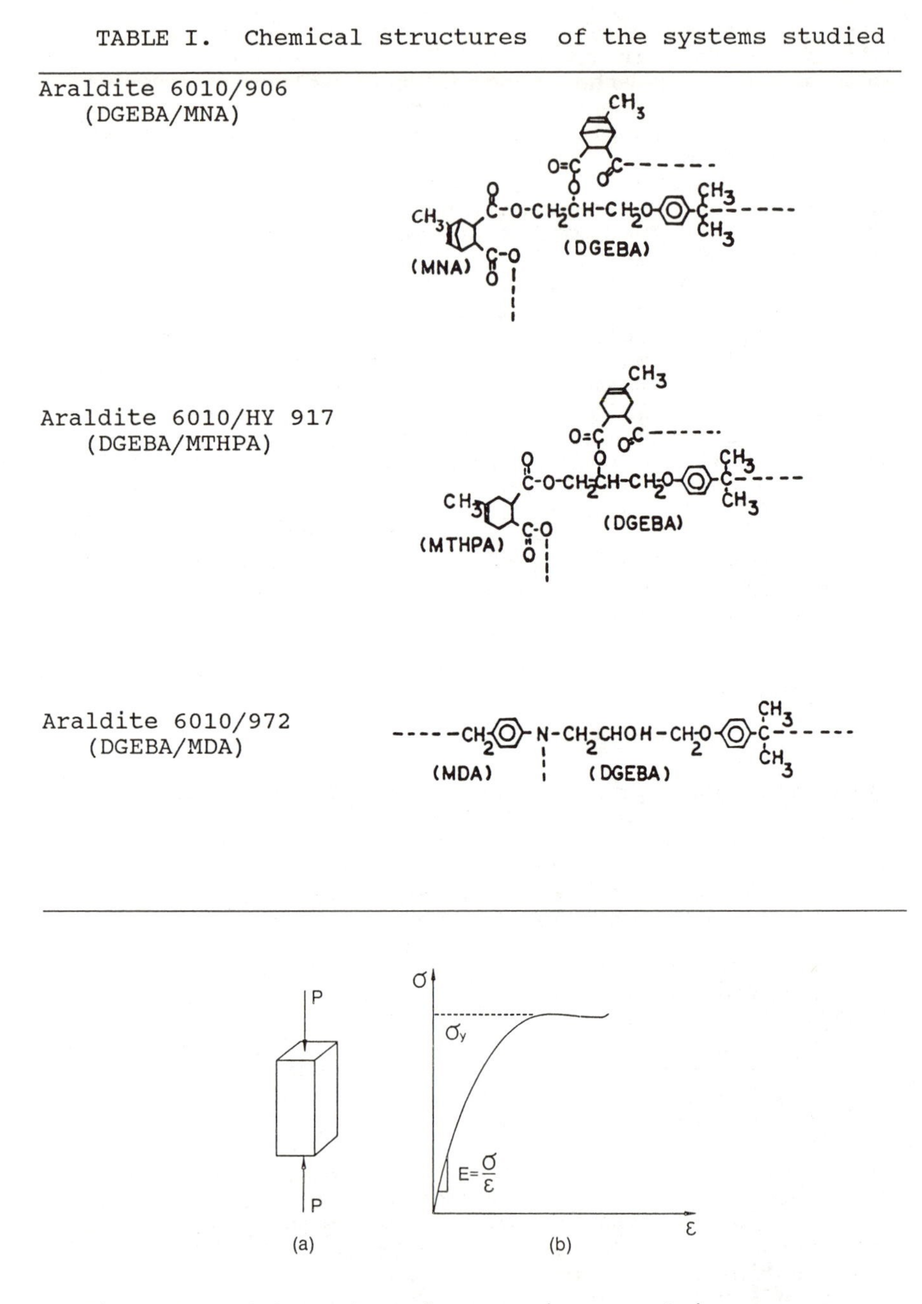

System	Structure
Araldite 6010/906 (DGEBA/MNA)	(MNA) (DGEBA)
Araldite 6010/HY 917 (DGEBA/MTHPA)	(MTHPA) (DGEBA)
Araldite 6010/972 (DGEBA/MDA)	(MDA) (DGEBA)

Figure 1. (a) Rectangular specimen used in compression test with (b) stress-strain curve showing the elastic modulus E and yield stress σ_y.

modulus measurement. The yield stress σ_y was determined from the level-off load P_y in the load-deflection curve by also considering the increased cross-sectional area due to the total compression strain ε_y at yield, i.e.,

$$\sigma_y = \frac{P_y(1-\varepsilon_y)}{A_o} \qquad (1)$$

where A_o is the original cross-sectional area of the specimen.

Results and Discussion

The Argon Theory. The model proposed by Argon (1-4) is based on a description of incremental alignment of molecular chain by the thermally activated rotation of molecular segments. The local deformation process, as shown in Figure 2, is produced by the formation of a pair of kinks at the end of the segment along the molecular chain. Such a segment is, in principle, closely related to the stiff units between the natural flexible hinges in the molecular structures. The kinks are formed by rotating the ends of the segment by an angle ω (of the order of bond angle). The kink formation is opposed not only by the molecule itself but also, and mainly, by the surrounding molecules.

The primary mechanism of kink formation involves the interaction of the segment (Figure 2) with the neighboring molecules treated as an elastic continuum with bulk elastic properties. The intermolecular energy change associated with such molecular deformation can be determined from elastic energy consideration . The activation free enthalpy ΔG of the process can be directly related to the configuration of the molecular segment. For the deformation mechanism to nucleate under applied load, the controlling segments must have a critical configuration with length z^* and radius **a** that maximize ΔG. The critical configuration $(z^*/\mathbf{a})$ is found to be

$$\left(\frac{z^*}{\mathbf{a}}\right) = \left(\frac{45}{8(1-\nu)}\frac{\mu}{\tau}\right)^{1/6} \qquad (2)$$

where τ is the shear yield stress, μ the shear modulus and ν the Poisson's ratio. The corresponding critical activation free enthalpy ΔG^* is

$$\Delta G^* = \frac{3\pi\mu\omega^2\mathbf{a}^3}{16(1-\nu)}\left[1-8.5(1-\nu)^{5/6}\left(\frac{\tau}{\mu}\right)^{5/6}\right] \qquad (3)$$

The deformation is essentially a thermally activated process and the strain at the molecular level determines the overall macroscopic plastic deformation.

The original mathematical derivations of the yield stress in relation to the molecular deformation are somewhat involved (1,2). However, a rather simple linear relationship can be obtained as follow:

$$\left(\frac{\tau}{\mu}\right)^{5/6} = A - B\left(\frac{T}{\mu}\right) \quad (4)$$

wher T is the absolute temperature. The constants A and B are defined by

$$A = \left(\frac{0.077}{1-\nu}\right)^{5/6} \quad (5)$$

$$B = A\left[\frac{16(1-\nu)k}{3\pi\omega^2 a^3} \ln(\dot{\gamma}_0/\dot{\gamma})\right] \quad (6)$$

where k is the Boltzmann's constant. The shear strain rate is $\dot{\gamma}$ and $\dot{\gamma}_0$ is a pre-exponential factor taken to be 10^{13} sec^{-1}

<u>Comparison of Experiments and The Argon Theory</u>. The measured compression modulus E and yield stress σ_y were first converted into μ and τ by using the simple relations of $\mu = E/(2+2\nu)$ and $\tau = \sigma_y/\sqrt{3}$. The $(\tau/\mu)^{5/6}$ values of the materials are then plotted against (T/μ) as shown in Figure 3. For each system a straight line was fitted through the data in Figure 3 using the least square method. It can be seen that good linear relations exist between $(\tau/\mu)^{5/6}$ and (T/μ) for all the systems studied. This strongly suggests that the Argon model gives a reasonable description of the deformation mechanism for these systems at least below T_g which is the case here. (T_g's of the materials are above 115 degree C.) For temperature close to T_g the model is not expected to be accurate as observed for thermoplastics (3,4) and thermosets (5). The different levels and slopes of the straight lines in Figure 3 directly reflect the different molecular structures of the polymers.

One important feature of the Argon theory is to allow the parameters A and B to be determined from the straight lines in Figure 3. Such parameters in turn give the critical activated molecular segment dimension which can be compared with the known molecular structures of the materials. Equation 6 can be immediately rearranged to give

$$a^3 = \frac{A}{B}\,\frac{16(1-\nu)k}{3\pi\omega^2} \ln(\dot{\gamma}_0/\dot{\gamma}) \quad (7)$$

($\omega = 2$, $\nu = 0.35$ and the actual $\dot{\gamma} = 1.64 \times 10^{-4}$ sec^{-1})
The critical distance, z^*, between the kinks along the molecule can also be determined from (3-5)

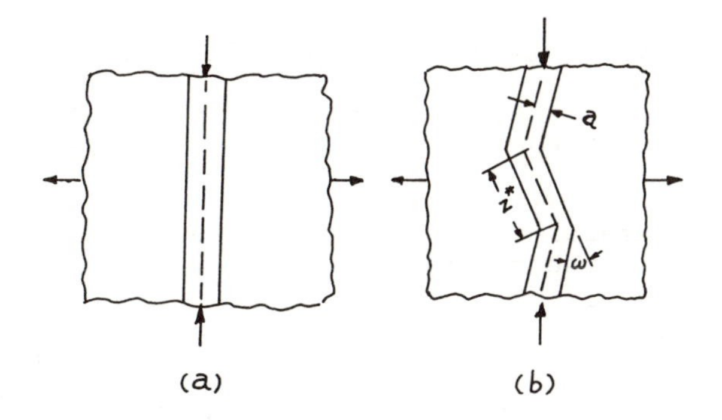

Figure 2. The physical model of local strain produced by nucleation of a pair of molecular kinks along the molecular chain (a) before and (b) after yielding.

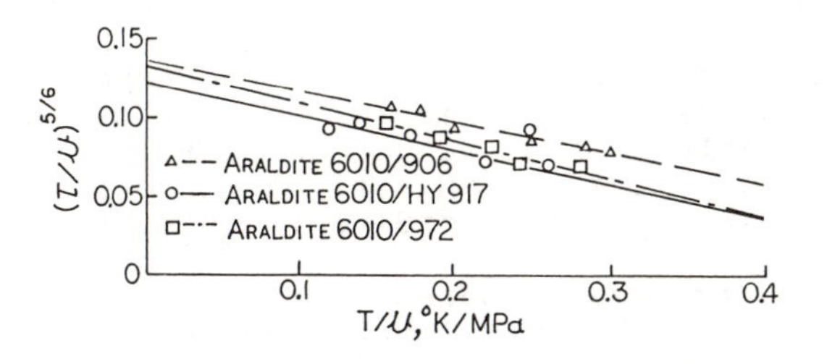

Figure 3. Experimental data $(\tau/\mu)^{5/6}$ plotted against (T/μ) with straight lines fitted according to the Argon theory.

$$z^{*} = a\left(\frac{45}{8(1-\nu)}\right)^{1/6} A^{-1/5} \quad (8)$$

The A, B, a and z^{*} values obtained from the experimental results based on the Argon theory are given in Table II. The molecular parameters should be related to the natural molecular hinges in the materials. For comparison, the mean spacing $\bar{l}$ of the natural molecular hinges are estimated (Table II) for the systems studied in the manner described in (3).

TABLE II. Measured and calculated parameters related to the plastic deformation in epoxy resins

Material	A	B (MPa/°K)	**a** (Å)	z^{*} (Å)	$\bar{l}$ (Å)	$(z^{*}/\bar{l})$
DGEBA/MNA	0.123	0.220	4.35	9.48	2.12	4.47
DGEBA/MTHPA	0.136	0.194	4.69	10.1	2.12	4.76
DGEBA/MDA	0.137	0.260	4.26	9.07	2.59	3.50
DGEBA/TETA(Ref(5))	0.120	0.258	4.10	8.97	----	----
PC (Ref(3))	0.169	0.129	5.77	11.6	2.8	4.14

The MDA cured system, with larger $\bar{l}$, has **a** and z^{*} smaller than the anhydride cured systems. This can also be explained by MNA and MTHPA with methyl side groups being bulkier than MDA. The two anhydride cured systems have very similar structures except that MNA has a slightly tighter ring than MTHPA. As a result, the MTHPA cured system has an activation configuration somewhat bulkier (larger **a** and z^{*}) than the MNA cured system.

All the systems studied appear to have similar A values, which are however lower than A=0.169 (ν= 0.35) according to Equation 5. This difference can be attributed to the real molecular cross-sections different from the circular one assumed in the Argon theory (3). In general, all the systems have molecular parameters (**a**, z^{*}) similar to the DGEBA/TETA systems studied by Yamani and Young (5) as shown in Table II. The structural similarity at least for these DGEBA based systems is evident.

It can be seen from Table II that $(z^{*}/\bar{l})$, which gives the number of segments in the activated configuration, is larger than one. Physically, this means that there are more than one natural segment between two kinks along the molecule. This is because the molecular radius $\mathbf{a}_o$ is comparable with or larger than the mean length $\bar{l}^{o}$ between hinges. The small aspect ratio $(\bar{l}/\mathbf{a}_o)$ necessitates a favorable activated molecular configuration with $(z^{*}/\bar{l}) > 1$. Similar observations were made by Argon on thermoplastic polymers (3,4). The trend of $(z^{*}/\bar{l})$

increasing with decreasing $\bar{l}$ observed here is also consistent with that for thermoplastics. For example, polycarbonate (PC) has a molecular structure similar to DGEBA. The $\bar{l}$ and $z^*/\bar{l}$ values of PC (3,4) and DGEBA based epoxies are quite comparable as can be seen from Table II.

The Argon theory, therefore, consistently interprets the yield behavior of both thermosets and thermoplastics. This indicates that crosslinks in thermosets do not introduce appreciable deviation to the kink formation process described. This point is also supported by Yamani and Young's finding of the molecular parameters, z^* and **a**, being insensitive to crosslinking density for DGEBA cured with different amount of TETA.

From the molecular mechanism described, it is clear that the energy associated with the kink formation is not localized in nature. Such an energy barrier primarily results from the deformed segment interacting with the surrounding molecules where the stored energy extends to far field. The crosslinks in the vincinity of the deformed segment will, therefore, not significantly contribute to the intermolecular energy. This, however, does not mean that crosslinks have no effect on the yield stress.

Actually, crosslinks control the molecular packing and indeed significantly affect the elastic modulus of the material. As the intermolecular energy of kink formation is also determined by elastic modulus, the yield stress will definitely vary with modulus and thus the cross -linking density. In other words, crosslinks may not seriously affect the activation segment configuration in the molecular chain but will indirectly control the yield stress.

Conclusions

The plastic deformation in several amine and anhydride cured epoxy resins has been studied. The experimental results have been reasonably interpreted by the Argon theory. The molecular parameters determined from the data based on the theory reflect the different molecular structures of the resins studied. However, these parameters are in similar enough range to also show the structural similarity in these DGEBA based systems. In general, the mechanisms of plastic deformation in epoxy resins below T_g are essentially identical to those in amorphous thermoplastics. The yield stress level being related to the modulus that controls the intermolecular energy due to molecular deformation will, however, be affected by the crosslinks in the thermosets.

Literature Cited

1. Argon, A. S. Phil. Mag. 1973, 28, 839.

2. Argon, A. S. In Polymeric Materials; American Society for Metals, Metals Park, Ohio 1975; p.411.
3. Argon, A. S.; Bessonov, M. I. Phil. Mag. 1977, 35, 917.
4. Argon, A. S.; Bessonov, M. I. Polymer Eng. Sci. 1977, 17, 174.
5. Yamani, S.; Young, R. J. J. Mater. Sci. 1980, 15, 1814.
6. Lee, S. M.; Schile, R. D. J. Mater. Sci. 1982, 17, 2095.
7. Lee, S. M. J. Mater. Sci. 1984, 19, 2278.
8. Lee, S. M. J. Composite Mater. 1986, 20, 185.

RECEIVED November 17, 1987

Chapter 12

Fractoemission from Epoxy and Epoxy Composites

J. T. Dickinson and A. S. Crasto

Department of Physics, Washington State University, Pullman, WA 99164-2814

Fracto-emission (FE) is the emission of particles (electrons, positive ions, and neutral species) and photons, when a material is stressed to failure. In this paper, we examine various FE signals accompanying the deformation and fracture of fiber-reinforced and alumina-filled epoxy, and relate them to the locus and mode of fracture. The intensities are orders of magnitude greater than those observed from the fracture of neat fibers and resins. This difference is attributed to the intense charge separation that accompanies the separation of dissimilar materials (interfacial failure) when a composite fractures.

When stressed, a material releases various types of emission prior to, during, and subsequent to ultimate failure. This emission includes electrons, positive ions, neutral molecules, and photons - including long wavelength electromagnetic radiation (radio waves), which we have collectively termed fracto-emission (FE). FE data have been collected from the fracture of a wide variety of single and multi-component solids, ranging from single crystals of molecular solids to fiber-reinforced composites, and also from the peeling of adhesives (1-16). In this paper, we will restrict our attention to FE arising from the failure of polymer composites (fibrous and particulate), and the individual components thereof (fibers and matrix resins).

Composites have a complex microstructure, and characterizing and monitoring the various fracture events when a sample is stressed to failure is not straightforward. In contrast to homogeneous materials where failure usually proceeds via the propagation of a single well-defined crack, composite materials exhibit several damage mechanisms, which may progress independently or in interaction with each other. These mechanisms include fiber breakage, interfacial debonding, and matrix cracking, well before ultimate failure. To aid our understanding of the performance of composite structures and the design of these products, it is essential to detect these pre-failure events and their sequence, which lead to crack growth.

Acoustic emission (AE) is a technique that has been successfully employed to study fracture events in composites, where potentially, each failure mechanism has a unique acoustic signature (17-19). FE is another technique, which can be used in parallel with AE, and offers better sensitivity to the various microfracture processes. We have shown that interfacial failure between fiber and matrix in a composite produces significantly more intense emission and longer lasting decay

0097-6156/88/0367-0145$06.75/0

compared to cohesive failure in the matrix (2,7,9). Analysis of the emission may prove useful in determining the mechanisms involved in delamination, fatigue, and/or environmentally induced failure. This technique could aid the materials scientist in understanding how, when, and where cracks originate and grow in a wide variety of materials. Other workers employing fracto-emission to investigate composite failure include Udris et al. (20), Kurov et al. (21), and Nakahara et al. (22).

Experimental Procedure

Equipment, sample preparation, and data acquisition have been described in detail elsewhere (1-16), and will only be briefly reviewed here. Detection of charged particles and gaseous species accompanying fracture events is necessarily carried out in vacuum. While pressures below 10^{-6} torr are sufficient for electron emission (EE) and positive ion emission (PIE), mass spectroscopy studies require pressures below 10^{-8} torr. Charged particles are detected with continuous dynode electron multipliers, usually positioned about 1 cm from the sample. The front surface is biased typically +300V (for electrons) and -2500V (for positive ions), and typical background noise is 1 and 10 counts per second, respectively. Photon emission (phE) is usually detected with various photomultiplier tubes, with single photon sensitivity. For experiments performed in air, cooled housings have been used, which reduce the background counts to less than 10 per second. When intensity comparisons are made, care is taken to acquire data at the same detector gains. Acoustic emission (AE) accompanying failure is frequently used to correlate the emission signals with various microfracture events, or for triggering a multichannel scaler for pulse counting. The signal is detected by attaching a standard AE transducer directly to the specimen, or the grip. A sketch of the experimental set-up is shown in Figure 1.

To correlate the occurrence of macroscopic crack growth in a specimen with the observed emission, we have employed a technique involving resistive changes. A grid of thin, parallel gold strips (0.5 mm thick and 0.5 mm apart) is deposited onto one face of the specimen, perpendicular to the crack path. As the crack grows these strips break, and the resulting resistance change is converted to a voltage and digitized. Thus, the approximate crack tip position is determined as a function of time, and periods of crack growth compared in time with the corresponding emission curve.

Standard techniques were used for sample preparation. Epoxy matrices were employed, including MY720, cured with diaminodiphenyl sulfone (TGDDM/DDS), and Epon 828, cured with Z-hardener (an aromatic amine eutectic from Shell Chemical Co.), m-phenylene diamine or Jeffamine polyoxyalkyleneamines. Cure schedules are summarized in Table I. Neat resins were cast in silicone rubber molds and cured to give dog-bone shaped specimens, as were model composites containing small volume fractions of aligned, continuous fibers. Unidirectional composite bars were also molded from prepregs. Glass, graphite, aramid and boron filaments, and alumina particles, were used to reinforce the above epoxy compositions. Notched and unnotched samples were fractured at various loading rates, and unless otherwise mentioned, in tension.

Results and Discussion

Neat Epoxy Resin. Neat epoxy resin (TGDDM/DDS) was found to be a relatively weak emitter of photons, electrons and positive ions. The general shape of all the emission curves consists of a relatively rapid burst, followed by a very low intensity decay which lasts approximately 100 µs. We frequently observed that during

Table I. Stoichiometry and Curing Schedules for Matrix Resins

Resin	Hardener	R/H Ratio	Cure Schedule
MY 720	Diaminodiphenyl sulfone	100/27	1h @ 135C, 2h @ 177C
Epon 828	Z-hardener	100/20	4h @ 55C, 16h @ 95C
	m-phenylene-diamine	100/14.5	1h @ 80C, 0.5h @ 150C and 150 psi, 4h @ 155C
	Polyoxyalkylene-amines D400:D2000 :: 53:10	100/63	2h @ 75C, 3h @ 135C

fracture, our detectors were saturating, ie. the instantaneous count rates for time periods of a few microseconds were greater than 10^8 counts/s. The time distributions of EE and PIE show nearly identical behavior, and this has been explained with an ion emission mechanism based on electron stimulated desorption (15). The total counts detected for these various emission components were 100-150.

It was frequently observed that notched specimens, and those broken under high strain rates, showed smoother, glassy fracture surfaces, compared to the rough surfaces produced with either unnotched specimens or low strain rates. Figure 2 compares the phE and surface characteristics of two notched specimens strained to fracture at 2%/s (data acquired at 100 ns/ch.). One specimen displays a smooth fracture surface and relatively low phE, while the other, with a considerably rougher fracture surface, yields much more intense phE. This correlation of emission intensity with surface roughness indicates that regions where significant plastic deformation and crack branching occur yield the highest emission rates. The molecular motion and bond alterations associated with this deformation may be important parts of the emission mechanisms.

To correlate (in time) the crack motion (from initiation to specimen failure) with the simultaneous emission curve, the voltage drop is continuously monitored across a grid of gold strips deposited on the sample surface. The crack breaks strips as it progresses, and the corresponding voltage drops map the crack motion. In general, both PIE and phE are most intense near the start or middle of fracture, with a decrease in emission prior to final surface separation. After separation, the emission, though weak, continues, decaying within 50-100 μs. This after-emission is attributed to thermal relaxation processes stimulated by fracture. The fracture surfaces are "glowing" after formation, similar to phosphorescence. Occasionally, small clusters of counts are seen (usually in charged particle emission which has a better signal/noise ratio) several milliseconds prior to the onset of crack growth. Since these experiments are conducted in a vacuum at room temperature, we do not expect to see the type of photon signals seen by Fanter and Levy (23) in an oxygen atmosphere at elevated temperature, where stress-induced chemiluminescence was observed well before failure. Thus, the occasional discrete bursts of emission we see are more likely due to microcracking and crack initiation, occurring prior to catastrophic crack propagation.

Typical results with such crack growth measurements are shown in Figure 3. In this example a notched specimen was used, with the first gold strip located immediately below the notch, and the phE recorded. The voltage is steady between

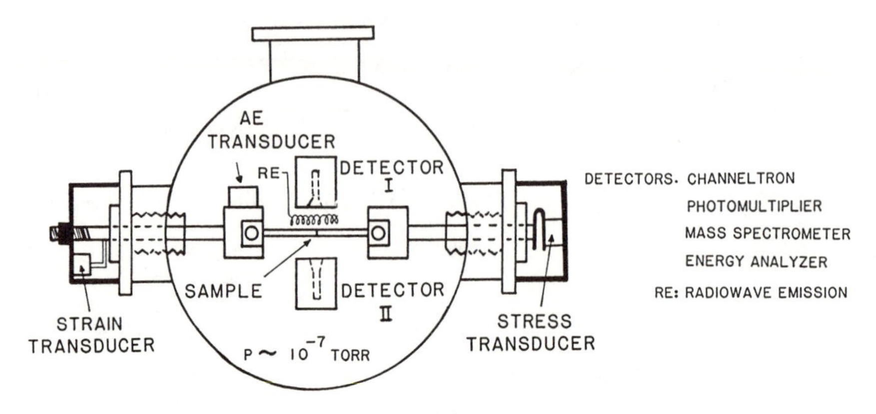

Figure 1. Schematic design of the experimental arrangement for fractoemission investigation. (Reproduced with permission from Ref. 7. Copyright 1984 Plenum Press.)

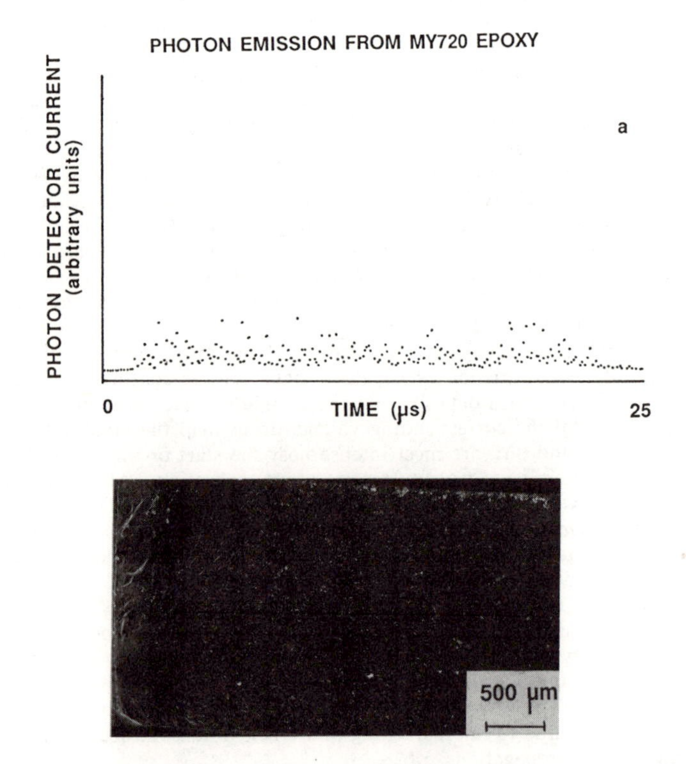

Figure 2a. Photon detector current as a function of time and corresponding fracture surface micrographs for smooth fracture surfaces of TGDDM/DDS.

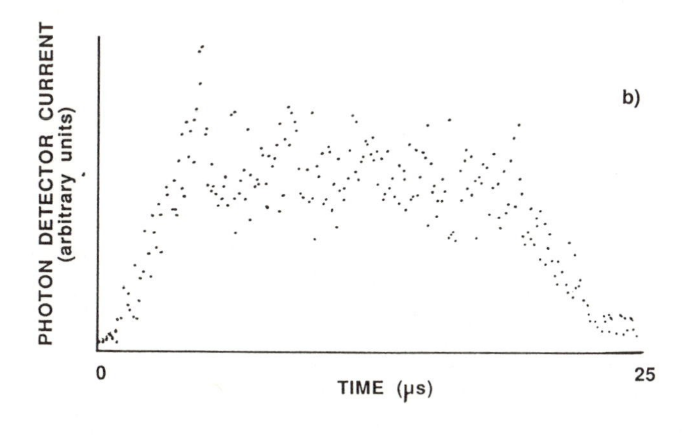

Figure 2b. Photon detector current as a function of time and corresponding fracture surface micrograph for rough fracture surfaces of TGDDM/DDS.

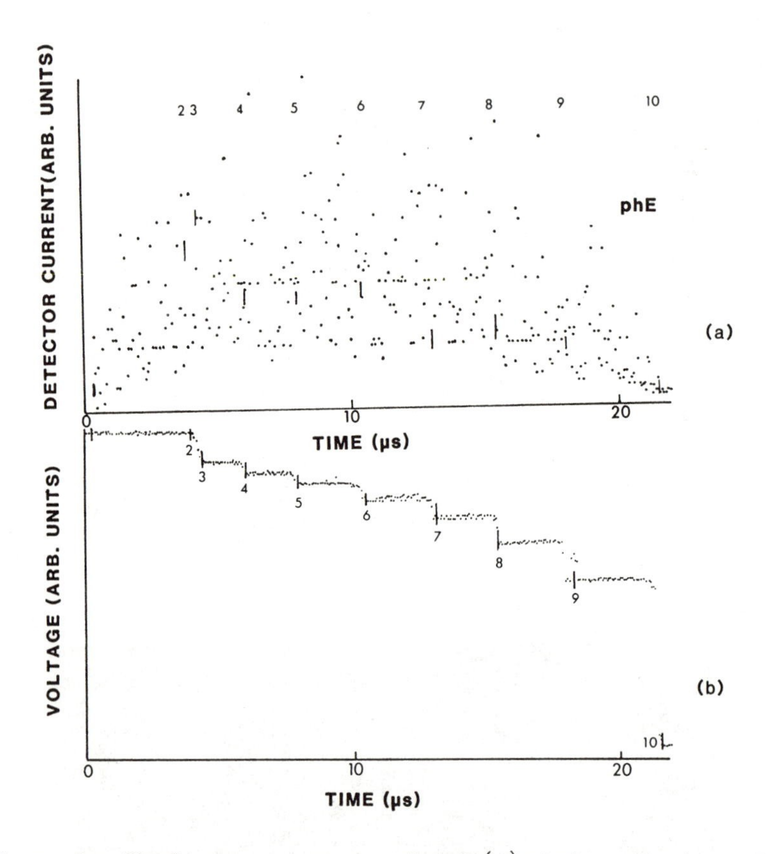

Figure 3. Simultaneous measurements of (a) photon emission and (b) voltage change across a conductive grid from the fracture of a TGDDM/DDS specimen.

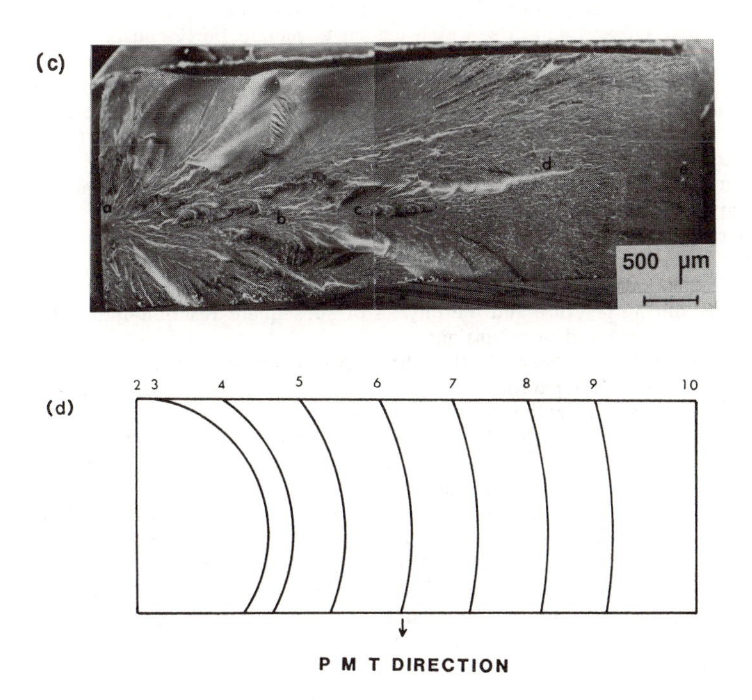

Figure 3 Continued. (c) SEM micrograph of the fracture surface. (d) Estimates of crack tip position corresponding to breaks in the conductive grid.

1 and 2 (Figure 3a), and the first voltage drop (between 2 and 3) corresponds to breakage of this strip and initial propagation of the crack. Comparison with the emission curve reveals phE (starting at 1) prior to any detectable crack growth, while the crack is being initiated. The location of the initiation zone is seen in the accompanying SEM micrograph. Using this zone as an epicenter, the approximate shapes of the crack front are also shown. From a knowledge of the grid spacing, the average crack speed was calculated for various regions. In the initiation zone and the region immediately surrounding it (about 10% of the entire crack path), this varied (typically) between 70 and 150 m/s. Beyond this region, the crack achieved a speed between 350 and 450 m/s, which tended to remain fairly constant.

The EE and phE mechanisms for neat polymers proposed by ourselves and others all involve the consequences of breaking bonds during fracture. Zakresvskii et al. (24) have attributed EE from the deformation of polymers to free radical formation, arising from bond scission. We (1) as well as Bondareva et al. (25) hypothesized that the EE produced by the electron bombardment of polymers is due to the formation of reactive species (e.g., free radicals) which recombine and eject a nearby trapped electron, via a non-radiative process. In addition, during the most intense part of the emissions (during fracture), there are likely shorter-lived excitations (e.g., excitons) which decay in a first order fashion with sub-microsecond lifetimes. The detailed mechanisms of how bond scissions create these various states during fracture and the physics of subsequent reaction-induced electron ejection need additional insight.

In filled polymers, we and others have shown that interfacial failure produces intense charge separation on the fracture surfaces as the crack propagates, which plays an important role in producing the observed intense and long lasting emission. In unfilled materials, charge separation may also play a role. Charge separation during fracture of piezoelectric materials is a reasonable expectation, due to the inherent stress-induced polarization of the material. In non-piezoelectric materials such as amorphous polymers, impurities and defects can lead to pseudo-piezoelectric behavior (26), which could contribute to charge separation. Total charge measurements on the TGDDM/DDS system found average charge densities on the order of 10^{-11} C/cm^2 of new fracture surface, usually positive. From the size of the samples fractured this corresponds to a net deficit of 10^8 electrons, which is many orders of magnitude larger than we observe. This suggests that the surface charge seen is not merely due to charge lost by emission. Further, when scanned with a spatial resolution of about 1 mm^2, considerable variation in charge density was observed over the fracture surface, of positive and negative sign. These results are consistent with the occurrence of charge separation. Clearly, a number of questions concerning emission mechanisms remain open.

Neutral emission (NE) from the fracture of tiny coupons of this neat resin under three-point bend was also investigated, in a chamber evacuated to 10^{-8} torr. Figure 4a shows the total pressure change as a function of time, arising from the emission of non-condensable gases only. Using a quadrupole mass spectrometer, individual masses can be monitored prior to and during fracture. The emission of water molecules (mass 18) is shown in Figure 4b, and comprises the major emission, consistent with the observations of Grayson and Wolf (27). On a similar scale is shown a much faster burst at mass 64, corresponding to the release of SO_2. Other components of the cracking fraction of SO_2 (e.g., masses 48 and 32) behave in exactly the same manner. Similar emission was detected by Grayson and Wolf, and shown to arise from fracture of the sulfone bond in the curing agent, and not from absorbed gases. Analysis of this burst, taking into account the time-of-flight from the sample to the mass spectrometer ionizer, indicates that the SO2 is evolved during the fracture event only. NE therefore provides a view of fracture at the molecular level and appears to be probing the mechanically induced decomposition of the epoxy during fracture. Currently, the effect of non-stoichiometric amounts

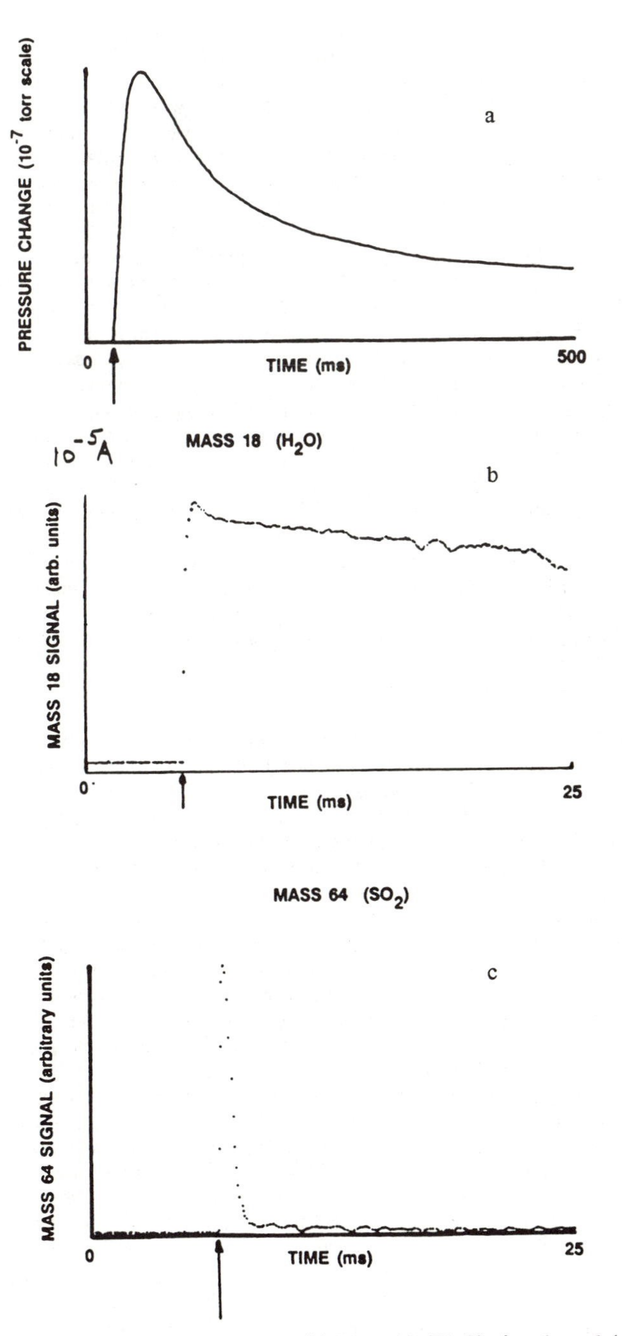

Figure 4. (a) Total pressure change, (b) Mass 18 (H_2O) signal, and (c) Mass 64 (SO_2) signal from the fracture of TGDDM/DDS.

of curing agent on the NE from neat epoxy is being studied, as well as the dependence of SO_2 emission on crack propagation parameters.

Single Fiber Emission. Graphite, Kevlar and glass fibers were used as reinforcements in composites tested, and so FE characteristics of individual filaments were also examined. Graphite and glass yielded similar electron emission (EE) curves, as shown in Figure 5. The data were acquired at 10 μs/channel and represent the accumulated EE from a number of fibers. The emission rises in one channel, which is consistent with the fracture of a brittle material with such a small cross-section. An important observation is the extremely rapid decay, with time constants of the order of 10-100 μs. The EE intensity (charge released per unit area) however, is high, of the order of 10^8 electrons/cm^2 of crack wall. There are two plausible explanations for this high intensity. When these thin filaments fracture, the crack surfaces separate instantaneously. There is less chance of particles from one crack surface hitting the other (which increases with increasing cross-section), and a greater probability of them reaching the detector. Secondly, filaments are known to have significantly higher tensile strengths than the bulk material, thus storing more elastic strain energy prior to fracture. This could lead to more energetic fracture and an increasing amount of excitation of the surfaces producing EE. Similar trends were noted in the positive ion emission from single fibers.

When Kevlar fibers were stressed in tension, a stranded form of fracture was observed, which resulted in multiple EE/PIE peaks spread over several hundred microseconds. The total emission from the entire fracture event correlated with the extent of "damage" to the fiber in fracture. Examination of the shape and intensity of the EE/PIE bursts could provide indications of the mechanism of failure, by differentiating between fibril formation and pull-out.

Fiber-Epoxy Strands. We have seen that reinforcing fibers and pure epoxy produce FE with a simple decay curve, with time constants of the order of 10-1000 microseconds. However, if we examine the FE accompanying the fracture of fiber/epoxy strands made from the same filaments and epoxy, we find emission curves that differ considerably from those of the pure materials. Figure 6 shows typical EE curves for a number of fiber/epoxy systems on a much slower time scale, compared to emission from the pure materials. There is a rapid rise in emission reaching a peak near the instant of rupture, then a decay with a non-first order time dependence. For all cases, the emission lasts much longer than for the pure materials. Time constants assigned to various portions of the decay curve vary from milliseconds to 5 minutes or longer. If the initial portions of the curves are examined on faster time scales, events are seen prior to catastrophic failure, sometimes with decay constants associated with the fracture of the pure components. The systems displayed in Figure 6 have different emission curves. Emission with E-glass is considerably more intense and longer lasting than with S-glass. Microscopic examination of the ruptured surfaces reveals more delamination and filament separation in the former, the latter showing very few resin-free fibers. While the EE from glass/epoxy strands follows relatively smooth curves, the EE of a graphite/epoxy strand, though intense, is very erratic. In this case the emission was so intense, that the detector was repeatedly saturating. By lowering the gain of the electron multiplier we were able to obtain a smooth decay curve, as seen in Figure 6d.

The above observations may be interpreted as follows. Prior to rupture in the glass/epoxy system, the stressed specimen suffers minor failures (primarily fiber breakage and epoxy cracking) which produce EE similar to that of the pure components. These failures accumulate, leading to strand rupture and a large amount of interfacial failure between filaments and epoxy. It is this latter form of

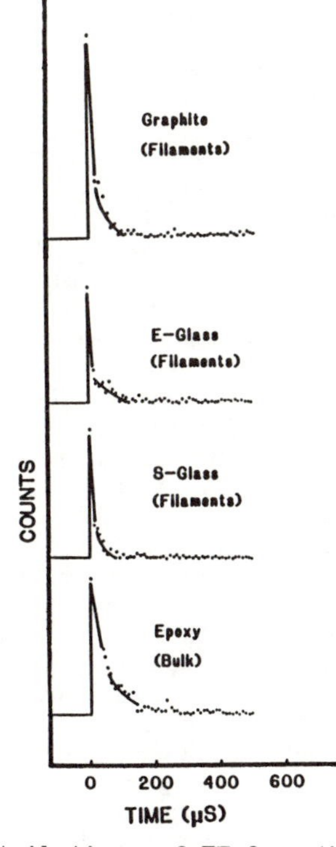

Figure 5. Time distributions of EE from the fracture of graphite and glass filaments, and neat epoxy (Epon 828/Z-hardener). (Reproduced with permission from Ref. 1. Copyright 1981 Chapman & Hall.)

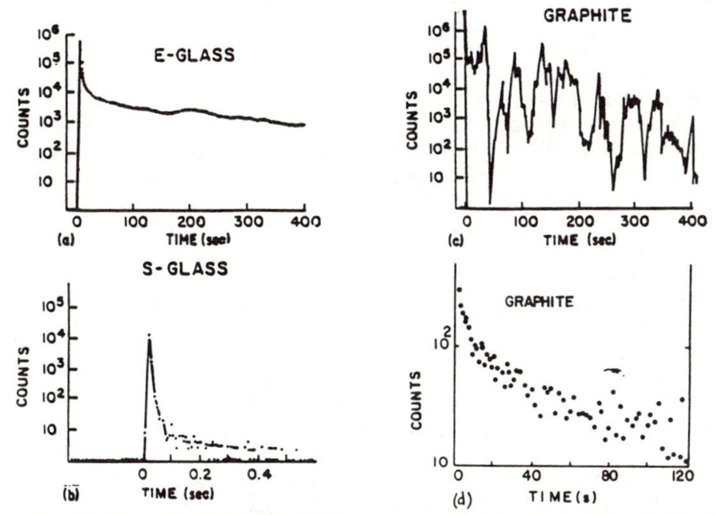

Figure 6. EE from the fracture of fiber/epoxy strands made with (a) E-glass, (b) S-glass, and (c & d) graphite filaments, in Epon 828. (Detection in (d) is at a lower gain of the electron multiplier compared to (c)).

failure that we believe is responsible for the major FE component with the slow decay.

When EE and PIE decay curves are compared over several samples, the total emission, on the average, is nearly the same. When the PIE from one sample is normalized to the EE data from another, within the fluctuations of the observed particle counts, the two curves are indistinguishable. This suggests the occurrence of a common rate-limiting step in both EE and PIE. By placing fine mesh grids between the sample and the detector, a retarding potential energy analysis (28) is possible for the EE and PIE accompanying fracture. The derivative of the count rate against retarding grid potential is the energy distribution of the emitted particles. Figure 7 shows the results for both EE and PIE from the fracture of two different E-glass/epoxy strands. Both charges seem to have very similar energy distributions peaking near 0 eV, with a significant quantity of higher energy particles tailing off in the range of a few 100 eV. The presence of these higher energy particles suggests that charging of the fracture surface is playing a role in their ejection from the surface. The similarity of the EE and PIE energy distributions again suggests that they share a crucial mechanistic step. EE and PIE were also simultaneously detected from a glass/epoxy composite, as well as a number of other filled polymers, to determine if they correlated in time. Coincidences were observed (4,15), consistent with a PIE mechanism involving electron stimulated desorption.

Previous publications (5, 11-15) have outlined a physical model for the emission of charged particles following fracture of composite materials in a vacuum. Fracture produces charge separation from the rapid separation of dissimilar materials that had previously been in intimate contact. In addition to long lasting EE and PIE, easily detectable phE and RE are also observed during fracture, indicating the occurrence of a microdischarge at the crack tip. The necessary gas in the region where breakdown is occurring consists of molecules evolving from the sample as the fracture occurs. These gases come either from occluded volatiles or from actual fracture fragments produced by bond scissions. The presence of these gases and an electric field due to charge separation (in the case of dissimilar materials a probable cause is contact charging) results in gaseous breakdown in the relatively close quarters of the crack tip during fracture.

This microdischarge causes ionization of the gases in the crack tip yielding high concentrations of electrons and positive ions, which are attracted to and strike the crack walls. Bombardment of the fresh crack walls creates primary excitations, usually explained in insulators in terms of electron-hole production, raising electrons into traps near conduction band energy levels. The recombination of electrons and holes is a thermally stimulated process and can yield an emitted electron (thermally stimulated electron emission), sometimes referred to as exoemission, via an Auger process, or a photon, eg. thermally stimulated luminescence. Furthermore, a portion of the electron emission may be attracted back to the fracture surface because of variations in the density and sign of the charge distributions (ie., charge patches), strike the surface, and thereby produce positive ion emission via an electron-stimulated desorption mechanism. In addition, the positive ions can be neutralized as they escape from the fracture surface and leave in an excited state (eg. a metastable molecule), forming a component of fracto-emission we call excited neutral emission. The decay curves are non-exponential, due to competition between retrapping of mobile species (probably electrons) and final annihilation, which produces the emission.

Alumina-Filled Epoxy. A different interfacial geometry and failure mechanisms are provided by particulate fillers. Alumina and silica are incorporated into plastics primarily because of their low cost, and may also improve properties to some extent. In our studies, the EE, phE and RE from neat and alumina-filled Epon 828

(Z-hardener) were investigated, the samples being fractured in flexure. All three signals showed a simultaneous rise (within microseconds of each other) coincident with the brittle fracture event, and the phE exhibited a tail that closely followed the decay of EE. Microscopic studies of the fracture surfaces showed extensive interfacial failure at the filler surfaces. In the neat resin where no interfacial failure occurred, the emission, as expected, was considerably weaker. Figure 8 shows the EE plotted for both unfilled and alumina-filled Epon 828, on the same graph. Several orders of magnitude difference in intensities are observed, the larger and longer lasting emission occurring when interfaces fail.

The emission intensity is strongly influenced by the concentration of filler particles. The counts in the first channel (0.8 s) were taken as a measure of the initial count rate and plotted against the alumina/epoxy ratio (α), and this dependence is shown in Figure 9. The total emission (over several hundred seconds) follows essentially the same curve. The EE intensity rises rapidly from a small value for the neat resin ($\alpha=0$) to a maximum near $\alpha=1$, followed by a slower decline. The amount of interfacial failure increases as α increases. However, as shown experimentally by Maxwell et al. (29), the fracture energy as a function of filler concentration goes through a maximum at precisely the same value as our data. Thus as the filler concentration goes up, the rate of separation at the particle/matrix interface tends to decrease. This reduces the intensity of the discharges due to charge separation, thereby reducing the emission producing excitation of the surfaces. Similar rate dependences have been seen in the phE and EE from the peeling of pressure sensitive adhesive tapes from rigid substrates (7).

Graphite/Epoxy Composite. Hitherto, fiber/epoxy combinations tested were restricted to impregnated strands. Unidirectional $(0)_{16}$ and angle-plied $(\pm 45)_{16}$ composites made from Thornel 300 graphite fibers and NARMCO 5208 epoxy resin were loaded in flexure, and the accompanying EE and AE examined. Load and deflection were simultaneously monitored and related to the observed emission. The results are displayed in Figure 10. In general, the AE shows a rapid rise from zero as the load is applied. The count rate builds up steadily prior to failure, and then there is a large burst followed by a drop in count rate at final failure. Similar trends were noted by Becht et al. (30) in the flexure of notched graphite/epoxy composites. Barnby and Parry (31) also noticed an AE buildup immediately following load application in a (0,90) glass/epoxy composite, but in unidirectional specimens, the onset of AE coincided with the large load drop (specimen failure).

In Figure 10a, the buildup of AE(in the unidirectional specimen) is attributed to a combination of single fiber fracture, and at higher loads, internal delamination. The rise of EE preceding first ply failure (first load drop) arises from damage occurring on the outer surface, eg. matrix cracking and fiber fracture. The two bursts of EE around 3 mm deflection occurred at the same time tiny bundles of fibers were seen to separate from the matrix on the surface in tension. This EE is a much more sensitive and discriminating sensor of front surface damage than AE, primarily because of the large rising background of AE emanating from the bulk of the sample. Finally, failure of the outer ply leads to large simultaneous bursts of both AE and EE.

In the angle-plied laminate (Figure 10b), interlaminar fracture is the main failure mechanism. The AE grows rapidly because of interlaminar shear and then levels off, due to continuous relative motion of the plies. The corresponding loss of modulus is apparent in the curvature of the load trace. The average EE count rate rises only slightly above the background level, since most of the damage is internal, and the outer ply never fractures at all. The frequent small EE bursts are probably due to fracture events on the edge of the sample, from where electrons can escape to the detector.

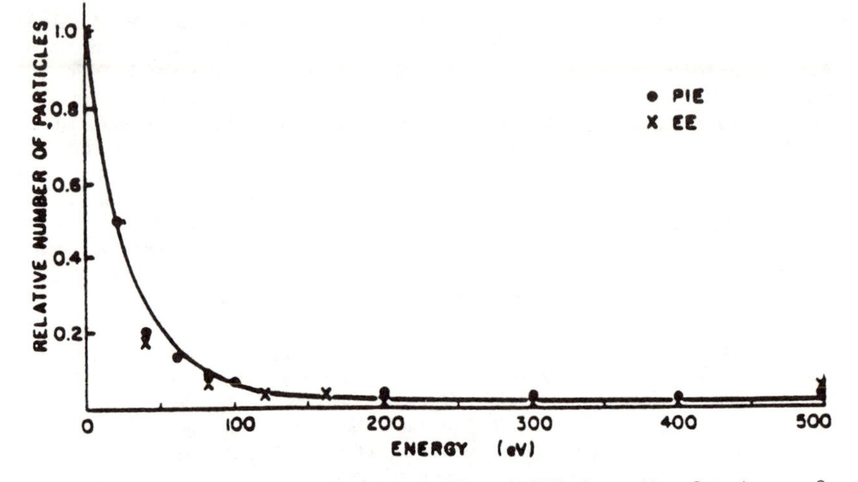

Figure 7. Energy distributions of EE and PIE from the fracture of E-glass/epoxy strands (Epon 828/Z-hardener). (Reproduced with permission from Ref. 1. Copyright 1981 Chapman & Hall.)

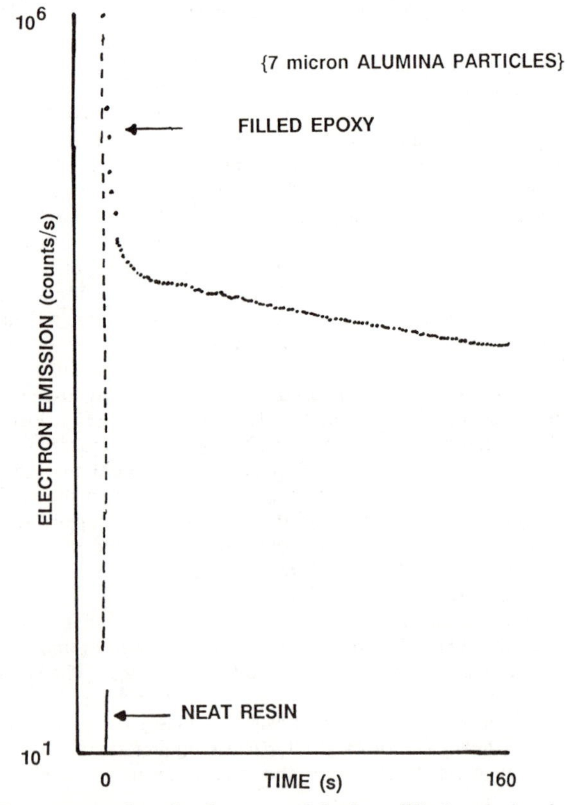

Figure 8. EE accompanying the fracture of both unfilled and alumina-filled Epon 828, cured with Z-hardener.

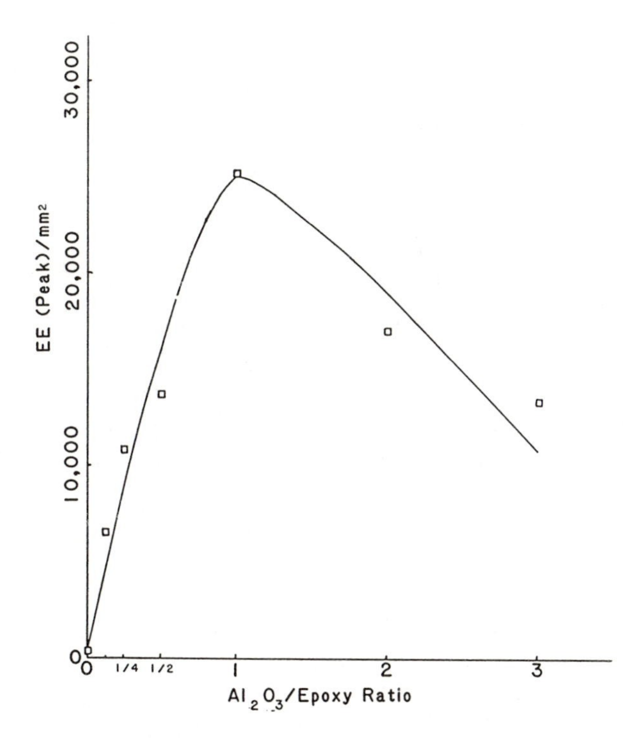

Figure 9. Peak EE from the fracture of alumina-filled Epon 828 (Z-hardener) as a function of the alumina/epoxy ratio (α). (Reproduced with permission from Ref. 7. Copyright 1984 Plenum Press.)

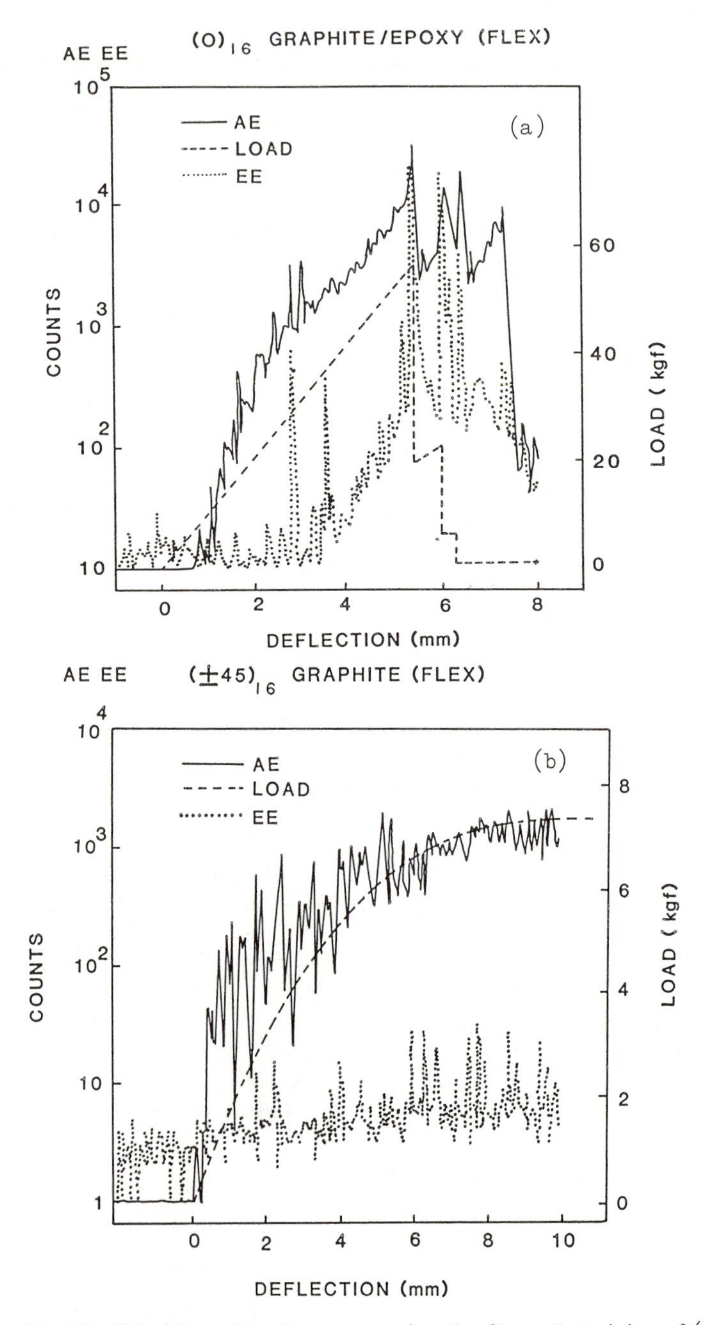

Figure 10. The EE, AE, and load accompanying the flexural straining of (a) $(0)_{16}$ and (b)$(\pm45)_{16}$ graphite/epoxy laminates (Thornel 300/NARMCO 5208).

The results of these experiments indicate that EE may be used to detect microfracture events on the composite surface, and signal the early stages of composite failure. It also clarifies the source of AE as a function of strain by the presence or absence of AE-EE correlations, and allows more details of failure mechanisms and composite characterization to be obtained.

To detect EE from interply failure, the newly-formed fracture surfaces have to be exposed to the detector, requiring a change of sample and testing configuration. Such a study is currently underway, with unidirectional composites made from Hercules AU (untreated) and AS4 (surface treated) graphite fibers. Fiber tows are impregnated with Epon 828 (cured with m-phenylene diamine) and the prepreg cut into plies and compression molded, with 16 plies to a composite. Thin Teflon strips are inserted between plies 8 and 9 prior to molding, which initiate an interlaminar crack when the bar is loaded as a double cantilever beam. A sketch of the arrangement is shown in Figure 11.

Preliminary results indicate at least two orders of magnitude difference in the EE from the two specimen types, the untreated (and consequently poorly bonded) fibers giving the more intense emission. A comparison is shown in Figure 12. SEM micrographs of the fracture surfaces reveal interfacial failure with an abundance of resin-free fibers in the AU sample, while good fiber-matrix adhesion in the AS4 specimen is evidenced from the large extent of cohesive matrix failure. The intensity of emission therefore correlates directly with the extent of interfacial failure. Efforts are underway to use specimens of similar geometry to evaluate composite fracture toughness, G_{IC}, by a hinged double cantilever beam technique (32), and relate it to the observed EE. G_{IC} may in turn be varied with appropriate fiber surface treatments.

Model Glass and Boron Fiber Composites. To detect charged particles, newly created fracture surfaces must be in communication with the surroundings, and further, a vacuum is necessary to enable the particles to reach the appropriate detectors. Photon detection is not as restricted, in that fracture events internal to relatively clear specimens may be detected prior to ultimate failure, and the experiments may be performed in air. Model unidirectional composites were therefore fabricated containing a single boron filament (100 μm) or 50-300 aligned E-glass filaments, embedded in a transparent matrix of Epon 828 (cured with Jeffamines D2000 and D400). These amine hardeners have different molecular weights, and by varying the ratio used, the modulus and failure strain of the matrix can be altered. The matrix was formulated to achieve a high failure strain (>30%) so that the filament(s) underwent multiple fracture prior to ultimate composite failure. These specimens were strained in tension at a rate of 15% /min, and phE and AE recorded (Crasto, A. S.; et al., Compos. Sci. Technol., in press).

In general, the AE bursts correlated with breakages of the fiber bundle (or single filament) at various intervals along the length, and ultimate specimen failure. The phE bursts did not coincide with those of AE, but occurred a fraction of a second later. Results for specimens made from untreated and commercially sized (AR-120-AA) glass fibers are displayed in Figure 13. The total counts from the former are an order of magnitude greater than those from the latter. When the counts for ultimate specimen failure were isolated (for the untreated fibers), they were found to be a small fraction of the total count, indicating that the major emission occurred before ultimate failure. SEM examination of the tensile fracture surfaces revealed the untreated glass fibers protruding from the fracture plane to be clean and devoid of resin (Figure 13c), attesting to interfacial failure as a result of poor adhesion. On the other hand, the sized fibers, also exhibiting pull-out (Figure 13d), were coated with resin, indicating the occurrence of cohesive failure in the matrix. Degradation of the glass-epoxy bond in an aqueous environment was also easily discerned, from the sizable increase in the resulting FE. Consequently, this

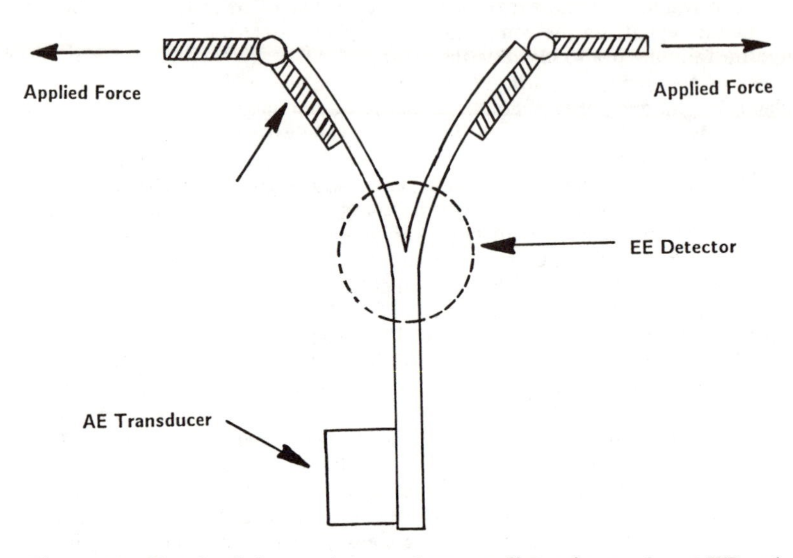

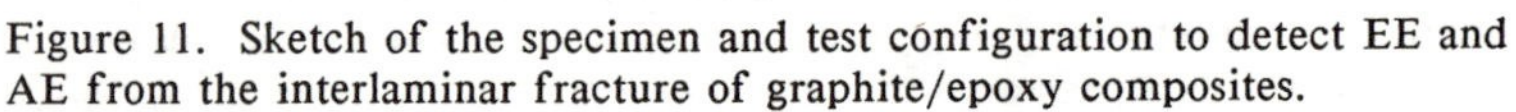

Figure 11. Sketch of the specimen and test configuration to detect EE and AE from the interlaminar fracture of graphite/epoxy composites.

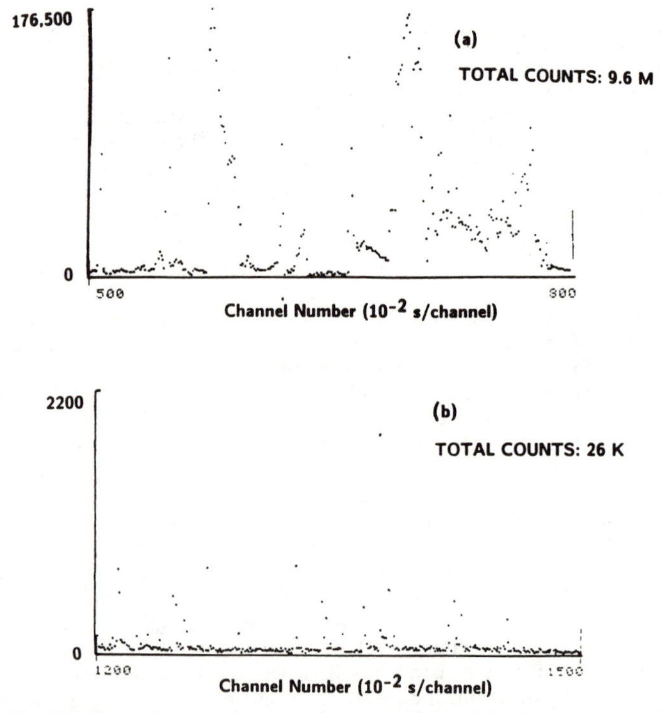

Figure 12. EE from the interlaminar fracture of graphite/epoxy composites made with (a) untreated, and (b) surface treated fibers (Epon 828/mpda).

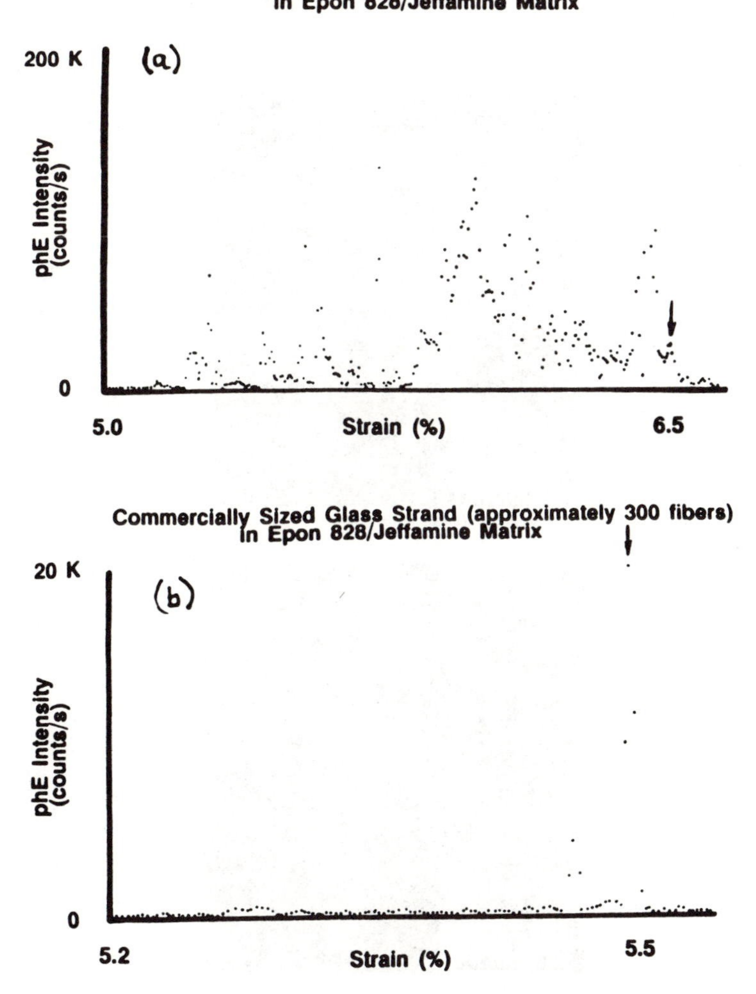

Figure 13. Photon emission from the tensile loading of model unidirectional epoxy composites made with (a) unsized and (b) sized E-glass fibers in Epon 828/Jeffamine. (Reproduced with permission from Ref. 33. Copyright 1988 Elsevier Applied Science.) (Continued on next page.)

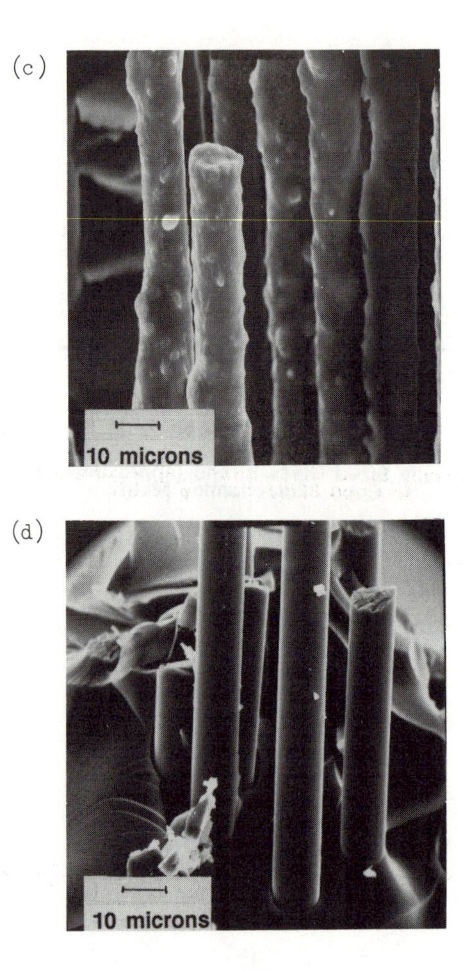

Figure 13 Continued. (c) and (d) Fracture surfaces that correspond to Figures 13a and 13b. (Reproduced with permission from Ref. 33. Copyright 1988 Elsevier Applied Science.)

technique can be used to assess bond deterioration as a result of environmental attack.

When these specimens are stressed in tension, the fibers fracture first. If the fiber-matrix bond is weak, the resulting stress is relieved by debonding on either side of the break. These debonded fibers slide back into their holes in the resin, against considerable friction, and the load is built up again in the fibers. This motion is most likely accompanied by considerable "making and breaking" of surface contact at the interface. This activity, as well as surface separation in debonding, with accompanying charge separation, contribute to the observed phE. A significant difference between the interfacial events occurring here and those in previous studies, is that the earlier work always involved macroscopic separation of surfaces. Here, the fiber and matrix remain in relatively intimate contact after debonding, which very likely influences the intensity and time dependence of the resulting emission. For example, we expect possible quenching mechanisms involving the nearby surfaces and gases in the narrow void created by the broken interface, which would tend to reduce the intensity and duration (decay) after a separation event.

With a strong interfacial bond, when a fiber fractures, the high stresses in the matrix near the broken ends are relieved by the formation of a short radial crack in the resin. There is no interfacial debonding and corresponding friction at a sheared interface, but rather, the load is transferred to the fiber by elastic deformation of the resin. The lack of adhesive failure in this case is responsible for the relatively low emission observed.

The large diameter single filaments of boron yield intense emission when they fracture. Easily detectable AE was used in the interpretation of the phE arising from the straining of single-filament composites made with untreated and lubricated boron filaments. The AE data are similar, with sharp bursts occurring at each filament break (typically 4-8 breaks/filament). On the average, the untreated (and better bonded) filament fractured more than a lubricated (poorly bonded) filament, yet the latter displayed two orders of magnitude greater phE (see Figure 14). Almost all of this emission occurred prior to catastrophic failure of the specimen. When these specimens were stressed under an optical microscope, large debonded lengths were clearly visible in the lubricated-filament samples. This reinforces the contention that the major emission from composite fracture arises from interfacial failure.

Conclusions

In general, the use of FE signals accompanying the deformation and fracture of composites offer elucidation of failure mechanisms and details of the sequence of events leading upto catastrophic failure. The extent of interfacial failure and fiber pull-out are also potential parameters that can be determined. FE can assist in the interpretation of AE and also provide an independent probe of the micro-events occurring prior to failure. FE has been shown to be sensitive to the locus of fracture and efforts are underway to relate emission intensity to fracture mechanics parameters such as fracture toughness (G_{IC}). Considerable work still remains to fully utilize FE to study the early stages of fracture and failure modes in composites.

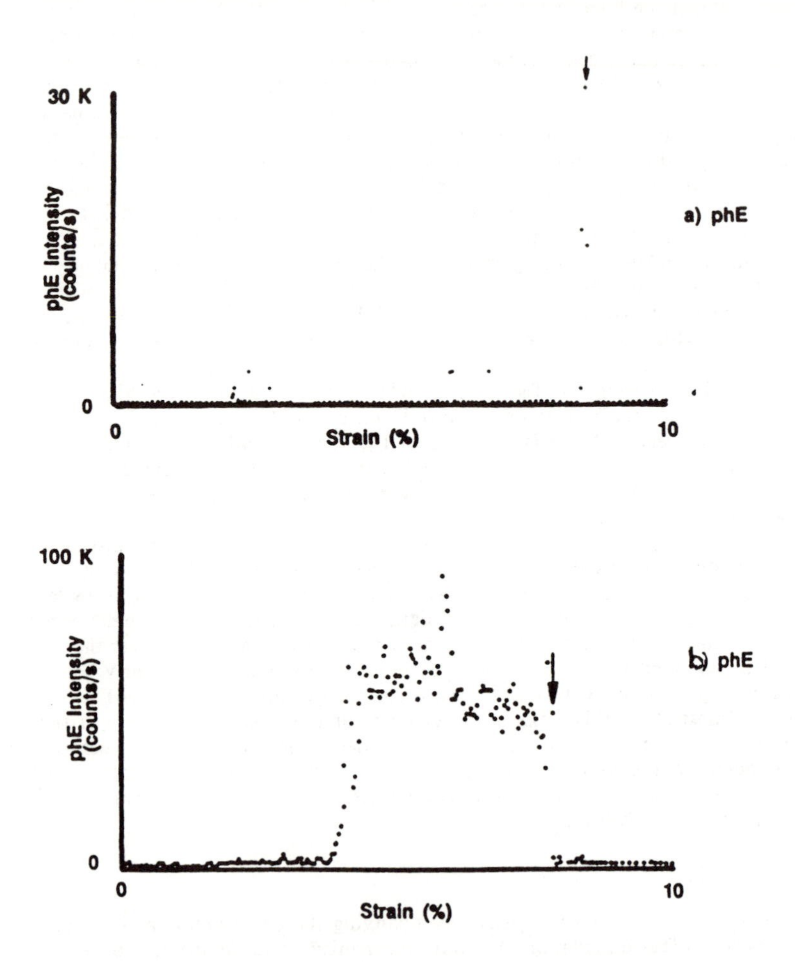

Figure 14. Photon emission from the tensile loading of single-filament boron/epoxy composites made with (a) untreated and (b) lubricated filaments in Epon 828/Jeffamine. (Reproduced with permission from Ref. 33. Copyright 1988 Elsevier Applied Science.)

Acknowledgments

The authors wish to thank Les Jensen, Russ Corey, and R. V. Subramanian of Washington State University, and Clarence Wolf of McDonnell Douglas, for their contributions to this work. These studies were supported by McDonnell Douglas Independent Development Fund, the Office of Naval Research (Contracts No. N00014-80-C-0213, N00014-87-K-0514), the Washington Technology Center, and the Graduate School at Washington State University.

Literature Cited

1. Dickinson, J. T.; Park, M. K.; Donaldson, E. E. J. Mater. Sci. 1981, 16, 2897.
2. Dickinson, J. T.; Park, M. K.; Donaldson, E. E.; Jensen, L. C. J. Vac. Sci. Technol. 1982, 20, 436.
3. Dickinson, J. T.; Jensen, L. C. J. Polym. Sci.: Polym. Phys. Ed. 1982, 20, 1925.
4. Dickinson, J. T.; Jensen, L. C.; Park, M. K. Appl. Phys. Letts. 1982, 41, 443, 827.
5. Dickinson, J. T.; Jensen, L. C.; Park, M. K. J. Mater. Sci. 1982, 17, 3173.
6. Dickinson, J. T.; Jensen, L. C.; Jahan-Latibari, A. J. Mater. Sci. 1984, 19, 1510.
7. Dickinson, J. T. In Adhesive Chemistry: Developments and Trends; Lee, L. H., Ed.; Plenum: New York, 1984; pp 193-244.
8. Dickinson, J. T.; Jensen, L. C.; Jahan-Latibari, A. Rubber Chem. Technol. 1984, 56, 927.
9. Dickinson, J. T.; Jensen, L. C.; Jahan-Latibari, A. J. Vac. Sci. Technol. 1984, 17, 1112.
10. Dickinson, J. T.; Jensen, L. C.; Jahan-Latibari, A. J. Mater. Sci. 1985, 20, 1835.
11. Dickinson, J. T.; Jensen, L. C. J. Polym. Sci.: Polym. Phys. Ed. 1985, 23, 873.
12. Dickinson, J. T.; Jahan-Latibari, A.; Jensen, L. C. In Polymer Composites and Interfaces; Kumar, N. G.; Ishida, H., Eds.; Plenum: New York, 1985; pp 111-131.
13. Dickinson, J. T.; Jensen, L. C.; Bhattacharya, S. K. J. Vac. Sci. Technol. 1985, A3, 1398.
14. Dickinson, J. T.; Jensen, L. C.; Williams, W. D. J. Am. Ceram. Soc. 1985, 68, 235.
15. Dickinson, J. T.; Dresser, M. J.; Jensen, L. C. In Desorption Induced by Electronic Transitions DIET II; Brenig, W.; Menzel, D., Eds.; Springer-Verlag: New York, 1985; pp 281-289
16. Dickinson, J. T.; Jensen, L. C.; McKay, M. R. J. Vac. Sci. Technol. 1986, A4, 1648.
17. Rotem, A.; Altus, E. J. Testing Eval. 1979, 7, 33.
18. Rooum, J.; Rawlings, R. D. J. Mater. Sci. 1982, 17, 1745.
19. Jeffery, M. R.; Sourour, J. A.; Schultz, J. M. Polym. Compos. 1982, 3, 18.
20. Udris, A. O.; Upitus, Z. T.; Teters, G. A. Mekhanika Kompozitnykh Materialov 1984, 5, 805.
21. Kurov, I. E.; Muravin, G. B.; Movshovich, A. V. Mekhanika Kompozitnykh Materialov 1984, 5, 918.
22. Nakahara, S.; Fujita, T.; Sugihara, K. Proc. 8th Exoelectron Emission Symp., 1985.

23. Fanter, D. L.; Levy, R. L. In Durability of Macromolecular Materials; Eby, R. K., Ed.; ACS Symposium Series No. 95; American Chemical Society: Washington, DC, 1979; pp 211-217.
24. Zakresvskii, V. A.; Pakhotin, V. A. Sov. Phys. Solid State 1978, 20, 214.
25. Bondareva, N. K.; Krylova, I. V.; Golubev, V. B. Phys. Stat. Sol. 1984, 83, 589.
26. Wada, Y. In Electronic Properties of Polymers; Mort, J.; Pfister, G., Eds.; John Wiley: New York, 1982; pp 109-160.
27. Grayson, M. A.; Wolf, C. J. J. Polym. Sci.: Polym. Phys. Ed. 1985, 23, 1087.
28. Edland, J. H. D. Photoelectron Spectroscopy; Butterworths: London, 1974; pp 34-37.
29. Maxwell, D.; Young, R. J.; Kinloch, A. J. J. Mater. Sci. Lett. 1984, 3, 9.
30. Becht, J.; Schwalbe, H. J.; Eisenblaetter, J. Composites Oct. 1976, p 245.
31. Barnby, J. T.; Parry, T. J. Phys. D: Appl. Phys. 1976, 9, 1919.
32. Standard Tests for Toughened Resin Composites, NASA Reference Publication 1092, 1983.
33. Crasto, A. S.; Corey, R.; Dickinson, J. T.; Subramanian, R. V.; Eckstein, Y. Composites Sci. Technol. 1988, in press.

RECEIVED October 20, 1988

Chapter 13

Fatigue Behavior of Acrylic Interpenetrating Polymer Networks

Tak Hur[1], John A. Manson[1,2], and Richard W. Hertzberg[3]

[1]Polymer Science and Engineering Program, Lehigh University, Bethlehem, PA 18015
[2]Department of Chemistry, Lehigh University, Bethlehem, PA 18015
[3]Department of Materials Science and Engineering, Lehigh University, Bethlehem, PA 18015

Interpenetrating polymer networks (IPNs) continue to excite both fundamental and technological interest. By combining elastomeric and brittle glassy phases it is often possible to obtain improved properties over a range of temperature and frequency. However, relatively little attention has been given to fatigue in IPNs, and to energy absorption in polyurethane rubber/poly(methyl methacrylate) (PU/PMMA) systems. In this paper it is shown that simultaneous interpenetrating networks based on an energy-absorbing PU are transparent, with a single broad glass-to-rubber transition indicating a microheterogeneous morphology. Fatigue resistance increases with the [PU] up to 50%, while energy absorption determined from dynamic properties and pendulum impact tests varies directly with the [PU]. The micromechanism of failure involves the generation of discontinuous growth bands associated with shear yielding rather than crazing.

Interpenetrating polymer networks, both sequential (IPN) and simultaneous (SIN), continue to excite both fundamental and technological interest (1-21). For example, by combining elastomeric and brittle glassy phases, it is possible to obtain improved properties such as fatigue resistance, as well as tensile, adhesive, and impact strengths. Although much research has emphasized crosslinked polystyrene as the glassy component, attention has also been given to crosslinked PMMA (4,5,8-13). Elastomeric phases have included a variety of polymers such as rubbery acrylics, polyesters, and polyurethanes (4,6,10). Among these latter types, polyurethanes have become important as energy-absorbing materials in orthopedic applications such as athletic footwear, and sound and vibration damping (22-25).

In such polyurethanes it is desirable to modify properties over a wide range of hardness and resilience; the polyurethanes themselves are of inherent interest as toughening phases for brittle

0097-6156/88/0367-0169$06.00/0

matrices. At the same time, although a fundamental understanding of fatigue behavior is important to both scientific inquiry and engineering applications (26), little has been published on fatigue in interpenetrating networks (21). For these reasons, new SINs based on a polyurethane (PU/PMMA) system were synthesized, and characterized with respect to properties including dynamic mechanical response, energy absorption, and fatigue crack propagation (FCP) behavior as a function of composition. In this paper, useful combinations of these properties, and correlations between energy absorption and fatigue behavior, are demonstrated; the micromechanism of failure is also shown to involve discontinuous growth bands associated with shear yielding rather than crazing.

Experimental

The polyurethane formulation involved a proprietary crosslinkable system based on poly(propylene glycol) and methylene diisocyanate (NCO/OH ratio = 1.0). For studies of viscoelastic, energy absorption, and fatigue behavior, the weight fractions of PMMA were 0, 0.25, 0.50, 0.75, and 1.0; for studies of tensile and tear strength, the ratios were 0, 0.10, 0.20, 0.25, 0.30, and 0.40. Reactants were mixed at room temperature, degassed, poured into a mold, and cured at 60°C for 48 hr.

Dynamic mechanical spectra were obtained at 110 Hz using an Autovibron unit, model DDV-IIIC; values of T_g were obtained from the temperatures corresponding to the maxima in the loss modulus (E") peaks. The shear modulus G was also measured at room temperature using a Gehman torsional tester. Fatigue crack propagation tests were conducted as before (21) using a servohydraulic test machine under ambient atmosphere, compact-tension specimens, a sinusoidal frequency of 10 Hz, and an R value of 0.1 (R = min/max load). The crack length a was measured using a traveling microscope, and values of log da/dN, the crack growth rate per cycle, were plotted against log ΔK, where ΔK is the range of the stress intensity factor K, a measure of the driving force for crack extension. This plot reflects the Paris equation (26,27): da/dN = A ΔK exp(n), where A and n are constants. ΔK was calculated as $\Delta K = Y\Delta\sigma\sqrt{a}$, where Y is a known geometrical factor and $\Delta\sigma$ is the range in applied stress.

The % energy absorption

$$\%\ \varepsilon(\mathrm{abs}) = \frac{100\ \varepsilon(\mathrm{abs})}{[\varepsilon(\mathrm{abs}) + \varepsilon(\mathrm{recovered})]} \qquad (1)$$

was determined in several ways: (1) using a Zwick pendulum tester to obtain the % rebound resilience R (% energy absorbed = 100-R); (2) using a computerized Instron tester (model 1332) to obtain hysteresis loops under cyclic compression at 10 Hz; and (3) using dynamic mechanical data to calculate the ratio of energy absorbed to energy input (per quarter cycle), given by $\pi \tan\delta/(\pi \tan\delta + 2)$ (28, p. 606).

Tensile and tear strengths were determined using ASTM standards D412 and D1004, respectively, at a crosshead speed of 0.42 mm/s (1 in/min); values reported are the average for 3 specimens. The elastic and inelastic (plastic) components of the total elongation at break were determined by measuring the retraction of the specimens after rupture; an extensometer was used to ensure accuracy of the strains involved.

Results and Discussion

Molecular and Viscoelastic Characterization. The SIN samples were relatively miscible, as evidenced by transparency to the eye. Densities were 0.6-0.7% higher than predicted by a rule of mixtures; such densification has been noted before (5). However, although the dynamic mechanical data shown in Figure 1 show a single transition, the transitions of the SINs are broader than for the homopolymers. As shown in Table 1 and Figure 1, the slopes of the complex modulus (E*) plots are lower for all the SINs than for the controls; the slope for the PU is also lower than for the PMMA. The breadths of the tan δ and E" curves (not shown) follow the same trend; as expected, the slopes and breadths of the tan δ peaks are inversely correlated. High transparency and a single transition have been noted for PU/polyacrylate SINs (19); in some other PU/PMMA SINs, two loss peaks shifted significantly towards each other were noted (5). Interestingly, the widths of the tan δ peaks as a function of [PU] exhibit a maximum at a PU/PMMA ratio of 50/50; the maximum value is very close to that reported by others (10), although at lower PU concentrations, values of tan δ for the IPNs of this study are higher.

Table 1. Properties of PU/PMMA SINs

PU/ PMMA	T_g (°C)	tanδ (25°C)	$E^a \times 10^{-3}$ (MPa)	dlog E^{*a}/dT (°C^{-1})	ΔT,[e] °C	ε(abs) %[b]	%[c]	%[d]
0/100	121	0.07	3.0	0.055	37	9.9	54	12
25/75	74	0.09	1.1	0.029	63	12.4	59	14
50/50	35	0.11	0.34	0.026	81	14.7	63	17
75/25	3	0.27	0.020	0.026	76	29.8	71	21
100/0	-24	0.93	0.0028	0.048	56	59.4	85	59

[a] E and E* represent the torsional and complex modulus, respectively, at 25°C.
[b] From dynamic spectra.
[c] From hysteresis loops.
[d] From pendulum impact test.
[e] Breadth of tan δ peak at tan δ(max)/2.

Thus the SINs presumably have a fine-scale, microheterogeneous character, while the PU itself has a relatively broad distribution of relaxation times relative to that of the PMMA. This overall

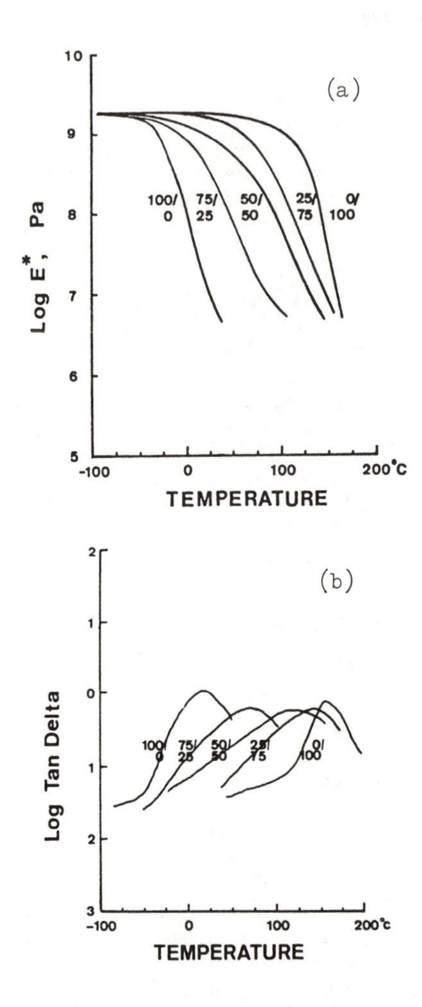

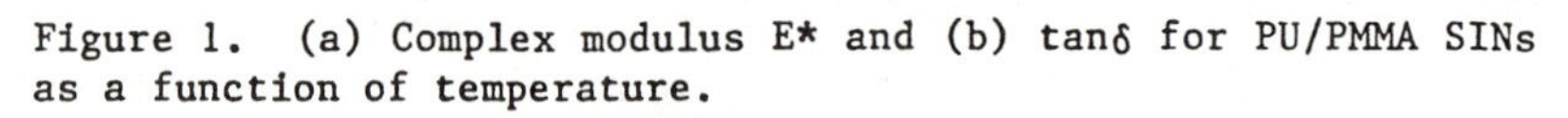
Figure 1. (a) Complex modulus E* and (b) tanδ for PU/PMMA SINs as a function of temperature.

partial miscibility of a system that may not be inherently fully miscible on a segmental scale may well reflect the favorable effects of H-bonding between the two constituents, similar solubility parameters (4), and of crosslinking-induced network constraints on gross phase separation. (The detailed morphology and the nature of the interactions between the PU and PMMA phases are under investigation by scanning and transmission electron microscopy and Fourier-transform infrared spectroscopy, respectively. So far, the size of the PU component appears to be in the range of ∿ 10-20 nm, a range consistent with the nearly-single-phase behaviors discussed above.)

As expected, the T_g varies regularly with composition (Table 1, Figure 2); values fell between those predicted by the Fox and Pochan equations ($1/T_g = w(A)/T_g(A) + w(B)/T_g(B)$, and $\ln T_g = w(A) \ln T_g(A) + w(B) \ln T_g(B)$, respectively, where w is the wt fraction and A and B refer to the two homopolymers involved). The modulus at 25°C also varies with composition, and is compared in Figure 3 with the predictions of several models. These include models of Paul (29) (series-parallel type), Davies (30) (especially suited to dual-phase continuity), Budiansky (31) (phase interactions considered), and Hourston and Zia (14) (a modified Davies equation). The Davies and modified Davies equations are given by: $G(\exp a) = \phi(A)\ G(A)\ (\exp a) + \phi(B)\ G(B)\ (\exp a)$, where G is shear modulus, ϕ is the volume fraction of components A and B, and a has the values of 1/5 (30) or 1/10 (14), respectively. As shown in Figure 3, the Davies and modified Davies equations give the best fit to the data available, the former being best at high, and the latter at low, PMMA concentrations. Thus the behavior is consistent with the existence of a significant degree of fine-scale dual phase continuity in this system.

Tensile and Tearing Behavior. As shown in Figure 4, the tensile and tear strengths increased with increasing PMMA content; measurements of tear strength were not feasible at PMMA contents > 30%. In any case, the incorporation of PMMA at even relatively low levels greatly improves the rather low strengths of the unmodified PU.

As shown in Figure 5, the overall elongation at break exhibited a maximum at a concentration of 20% PMMA. At higher concentrations of PMMA, the relative contribution of plastic (irreversible) deformation increased with increasing PMMA content, while that of elastic (reversible) deformation decreased.

Energy Absorption. Energy absorption [ε(abs)] data are presented in Table 1. Clearly the energy absorption values increase directly with the PU content, and the values for the PU are very high indeed at room temperature. One would not expect absolute values of ε(abs) obtained from such widely different tests involving quite variable states of stress and both linear and nonlinear viscoelastic behavior to agree. Even so, the values relative to that of the PU agree fairly well (within ± 7%) up to a PU/PMMA ratio of 50/50, and good correlations between tests are obtainable between results obtained over the whole range of composition. It is especially interesting that the value of ε(abs) obtained from measurements of tan δ (in the range of linear viscoelasticity) agree quite well with values

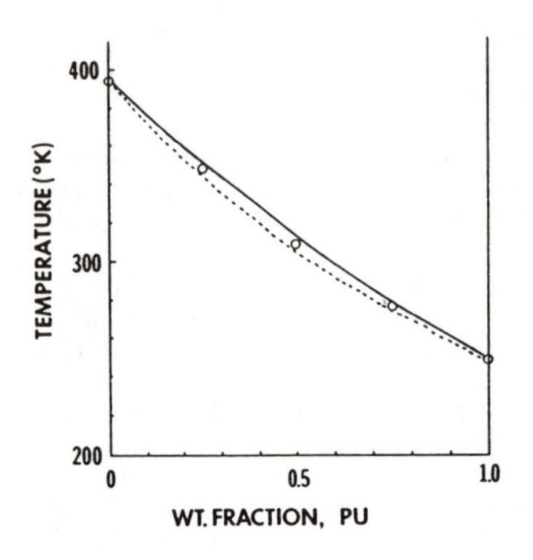

Figure 2. Glass transition temperature of PU/PMMA SINs as a function of composition: —, Pochan equation; - - -, Fox equation.

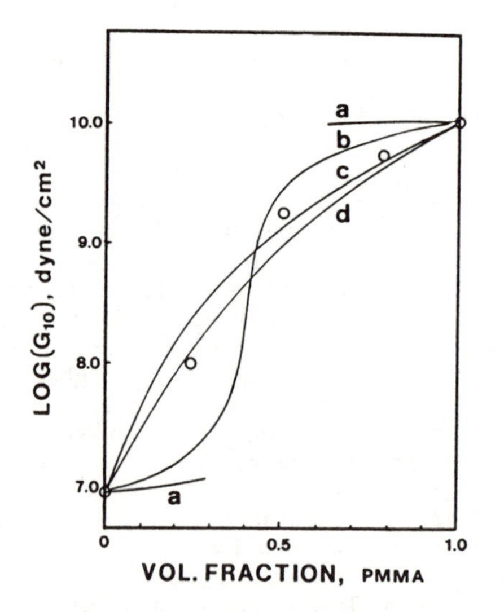

Figure 3. Shear modulus as a function of composition for PU/PMMA SINs. Models of (a) Paul (*29*); (b) Budiansky (*31*); (c) Davies (*30*); and (d) Hourston and Zia (*14*).

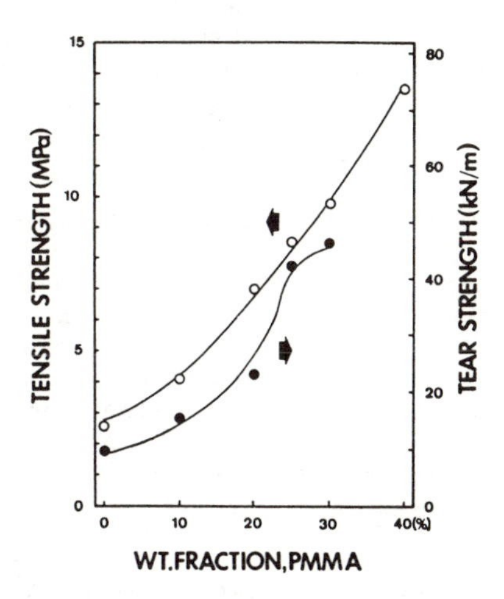

Figure 4. Tensile and tear strength of PU/PMMA SINs as a function of composition.

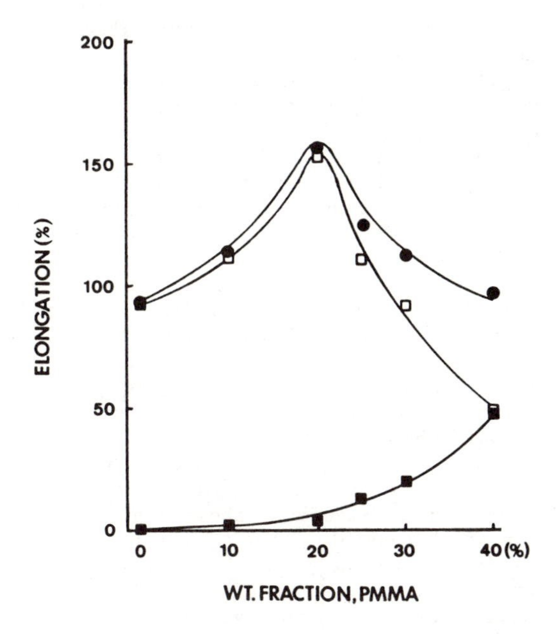

Figure 5. Elongation behavior of PUR/PMMA SINs. Total elongation at break (●) is broken down into both irreversible (■) and reversible (□) contributions.

obtained from the pendulum impact tests. Thus, as shown in Figure 6, the values of ε(abs) obtained from hysteresis loops parallel those obtained by averaging the results for the pendulum and dynamic mechanical tests. In each case the dependence of ε(abs) on composition follows a lower-bound rule of mixtures (Equation 2) rather well.

$$\varepsilon(\text{abs, SIN}) = \frac{\varepsilon(\text{abs, PU}) \times \varepsilon(\text{abs, PMMA})}{[\phi(\text{PMMA})\ \varepsilon(\text{abs, PU}) + \phi(\text{PU})\ \varepsilon(\text{abs, PMMA})]} \quad (2)$$

<u>Fatigue Crack Propagation</u>. FCP data were obtainable only with PU/PMMA ratios of 0/100, 25/75, and 50/50, at least with the specimen and test conditions used; at higher concentrations it was not possible to obtain stable crack growth. This is not unexpected, for the FCP behavior of rubber-toughened plastics is believed to be dominated by a dynamic balance between the ability of the rubber to generate energy-dissipating processes at the crack tip and softening of the bulk polymer (26). In this case, the significant decrease in modulus as the PU content increases evidently overcame the beneficial effects of energy dissipation. Specimens specifically designed for elastomers will have to be used to evaluate the highly elastomeric systems.

With respect to the three specimens that could be tested, the FCP behavior followed the Paris law (Figure 7) over the range of ΔK studied, and the FCP rates at constant ΔK varied inversely with the PU content, while the maximum value of ΔK attainable (K'_c) varied directly (Figure 8). Thus at a constant value of ΔK, the FCP rate of PMMA was reduced by more than an order of magnitude for the 50/50 SIN (at ΔK = 0.6 MPa $\sqrt{m}$) and over half an order of magnitude as final failure was approached. At the same time, the driving force for crack extension ΔK^* (the value of ΔK corresponding to an arbitrary crack speed, in this case 10^{-3} mm/cycle) was increased from 0.7 to 0.9 as the PU content was increased to 50%. Although the values of ε(abs) do not correspond to values of the strain energy release rate G (where $K^2 = EG$), the fatigue parameters ΔK^* and da/dN at constant ΔK correlate well with ε(abs) (Figure 9). Also, division of $(\Delta K^*)^2$ by E gives a figure of merit analogous to the strain energy release rate that increases from 0.17 to 2.36 as the PU content increases, in qualitative agreement with the increase in ε(abs).

With respect to micromechanisms of failure, at low values of ΔK, discontinuous growth bands whose spacings correspond to many cycles of loading were observed (26,32). Figure 10 shows the effect of composition on r, the spacing of the bands, the yield stress σ_y (estimated from the Dugdale relationship, $r = \pi K^2_{max}/8\sigma^2_y$), and the number of cycles per band. Figure 11 shows that the trend of band size as a function of ΔK resembles that of a typical ABS (acrylonitrile-butadiene-styrene terpolymer), and that the slope of the curve conforms to the Dugdale prediction. While such bands are often associated with crazing, shear yielding is expected in a crosslinked system. Indeed the estimated value of σ_y for the PMMA is 77 MPa, a value more typical of yield, rather than crazing, stresses. At high values of ΔK, diamond and feather markings are seen; such markings have been related to the occurrence of multiphase fracture and crack forking in a heterogeneous polymer

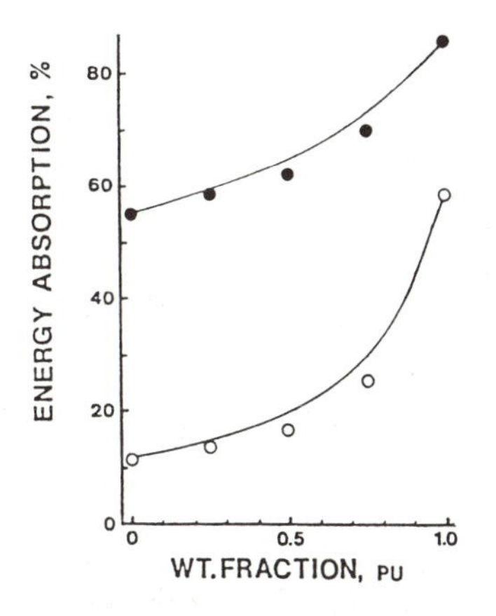

Figure 6. Energy absorption in PU/PMMA SINs as a function of composition: ●, from hysteresis loops; ○, average of results from pendulum and dynamic mechanical tests; –, lower-bound theoretical curve.

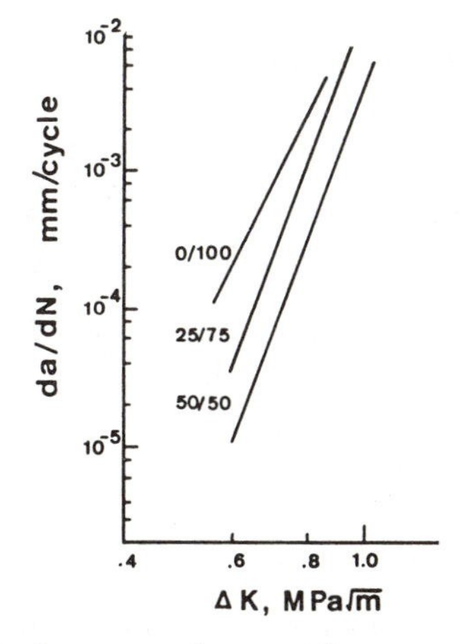

Figure 7. Fatigue crack growth rates as a function of composition for PU/PMMA SINs. Ratios are for PU/PMMA contents.

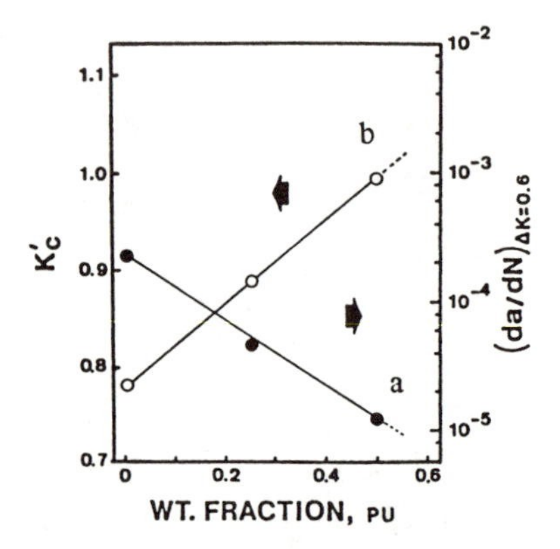

Figure 8. Effects of composition of PU/PMMA SINs on (a) FCP rate (at $\Delta K = 0.6$ MPa) and (b) K'_c (at $da/dN = 3X\ 10^{-3}$ mm/cycle).

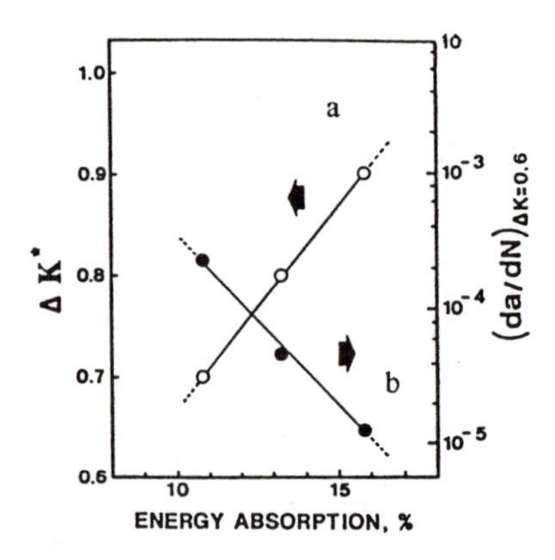

Figure 9. Effects of % energy absorption $\varepsilon(abs)$ on FCP parameters (a) ΔK^* (at $da/dN = 10^{-3}$ mm/cycle) and (b) da/dN ($\Delta K = 0.6$ MPa).

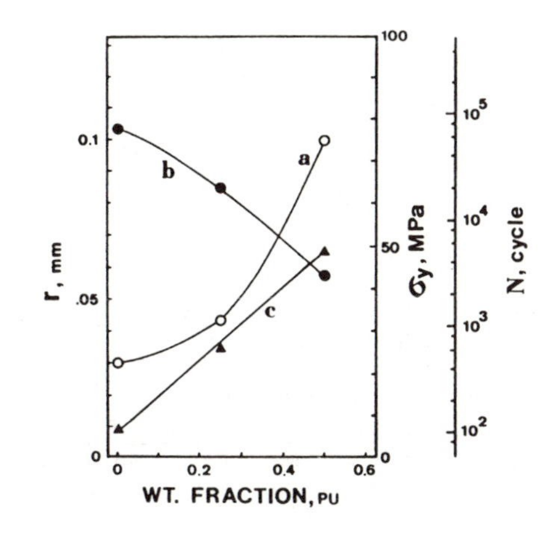

Figure 10. Effects of composition of PUR/PMMA SINs on: (a) r, the size of DGB spacings; (b) σ_y, yield stress; and (c) log N, the number of cycles per band.

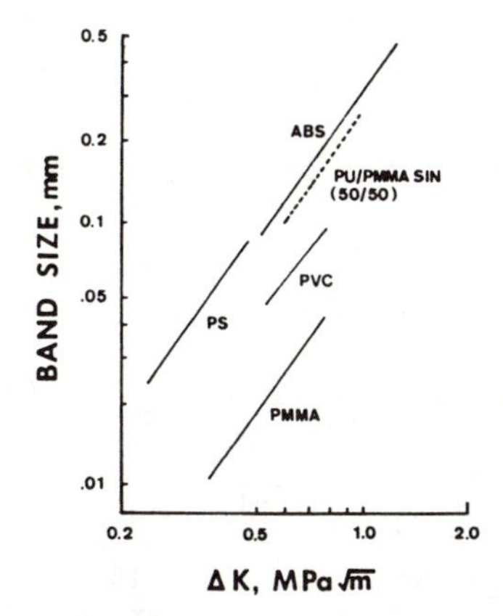

Figure 11. Comparison of r, the length of discontinuous growth bands as a function of ΔK, for PU/PMMA SINs with values for several other polymers (26).

subjected to more applied energy than is required to propagate a stable crack (33).

Acknowledgment

The authors wish to acknowledge partial support from the Ben Franklin Program, NET/ATC (State of Pennsylvania) and from the National Science Foundation, Grant No. 8106489. Assistance with polymer formulation was provided by Dr. Fritz Hostettler, Polymer Dynamics, Inc., Bethlehem, PA.

Literature Cited

1. Manson, J. A.; Sperling, L. H. Polymer Blends and Composites; Plenum: New York, 1976; Chap. 3.
2. Sperling, L. H. Interpenetrating Polymer Networks and Related Materials; Plenum: New York; 1981.
3. Sperling, L. H., ACS In Multicomponent Polymer Materials; Paul, D. R.; Sperling, L. H., Eds.; ACS Advances in Chemistry Series No. 211, American Chemical Society: Washington, D.C., 1986; p. 21.
4. Kim, S. C.; Klempner, D.; Frisch, K. C.; Radigan, W.; Frisch, H. L. Macromolecules 1976, 9, 258.
5. Kim, S. C.; Klempner, D.; Frisch, K. C.; Frisch, H. L. Macromolecules 1976, 9, 263, 1187.
6. Allen, G.; Bowden, M. J.; Todd, S. M.; Blundell, D. J.; Jeffs, G. M.; Davies, W. E.A. Polymer 1974, 15, 28.
7. Kim, S. C.; Klempner, D.; Frisch, K. C.; Frisch, H. L. Macromolecules 1977, 10, 1187.
8. Djomo, H.; Morin, A.; Damyanidu, M.; Meyer, G. C. Polymer 1983, 24, 65.
9. Djomo, H.; Widmaier, J. M.; Meyer, G. C. Polymer 1983 24, 1415.
10. Morin, A.; Djomo, H.; Meyer, G. C. Polym. Eng. Sci. 1983, 23, 394.
11. Hermant, S.; Damyanidu, M.; Meyer, G. C. Polymer 1983, 24, 1419.
12. Hermant, J.; Meyer, G. C. Eur. Polym. J., 1984, 20, 85.
13. Lee, D. S.; Kim, S. C. Macromolecules, 1984, 17, 268, 2193.
14. Hourston, D. J.; Zia, Y. J. Appl. Polym. Sci. 1983, 28, 3745.
15. Hourston, D. J.; Zia, Y. J. Appl. Polym. Sci. 1984, 29, 2951.
16. Kim, S. C.; Klempner, D.; Frisch, K. C.; Frisch, H. L.; Ghiradella, H. Polym. Eng. Sci. 1975, 15, 339.
17. Frisch, H. L.; Frisch, K. C.; Klempner, D. Polym. Eng. Sci. 1974, 14, 646.
18. Meyer, G. C.; Mehrenberger, P. Y. Eur. Polym. J. 1977, 13, 383.
19. Frisch, K. C.; Klempner, D.; Migdal, S.; Frisch, H. L.; Ghiradella, H. Polym. Eng. Sci. 1974, 14, 76.
20. Devia, N.; Sperling, L. H.; Manson, J. A.; Conde, A. Polym. Eng. Sci. 1979, 19, 878.

21. Qureshi, S.; Manson, J. A.; Sperling, L. H.; Murphy, C. J. In Polymer Applications of Renewable Resource Materials; Carraher, C. E.; Sperling, L. H. Eds.; Plenum: New York, 1983.
22. Light, L. H.; McLellan, G. E.; Klenerman, L. J. Biomech. 1980, 13, 477.
23. Lin, J. S. Ph.D. dissertation, Lehigh University, 1986.
24. Lin, J. S.; Manson, J. A. In Proc. 45th Ann. Tech. Conf., Society of Plastics Engineers: Los Angeles, May, 1987, p. 478.
25. Newton, C. H.; Connelly, G. M.; Lewis, J. E.; Manson, J. A. In Proc. 39th ACEMB Conference, Baltimore, September, 1986, p. 38.
26. Hertzberg, R. W.; Manson, J. A. Fatigue in Engineering Plastics; Academic: New York, 1980.
27. Paris, P.C. In Proc. 10th Sagamore Army Res. Conf.; Syracuse University Press: Syracuse, 1964; p. 107.
28. Ferry, W. D. Viscoelastic Properties of Polymers, 2nd ed., Wiley: New York, 1970.
29. Paul, B. Trans. AIME 1980, 218, 36.
30. Davies, W. E. A. J. Phys. (D) 1971, 4, 318.
31. Budiansky, B. J. Mech. Phys. Solids 1965, 13, 223.
32. Skibo, M. D.; Manson, J. A.; Webler, S. M.; Hertzberg, R. W.; Collins, E. A. In Durability of Macromolecular Materials; Eby, R. K.; Ed.; ACS Advances in Chemistry Series No. 95, American Chemical Society: Washington, D.C., 1979; p. 311.
33. Andrews, E. H. Fracture of Polymers; American Elsevier: New York; 1968.

RECEIVED February 8, 1988

Chapter 14

Properties of Epoxy Resins Cured with Ring-Alkylated *m*-Phenylene Diamines

Robson F. Storey, Sudhakar Dantiki, and J. Patrick Adams

Department of Polymer Science, University of Southern Mississippi, Hattiesburg, MS 39406

Diglycidyl ether of bisphenol-A resin was cured, individually, with m-phenylene diamine (MPD) and ring-alkylated MPD's, i.e., toluene diamine, diamino ethyl benzene, diamino isopropyl benzene and diamino tert-butyl benzene. For each diamine, glass transition temperature (DSC) of the cured resin was studied as a function of epoxide/amine stoichiometry. Room temperature tensile properties of resins cured at the stoichiometric ratio, using two cure cycles, were determined. Alkyl substitution generally decreased tensile strength relative to MPD-cured resins. Thermogravimetric analysis indicated slightly higher thermal stability for resins cured with ring-alkylated MPD's compared to MPD. Dynamic mechanical analysis was used to detect the molecular relaxations in the resins cured with ring-alkylated MPD's. The concept of nodular morphology was invoked to explain the shift in the maximum in the beta relaxation observed for resins cured with ring-alkylated MPD's.

Aromatic diamines were introduced into epoxy resin curing technology to improve heat and chemical resistance over that attained with aliphatic diamines. They have been used successfully in laminating applications and to a certain extent in casting and adhesive applications. The most widely used aromatic diamines are m-phenylene diamine (MPD), methylene dianiline (MDA) and diaminodiphenyl sulfone (DADS). DADS is generally viewed to be separate from the other two, being a higher priced, high-performance diamine yielding an elevated glass transition temperature (Tg) of the cured resin. Due to its low basicity DADS requires higher temperature cure schedules than MPD/MDA. Thus, MPD and MDA have for many years been the workhorses in the aromatic diamine-cured epoxy industry. However MDA, and to a lesser extent MPD, have recently come under

0097-6156/88/0367-0182$06.00/0

increasing attack due to their high toxicity. In part, this is due to the recent finding that MDA causes cancer in laboratory rats (1). Since liquid curing agents are desirable, MPD and MDA are most commonly sold as proprietary eutectic blends containing lesser amounts of other diamines. Thus, there is a need for new aromatic diamines which can serve as replacements for MDA and/or MPD/MDA eutectic solutions.

As we reported previously (2), ring-alkylated MPD's appear promising as candidate replacements for MDA. Certain of these compounds are neat liquids at room temperature, and others form eutectic liquids with MPD. It is also thought that the presence of the alkyl group on the ring might produce a less toxic, more easily metabolized compound.

We have undertaken a systematic investigation of the effect on cured resin physical properties of various alkyl groups on the MPD aromatic ring. In this report we present static and dynamic mechanical properties, density measurements, and glass transition temperature (Tg) measurements of diglycidyl ether of bisphenol-A (DGEBA) cured with various alkylated MPD's.

In a thorough study on the temperature dependence of mechanical properties of MPD-cured epoxy resins, Gupta et al. (3) concluded that in the glassy state, high strain properties such as tensile strength, elongation, and toughness are affected by intermolecular packing, molecular architecture, and molecular weight between crosslinks (M_c). In the rubbery state, crosslink density was reported to be the important factor. Gupta et al. undertook these investigations because of conflicting observations reported in the literature on the effect of stoichiometry, and thus crosslink density, on room temperature mechanical properties of MPD-cured epoxy resins. We were also concerned about the effects of stoichiometry on the optimization of properties using ring-alkylated MPD's, and we have viewed our results in terms of the steric effect of large alkyl groups within the crosslinked structure.

In addition, we were interested in the effect of phase-morphology of the crosslinked structure on the ultimate mechanical properties of epoxy resins. A stimulating controversy exists regarding the possible micro-phase separation of crosslinked epoxy networks (4). Several studies which focus on this point have used aliphatic amine curing agents (5-8). These studies have concluded that cured epoxy resins consist of higher crosslink density nodules imbedded in a lower crosslink density matrix. The concept of nodular morphology was applied in the present investigation to explain the differences in the dynamic mechanical relaxations of epoxy resins cured with stoichiometric amounts of different alkylated aromatic diamines.

EXPERIMENTAL

Materials. MPD was obtained from Aldrich Chem. Co. and used as received.

The various alkylated diamines were experimental quantities kindly supplied by Dr. Arthur Bayer of First Chemical Corp. The various diamines studied are indicated below:

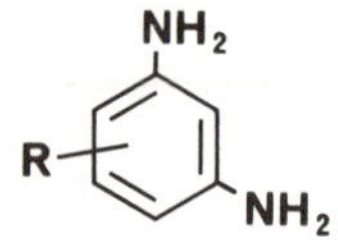

R = -H, m-phenylene diamine (MPD)
= $-CH_3$, toluene diamine (TDA)
= $-CH_2CH_3$, diaminoethyl benzene (DAEB)
= $-CH(CH_3)_2$, diaminoisopropyl benzene (DAIPB)
= $-C(CH_3)_3$, diamino-tert-butyl benzene (DATBB)

The epoxy resin, diglycidyl ether of bisphenol A (DGEBA) (DER 332, Dow Chemical Co.), was used without purification. According to the manufacturer, it is pure DGEBA without appreciable amounts of higher molecular weight oligomers and has the following structure:

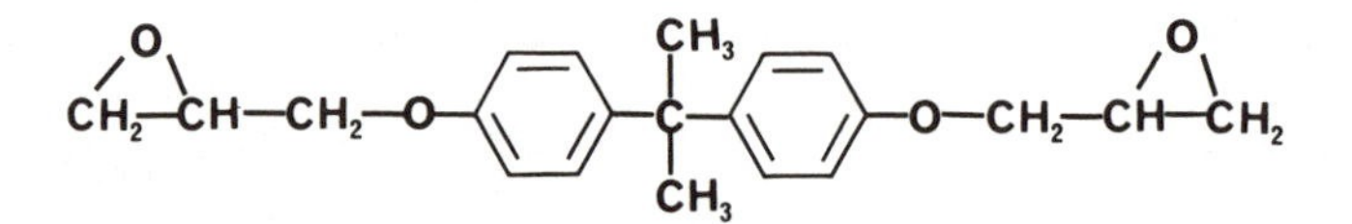

Procedures. The equivalent weight of DGEBA was determined to be 172.1, by titrating epoxide groups using the hydrogen bromide/acetic acid method (9).

The procedure for curing epoxy resins in aluminum molds was as follows: An excess amount of DGEBA was heated in a paper cup using a circulating air oven at 85°C. Meanwhile, an excess amount of the diamine was heated just to melting in a test tube using a hot oil bath. As the size of the alkyl substituent increases, the melting point of the compound decreases such that DAIPB is a wet solid and DATBB is a liquid at room temperature.

Appropriate amounts of epoxy resin and liquified amine were mixed together and evacuated at 60°C in a vacuum oven for ten minutes to remove air bubbles. The reaction mixture was poured into a hot (85°C) aluminum mold which was prepared in advance by lightly spraying mold release agent (MR 515, Green Chem. Products Inc.) to the inner surfaces of the mold. Samples were cured two hours at 85°C followed by two hours at 150°C (standard cure). In some cases samples were subjected to an alternate, high temperature cure of two hours at 85°C and two hours at 175°C (HT cure). After curing, samples were stored in a desiccator until use.

Tensile properties were determined according to ASTM D638 using an Instron tensile tester equipped with a 500 kg load cell.

Glass transition temperatures (Tg's) were determined using a Dupont DSC 910 attached to a 9900 data analysis system. For off-stoichiometric studies, epoxy resin and diamine were cured in situ within a hermetically sealed DSC pan (sample taken from 25°C - 300°C at 10°C/min), then cooled rapidly back to 25°C, and finally scanned from 40°C - 220°C to record the Tg. All samples were scanned under nitrogen atmosphere at a rate of 10°C/min.

The dynamic mechanical spectra were recorded using a Dupont 982 dynamic mechanical analyzer attached to a 9900 data analysis system. The amplitude was held constant at 0.2mm. All thermal scans were obtained at a scanning rate of 5°C/min under nitrogen atmosphere. Molecular weight between crosslinks, Mc, was calculated from the rubbery modulus, G', in the dynamic mechanical spectrum in the region where G' is independent of temperature. The following empirical equation given by Nielsen (10) was used to compute the Mc values:

$$\log_{10} G' = 7 + 293\ \rho/Mc$$

where ρ is the density at 293°K.

Densities of cured resins were obtained by accurately weighing rectangular solids which were precision-machined from expended tensile specimens. Approximate dimensions of the specimens were 1 cm x 1 cm x 0.32 cm; accurate dimensional measurements were obtained using a micrometer.

RESULTS AND DISCUSSION

In our initial studies (11), we examined several series of cured resins using ortho and meta isomers of phenylene diamine. p-Phenylenediamine was not considered for study due to its known carcinogenic nature. As shown in Table I, o-phenylenediamine (OPD) when used alone was found to impart inferior properties to the crosslinked epoxy resin. We suspect the poor performance of OPD is attributable to intramolecular hydrogen bonding, as shown below:

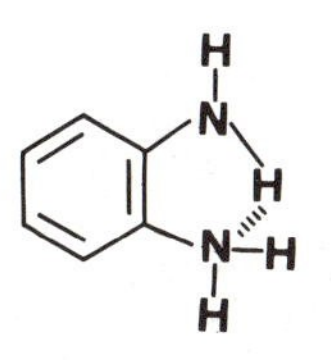

OPD is the only isomer which can form such a structure, and its presence may interfere with complete curing.

Interesting results were obtained when MPD/OPD blends were used as curing agents. When an 80/20 (wt/wt) MPD/OPD blend was used in the standard cure procedure, a significantly higher glass transition temperature (Tg) was observed compared to resins cured with MPD alone. However, for samples cured with the HT cure schedule, MPD and 80/20 MPD/OPD yielded resins of essentially the same Tg. The increased Tg observed in the case of epoxy resins cured with MPD/OPD blends under standard cure conditions may be due to the molecular architecture and/or intermolecular packing, but at this time we do not have a satisfactory explanation for the observed Tg's of resins cured with these blends. Thermal stability was evaluated for resins cured using standard cure. The temperature at which 5% weight loss occurred (Td) was taken as the onset temperature of decomposition. There was no appreciable difference in the thermal stability of the resins cured with MPD/OPD blends compared to MPD alone.

Tensile strength of the MPD cured resins decreased with a change to HT cure, in spite of the higher Tg. Gupta et al. (3) have observed similar behavior in MPD cured epoxy resins and attribute this to an increase in free volume of the sample. The increase in crosslink density is apparently less significant in this case. HT cure increases the tensile modulus in MPD cured samples; whereas it did not show much impact in epoxies cured with MPD/OPD blends. This confirms that the intermolecular packing is different in epoxies cured with different curing systems. In all cases, cured products with excess epoxy have higher modulus than the stoichiometric products. This increased stiffness results because fewer diamine molecules are involved in the final network structure.

Further efforts involved a study of the effect of alkyl substituents on the MPD aromatic ring. We have chosen methyl, ethyl, isopropyl, and tert-butyl substituents. Not only are the branched alkyl derivatives easier to synthesize, but also it was felt that straight-chain substituents, e.g., n-butyl, might lower the Tg of the cured epoxy resin unfavorably. It is known that with longer n-alkyl groups, the Tg of methacrylate homopolymers is lowered substantially due to internal plasticization (12); however it has been reported that in both free radically (13) and anionically (14) prepared poly (alkyl methacrylates), replacing the n-alkyl group with the corresponding iso or tert-alkyl group shifted the glass transition back to higher temperatures.

Because stoichiometry plays an important role in the network formation of epoxy resins, there was some concern that large alkyl groups on the ring might prevent full reaction of the diamine with four equivalents of epoxy. The presence of bulky groups in the vicinity of the amine group are known to preferentially slow the reaction rate of the secondary amine hydrogen relative to the primary amine hydrogen. In their molecular model studies, Morgan et al (15). have found that the methyl group adjacent to the amine group in an aliphatic polyether triamine would produce sufficient steric interference to significantly and preferentially slow the rate of reaction of the secondary amine with epoxide. Thus it appeared possible that a theoretically stoichiometrically balanced system might in fact be deficient in amine, and the highest crosslink density might only be obtained for amine-rich systems. To test this possibility, epoxy resin was cured with varying amounts of amine, on either side of stoichiometry, and the Tg (DSC) of the cured resin was taken to be a direct measure of crosslink density. Figure 1 shows the effect of varying stoichiometry on the Tg of cured resins. Note that except for TDA, all of the diamines show a maximum crosslink density at 1:1 stoichiometry. Thus it is felt that steric bulk of the ring substituent has very little effect on the final extent of cure obtained under these conditions. As will be discussed below, TDA tends to also yield anomalous behavior in other respects.

Although Tg is an acceptable indicator of crosslink density when compared using the same curing agent, it may be unwise to attempt to correlate Tg and crosslink density among several different curing agents. The trend of the data in Figure 1 is apparent, however. DATBB yields the highest Tg resins, followed

Table I

PROPERTIES OF EPOXY RESINS CURED WITH MPD/OPD BLENDS

Curing Agent	Epoxy: Amine	DSC Tg, °C	TGA Td, °C	T.S. psi, $x10^{-3}$	Elonga-tion, %	Modulus psi, $x10^{-5}$
STANDARD CURE						
MPD	1:1	146	376	11.9	5.74	2.8
MPD	1:0.75	112	375	8.2	2.84	3.4
MPD/OPD 80/20	1:1	160	378	8.8	4.02	2.9
MPD/OPD 80/20	1:0.75	93	-	10.3	3.3	3.4
HT CURE						
MPD	1:1	168	-	10.9	3.9	3.8
MPD/OPD 80/20	1:1	170	-	8.8	2.9	3.4
MPD/OPD 80/30	1:1	167	-	8.8	4.2	2.9
MPD/OPD 50/50	1:1	168	-	8.3	3.7	3.2
MPD/OPD 20/80	1:1	167	-	5.2	1.9	3.0
OPD	1:1	152	-	7.6	3.3	3.0

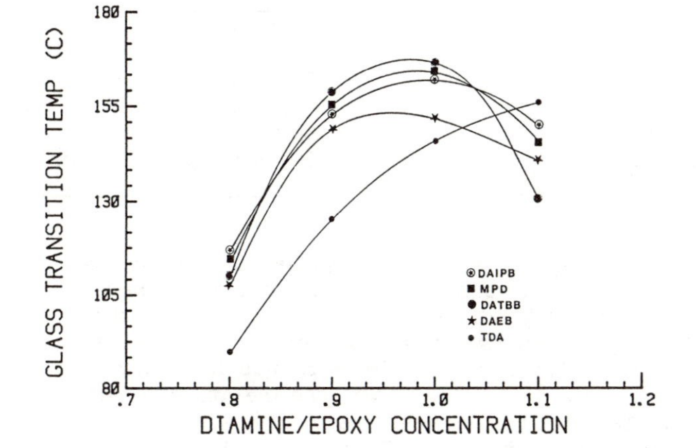

Figure 1 - Effect of stoichiometry on glass transition temperature (DSC).

by MPD itself and then DAIPB > DAEB > TDA. Except for TDA, which shows anomalous behavior, this ranking of Tg's is perfectly consistent with, for example, the family of poly (4-alkylstyrenes) (16) as shown in Table II. Thus it appears that alkyl groups on the aromatic ring create two conflicting effects. The group increases the bulk of the aromatic ring and thus decreases the mobility of that chain segment to which it is attached. This increases Tg and is the predominate effect for the rigid tert-butyl group. The presence of the alkyl group also creates free volume by decreasing the packing efficiency of neighboring molecules. This effect decreases Tg and appears to be the predominate effect for the less bulky isopropyl and ethyl substituents. The anomalous behavior of methyl-substituted TDA escapes satisfactory explanation at this time.

TABLE II

Glass Transition Temperatures of Poly (4-Alkyl Styrenes)

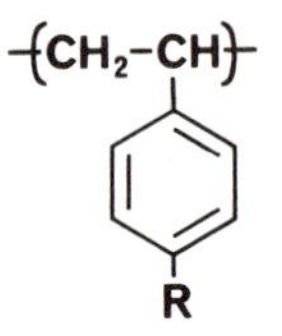

R	Tg
-tert-butyl	403
-H	373
$-CH_3$	366, 374 (conflicting data)
-iso-propyl	360
$-CH_2CH_3$	300, <351 (conflicting data)

Room temperature tensile properties and Tg's (DMA) of cured epoxy resins using standard and HT cures are given in Tables III and IV, respectively. It should be noted that Tg values obtained for MPD cured resins agree perfectly with values reported elsewhere (3). Although DMA yields higher Tg values than DSC, as expected, it is satisfying that the relative ranking of Tg's shown in Table III agrees precisely with those shown for stoichiometric samples in Figure 1. It is clear that the presence of large, branched alkyl groups does not significantly depress the Tg of MPD-cured resins in any case.

Table III

Mechanical properties - standard cure

Diamine	DMA Tg°C	Tensile Property: T.S psi,x10^{-3}	Elong. %	Modulus psi,x10^{-5}
MPD	167	11.9	5.7	2.8
TDA	160	8.3	3.1	3.5
DAEB	163	8.5	3.5	3.0
DAIPB	167	7.8	3.4	3.0
DATBB	168	7.3	2.6	2.5

Table IV

Mechanical properties - HT cure

Diamine	DMA Tg°C	Tensile Property: T.S psi,x10^{-3}	Elong. %	Modulus psi,x10^{-5}
MPD	174	10.9	3.9	3.8
TDA	164	7.5	4.2	2.0
DAEB	170	12.0	6.5	2.9
DAIPB	175	5.9	2.7	2.5
DATBB	171	7.8	3.7	2.3

The most significant deleterious effect of the alkyl substituents is a lowering of the tensile strength of the cured epoxy resins, e.g., by about 30% compared to MPD-cured samples using the standard cure. In general, a decrease in tensile strength is accompanied by a reduction in the strain at break. Surprisingly, use of the HT cure dramatically improves the tensile strength of DAEB-cured resins, but actually lowers the tensile strength of other resins, notably those cured with DAIPB. Many of our further studies have been directed toward explaining these observations, which are still poorly understood at this time.

As shown in Tables III and IV, for both cure schedules, modulus and Tg display inverse relationships upon varying the diamine curing agent. This is in agreement with a previous report (3). In standard cured samples, thermal stability of the resins cured using alkylated aromatic diamines was found to be independent of the size of the alkyl group and about 10° higher than the MPD cured resin.

Room temperature densities of the epoxy resins cured with stoichiometric quantities of diamines are shown in Table V for both standard and HT cures. Molecular models of crosslinked epoxy resins indicate that as the crosslink density of the network increases, the packing efficiency decreases, and the density goes down (15). In agreement with this, Gupta et al. (3) found an inverse relationship between Tg and room temperature density for MPD-cured epoxy resins. It can be seen from Table V that use of the HT cure generally decreases the density of the resulting network, but only slightly. Since the HT cure tends to increase the crosslink density, this behavior is expected. It is also noteworthy that DAEB-cured resins, which have a relatively low Tg, display the highest density using the standard cure.

Table V

Densities of cured epoxy resins

Curing agent	Standard cure g/ml	HT cure g/ml
MPD	1.159	1.186
TDA	1.150	1.142
DAEB	1.167	1.163
DAIPB	1.130	1.128
DATBB	1.131	1.127

DMA was used to characterize the mechanical properties of cured resins obtained using the various substituted MPD's. In the dynamic mechanical spectrum of a diamine-cured epoxy resin, three different transitions can be observed. The highest temperature transition is the alpha or glass transition. As such, the peak temperature in mechanical loss (or tan δ) is taken as the Tg. Moving lower in temperature, the beta transition occurs from -50°C to 50°C and arises from rotational motion of the $-OCH_2-CHOH-CH_2O-$ linkages produced by the cure reaction. Thus, the beta relaxations may be correlated to the concentration of reacted epoxy group, i.e., to the degree of cure (16). A gamma relaxation is often observed around -150°C and is due to motion of unreacted or one-end-reacted epoxy molecules (dangling chain ends). Usually the gamma transition is not observed in resins which are fully cured and stoichiometrically balanced.

Dynamic spectra for the standard-cured and HT-cured epoxy resins are shown in Figures 2-5. Though the concept of inhomogeneity in crosslinked structures is not beyond dispute, we believe that enough evidence exists to support the heterogeneity

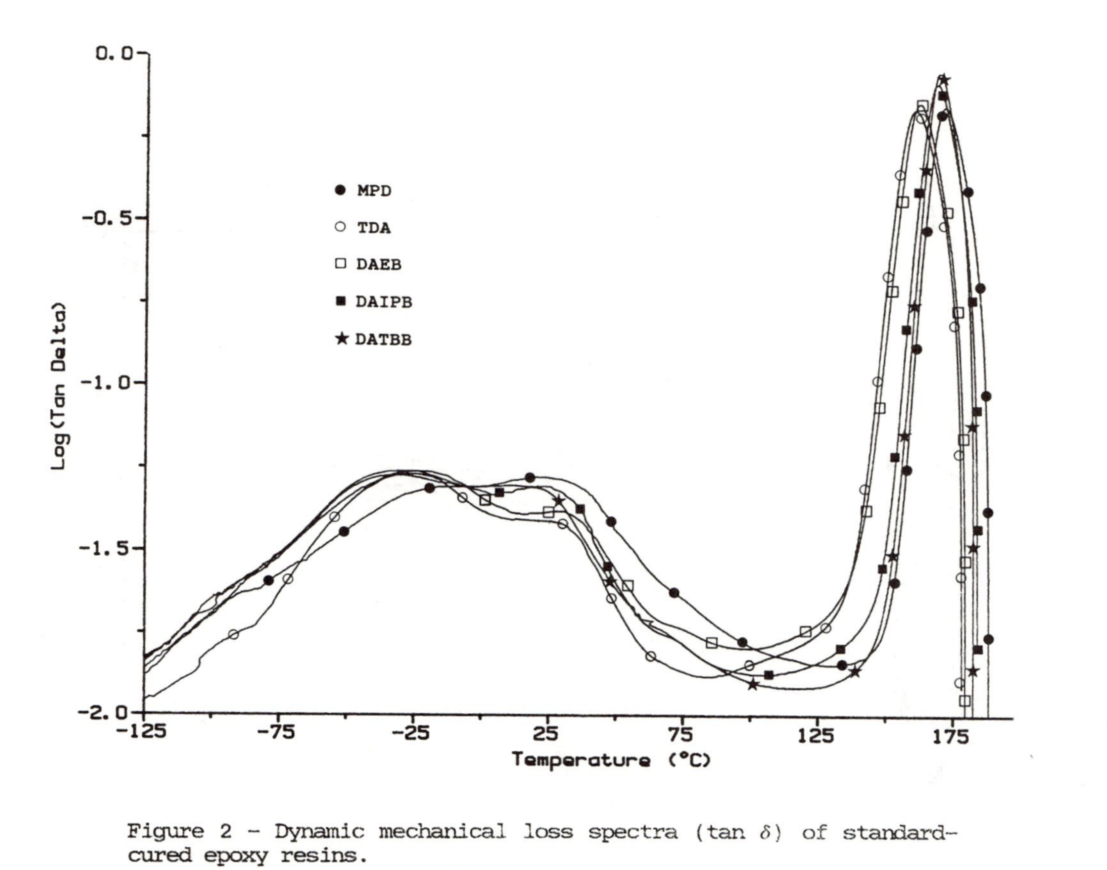

Figure 2 - Dynamic mechanical loss spectra (tan δ) of standard-cured epoxy resins.

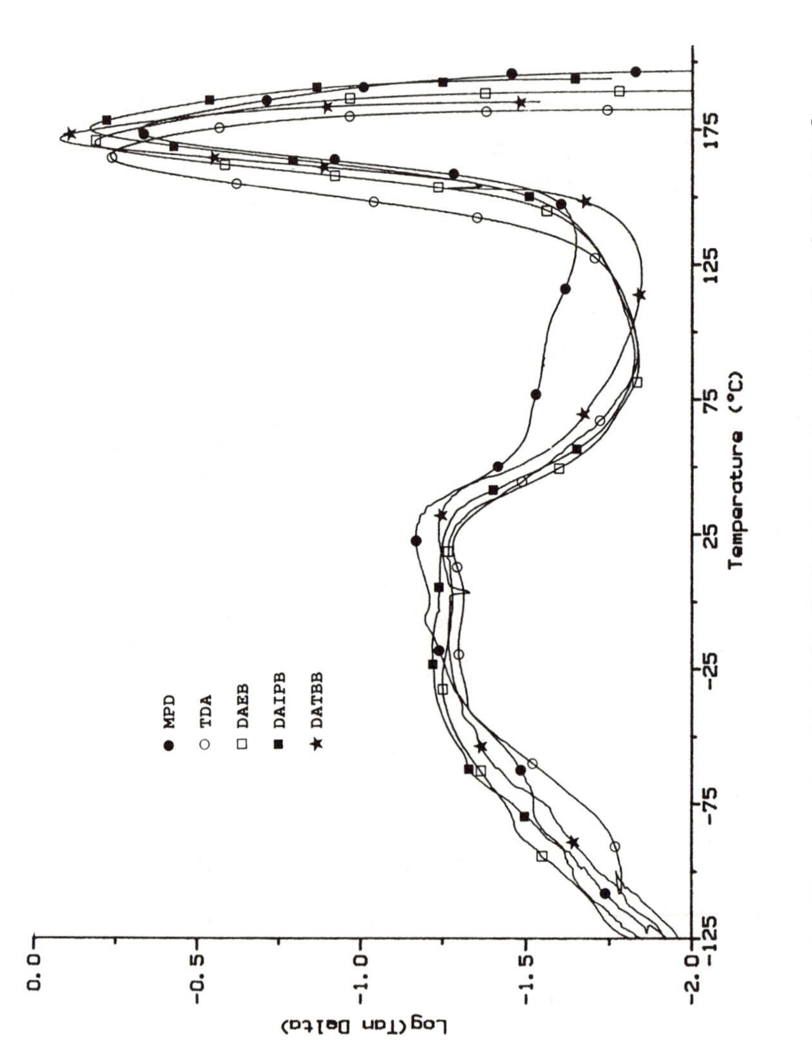

Figure 3 - Dynamic mechanical loss spectra (tan δ) of HT-cured epoxy resins.

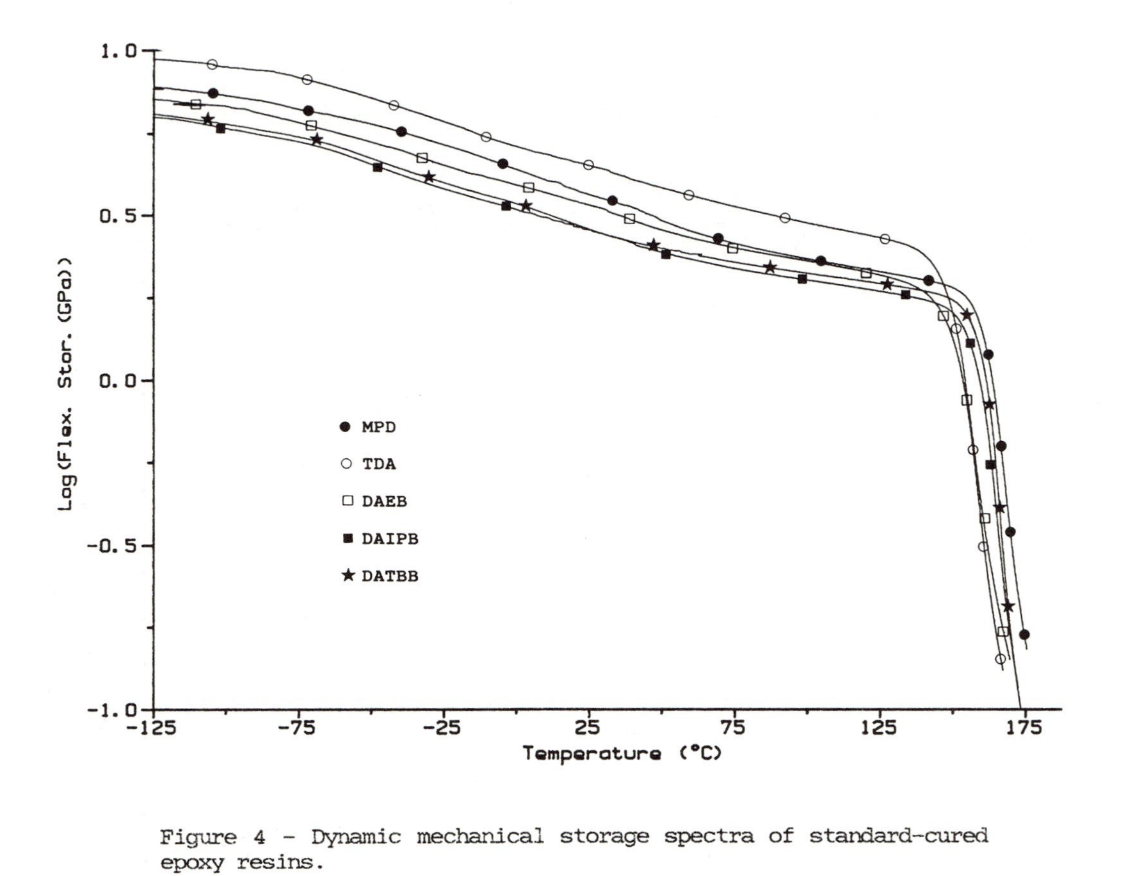

Figure 4 - Dynamic mechanical storage spectra of standard-cured epoxy resins.

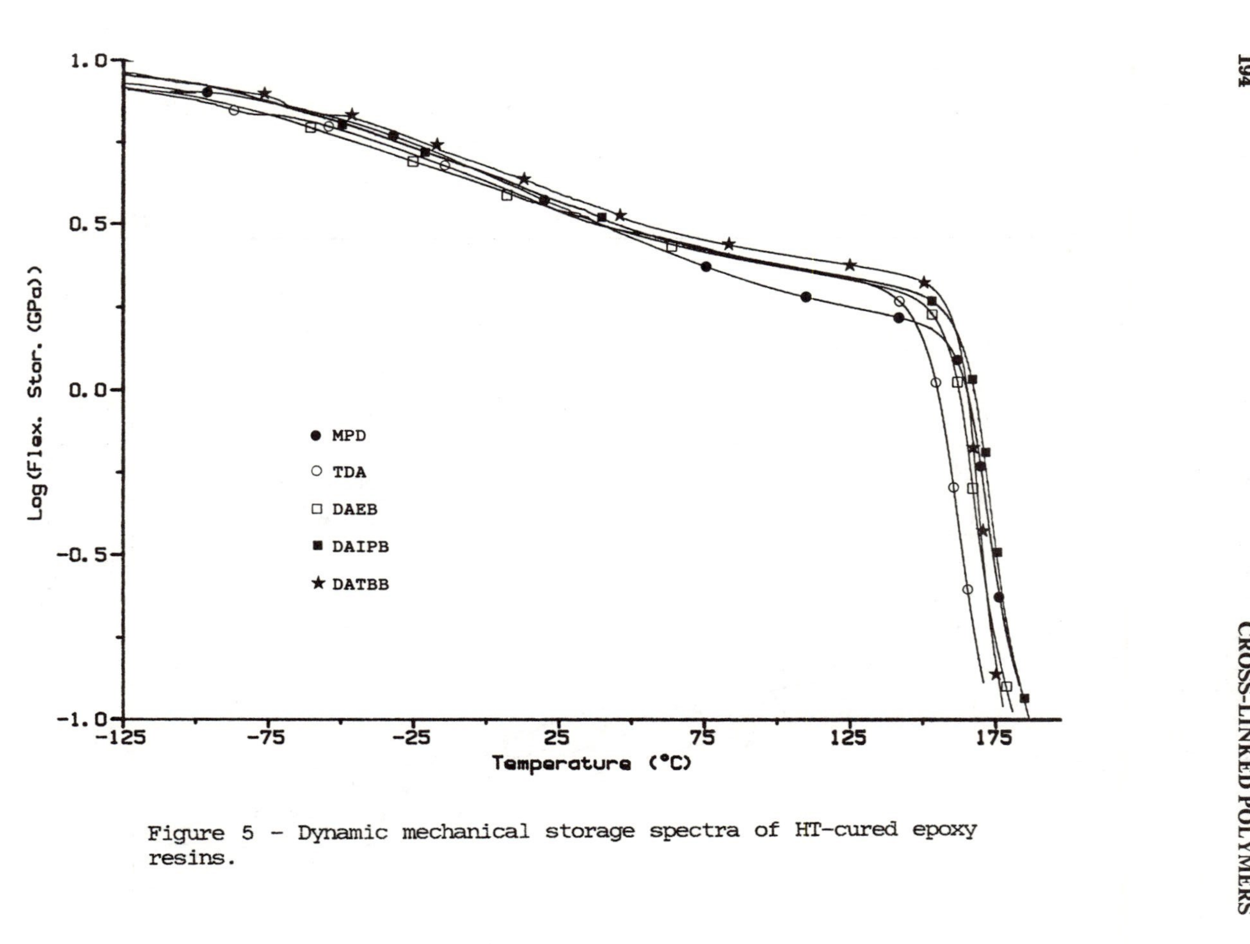

Figure 5 - Dynamic mechanical storage spectra of HT-cured epoxy resins.

of the crosslinked epoxy structure. We invoke the nodular morphology concept (5-8) to account for the differences in the mechanical relaxations of the epoxy resins cured using different alkylated aromatic diamines. In the popular terminology, nodules are micro-phase-separated domains of relatively high crosslink density surrounded by a continuous matrix of lesser crosslink density. According to Mijovic (5,6), the formation of nodules commences at random spots in the resin/curing-agent mixture, as shown in Figure 6. An increase in curing agent concentration, up to the stoichiometric ratio, leads to the formation of smaller nodules of higher crosslink density at the expense of the relatively weaker matrix. The elastic storage modulus in the glassy state is determined primarily by the internodular matrix, whereas the onset of molecular motion, corresponding to the glass transition, depends directly upon the intranodular crosslink density.

Looking first at the tan δ curves, Figures 2 and 3, it should be noted that the breadth of the alpha peak is influenced very little by the presence of the alkyl substituents. Imperfections in the network structure are reported to lead to a broadening of the alpha peak (15). As mentioned earlier, the peak temperature of the alpha peak, a measure of Tg, follows exactly the trend in Tg's as measured by DSC.

The beta transition is the mechanical relaxation process which contributes to the outstanding toughness of diamine cured epoxies, permitting their high modulus without the inherent brittleness of a typical glass. Beta transitions appear bimodal for all samples studied. The two overlapping peaks appearing in the beta relaxations are apparently due to the relaxations of matrix and nodules respectively. Among the standard cured samples, the MPD cured resin shows a maximum in the beta region occurring around 25°C whereas all other resins show a maximum around -40°C. According to Mijovic and Tsay (5) the more hindered groups within highly crosslinked nodules will display more difficulty in initiating the crankshaft motion and will do so only at a relatively higher temperature. When compared to MPD cured samples, the alkylated aromatic diamine cured epoxies have less hindered groups in the crosslinked structure which can initiate the crankshaft motion at a relatively lower temperature. The breadth of the beta transition did not vary much in changing from standard to HT cure, in agreement with a previous report (15). However, the bimodality in the beta relaxation has, in general, decreased in the HT cured samples indicating the predominance of internodular matrix due to greater crosslinking.

Examination of the dynamic storage moduli in Figures 4 and 5 shows that in the standard cure, TDA-cured resins display the highest moduli and DAIPB and DATBB-cured resins display the lowest. For all samples, the alpha peak appears to narrow perceptively upon changing to the HT cure. As mentioned earlier, elastic storage modulus in the glassy state is determined primarily by the weak internodular matrix. On changing to HT cure, all resins display nearly the same storage modulus but higher temperature glass transitions. This indicates that in the HT cured samples the intranodular matrix has increased at the expense of the internodular matrix.

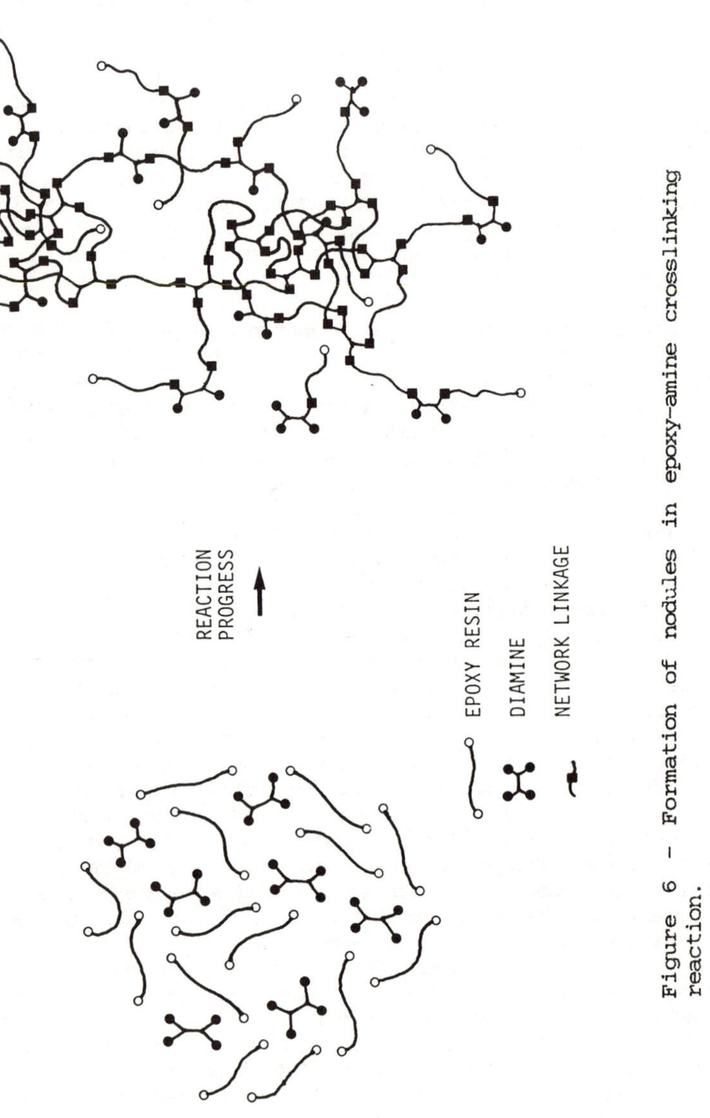

Figure 6 - Formation of nodules in epoxy-amine crosslinking reaction.

Gupta et al. (3) reported that elastically effective molecular weight between crosslinks, Mc, was lowest at the stoichiometric ratio of MPD and epoxy, and post curing resulted in a considerable decrease in Mc only in epoxy rich samples (3). As presented in Table VI, for stoichiometric systems we found that Mc decreases slightly in all samples with HT cure. In addition, the effective Mc's were higher than those predicted based on the molecular weights of the epoxy resin and diamine crosslinkers. This indicates some degree of incompleteness of cure. Note that Mc increases with increasing size of the alkyl substituent in standard cured resins, as expected. However, the increase is much less uniform with HT cure, indicating that in this case Mc's are being significantly influenced by differing extents of cure.

TABLE VI

MOLECULAR WEIGHTS BETWEEN CROSSLINKS IN CURED EPOXY RESINS

DIAMINE	STANDARD CURE	HT CURE
MPD	324	284
TDA	333	304
DAEB	323	291
DAIPB	335	282
DATBB	366	300

CONCLUSIONS

A systematic investigation of isomeric phenylene diamines and ring-alkylated MPD's as epoxy curing agents was performed. Initial studies on blends of phenylene diamine isomers showed that 20% substitution of OPD in MPD increased the Tg of the cured resin but decreased the tensile strength and elongation. When used alone, OPD resulted in resins of inferior properties.

Ring alkylated-MPD's yielded highest crosslink density at the theoretical 1:1 stoichiometry. Thus, large branched alkyl groups do not change the optimum stoichiometry. In both cure schedules studied, ring alkylation of MPD decreased the tensile strength of the cured resins, but slightly increased the thermal stability.

Dynamic mechanical analysis was discussed in terms of the nodular morphology concept in crosslinked structures. Beta relaxations in all the cured resins were bimodal in appearance. But, while MPD-cured resins showed a maximum at 25°C with a smaller shoulder at -40°C, TDA and DAEB-cured resins had maxima at -40°C with a less significant peak at 25°C. For DAIPB and DATBB-cured resins the two peaks were approximately equal in magnitude. The two overlapping peaks at -40 and 25°C were attributed to crankshaft motions in the matrix and nodules, respectively. The overall breadth of the beta transition region did not vary much in changing from standard to HT cure. Tg's of the cured resins, as measured from the peak position of the alpha transition, were as follows for either cure schedule:

$$\text{DATBB} > \text{MPD} \cong \text{DAIPB} > \text{DAEB} > \text{TDA}$$

In analysis of the dynamic storage modulus, toluene diamine-cured resins exhibited the highest modulus in standard-cured samples, whereas all the HT-cured resins showed nearly similar elastic moduli.

ACKNOWLEDGMENT

Financial support for this research by First Chemical Corporation, Pascagoula, MS, is gratefully acknowledged.

LITERATURE CITED

1. E. Helmes, J. Bakker and J. Ahimosata, Epoxy Resins Marketing Research Report, Chemical Economics Handbook (1984).
2. R. F. Storey, D. Sudhakar and L. V. Bui, ACS Polym. Mat. Sci. Eng., 56, 357 (1987).
3. V. B. Gupta, L. T. Drzal, C. Y-C. Lee and M. J. Rich, Polym. Eng. Sci., 25, 812 (1985).
4. J. Mijovic and L. Tsay, Polymer, 22, 902 (1981).
5. J. Mijovic and J. A. Koutsky, Polymer, 20, 1095 (1979).
6. E. Kontou, G. Spathis and P. S. Theocaris, J. Polym. Sci., Polym. Chem. Ed., 23, 1493 (1985).
7. J. L. Racich and J. A. Koutsky, in "Chemistry and Properties of Crosslinked Polymers," S. S. Labana, ed., Academic Press, New York, 1977, pp. 303.
8. A. S. Kenyon, and L. E. Nielsen, J. Macromol. Sci., (A) 3, 275 (1969).
9. C. A. May and Y. Tanaka, ed., Epoxy Resins, Marcel Dekker Inc., New York, 1973.
10. L. E. Nielsen, J. Macromol. Sci., C-3, 69 (1969).
11. D. Sudhakar, R. F. Storey and D. L. Smith, ACS Polym. Mat. Sci. Eng., 55, 38 (1986).
12. N. G. McCrum, B. E. Read, and G. Williams, "Anelastic and Dielectric Effects in Polymeric Solids". Chap. 8, pp. 256, John Wiley and Sons, New York, 1967.
13. Y. Ishida and K. Yamafuji, Kolloid-Z., 177 (4), 97, (1961).
14. J. W. Walstrom, R. Subramanian, T. E. Long, J. E. McGrath and T. C. Ward, ACS Div. Polym. Chem. Polym. Preprs., 27 (2), 135 (1986).
15. R. J. Morgan, F-M. Kong and C. M. Walkup, Polymer, 25, 375 (1984).
16. W. A. Lee and R. A. Rutherford, "Glass Transition temperatures of Polymers," in Polymer Handbook, J. Brandrup and E. H. Immergut, eds., John Wiley and Sons, New York, 1975, pp. III-152-154.

RECEIVED November 17, 1987

Chapter 15

Cross-Linking of an Epoxy with a Mixed Amine as a Function of Stoichiometry

Dynamic Mechanical Spectroscopy Studies

D. Wingard, W. Williams [1], K. Wolking, and C. L. Beatty

Department of Materials Science and Engineering, University of Florida, Gainesville, FL 32611

Variation of the epoxy/curing agent ratio for a system containing a diglycidyl ether of bisphenol A (DGEBA) and a mixed aromatic amine was found to have a significant effect on the cure kinetics and final dynamic mechanical properties of both the neat resin and glass reinforced epoxy. Curing of the mixed system followed first-order kinetics at about 60-100°C and an epoxy/curing agent ratio ranging from -10 wt.% to +5 wt.% of the stoichiometric ratio. For the postcured resin, the molecular weight between crosslinks passed through a minimum and the glass transition temperature passed through a maximum at the stoichiometric ratio.

Epoxy resins are an important engineering matrix material for polymeric composites where a high modulus and glass transition temperature are desirable. The mechanical properties and curing conditions of a number of these thermoset resins have been well studied (1-5). Accurate determination of the ratio of epoxy to curing agent is important because a variation of more than ±2 wt.% of the stoichiometric ratio of epoxy to curing agent can have a significant effect on the final mechanical properties of the cured resin.

This work focuses on the use of dynamic mechanical spectroscopy to characterize the curing behavior of an epoxy-amine system analogous to previous dynamic mechanical studies of epoxy crosslinking (6,7). Murayama and Bell (6) varied the epoxy/curing agent ratio for a DGEBA resin (Shell Epon 828) cured with methylene dianiline (MDA) and determined its effect on crosslink density, modulus as a function of temperature, and glass transition temperature. Galy et al. (8) performed similar work for several DGEBA/anhydride and DGEBA/diamine systems using differential scanning calorimetry (DSC). Although some workers have reported the effect of the epoxy/amine ratio on cure kinetics by DSC and FTIR methods (9-11), the effect of

[1] Current address: Naval Air Rework Facility, Naval Air Station (NASJAX), Jacksonville, FL 32212

0097-6156/88/0367-0199$06.25/0

the epoxy/amine ratio on cure kinetics using dynamic mechanical spectroscopy is relatively non-existent in the literature.

For this work, the epoxy/amine ratio was varied and the dynamic mechanical response was determined for:

1. Conversion of the mixed system to the B-stage cure condition to yield isothermal cure kinetics.
2. Isothermal cure of the B-stage neat resin to the C-stage cure.
3. C-stage cure of the neat resin as a function of temperature and frequency.

Experimental

The dynamic mechanical properties were studied via a Polymer Laboratories Dynamic Mechanical Thermal Analyzer (DMTA) used with:

(a) Universal Temperature Programmer (UTP);
(b) Hewlett Packard Model 85 Desktop Computer and Model 7470-A Plotter;
(c) Flexural Deformation Head.

All samples were mounted in the flexural deformation head for single cantilever bending at a strain amplitude of 0.016 mm. For the cure kinetics, isothermal temperatures from 59-159°C were used. A heating rate of 12°C/minute was used to ramp from room temperature to the desired isothermal temperature. Each sample was run at a single frequency, ranging from 0.33 to 90 Hz. The mixed resin was mounted between two pieces of stainless steel shimstock, the geometry of which is shown in Figure 1. Two types of samples were prepared for the cure kinetics studies: the neat resin enclosed within a glass cloth border, and a composite sample made by rolling the neat resin into the glass cloth with a glass pipette. The latter sample contained an average of 50% glass by weight. The glass cloth used was Owens-Corning Fiberglas glass yarn (0.003" thick). For the isothermal cure of the B-stage resin to the C-stage cure, a ramp rate of 5°C/minute was used and the DMTA was run manually at frequencies of 3, 10 and 30 Hz for each sample. These samples were prepared in an aluminum mold, the sample geometry of which is shown in Figure 2. For the post C-stage curing of the neat resin, samples were analyzed from -100°C to 200°C at a heating rate of 4°C/minute and at frequencies of 3, 10 and 30 Hz. The samples were prepared by the geometry shown in Figure 2.

The epoxy resin used was a diglycidyl ether of bisphenol A (DGEBA), Shell Epon 828, and the aromatic amine, 360L, was manufactured by Magnolia Plastics Company. The aromatic amine is primarily a 50/50 wt.% blend of methylene dianiline (MDA) and m-phenylene diamine (m-PDA) with other oligomeric fractions included. Both the resin and curing agent were used without further purity. The recommended resin to curing agent ratio was 81 wt.% to 19 wt.%. The recommended B-stage cure was 129°C for 30-45 minutes, followed by a C-stage cure of 149°C for 2 hours. Studies were conducted by varying the concentration of Epon 828 by ±5, ±2 and -10 wt.% from the stoichiometric value of 81 wt.%. Changing the concentration of resin by a value of +10 wt.% resulted in samples too brittle for accurate analysis. All samples were hand mixed for 2 minutes and degassed under 29 inches of Hg vacuum for 15-20 minutes prior to heat curing.

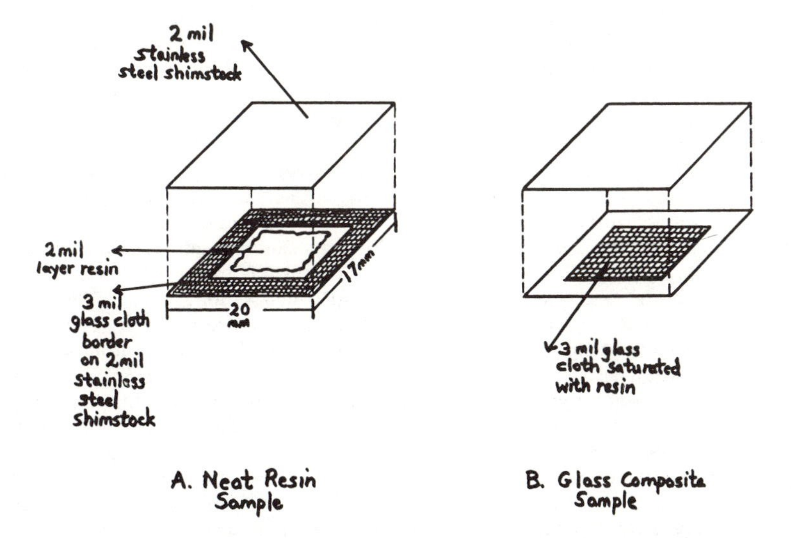

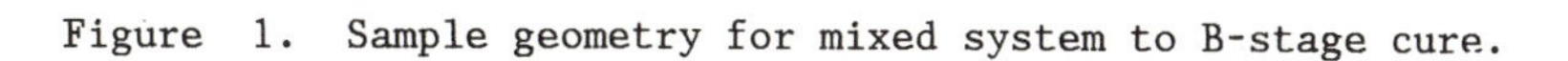
Figure 1. Sample geometry for mixed system to B-stage cure.

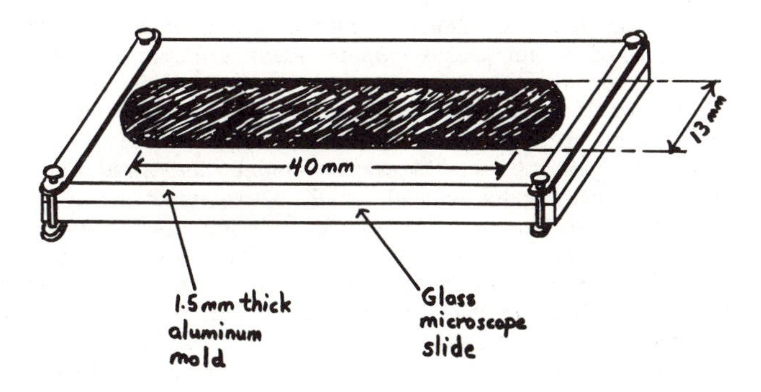

Figure 2. Sample geometry for B-stage to C-stage and post C-stage cure.

Results and Discussion

Cure Kinetics. The cure kinetics of the mixed system to the B-stage were determined by the method outlined by Senich, MacKnight and Schneider (7) for two epoxy resins cured with dicyandiamide by dynamic spring analysis (DSA). Senich et al. (7) used the elapsed time to the loss peak maximum of tan delta as a measure of the rate of the reaction at each temperature and for each frequency. The slope of an Arrhenius plot of ℓn (tmax) vs. 1/T was then used to determine the activation energy.

Some of the data for our work is shown in Figures 3 and 4 for tan delta and flexural modulus E' vs. time for several isothermal temperatures. There is a shoulder at the beginning of the tan delta peak at 119°C. At 149°C the shoulder has become an initial smaller peak. Enns and Gillham (12) also observed this behavior with increasing cure temperature for torsional braid analysis (TBA) of an amine-cured epoxy. They attributed the first peak to a liquid-to-rubber transformation (gelation) and the second peak to a rubber-to-glass transformation (vitrification). Figure 5 shows an Arrhenius plot for the neat resin and glass cloth composite at the stoichiometric ratio. The data shows a linear relationship over the temperature range of 59-129°C. The data also shows that the glass cloth has little effect on the activation energy of the resin. However, this data does not take into account the effect of curing that takes place during ramping to the isothermal temperature. A computer program was developed by K. Wolking (this work) to correct for the ramping effects, and the corrected data showed a significant curvature at higher temperatures in Figure 6. Other kinetic data was obtained at different epoxy/curing agent ratios (+5 and -10 wt.%) for glass cloth composites and is compared with the stoichiometric ratio (0 wt.%) in Figure 7. The activation energies calculated from Figure 7 are tabulated in Table I and are compared to data obtained by Mijovic (10) for an amine-cured tetrafunctional epoxy. Molar amine/epoxy ratios were calculated to be 0.8, 1.2 and 2.0 for the +5, 0 and -10 wt.% samples, respectively. Table I shows that the activation energy increases with increasing amine/epoxy ratio for the glass cloth composites. Mijovic showed the same trend for DSC data when glass microspheres (40 wt.%) were used as a filler. He also showed that the activation energy of the neat resin decreased with increasing amine/epoxy ratio, implying that the glass filler restricted molecular mobility in the resin. Table II shows that the activation energy for curing of the neat resin at the stoichiometric ratio for this work is slightly lower than some literature data for other amine-cured epoxies. Table II also shows that some workers (11,13) used an epoxy-amine system very similar to the one used for this work. They postulated an autocatalytic mechanism involving two rate constants (k_1 and k_2), which correspond to the consumption of primary amine and the formation of secondary amine, respectively (13). Presently, DSC data is being obtained on the epoxy-amine system used in this work and a similar autocatalytic behavior has been observed. Also, future work will include the DMTA and DSC analysis of the Epon 828-MDA and Epon 828-m-PDA systems. By analyzing data for the reactions of the epoxy with these individual components in the mixed amine, perhaps a better understanding of the

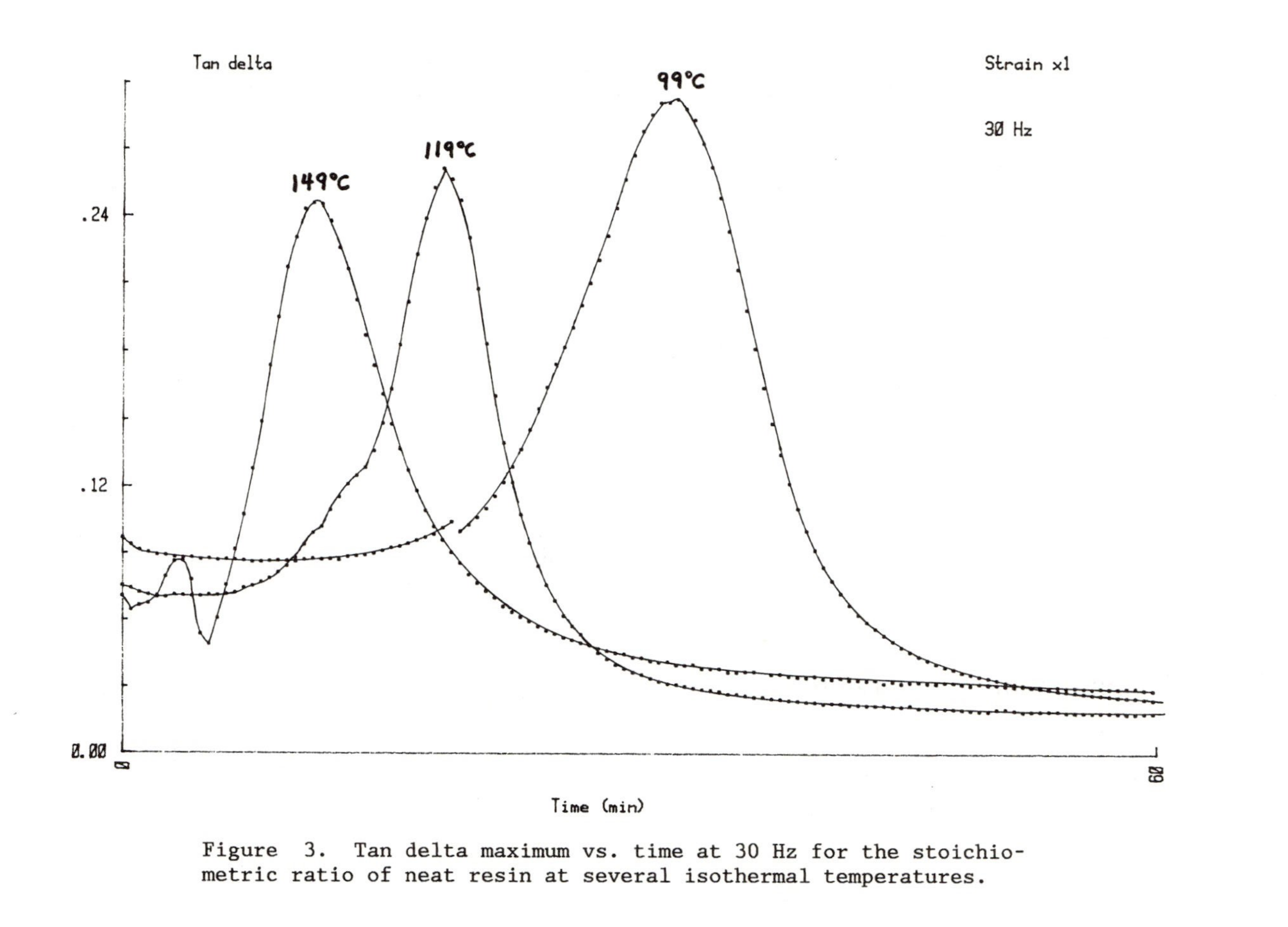

Figure 3. Tan delta maximum vs. time at 30 Hz for the stoichiometric ratio of neat resin at several isothermal temperatures.

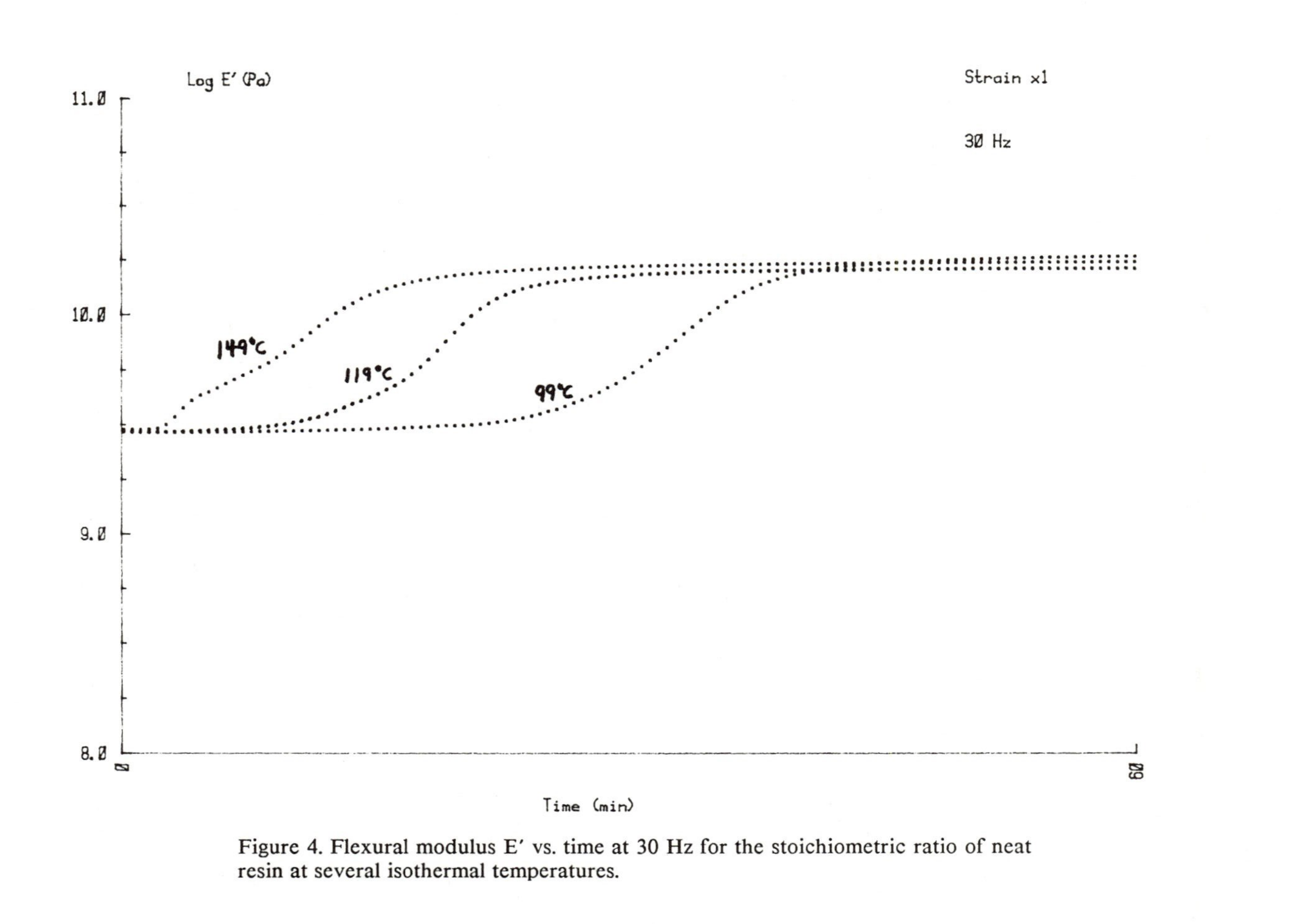

Figure 4. Flexural modulus E′ vs. time at 30 Hz for the stoichiometric ratio of neat resin at several isothermal temperatures.

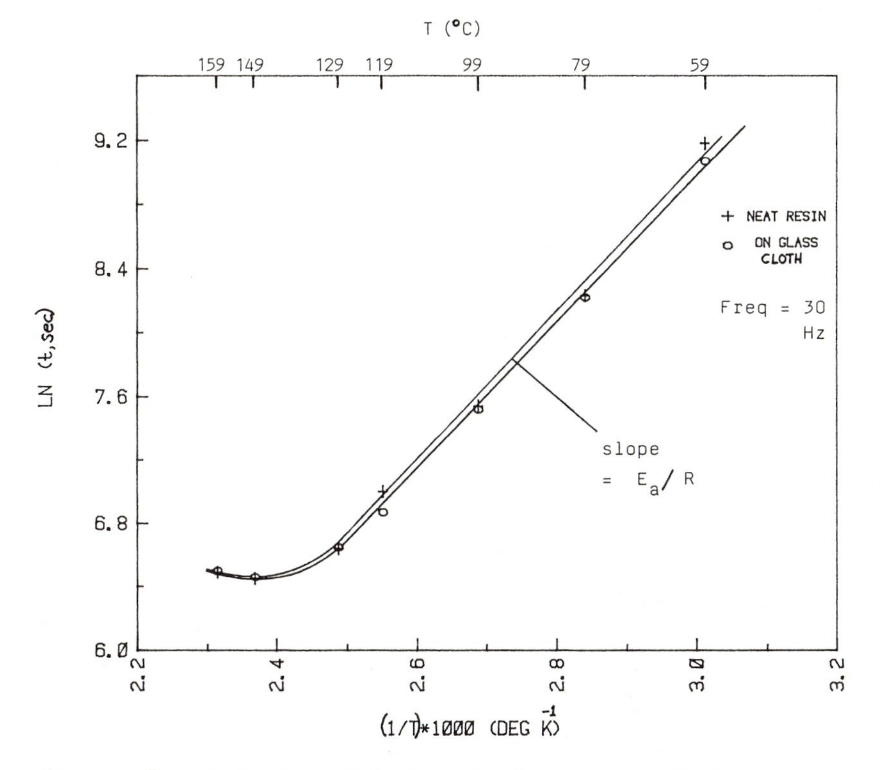

Figure 5. Arrhenius plot for mixed system to B-stage cure: neat resin and glass cloth composite at the stoichiometric ratio.

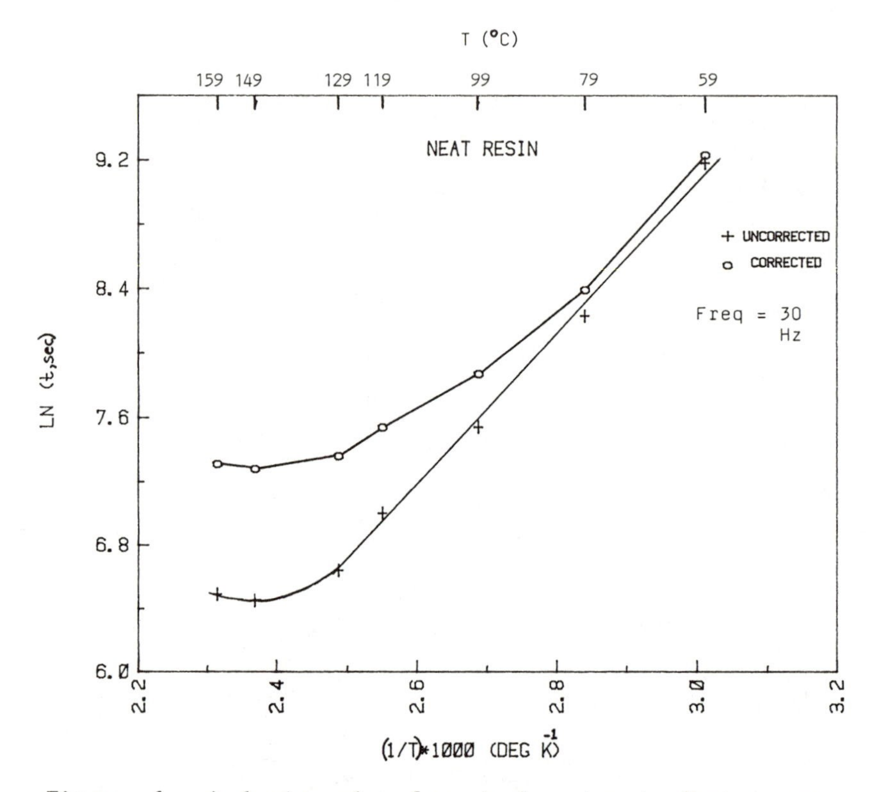

Figure 6. Arrhenius plot for mixed system to B-stage cure: uncorrected vs. corrected plots for curing of the neat resin during temperature ramping.

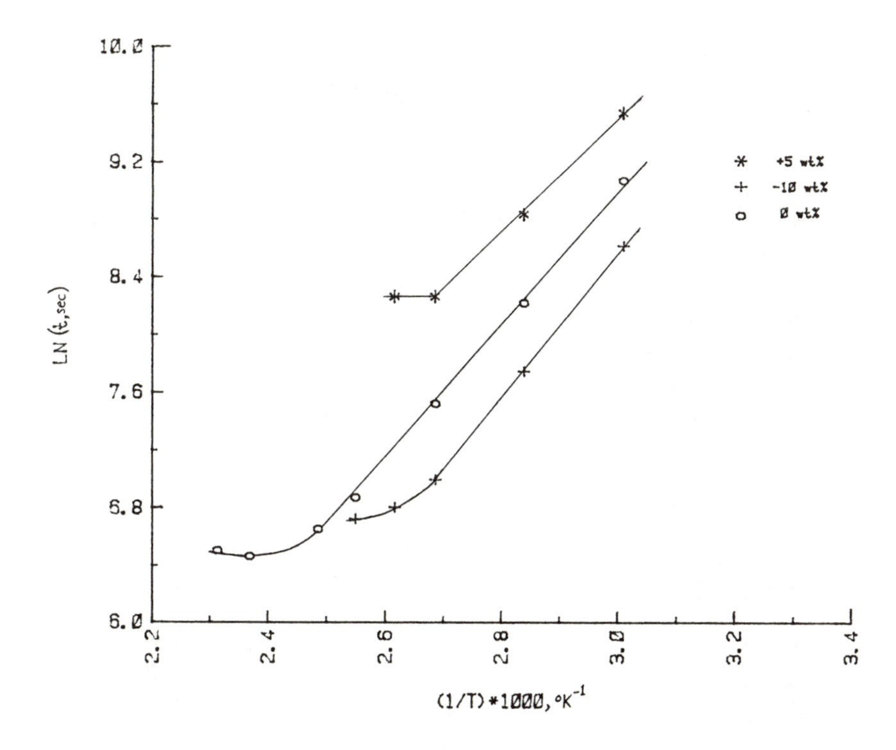

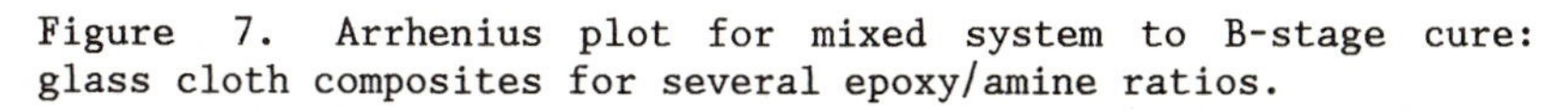
Figure 7. Arrhenius plot for mixed system to B-stage cure: glass cloth composites for several epoxy/amine ratios.

Table I. Activation Energy of Cure as a Function of Amine/Epoxy Ratio

Epoxy	Curing Agent	Amine/ Epoxy Ratio	Technique	Reinforcement	Ea, kcal/mole	Ref.
DGEBA	MDA/m-PDA	0.8	DMTA	Glass cloth	7.9	This work
		1.2		Glass cloth	9.2	
				None	9.4	
		2.0		Glass cloth	9.9	
TGDDM	DDS	0.44	DSC	None	15.9	10
				Glass micro-spheres	12.3	
		0.71		None	11.9	
				Glass micro-spheres	13.9	

TGDDM = tetraglycidyl-4, 4'-diaminophenyl methane
DDS = diaminodiphenyl sulfone

Table II. Activation Energy of Cure for Neat Resin at the Stoichiometric Ratio

Epoxy	Curing Agent	Technique	Kinetic Experiment	Measurement Frequency Range, Hz	Ea, kcal/mole	Ref.
DGEBA	MDA/m-PDA	DMTA	I	30	9.4	This work
DGEBA/ DGEBD (80/20)	MDA/m-PDA	FTIR	I (50% conv)	--	11.9	11
		DSC	I	--	13.5 (k_1), 12 (k_2)	13
		DSC	D	--	22.3	
DGEBA	MDA	TBA	I (gel time)	1.0	9	14
DGEBA	m-PDA	TBA	I (gel time)	1.0	11	14
			I	--	15.4 (k_1), 10.9 (k_2)	15
DGEBA	polyamide	dielectric	I	1, 5, 10 kHz	8 (22-45°C) 10 (45-60°C) 13 (60-75°C)	16

I = isothermal
D = dynamic

autocatalytic mechanism for the mixed amine can be obtained. Table II shows that literature data is available for the cure kinetics of DGEBA-MDA and DGEBA-m-PDA systems.

Data for determining rate constants for this work was also obtained by the method used by Senich et al. (7) for dynamic spring analysis (DSA). Their experiments were conducted at four frequencies (3.5-110 Hz) and, since the viscosity frequency product is constant at the loss modulus maximum, four reaction times to a known viscosity level were available. From this data, a plot of logarithmic viscosity vs. reaction time for each isothermal temperature was constructed. The same kind of plot was made for this work on DMTA frequencies of 0.33-90 Hz and the results are shown in Figure 8 for the stoichiometric ratio of epoxy to amine on glass cloth. The linear relationship indicates that the reaction is first order in the temperature range of about 60-100°C and the slope is equal to the overall rate constant. Rate constants were determined as a function of the amine/epoxy ratio at three isothermal cure temperatures and the results are shown in Table III. The rate constants were determined by a linear least squares fit and the correlation coefficients are also shown. Although there is some scatter in the data and the data in Table III is not complete, some trends can be observed. The rate constant increases with increasing cure temperature at a constant amine/epoxy ratio. Also, with the exception of the 99°C cure, the rate constant increases with increasing amine/epoxy ratio. This is in agreement with the work of Golub and Lerner (11) for analysis of a similar DGEBA/DGEBD and MDA/m-PDA system by FTIR. The rate constants shown in Table III are for the resin on glass cloth; no neat resin data was obtained. Mijovic (10) showed that the rate constant for an amine-cured tetrafunctional epoxy increased with an increasing amine/epoxy ratio for both the neat resin and glass-reinforced resin. However, the rate constant was slightly lower for the reinforced resin. The rate constants in Table III are very similar in magnitude to those obtained by Senich et al. (7) for the DSA analysis of two epoxy resins cured with dicyandiamide.

<u>B-Stage to C-Stage Cure</u>. Samples that were B-stage cured at the recommended conditions (129°C for 45 minutes) were isothermally cured in the DMTA at the recommended C-stage conditions (149°C for 2 hours). Each sample was run manually at three frequencies (3, 10 and 30 Hz) and some dynamic mechanical data at 30 Hz for different epoxy/amine ratios are shown in Figure 9. This plot shows that the kinetics for the B-stage to C-stage cure is very rapid. For the -5 wt.% and -10 wt.% samples, log E' after 120 minutes was only 9.0 and 8.0, respectively. For all other samples (0, ±2, +5 wt.%), the final log E' was 9.3-9.7. Tan delta showed a decrease with increasing cure time for all epoxy/amine ratios except the -5 wt.% and -10 wt.% samples, which showed a gradual increase in tan delta with increasing cure time.

<u>Post C-Stage Cure (Dynamic)</u>. Samples that were C-stage cured at 149°C for 2 hours were scanned in the DMTA as a function of temperature (-100°C to 200°C). Raw data of tan delta and E' vs. temperature at 30 Hz for the -5 and +5 wt.% samples is shown in Figure 10.

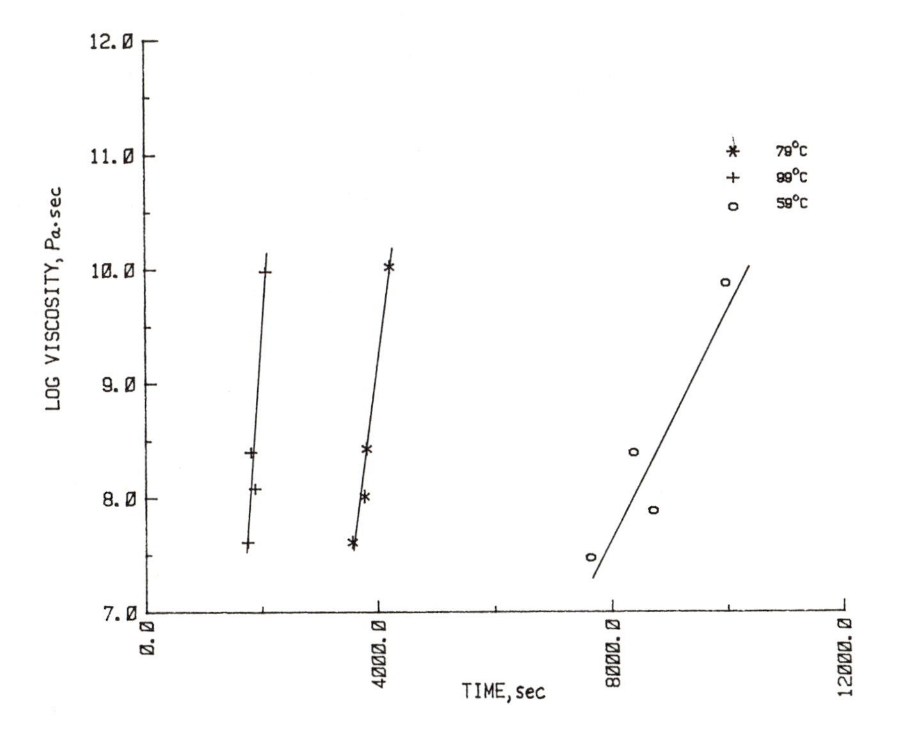

Figure 8. A plot of log viscosity vs. time for determining the overall rate constant of the curing reaction.

Table III. Overall Rate Constants for Mixed System Cure of Resin on Glass Cloth

	Rate Constant, $k(10^3)$, sec^{-1} *		
Amine/epoxy ratio	59°C	79°C	99°C
0.8 (+5%)	0.59 (0.977)	----	----
1.2 (0%)	1.0 (0.933)	3.8 (0.987)	6.7 (0.946)
2.0 (-10%)	----	4.1 (0.919)	5.5 (0.978)

*Calculated from time to reach tan $(\delta)_{max}$; correlation coefficients of linear least squares fit shown in parentheses.

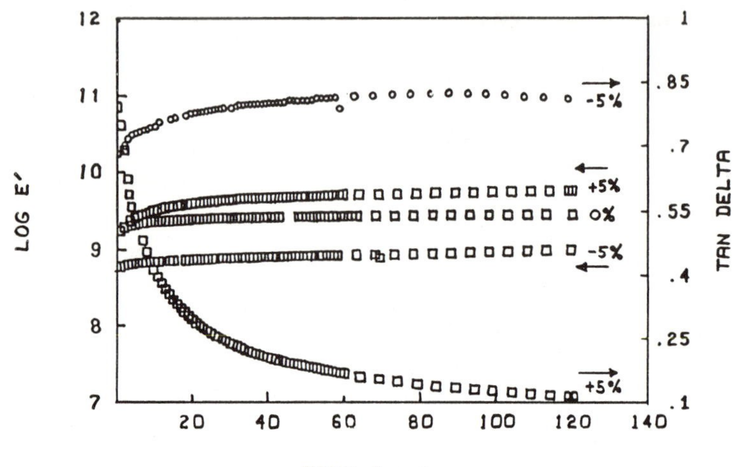

Figure 9. Log E' and tan delta vs. time at 30 Hz for several epoxy/amine ratios during the B-stage to C-stage cure.

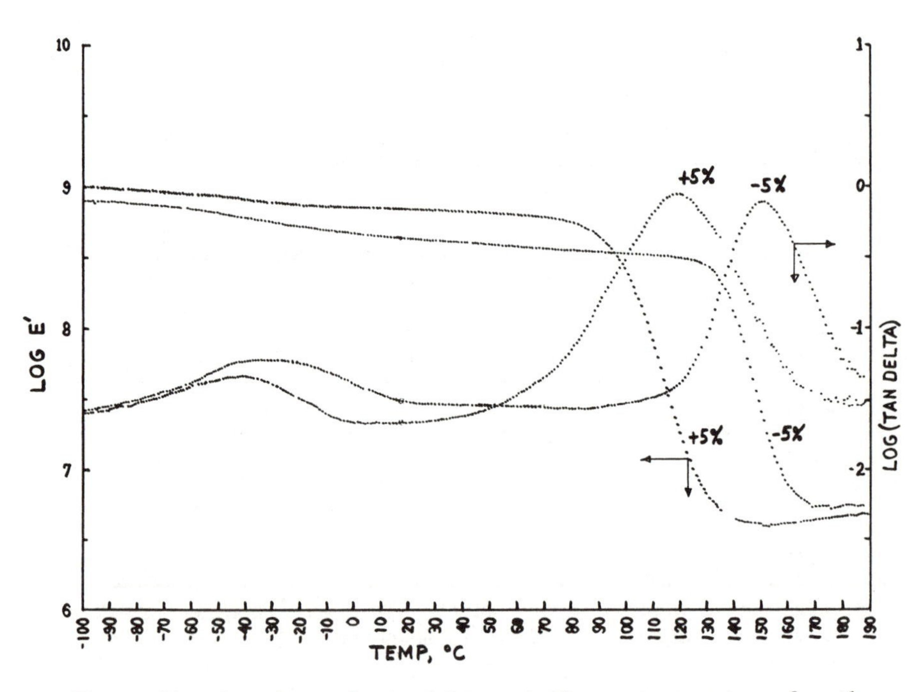

Figure 10. Raw data of tan delta and E' vs. temperature for the -5 and +5 wt.% epoxy/amine ratios.

The sharp drop in modulus, accompanied by a peak in tan delta, indicates the glass-to-rubber transition of the epoxy resin. From the maximum in tan delta, the glass transition temperature (Tg) can be obtained. From the smaller tan delta peak at -40 to -30°C in Figure 10, the sub-Tg can be determined. The sub-Tg is a secondary transition believed to be attributed to the crankshaft rotation of the glycidyl amine linkage ($-CH_2-CH(OH)-CH_2-O-$) after reaction of the epoxide ring and the amine (17). For all the epoxy/amine ratios tested, both the Tg and sub-Tg increased with increasing frequency. This is shown in Figures 11 and 12 by the tan delta peak shifting to increasing temperature with increased frequency.

From the variation of Tg and sub-Tg with frequency, transition maps of ℓn (frequency) vs. 1/T were made from which the activation energy could be determined. Table IV shows that the activation energy (Tg) of the fully cured resin at the stoichiometric ratio agrees fairly well with that reported by Senich et al. (7) for two amine-cured epoxy resins.

The molecular weight between crosslinks (Mc) was determined for each epoxy/amine ratio of the neat resin from the rubbery plateau region of the modulus curve following the Tg region. This can be seen in Figure 13 for several epoxy/amine ratios. The Mc values were calculated from the following equation:

$$Mc = dRT/G \tag{1}$$

where d is the polymer density, R is the gas constant, T is the absolute temperature, and G is the rubbery plateau shear modulus (E'). The Mc was calculated at 185°C for each sample because this temperature was clearly in the rubbery plateau region. The density at 185°C was determined by preparing a block of crosslinked epoxy and measuring its dimensions and weight. Since the density was nearly invariant for all epoxy/amine ratios, an average value of 1.13 g/cm^3 was used in calculating Mc for all the samples. The Mc and Tg were plotted vs. epoxy/amine ratio of neat resin in Figure 14 and it shows that the molecular weight between crosslinks passes through a minimum and the glass transition temperature passes through a maximum at the stoichiometric ratio. Table V shows that the Mc values for this work agree fairly well with some reported in the literature for other DGEBA-amine systems.

Summary

Dynamic mechanical spectroscopy was utilized to characterize the cure kinetics of a mixed epoxy to a B-stage cure as well as for a B-stage to C-stage cure for unfilled epoxy and epoxy composites. A technique for correcting for the partial curing the occurs during ramping to the isothermal cure temperature was developed. The effect of epoxy to amine ratio on cure kinetics and the final dynamic mechanical properties of unfilled epoxy and epoxy composites was determined. The molecular weight between crosslinks passed through a minimum and the glass transition temperature passed through a maximum at the stoichiometric ratio for the DGEBA/MDA-m-PDA system tested. Future work will involve DSC, FTIR and dielectric relaxation studies of the cure kinetics of the same epoxy system.

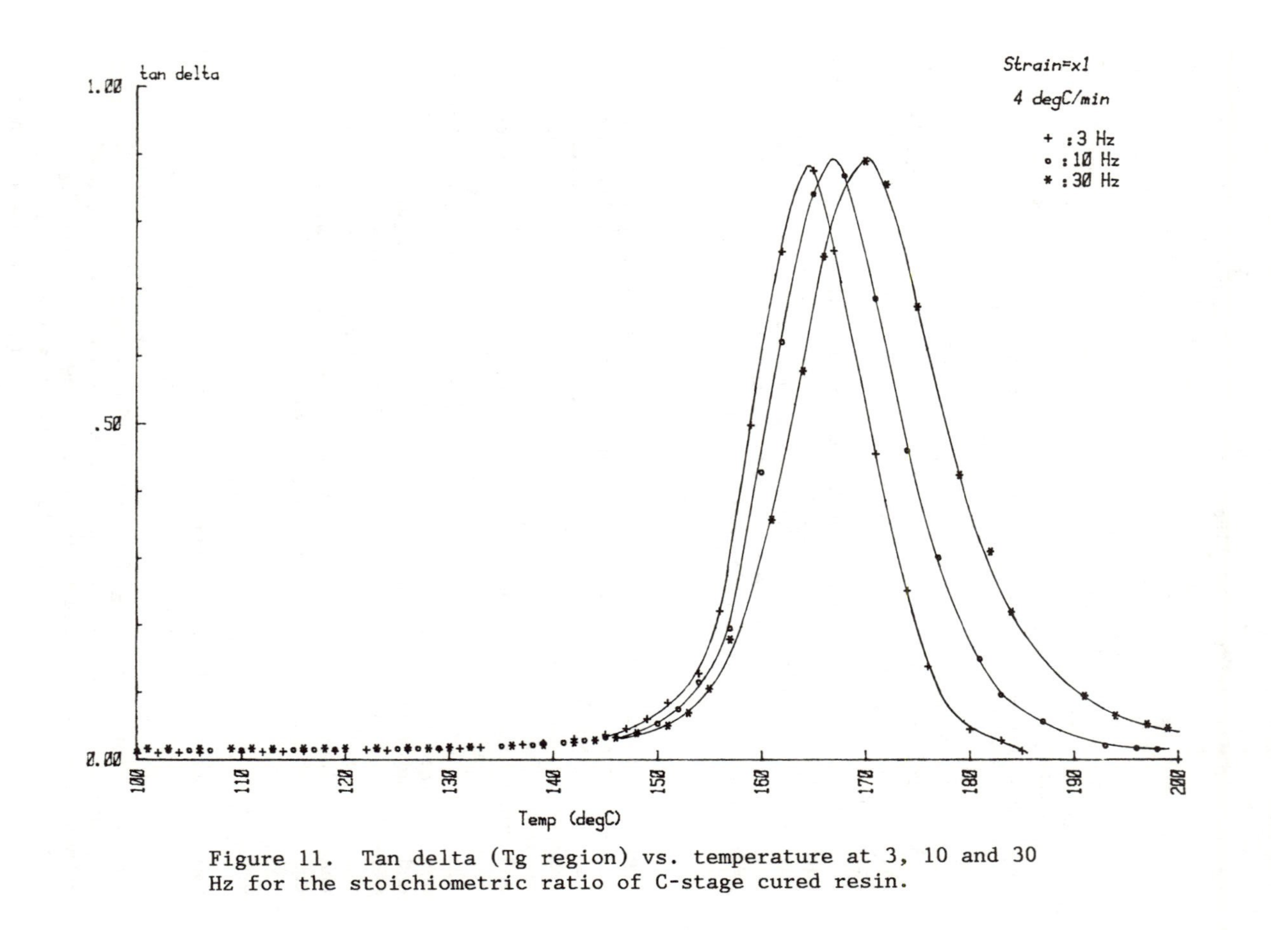

Figure 11. Tan delta (Tg region) vs. temperature at 3, 10 and 30 Hz for the stoichiometric ratio of C-stage cured resin.

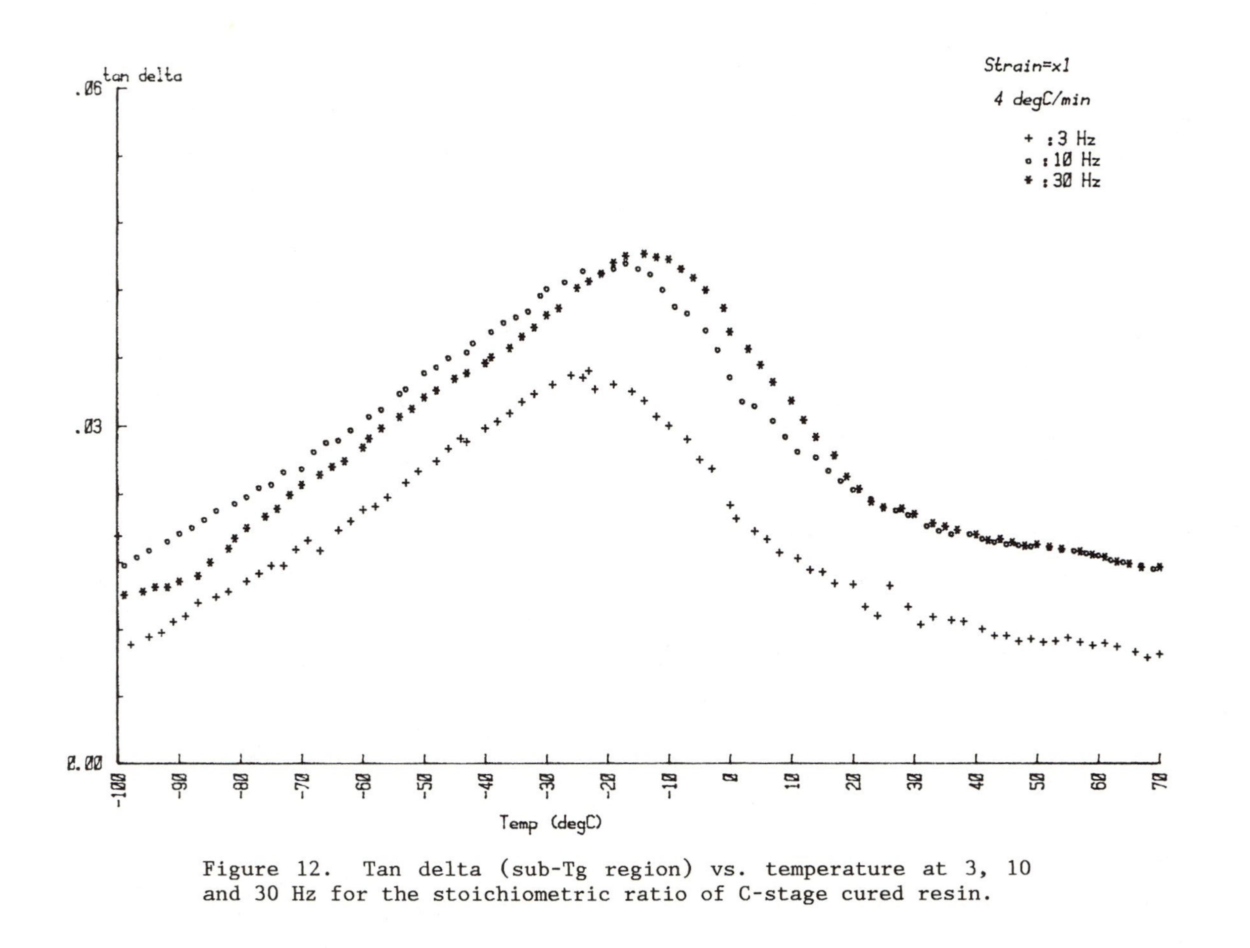

Figure 12. Tan delta (sub-Tg region) vs. temperature at 3, 10 and 30 Hz for the stoichiometric ratio of C-stage cured resin.

Table IV. Post C-Stage Curing of Neat Resin at the Stoichiometric Ratio of Epoxy to Amine

Epoxy	Curing Agent	Technique	State of Stress	Measurement Frequency Range, Hz	Ea, kcal/mole		Ref.
					Tg	sub-Tg	
DGEBA	MDA/m-PDA	DMTA	Flexure	3 - 30	175	27.9	This work
Difunctional	dicyandiamide	DSA	Tension	3.5 - 110.0	119-155	----	7
Tetrafunctional	dicyandiamide				119-155		

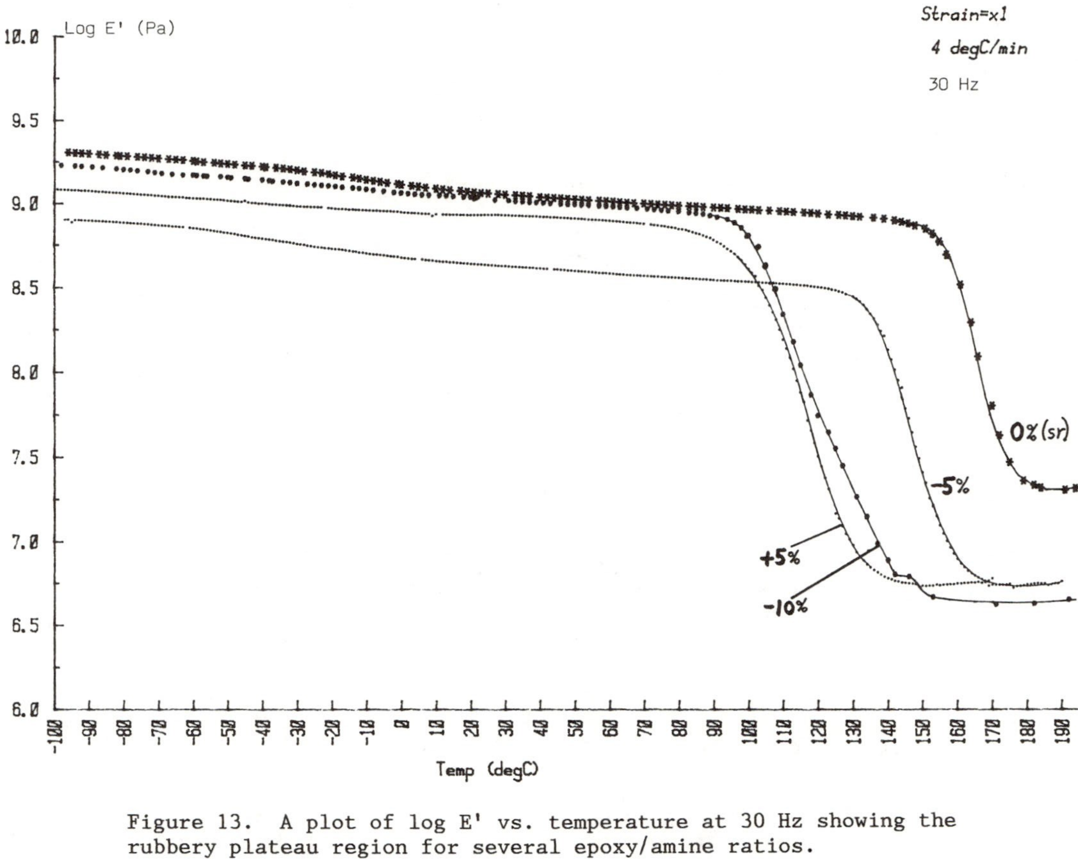

Figure 13. A plot of log E' vs. temperature at 30 Hz showing the rubbery plateau region for several epoxy/amine ratios.

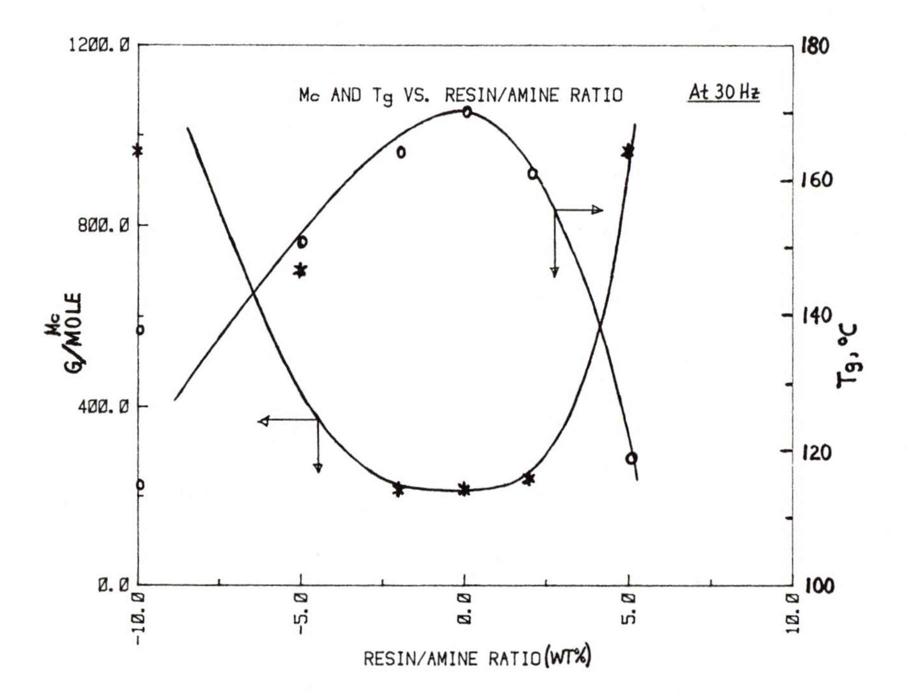

Figure 14. Glass transition temperature (Tg) and molecular weight between crosslinks (Mc) as a function of epoxy/amine ratio for C-stage cured neat resin.

Table V. Molecular Weight Between Crosslinks, Mc, for Several Epoxy/Amine Ratios

Epoxy	Curing Agent	Stoichiometric Ratio (wt.%)	Technique	State of Stress	Measurement Frequency Range, Hz	Mc (g/mole)	Ref.
DGEBA	MDA/MPDA	+5	DMTA	Flexure	3 - 30	963	This work
		+2				242	
		0				215	
		-2				215	
		-5				700	
		-10				963	
DGEBA	MDA	+5.96	Vibron	Tension	11	393	6
		+2.41				372	
		-11.64				412	
DGEBA	IPD	0	Rheovibron	Tension	11	647	8

IPD = isophorone diamine

Similarly, other epoxy and polyester thermosetting systems will be investigated.

Acknowledgments

The authors wish to express appreciation to Mr. Howard Novak and McDonnell-Douglas Astronautics Company, Titusville, FL 32780, and the Microstructure of Materials Center of Excellence (CLO) for partial support of this research. The authors would also like to acknowledge the use of the Polymer Products and Services, Inc. dynamic mechanical instrument in this work.

Literature Cited

1. Garton, A.; Daly, J. Polymer Composites 1985, 6(4), 195.
2. Takahama, T.,; Geil, P. J. Polym. Sci.: Polym. Phys. Ed. 1983, 21, 1237.
3. Manizone, L.; Gillham, J.; McPherson, C. J. Appl. Polym. Sci. 1981, 26, 889.
4. Walker, J.; Richardson, W.; Smith, C. Modern Plastics 1976, May, 34.
5. Garton, A. J. Polym. Sci.: Polym. Chem. Ed. 1984, 22, 1495.
6. Murayama, T.; Bell, J.P. J. Polym. Sci.: Part A-2 1970, 8, 437.
7. Senich, G.; MacKnight, W.J.; Schneider, N.S. Polym. Eng. Sci. 1979, 19(4), 313.
8. Galy, J.; Sabra, A.; Pascault, J.P. Polym. Eng. Sci. 1986, 26(21), 1514.
9. Horie, K.; Hiura, H.; Sawada, M.; Mita, I.; Kambe, H. J. Polym. Sci.: Part A-1 1970, 8, 1357.
10. Mijovic, J. J. Appl. Polym. Sci. 1986, 31, 1177.
11. Golub, M.A.; Lerner, N.R. J. Appl. Polym. Sci. 1986, 32, 5215.
12. Enns, J.B.; Gillham, J.K. J. Appl. Polym. Sci. 1983, 28, 2831.
13. Moroni, A.; Mijovic, J.; Pearce, E.M.; Foun, C.C. J. Appl. Polym. Sci. 1986, 32, 3761.
14. Babayevsky, P.G.; Gillham, J.K. J. Appl. Polym. Sci. 1973, 17, 2067.
15. Ryan, M.E.; Dutta, A. Polymer 1979, 20, February, 203.
16. Lane, J.W.; Seferis, J.C. J. Appl. Polym. Sci. 1986, 31, 1155.
17. Chu, H.W.; Seferis, J.C. Polymer Composites 1984, 5(2), 124.

RECEIVED November 17, 1987

Chapter 16

Thermal Stress Development in Thick Epoxy Coatings

D. King[1] and J. P. Bell

Chemical Engineering Department and Institute of Materials Science, Box U-136, University of Connecticut, Storrs, CT 06268

The thermal stress development in thick (60 mils), solventless, amine-cured epoxy coatings on aluminum was investigated. The stress level as a function of temperature was measured by the bending beam technique. The results obtained showed increased deviation from the behavior of thinner coatings with increasing Tg and elastic modulus epoxies. The measured stress was greatly affected by increasing coating thickness up to eleven mils thick, with only slight effect after eleven mils. The curing schedule was found to control the temperature at which the epoxy-metal system is at zero stress. Plasticization by moisture was found to only slightly relieve the stress developed.

The thermal shock resistance of epoxies is an increasingly important field of research because of the widening use of epoxies in application areas where the conditions for thermal shock exist. The Handbook of Epoxy Resins (1) defines thermal shock as "the likelihood of stress cracking during cure or thermal cycling", and thermal shock is "a function of the elastic modulus of the resin, the thermal expansion characteristics, and the stresses imposed by the curing temperature". An understanding of the development of thermal stresses in epoxy-metal systems would aid in designing thermal shock-resistive epoxies.

The interesting recent work in thermal shock by McCoy (2-3) and Rauhut (4) gives qualitative results, but no indication of the

[1]Current address: A. O. Smith Corporation, Corporate Technology Center, 12100 West Park Place, Milwaukee, WI 53224-3006

0097-6156/88/0367-0221$06.00/0

stress levels achieved. Dewey (5) applied a linear finite-element analysis to the McCoy work, showing where the maximum area of stress was located and an indication of problem areas, but with no absolute values.

There has been extensive work in understanding the origin and level of stresses that develop with coating systems due to curing and/or a thermal gradient. For solventless cured epoxy systems, curing stresses are small (6-8); these can be neglected in the present study since the concern is with solventless cured epoxy systems.

Dannenberg (9) and also Shimbo, Ochi and Arai (10) present work on 2-5 mil thick coatings of epoxy on aluminum. Their work indicates that above the glass transition temperature, Tg, of the epoxy, there are no stresses present because the epoxy will relax to alleviate any stresses developed. Upon cooling to temperatures below Tg, the stresses increase because the epoxy is a glassy material contracting at a rate greater than the aluminum. The stress level achieved at some temperature T<Tg can be calculated from (10-11):

$$\text{Stress} = \sigma = E \int_{T}^{T_g} (\alpha_E - \alpha_S)\, dT \qquad (1)$$

where E is the elastic modulus of the epoxy, and α_E and α_S are the coefficients of expansion for the epoxy and metal substrate, respectively.

Croll's work (6,11-13) in this area is consistent with the results of Dannenberg and Shimbo et al. for thin coatings. However, for coatings of epoxy of thickness greater than five mils, Croll (13) found deviation from Equation (1), and also an expansive stress was present. Croll explained this by plasticization by moisture. He postulated that the thinner coatings absorbed more water and so were plasticized and able to release the stresses better than the thicker coatings, where non-equilibrium water absorption is a considerably smaller overall percentage. If the thicker coatings were plasticized as much as the thinner coatings, their response would be the same as the thin coatings. Croll presented no test of this theory.

Dannenberg's experiments were carried out in a dry environment, and he found results that satisfied Equation (1). These stresses were only slightly relaxed when the sample was placed in a wet environment, showing small plasticization by water, and also this effect was shown to be reversible. Therefore, Dannenberg's satisfaction of Equation (1) was for low plasticization by moisture, putting doubt on Croll's explanation for moisture as the reason for variation between thick and thin coatings.

The present study is concerned with the development of thermal stresses in thick epoxy coatings (60 mils) on aluminum when the sample is thermally cycled at 1°C/min between 200°C and -30°C. A typical amine-cured epoxy system, Epon 828 cured with methylene dianiline (MDA), was selected and the resin to curing agent ratio was varied to obtain different epoxy structures. Varying the resin to curing agent ratio has been shown by Bell (14-15) and Kim, et al. (16) to change the molecular weight between crosslinks and also the toughness of the epoxy.

Experimental

Sample Preparation. The sample series was prepared by varying the ratio of Epon 828 to the methylene dianiline (MDA) curing agent. The compositions tested contained 0%, 23.1%, 50.0% and 100% excess MDA over the stoichiometric amount. The MDA and diepoxide were heated to 100°C and mixed. The mixture was cast onto one side of a sheet of aluminum 5.5 cm. x 24.5 cm. x 5 mils, which had first been wiped with acetone to degrease the bonding surface. The final coating thickness was controlled at approximately 60 mils.

After curing for 2 hrs. at 80°C, 2 hrs. at 150°C and 2 hrs. at 180°C, the coated aluminum sheet was cut into strips 10 cm. x 1 cm. by use of a carbide-toothed cutting wheel. Samples were held, while being cut, with double-backed adhesive tape and were removed by sliding an eight mil thick steel strip between the tape and aluminum. A second identical sample was cut and a hole drilled through the epoxy coating for insertion of a thermocouple during testing.

Thermal Stress Determination. The method selected to determine the thermal stress developed at the epoxy-aluminum interface was the bending beam technique utilized by Dannenberg (9), Shimbo, et al. (10) and others (12-13,17). The exact apparatus configuration is that of Dannenberg's except that thicker coatings were applied to the beam.

The sample was supported at two points, each one quarter of the beam length from an end. This was to remove any curvature due to the weight of the center portion of the beam. The supports were on brass prisms one inch in height. A second identical sample was placed next to the first and a thermocouple was inserted into the hole in the epoxy coating. A large dish of drying agent was placed in the oven to remove any moisture. Liquid CO_2 coolant was introduced into the chamber through copper tubing. A propeller was present to give mechanical mixing of the oven atmosphere. A Gardner microscope cathometer was used to measure the deflection of the center of the beam from the horizontal of the prisms, to ±0.001 cm. A window in the oven and an exterior light allowed easy deflection measurement.

After the thickness of the sample was measured, each sample was heated to 200°C and held at that temperature for ten minutes. Next, the deflection of the beam was measured at ten degree intervals as the sample was cooled to -30°C at 1°C/min. The sample was also held at -30°C for ten minutes. The deflection was again measured at ten degree intervals as the sample was heated to 200°C at 1°C/min.

The deflection of the beam was mathematically converted to stress by Equations (2) and (3) presented by Dannenberg (9):

$$S = P/(bh_1) = E_2h_2^3/12h_1 * 8d/\ell H * F(m,n) \quad (2)$$

$$F(m,n) = ((1-mn^2)^3(1-m))/(1+mn^3) + ((mn(n+2)+1)^3 + m(mn^2+2n+1)^3)/(1+mn)^3 \quad (3)$$

Where S is the stress through the resin cross sectional area, P is the total force required to bend the beam, b is the beam width,

E_1 is Young's modulus of the resin, E_2 is Young's modulus of the metal (10.1×10^6 psi), h_1 is the thickness of the resin, h_2 is the thickness of the metal (5 mils), H is the thickness of the resin plus the metal, ℓ is the length of the beam between supports (5 cm), d is the deflection of the beam center, m is E_1/E_2, and n is h_1/h_2.

To determine the effect of humidity on the stress, one sample (0% excess MDA) was placed in the oven with desiccant at 200°C for 6 hours, and the deflection was measured during that time span. The sample was next cooled to room temperature, weighed, and returned to the oven at 200°C; the dish of desiccant was now replaced by a dish of water. The deflection as a function of time was again measured over a time span of 6 hours. The sample was cooled to room temperature and weighed again to determine percent weight gain.

To determine the effect of the curing schedule upon the thermal stress development, a 0% excess MDA sample was made by first B-stage curing (2 hrs at 80°C) the sample, cutting it into the beam dimensions and then placing it in the apparatus. The deflection of the beam was measured during the remainder of the curing schedule (2 hrs at 150°C and 2 hrs at 180°C).

The stress distribution through the thickness of the 0% excess MDA sample was analyzed by removing three-mil thick layers of the coating by use of a grinding wheel, measuring the resulting curvature after each layer was removed. The grinding technique was tested and was found to be essentially isothermal when done slowly; only one-mil thick layers of the coating were removed per pass of the wheel.

A glass mold treated with release agent was used to prepare 0.125" thick cured plates of the resin compositions. These cured plates were used to provide samples for tensile, differential scanning calorimetery and thermomechanical analysis.

Tensile Modulus. Tensile samples were cut from the 0.125 in. plates of the compositions according to Standard ASTM D638-68, into the dogbone shape. Samples were tested on an Instron table model TM-S 1130 with environmental chamber. Samples were tested at temperatures of -30°C, 0°C, 22°C, 50°C, 80°C, 100°C and 130°C. Samples were held at test temperature for 20 minutes, clamped into the Instron grips and tested at a strain rate of 0.02 in./min. until failure. The elastic modulus was determined by ASTM D638-68. Second order polynomial equations were fitted to the data to obtain the elastic modulus as a function of temperature for each of the compositions.

Thermal Analysis. The glass transition temperature, Tg, of each of the samples was determined by the use of an Omnitherm Q.C.25 Thermal Analyzer at a heating rate of 20°C/min. The midpoint of the inflection in the DSC output was taken as the Tg.

To measure the thermal expansion coefficients, a 0.125 in. thick sample was taken from the cured plates. The increase in length with temperatures was measured by use of a Dupont 941 Thermomechanical Analyzer (TMA), with a heating rate of 5°C/min. The instrument was calibrated with an aluminum standard. Three runs were made for each sample and standard deviations calculated.

Results and Discussion

Bulk Properties of Cured Resin. Table I presents glass transition temperature (Tg) and the coefficients of expansion values obtained for each sample.

Table I. Tg and Coefficient of Expansion

% Excess MDA	Tg(°C)	Expans. Coeff. x 10^6, $°C^{-1}$ <Tg	>T
0.0	163	62.3(1.7)	158(3)
23.1	156	74.0(8.0)	169(21)
50.0	132	74.1(3.5)	170(1)
100.0	99	59.9(4.9)	217(12)

Quantities in parentheses are standard deviations.

Since increasing the amount of curing agent increases the molecular weight between crosslinks, the Tg should correspondingly decrease. The results of Table 1 show this. The coefficients of expansion are not greatly affected by added excess curing agent, although at 100% excess there seems to be a large increase in the coefficient above Tg. Because of the increase in free volume, one would expect some increase in the coefficients. The values obtained in Table 1 will be used in the calculation of stresses by Equation 1.

The elastic moduli are presented in Table II.

Table II. Elastic Moduli

Modulus x 10^{-5}, lb/in^2

% Excess MDA	-30°C	0°C	22°C	50°C	80°C	100°C	130°C
0.0	3.80	3.24	3.08	2.64	2.34	2.16	1.92
23.1	3.70	2.47	3.10	2.61	2.17	2.17	1.64
50.0	3.64	3.21	3.72	2.88	2.44	2.34	0.04
100.0	4.47	3.82	3.84	3.37	2.43	0.16	0.02

The modulus decreases with increasing curing agent at higher temperatures, primarily due to the lower Tgs of these compositions.

The modulus data were fitted with a second degree polynomial equation, and these functions were used in the calculations of the thermal stresses from Equations 1 through 3. The polynomial coefficients and the correlation coefficient for each sample are given in Table III.

Table III. Elastic Modulus Functions

$E(lb/in^2) = AO + A1*T + A2*T^2 \quad T=(^oC)$

Composition	AO	A1	A2	Correlation Factor
	Coeffs. x 10^5			
0.0% Excess	3.315	-.0146	2.97×10^5	.997
23.1% Excess	3.082	-.0127	1.06×10^5	.893
50.0% Excess	3.616	.0013	-2.00×10^4	.955
100.0% Excess	4.126	-.0140	-1.57×10^4	.9600

Thermal Stress Behavior During Cooling and Reheating. The deflection of the epoxy-coated aluminum strips from the horizontal as the strips were cooled from 200°C to -30°C and after ten minutes reheated to 200°C is given in Figure 1. After the temperature cycle was completed, the strips did not return to the original deflection for any of the samples tested. The stress development during the cycle is not reversible in thick coatings, different from the results with thin coatings by Dannenberg (9).

Dannenberg's (9) rate of cooling was approximately 0.4°C/min, slower than the 1°C/min used in this study. The relatively faster cooling rate in our work is believed to have prevented the coatings from reaching their true equilibrium free volume at each temperature. This results in the entrapment of excess free volume upon cooling the sample, which is released when the sample is heated above its Tg if not enough time elapsed to achieve the equilibrium state. This is evident by the plateau region in the heating curves of Figure 1 at the Tgs of each composition. It is less evident as the Tg of the sample decreases and the heating and cooling curves become almost superimposable (Figure 1(d)). The lower Tg samples are able to relax and release the excess free volume faster than the higher Tg samples.

All of the curves show a change in slope at a temperature corresponding to the Tg of the respective sample. The change in slope is a result of the change in the coefficient of expansion at Tg.

All of the curves, except the 100% excess curing agent, show deflections greater than zero. A positive deflection is caused by an expansive stress through the thickness of the coating. Dannenberg's (9) and Shimbo's, et al. (10), deflections were always negative (contractive stresses) for thin coatings. Croll (13) did find positive deflections for thicker coatings and suggested the absence of plasticization by moisture in thicker samples as the cause.

The different behavior of the 100% excess sample is a result of the low elastic modulus and Tg. As the elastic modulus and Tg of the sample decrease, the deflections become more negative and contractive stresses begin to appear.

Effect of Humidity on Thermal Stress. To evaluate whether or not moisture causes the positive deflections for thick coatings (13) the deflection of a 0% excess curing agent sample was measured in both dry and wet environments at 200°C for 6 hrs. Figure 2 gives the results of this experiment. There is no evidence of appreciable plasticization/stress relaxation. Any stress relaxation must occur during the heating of the sample. The difference in starting deflection between the dry and wet environments is caused by the irreversibility of the development of the stress; the 0% RH treatment was carried out before the 100% RH treatment of the sample (see Experimental Procedure Section).

The constant-stress results of Figure 2 are different from the findings of Dannenberg and Shimbo for thin coatings. They reasoned that above Tg the epoxy is viscoelastic and will release any stresses that are formed upon heating. The present results for thick coatings suggest that some stress relaxation does occur when the sample is heated above Tg, but the three-dimensional network of the epoxy limits the amount of stress relaxation. Stresses then still exist above Tg, proportional to the modulus of the epoxy at that temperature.

The wet environment does cause a slight relaxation of the stress, but the extent of relaxation required to prove Croll's reasoning (13) is far from being achieved. The weight percent gained by the epoxy was found to be 0.1%. The slight stress relaxation by moisture was also found by Dannenberg, who also showed the effect to be reversible.

Effect of Composition (Amount of Curing Agent) on Thermal Stress. The deflections from the samples of various curing agent concentration were converted to the stress present in the epoxy coating by use of Equations 2 and 3. These results are given in Figure 3. The stress is a function of the coefficient of expansion and the elastic modulus of the epoxy. Both affect the stress curves and give them the shapes shown. The change in the coefficient of expansion at the Tg causes an abrupt change in the slope of the stress curves. The elastic modulus steadily decreases with an increase in temperature and causes the stress level to decrease with temperature. For low modulus and Tg samples (50% and 100% excess curing agent) where the modulus of elasticity goes to zero in the higher range of temperatures, the stress level will also go to zero and will distort partially, if not totally, the coefficient of expansion effect.

The form of the 100% excess curing agent curve is expected, based upon the work of Dannenberg and Shimbo, et al. They found that above the Tg of the epoxy, the stress was equal to zero. Contraction stresses would then continually build up upon cooling below Tg according to Equation 1. The results presented in Figure 3 show that agreement with Equation 1 is obtained when the Tg and the elastic modulus, at higher temperatures, decrease. All of the samples do show contraction of the epoxy upon cooling the strip, but they also show a difference in the zero stress temperature. Dannenberg and Shimbo suggest the Tg is the zero stress temperature, while present results suggest a variance with epoxy composition. For highly crosslinked systems, the stresses do not relax above Tg.

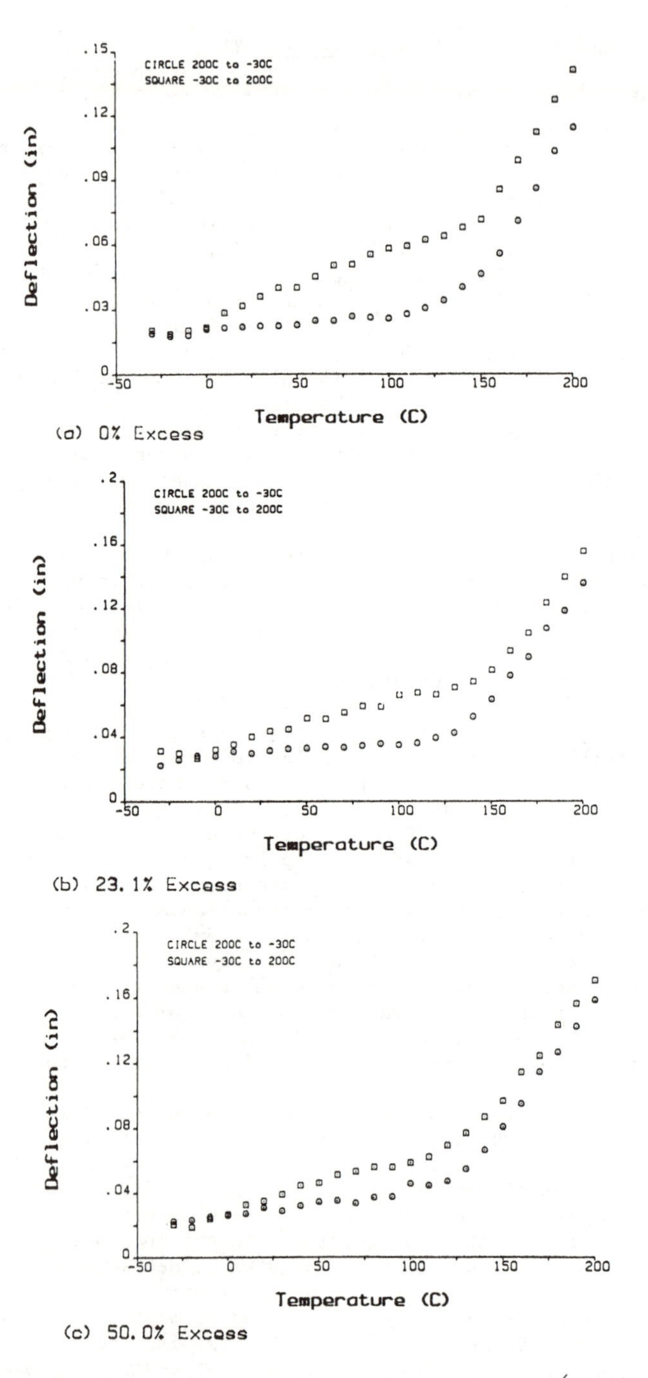

Figure 1. Deflection results vs. temperature. (Continued on next page.)

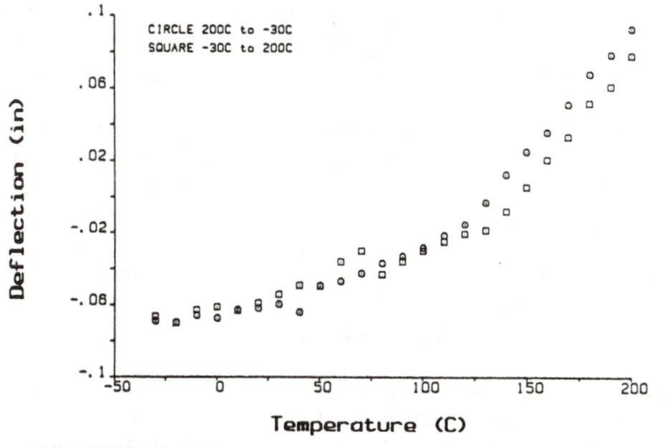

(d) 100% Excess

Figure 1 Continued. Deflection results vs. temperature.

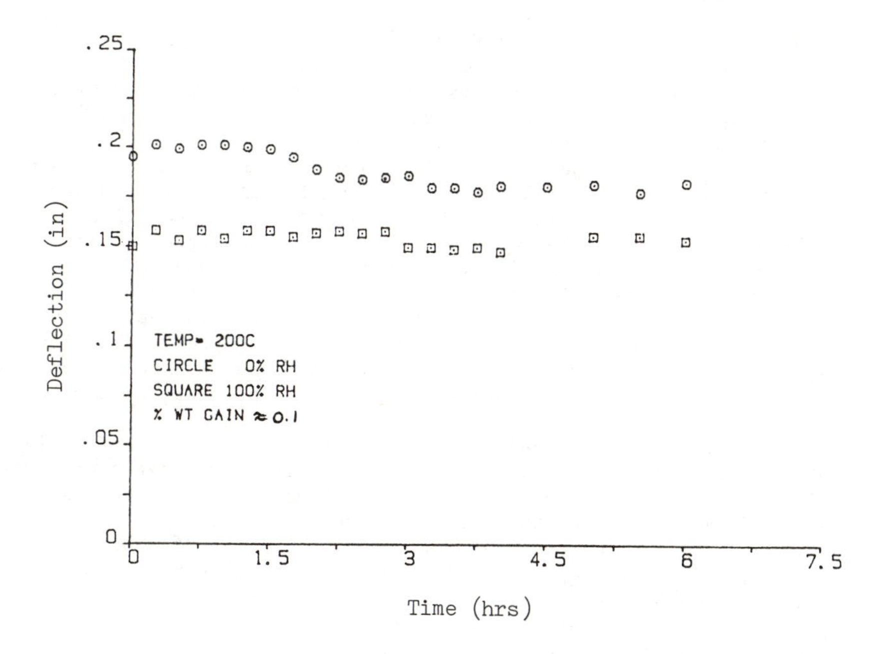

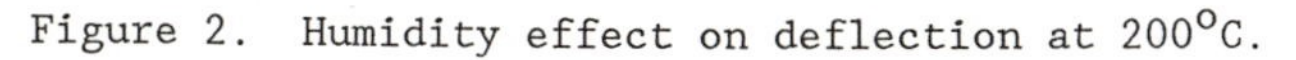

Figure 2. Humidity effect on deflection at 200°C.

If the zero stress temperature is the only variant between these results and those of Dannenberg and Shimbo, the derivative of Equation 1 should correspond to the change in stress with temperature for the results of Figure 3. By taking a given increment in stress and dividing by the corresponding increment in temperature, the derivative of Equation 1 can be compared to the experimental stress increment per unit temperature, shown in Figure 4.

The experimental results have approximately the same shapes but are generally lower in value than the results obtained from Equation 1. The shape is a result of the stress being a function of the coefficient of expansion and the elastic modulus of the epoxy. The lower values indicate that the stress change predicted from the equation is larger than that of the experimental data. Again, it is the samples with the lower Tgs and moduli at high temperatures that are closest to the equation results.

<u>Effect of Cure Schedule on Zero Stress Temperature</u>. The effect of the curing schedule on the zero stress temperature was determined by B-stage curing (2 hrs. at 80°C) a 0% excess curing agent sample and then measuring the deflection during the remainder of the curing schedule. The results are shown in Figure 5.

At each stage of the curing schedule, a certain percentage of the crosslinks are being formed, based on the final Tg of the sample. Lower Tg epoxies form the majority of their crosslinks at the first stages of the curing schedule. The 0% excess curing agent has a Tg of approximately 163°C and, as shown in Figure 5, crosslinks are being formed in an unstressed state at each stage of the process. At each temperature increase to the next stage, the network formed at that point will expand because of the increase in temperature. There is stress relaxation after each temperature increase but it is limited by the extent of the crosslink network formed. Because the crosslinks are in three different stages of stress at a given temperature, the zero stress temperature for the epoxy-metal system is determined by the temperature of formation and percent of the network each of the crosslinks represent. The zero stress temperature for the 0% excess sample is therefore at some location between 80°C and 180°C, and is based on the percent of crosslinks formed at each temperature of the curing schedule.

For the 100% excess curing agent sample, where the Tg is 99°C, the crosslinks are in only two different stages of stress at a given temperature. This is because most of the crosslinks were formed at the 80°C stage with the remainder formed at 150°C. No crosslinks were formed at the 180°C curing stage. The zero stress temperature for this system is therefore somewhere between 80°C and 150°C and is based on the percentage of crosslinks formed at each of those temperatures. The lower zero stress temperature for this epoxy-metal system is the reason the sample shows contractive stress at -30°C and the 0% excess sample does not.

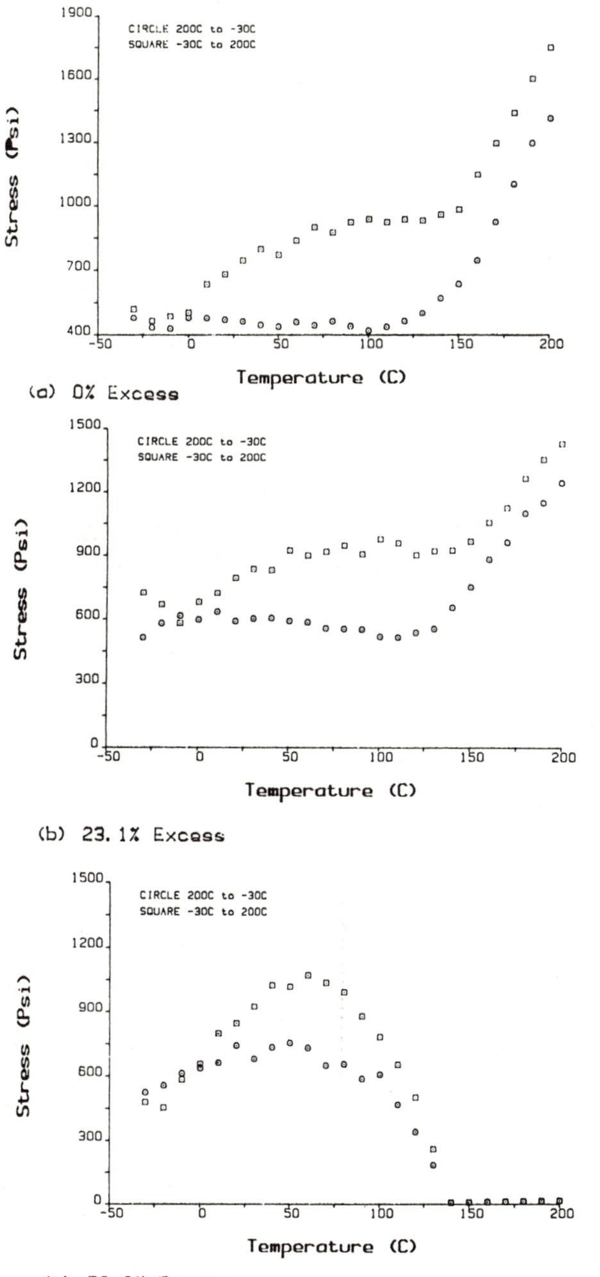

Figure 3. Stress vs. temperature. (Continued on next page.)

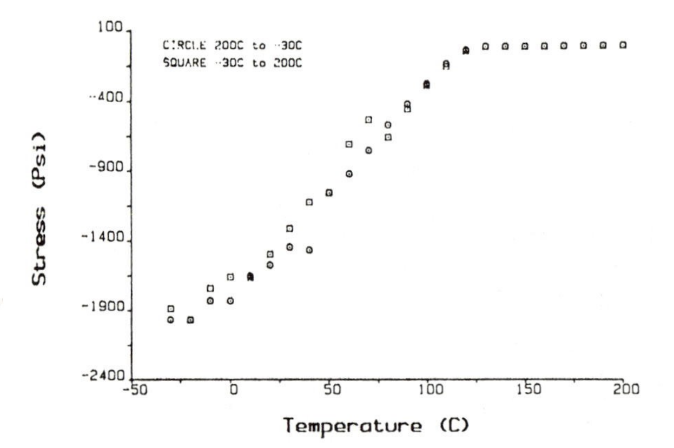

(d) 100% Excess

Figure 3 Continued. Stress vs. temperature.

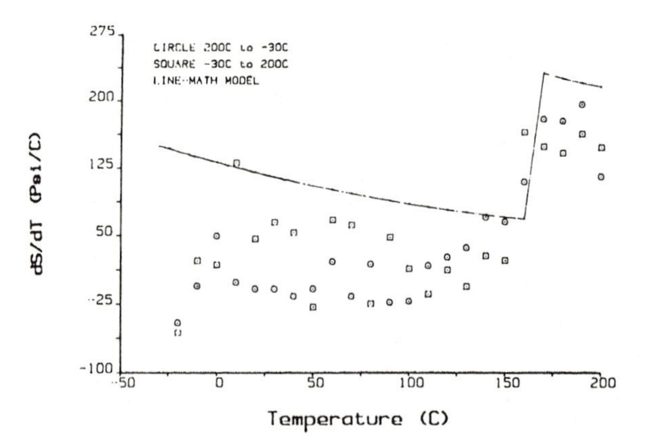

(a) 0% Excess

Figure 4. Comparison of Equation 1 with experimental results. (Continued on next page.)

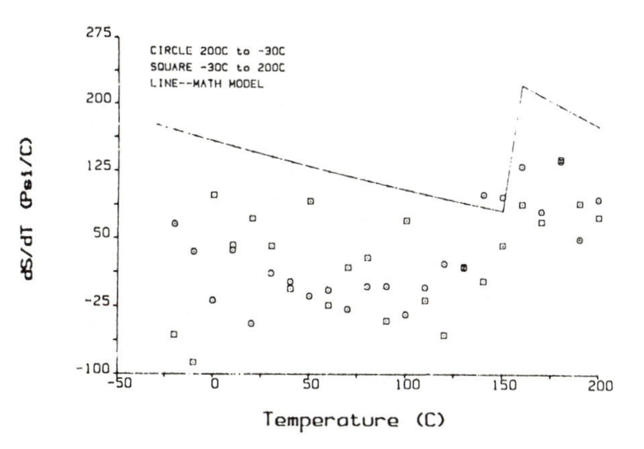

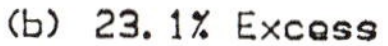
(b) 23.1% Excess

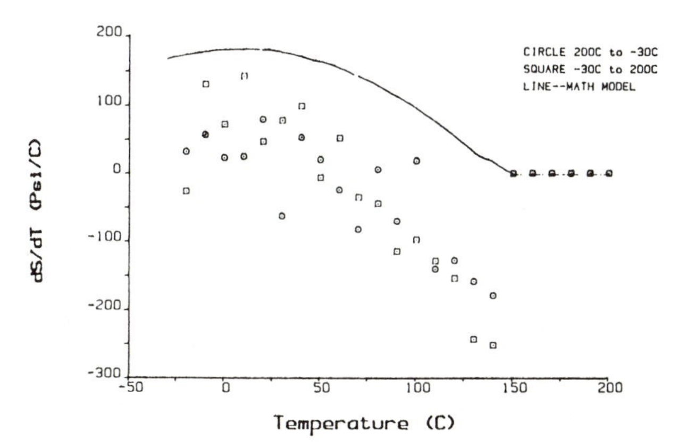

(c) 50.0% Excess

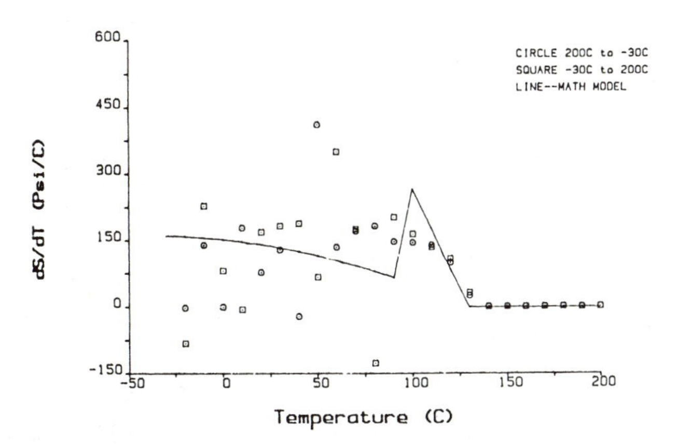

(d) 100% Excess

Figure 4 Continued. Comparison of Equation 1 with experimental results.

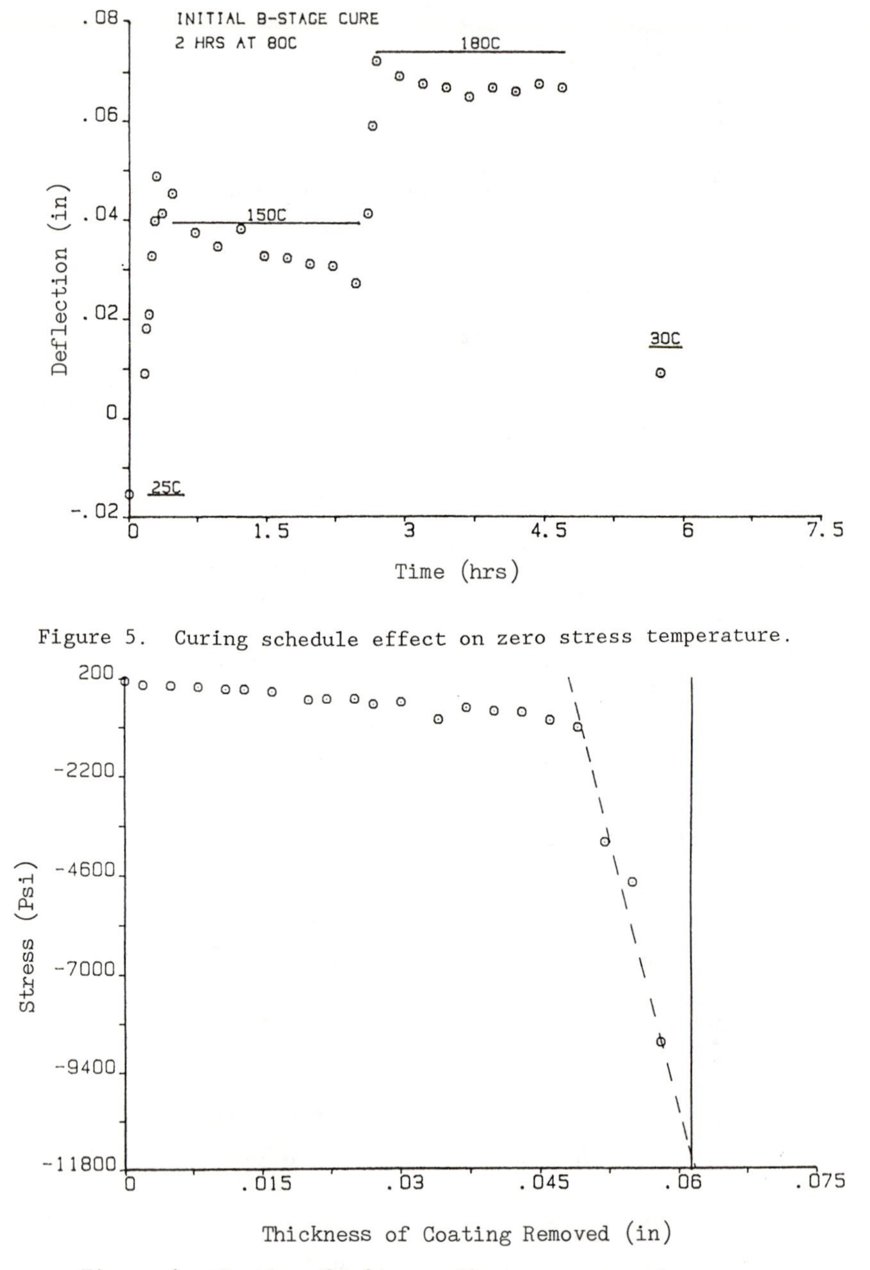

Figure 5. Curing schedule effect on zero stress temperature.

Figure 6. Coating thickness effect on measured stress.

Effect of Coating Thickness on Thermal Stress. The effect of coating thickness on the measured stress was determined by shaving off thin layers of the epoxy coating from the 0% excess curing agent sample and measuring the resulting beam reflection. The deflection was then converted to stress. The results are given in Figure 6.

The difference between the stress at the epoxy-metal interface (0.062 in. of material removed) and the overall measured stress increases with increasing coating thickness, up to a thickness of approximately 13 mils. For thickness >13 mils, the residual stress changes only slightly.

Equation 1 predicts an interfacial stress of only 3,000 psi when the sample is cooled from 180°C to 25°C, assuming the stress does not equal zero at T>Tg. This stress is apparent at a coating thickness of approximately 11 mils. By extrapolation of the last four data points of Figure 6, the interfacial stress (0.062 in. removed) is about 11,300 psi. This is greater than that predicted by equation 1 but is of the same magnitude as the breaking stresses from the tensile data at this temperature. Possible reasons for this include the concept that a significant thermal gradient occurs in thick coatings during cooling, and this gradient produces constraints near the interface that are not present for thin coatings (18).

Conclusions

The thermal stress development of thick epoxy coatings on metal was found to be different from that found with thinner coatings. This deviation was greatest for high Tg, rigid epoxy systems. As the Tg and elastic modulus of the epoxy resin decreased, the development corresponded more to that found with thin coatings. Plasticization by moisture was found to relieve some of the stress developed but was not great enough to explain the deviation found.

The curing schedule of the epoxy controls the temperature at which the epoxy-metal system is at zero stress. The controlling variables are the final Tg of the system and the time and temperature of each stage in the curing schedule. The variables determine the percent of crosslinks formed in an unstressed state at each temperature of the curing schedule.

The stress measured was found to decrease as the thickness of the coating increased. The equation used for thin coatings predicted the correct stress level at a coating thickness of 11 mils. The stress measured was lower than predicted at greater thickness and higher at smaller thickness.

Acknowledgments

The authors wish to gratefully acknowledge the financial support of Westinghouse Electric Corp. for this research.

References

1. H. Lee and K. Neville, Handbook of Epoxy Resins, McGraw-Hill, New York, 1967.

2. H. E. McCoy, Jr., "Evaluation of Polymeric Films for Electrical Insulation", Oak Ridge National Laboratory, Report ORNL-6134, Apr. 1985.
3. H. E. McCoy, Jr., and C. R. Brinkman, "Evaluation of Several Polymer Films for use as Electrical Insulators", Oak Ridge National Laboratory, Report CONF-8509141, Sept. 1985.
4. H. Rauhut, "Advanced Epoxy Encapsulates; Low-Stress and T-Shock resistant Semiconductor Device Encapsulation", Thermosets by Design, RETEC, Des Plaines, Ill., p.1-8, March 7-9, 1984.
5. B. R. Dewey, "Thermal Stress Analysis of an Aluminum-Mylar Transformer Coil", Oak Ridge National Laboratory, Report ORNL/TM-8095, Feb. 1982.
6. S. G. Croll, "Adhesion and Internal Strain in Polymeric Coatings", Adhesion Aspects of Polymeric Coatings, (K.L. Mittal, Ed.), pp. 107-129, Plenum Press, New York (1982).
7. I. S. Klaus and W. S. Knowles, "Reduction of Shrinkage in Epoxy Resins", J. Appl. Polym. Sci., 10, 887-889 (1966).
8. L. M. Vinogradova, Yu. V. Zherdev, A. Ya. Korolev, R. V. Simonekova, and R. V. Artamonova, "Adhesion and the Internal Stresses of Polymers", Vysokomol.,soyed., A12, No. 2, 348-354 (1970).
9. Hans Dannenberg, "Determination of Stresses in Cured Epoxy Resins", SPE Journal, 669-675, July 1965.
10. M. Shimbo, M. Ochi and K. Arai, "Internal Stress of Cured Epoxide Resin Coatings Having Different Network Chains", J. of Coatings Technology, Vol. 56, 713, 45-51, June 1984.
11. S. G. Croll, "Adhesion Loss Due to Internal Strain", J. of Coatings Technology,. Vol. 52, 665, 35-43, June 1980.
12. S. G. Croll, "An Overhanging Beam Method for Measuring Internal Stress in Coatings", J. Oil Col. Chem. Assoc., 63, 271-275 (1980).
13. S. G. Croll, "Residual Stress in a Solventless Amine-Cured Epoxy Coating", J. of Coatings Technology, Vol. 51, 659, 49-55, Dec. 1979.
14. J. P. Bell, "Structure of a Typical Amine-Cured Epoxy Resin", J. Polym. Sci., A-2, 6, 117, (1970).
15. J. P. Bell, "Mechanical Properties of a Glassy Epoxide Polymer: Effect of Molecular Weight Between Crosslinks", J. Appl. Polym._Sci., 14, 1901-1906 (1970).
16. S. L. Kim, M. D. Skibo, J. A. Manson, R. W. Hertzberg, and J. Janiszewski, "Tensile Impact and Fatique Behaviour of an Amine-Cured Epoxy Resin", Polym. Eng. and Sci., Vol. 18, 14, 1093-1100 (1978).
17. A. Saarnak, E. Nilsson and L. O. Kornum, "Usefulness of the Measurement of Internal Stresses in Paint Films", J. Oil Col. Chem. Assoc., 59, 427-432 (1976).
18. D. King, Factors Controlling the Thermal Shock Resistance of Epoxy-Metal Systems., M. S. Thesis, University of Connecticut, 1986.

RECEIVED October 7, 1987

Chapter 17

Recycling of Cured Epoxy Resins

Conchita V. Tran-Bruni and Rudolph D. Deanin

Plastics Engineering Department, University of Lowell, Lowell, MA 01854

Epoxy resin was cured with polyamine, polyamide, or anhydride; ground to pass a 0.5-mm screen; and 20% was recycled into the next batch of virgin resin, either directly or after pre-soaking in the liquid ingredients. When these "filled" resins were cured, hardness was generally higher than for virgin resins. For polyamine-cured resin, recycle of pre-soaked epoxy also increased both impact strength and heat deflection temperature. For polyamide-cured resin, recycle also increased volume resistivity. Pre-soaked recycle dramatically increased adhesion to aluminum, especially in the anhydride system.

Urbanization and industrialization of the world are producing increasing amounts of solid waste, and disposal of this solid waste is becoming a growing problem; the rapid growth in usefulness and production of plastics in our society is making plastics a growing portion of this total solid waste (1). Thermoplastic waste in manufacturing can generally be recycled by blending it homogeneously with virgin thermoplastic material, with little or no sacrifice in properties and usefulness (2-4); and even collection, purification, and recycling of post-consumer waste is becoming practical in selected fields such as polyethylene terephthalate and high-density polyethylene bottles and polypropylene battery cases (5).

Thermosetting plastics offer many advantages in end-use properties for high-performance applications (6,7). They suffer from greater difficulty in processing, and from the low utility of manufacturing scrap, which therefore becomes solid waste, and constitutes a double economic burden. The scrap can of

0097-6156/88/0367-0237$06.00/0

course be ground and blended with virgin resin; but since it is insoluble and infusible, this creates more problems than it solves. It invariably increases viscosity and makes processing more difficult. It can be used as a solid filler; because of its chemical similarity to the virgin resin, it might be hoped that it would give a stronger interfacial bond and thus reinforce properties; but in most cases, the thermoset scrap is so chemically resistant (i. e. unreactive) that the virgin resin cannot bond to it successfully. On the contrary, since grinding of the rigid glassy thermoset scrap produces particles with sharp edges and sharp corners, these act as stress concentrators in the cured virgin resin and thus only weaken and embrittle it. Therefore recycle of thermoset scrap has rarely been successful.

About 20% of vulcanized rubber scrap is recycled by thermal/chemical cleavage of cross-links to make it melt processable again (8). Experimenters have reported that they can grind and recycle moderate proportions of thermoset polyesters, polyurethane rubber, and phenolic resins (9-11) into virgin material, and sometimes even observe some reinforcement of properties. The present study was undertaken to explore the possibility of grinding and recycling cured epoxy resins back into virgin epoxy resin formulations, particularly by pre-soaking the ground scrap in the virgin liquids in the hope of penetrating the thermoset particles and thus bonding them more firmly into the virgin matrix during the ultimate cure reaction.

Experimental

Shell Epon 828 epoxy resin was cured with 13 PHR (parts per hundred of epoxy) of triethylene tetramine as a typical polyamine, 65 PHR of Shell Epon V-15 polyamide as a typical polyamide, or 130 PHR of dodecenyl succinic anhydride as a typical anhydride. Liquid formulations were cast in aluminum molds and oven-cured as follows:

Polyamine: 1 Hr./100°C
Polyamide: 1 Hr./100°C
Anhydride: 1% Benzyl Dimethyl Amine Accelerator, 1 Hr./120°C + 2.5 Hr./150°C.

This produced test specimens directly.

For recycle, cured specimens were cut up with a band saw, mixed with dry ice, and ground in a Fitzpatrick Homoloid WD-36-3 hammer mill to pass a 500-micron screen; 70-80% of the ground recycle was between 200-500 microns, the remainder finer. Twenty percent of recycle was mixed with 80% of the corresponding virgin formulation, and cured the same way as before. When the recycled resin was simply added to the liquid system just before casting and curing, it was referred to as "Dry" filler.

Alternatively, the recycled resins were pre-soaked in virgin liquids, 1 Hr./90°C + 4 Days/Room Temperature, in the hope of "activating" them, before mixing, casting, and curing the formulations. Polyamine- and polyamide-cured recycles were soaked in virgin epoxy resin; anhydride-cured recycle was soaked in virgin anhydride curing agent. These systems, containing "Soaked" fillers, were cured the same way as before.

Rockwell L hardness was measured according to ASTM D-785, flexural modulus and strength according to D-790, dart impact strength according to D-3763, heat deflection temperature according to D-1637 at 264 PSI, and volume resistivity according to D-257. Results are summarized in Tables I-III.

Discussion

Recycle of a thermoset powder would be expected to resemble the use of particulate fillers. Thus we might expect it to increase hardness, modulus, and heat deflection temperature, and decrease strength and impact resistance. On the other hand, if similar polarity and/or chemical reactivity produced interfacial bonding, we might hope that the recycle could act as a reinforcing filler, and help to retain or even improve strength and impact resistance; and we might further hope that pre-soaking the recycle in the virgin liquid system would improve interfacial bonding even more.

The experimental results indicated that the recycle did generally increase hardness and decrease strength as expected. Effects on modulus ranged from negligible to negative, suggesting that the recycle either inhibited cure or simply introduced weak interfaces.

Other effects were more selective. While recycle usually lowered impact strength and heat deflection temperature, pre-soaking polyamine recycle surprisingly improved both of these properties. While recycle usually lowered volume resistivity, polyamide recycle improved it. Finally, adhesion of epoxy formulations to the aluminum mold, in spite of wax and silicone mold release agents, was dramatically increased by the use of pre-soaked recycle, especially in the anhydride system, suggesting unexpected usefulness in epoxy adhesive formulations.

Such a variety of effects suggests that (1) Further analytical study would be required to understand them theoretically, but (2) Broader practical study would pinpoint formulations which would permit the recycler to optimize critical end-use properties. This would be useful primarily in epoxy moldings and castings, and in adhesive formulations.

Table I. Polyamine Cure

Recycle	None		Dry		Soaked	
Rockwell L Hardness	120	±2	129	±1	128	±1
Flexural Modulus, KPSI	475	±30	336	±12	325	±9
Flexural Strength, KPSI	16.2	±1.0	5.67	±0.46	6.67	±0.88
Dart Impact Energy, Ft-Lb	0.834		<0.834		1.67	
Heat Deflection Temperature, °C	103	±1	90	±2	108	±1
Volume Resistivity, 10^{15} Ohm-Cm	1.70	±0.17	1.34	±0.01	0.024	

Table II. Polyamide Cure

Recycle	None		Dry		Soaked	
Rockwell L Hardness	118	±3	121	±3	113	±5
Flexural Modulus, KPSI	383	±13	364	±13	236	±13
Flexural Strength, KPSI	12.2	±0.6	11.9	±0.3	5.76	±0.50
Dart Impact Energy, Ft-Lb	2.50		<0.834		0.834	
Heat Deflection Temperature, °C	60	±2	54	±2	46	±1
Volume Resistivity, 10^{15} Ohm-Cm	0.69	±0.03	2.30	±0.06	0.87	

Table III. Anhydride Cure

Recycle	None		Dry		Soaked	
Rockwell L Hardness	121	±1	122	±1	126	±2
Flexural Modulus, KPSI	358	±6	368	±13	244	
Flexural Strength, KPSI	12.0	±0.2	6.84	±0.66	7.31	
Dart Impact Energy, Ft-Lb	1.67		0.834		<0.834	
Heat Deflection Temperature, °C	70	±1	71	±1	65	
Volume Resistivity, 10^{15} Ohm-Cm	12.1	±0.6	7.19		8.01	

Acknowledgment

Taken from C. V. Tran-Bruni's M.S. Plastics Engineering thesis at the University of Lowell.

Literature Cited

1. Baum, B.; Parker, C. H. Solid Waste Disposal; Ann Arbor Science Publishers: Ann Arbor, Michigan, 1974. Plastic Waste Management; Manufacturing Chemists Association, 1974.
2. Leidner, J. Plastics Waste, Recovery of Economic Value; Dekker: New York, 1981.
3. Hawkins, W. L. SPE ANTEC 1981, 27, 477-510.
4. Kinstle, J. F. ACS Polym. Preprints 1983, 24 (2), 425-448.
5. Leaversuch, R. D. Mod. Plastics 1987, 64 (3), 44-47.
6. Brydson, J. A. Plastics Materials; Butterworths: London, 1982.
7. Modern Plastics Encyclopedia; McGraw-Hill: New York, Annual.
8. Brothers, J. E. In Rubber Technology; Morton, M., Ed.; Van Nostrand Reinhold: New York, 1973; 2nd Edition, Ch. 19.
9. Kruppa, R. A. Plastics Tech. 1977, 23 (5), 63.
10. Bauer, S. H. SPE ANTEC 1976, 22, 650.
11. Deanin, R. D.; Ashar, B. V. Org. Coatings & Plastics Chem. 1979, 41, 495.

RECEIVED October 7, 1987

HIGH-PERFORMANCE
POLYMER NETWORKS

Chapter 18

Semiinterpenetrating Networks Based on Triazine Thermoset and *N*-Alkylamide Thermoplastics

J. A. Feldman and S. J. Huang

Institute of Materials Science, University of Connecticut, Storrs, CT 06268

In this study a semi-interpenetrating network (SIPN), is developed using bisphenol-A dicyanate with polyethyloxazoline and polyvinylpyrrolidone. The resulting SIPN is characterized with thermal analysis, infrared spectroscopy and electron microscopy. A water etching process shows that the polyethyloxazoline system is more compatible than the polyvinylpyrrolidone system. Phase domains of the thermoplastics range in size from 0.5 μm to 5.0 μm, respectively. This is the effective size for thermoset toughening. The results also indicate that the material may have useful biomedical applications. The porous surface may enhance tissue-polymer interactions, thus increasing the compatibility of implant prostheses.

The end use of a material is often decided from an engineering viewpoint. The molecular level, although important, is not the only concern when the application requires bulk properties to be in a specific range. Many synthetic polymers are tailor made such that the physical properties will be optimum for a particular end use. In the past this meant developing completely new polymers, and although there is practically no limit to the number of polymers that can be made, others have found that by combining two known polymers, one can also obtain a specialized system. Combining various known systems together produces physical properties averaged according to the volume of each component. Conversely, polymers can be physically mixed with the retention of their properties.(1) In this way the desired mechanical properties and processing capabilities can be obtained. However, some problems exist from a thermodynamic standpoint. The degree of mixing is directly related to the compatibility between the two polymers. If the cohesive energy densities are quite different, then phase separation is likely to occur and the final product will not be homogenous in bulk composition.(1) A blend of this type may show anisotropy as a function of domain size and position. Whether this is beneficial or detrimental will depend on its application. If

0097-6156/88/0367-0244$07.00/0

the cohesive energy densities are similar, then the free energy of mixing will be optimized for a homogenous blend, better approximated as a polymer solution (2). The physical properties of a polymer solution are closer to a true average than a blend. There the two polymers are highly intertwined. The degree of mixing will be high but there is a greater dependency on the entropy of mixing. The change in enthalpy may enhance mixing, but the change in entropy needed to bring one polymer into contact with the other may be too large. This would limit the homogeneity of the polymer solution. One way to maximize the mixing of two polymers is to actually polymerize the monomers after they are mixed, forming an interpenetrating network, (IPN).

The free energy of mixing must be minimized according to the enthalpy and entropy of the system. The enthalpy of mixing between two chemicals can be minimized by matching the solubility parameters or cohesive energy densities. Monomers with the same general primary and secondary forces should exhibit similar thermodynamic properties. The enthalpy of mixing will be favorable in this case. One recognizes too, that the entropy of mixing should be maximized.(3) This may not be easily accomplished using high molecular weight polymers, but by mixing the monomers together first, one finds the total increase in entropy to be much higher. The result is a lower free energy of mixing for the monomers relative to the polymers. The same argument will hold, but to a lesser degree, for a two part system using a monomer and a polymer instead of two monomers. After the compounds are mixed, the final parameter to be concerned with is the diffusion rate. Assuming Fickian behavior, one expects the diffusion rate to be a function of the square root of time. Ample time should be allowed to obtain a homogenous system.

The first stipulation to be recognized in obtaining a true IPN is that each monomer should only polymerize with itself. Separate initiators may be needed to accommodate each monomer. As long as the polymerization mechanism is sufficiently different one should obtain an interpenetrating network, IPN. On a molecular level the polymers should be intertwined. There should be a high degree of entanglement but no covalent bonding between species. This type of interlocking has been coined "catenanes". (1)The observed mechanical properties should be closely approximated by rule of mixture theories. There is still controversy to the boundary between polymer blends, polymer solutions and interpenetrating networks. The work presented in this paper seems to have an interpenetrating morphology even though separate domains are observed.

The original definition of IPN requires both chemical species to be crosslinked, forming "catenanes". However if one component is linear while the other is crosslinked then the final system will only be partially, but selectively crosslinked. The semi-interpenetrating networks, SIPNs, are actually synthesized by sequential addition of a monomer to a polymer.(3) There are two types of SIPNs. The distinction arises from the order of addition. The original polymer is designated I and the additional monomer is designated II. The SIPN of the first kind is obtained when I is crosslinked and II is swollen into it, polymerized, but not crosslinked. The linear polymer is not formally crosslinked. However it will be highly entangled. The final material should be less rigid than a full IPN when used above the glass transition temperature of the linear polymer. It should be tougher since the linear molecules may be free to move and dissipate shock energy. A SIPN of the second kind results in a system of the same chemical nature as the first kind but with a different topography and morphology. This type of SIPN starts with a linear polymer to which a multifunctional monomer is added. The monomer is polymerized and crosslinked in situ.

In the last decade much interest has evolved in the properties and applications of polymer blends. It is often found that mixing two systems together produces a synergistic effect. An example is the addition of rubber particles to epoxies or polystyrenes(4,5,6). A toughened material is obtained if the rubber domain size is chosen correctly (7). As a result of polymer blend technology, materials conventionally regarded as being too brittle for certain applications can now be used. Two underlying principles are emphasized in this work. One is analogous to the toughening mechanism described above, while the second includes integration of the material for biomedical applications.

The thermoset included here is derived from bisphenol-A dicyanate. It can be thermally trimerized yielding a triazine or cyanurate network (8,9,10) as seen in the reaction scheme (Table 1). The critical molecular weight between crosslinks is relatively low, resulting in an extremely tight, brittle network. The material is usually used as a prepeg because a total cure produces a hard, infusible, and insoluble matrix. It possesses excellent adhesive properties and is currently used as a metal coupling agent. It offers many superior properties relative to conventional epoxies derived from bisphenol-A.

Bisphenol-A type epoxies are widely exploited as adhesives, sealants and coatings for a multitude of applications. It is also used as a matrix for fiber composites. One recognizes, though, that when an epoxy reacts there are alcohol moieties produced which significantly complicate future behavior. Improving properties of epoxies in different environments remains a major concern, range from large temperature fluctuations to large humidity fluctuations (5,11,12).

The epoxy matrix, containing multiple hydroxyl and amino groups, has a high tendency to adsorb water, a major cause of material failure in many applications. Finally, when one uses the epoxy as a fiber matrix, there is a distinct probability that voids will appear due to reaction by-products which do not escape during cure. Once the material vitrifies, the mobility of the molecules is reduced, making it difficult for the small molecular weight products to be eliminated.

The triazine thermoset has been developed to combat these situations. Reaction conversions approach 100% when the curing is done stepwise following a temperature cycle. There are few unreacted groups which can complicate material properties over long periods of time. Complications may arise as a result of side reactions if residual unreacted groups are present therefore complete reactions are desired. Another advantage is that during the course of reaction there are no secondary reactive groups produced, only triazine rings are formed, whereas epoxies produce alcohol groups. No reactive center is produced. Finally, it is seen that no volatile molecules are produced in the reaction.

TABLE 1 MATERIAL INFORMATION

MATERIAL	ABBREV	PRODUCER	MW (g/m)	STRUCTURE
polyethyloxazoline	PEOX	Dow	250,000	$-(CH_2CH_2N)_n-$ with $O{=}CCH_2CH_3$ on N
polyvinylpyrrolidone	PVP	Aldrich	360,000	$-(CH_2CH)_n-$ with pyrrolidone ring on CH
bisphenol-A dicyanate	BPADC	UCONN	278	$N{\equiv}C-O-C_6H_4-C(CH_3)_2-C_6H_4-O-C{\equiv}N$

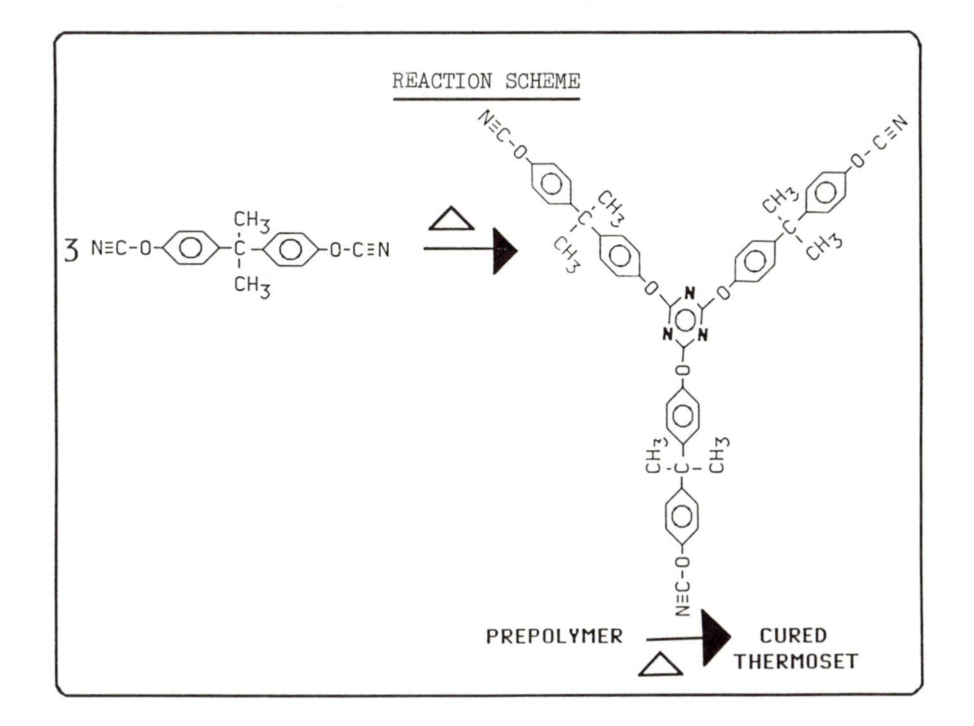

Three triple bonds react to give the aromatic cyanurate ring without any by-products. This is especially enticing for fiber matrices where void formation will be reduced.

The materials developed should also be useful for biomedical applications. Plastics are currently used in the body with a large range of applications, but many problems exist (13). It is imperative to evaluate the biocompatibility of any material brought into contact with the body whether one discusses polymeric drug delivery systems, intracorporeal synthetic prostheses or extracorporeal plastic pumps or tubing(14,15). Assorted polymer formulations have been tested to assess the degree of compatibility but no absolute analysis method has been established. Discussion persists in the literature on whether or not the surface of an artificial component must be smooth to maximize biocompatibility or rough to enhance polymer-tissue interactions(14). A smooth surface is believed to help blood components pass by without a great deal of friction which can induce thrombosis and hemostasis. This is somewhat of an ideal picture since a protein layer is deposited over the plastic soon after contact with the tissue. The healing process stabilizes the prosthesis in its position and segregates it from bodily fluids. However, there is no sound evidence that the protein layer is smooth. One thought is that the protein layer follows the contours of the substrate, resulting in a smooth surface (16). Tissue growth is quite extensive after the fibrin layer is deposited on a synthetic component. Others believe that a rough surface is needed which can interact with the tissue a great deal more(14). A better bond is realized if the pore size and porosity are chosen correctly. The rough surface will, however, induce some variations in blood flow leading to hemostasis (17). There is sufficient data to promote each ideology but there are many inherent complications in defining rough VS. smooth surfaces(15).

The technique used to determine the surface characteristics is critical but again no absolute method or standard exists. Much also depends on the interaction scale which is a function of the biological particle of interest. The blood components range in size from small molecules to protein molecules and larger blood cells. These can be selectively separated by choosing the correct pore size. Some species will preferentially interact with the prosthesis due to its dimensions. The extent of interaction between various components is measured using several methods. X-ray photoelectron spectroscopy and scanning electron microscopy are used to assess the degree of adhesion between the biological species and the plastic component generally providing qualitative results(13,15,18,19). These methods are often used to decide whether or not a material is biocompatible, which remains the primary concern when introducing a polymer to the body. Once a polymer is recognized as being biocompatible it still does not assure the system will be applicable for a particular application. Adhesion between the two components is very important.

Biocompatible acrylic cements have been widely employed to combat slippage or complete failure between the tissue and the polymer(18). These glues are said to be safe in specified amounts but the monomers are toxic and if excessive quantities are used over a period of time many undesirable side effects may arise. The system developed here may help to minimize many of the current problems. The composite discussed here may improve tissue-polymer stability. This is the main factor for choosing the thermoplastics. Polyethyloxazoline and polyvinylpyrrolidone are non-toxic up to relatively high doses. The former is an experimental polymer made by Dow Corporation. The glass transition temperature (Tg), is 68°C, but it remains very tough below this temperature. It is impossible to fracture even when cooled by liquid nitrogen. Polyvinylpyrrolidone is a polymer accepted by the FDA to be used in the body at moderate molecular weights (<20,000g/m)(15,18,20). It is biocompatible but not biodegradable. Molecular weights higher than this are used but it is known that the body cannot dispose of these as with lower molecular weight polyvinylpyrrolidone polymers[19]. The high molecular weight fractions may be stored in the body indefinitely. In any case, this polymer has found many applications in polymeric drug release matrices and blood plasma substitutes. The reported glass transition temperature varies greatly, ranging from 54°C to 185°C(21). A DSC thermogram shows the Tg at 67°C, but as will be shown later this value changes. The polymer is extremely hygroscopic and difficult to dry completely. Small amounts of water significantly influence the glass transition temperature. In this study, the thermoplastics are dispersed with the bisphenol-A dicyanate (BPADC), then the monomer is cured. A SIPN of the second kind is obtained.

Experimental

Bisphenol-A dicyanate (BPADC), was synthesized using bisphenol-A, cyanogen bromide and triethylamine following previous procedures(9,22). It was purified by successive precipitations from acetone with water, mp. 78-80°C. The compound cannot be recrystallized since it tends to react at relatively low temperatures while in solution. Gel permeation chromatographs (GPC), showed that a high degree of purity (98%) was attained. Polyethyloxazoline (PEOX), was used as obtained from Dow Chemical. The clear yellow material had a Tg at 68°C seen in the thermogram, Figure 1, run under nitrogen at 20°C/min. The molecular weight was reported to be approximately 250,000 g/m. Polyvinylpyrrolidone (PVP), was used as obtained from Aldrich Chemical. The molecular weight was reported to be 360,000 g/m having a Tg at 67°C under the same conditions, Figure 2. The molecular weight distributions of PEOX and PVP, relative to poly(ethyleneoxide) standards, were shown in Figures 3 & 4 respectively, using a Waters Aqueous GPC 150-C with a 0.1 N $NaNO_3$ aqueous carrier solution. 100 µl samples of 5 mg/ml solutions were injected and sent through the columns with a 1 ml/min flow rate at 30°C.

The samples were prepared follows. Solutions were made using methylene chloride and a 1:3 polymer to monomer weight ratio. A 2:1 volume to weight ratio of solvent to solids was used. These were mixed for several hours to attain homogeneity. The polyethyloxazoline/bisphenol-A dicyanate solution (I) was clear, indicating high compatibility. The polyvinylpyrrolidone/bisphenol-A dicyanate solution (II) was slightly turbid. Next the solutions were thoroughly dried under vacuum for one day. Zinc stearate, in the

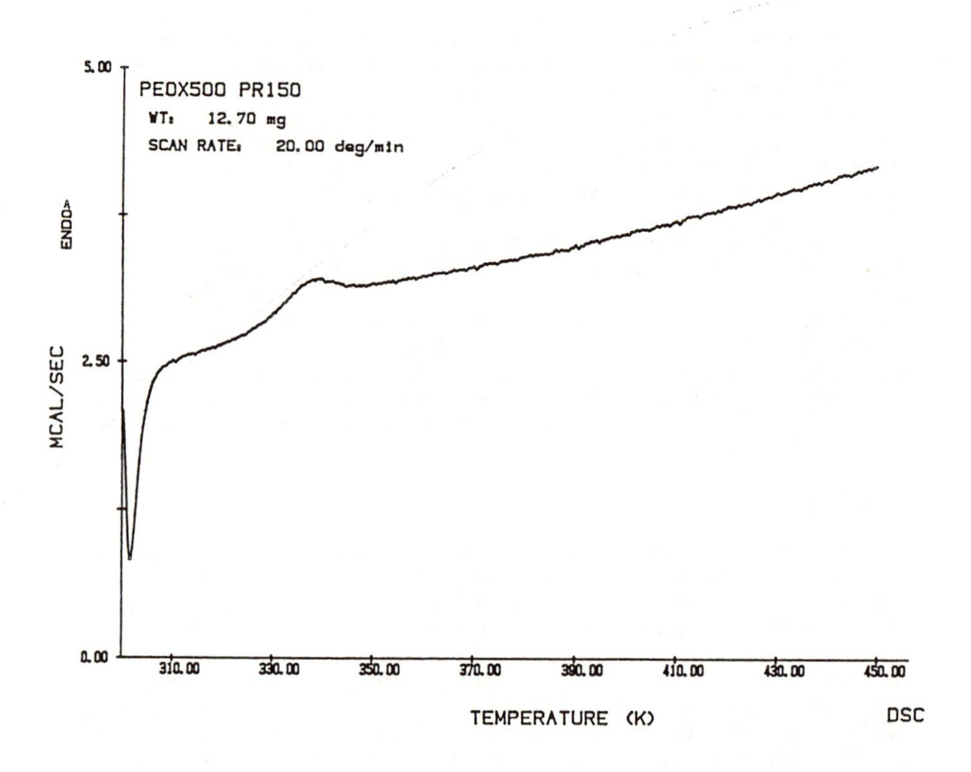

Figure 1. DSC thermogram of PEOX hot pressed at 150°C

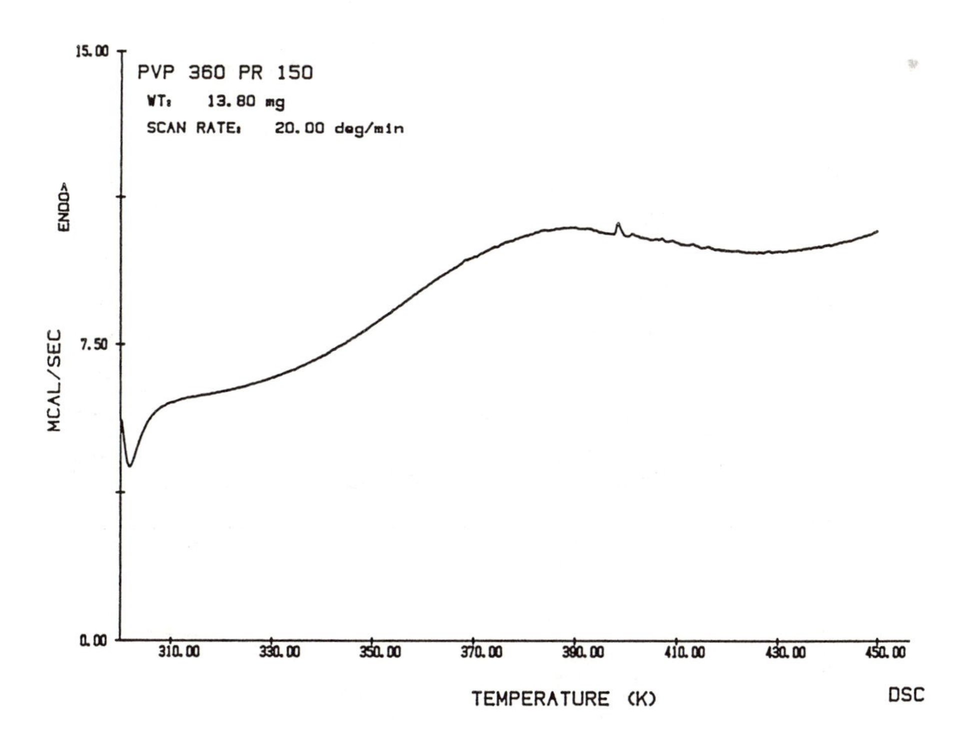

Figure 2. DSC thermogram of PVP hot pressed at 150°C

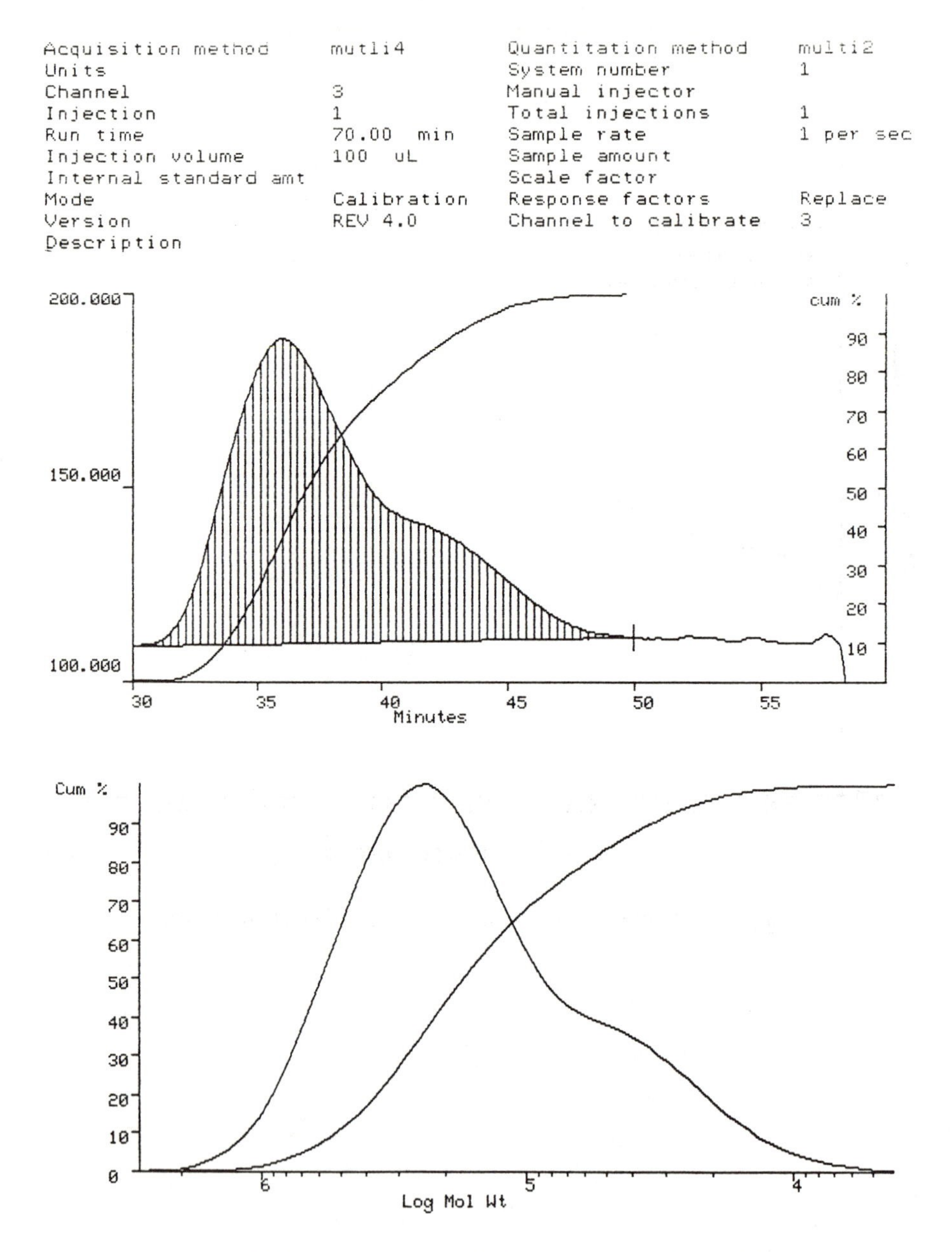

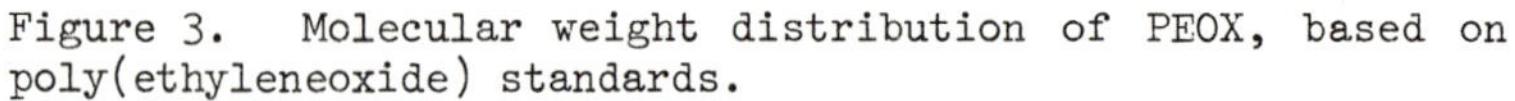

Figure 3. Molecular weight distribution of PEOX, based on poly(ethyleneoxide) standards.

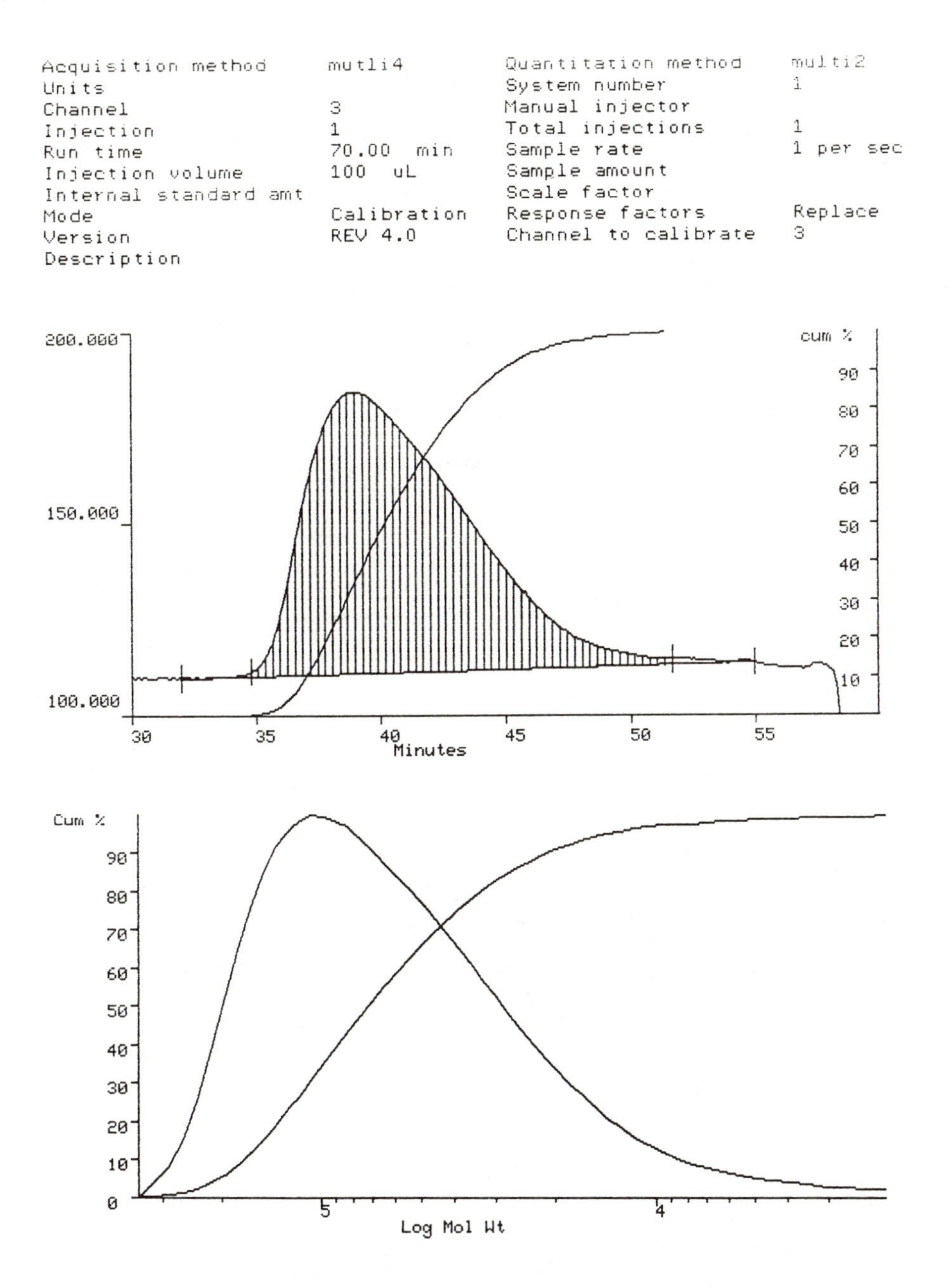

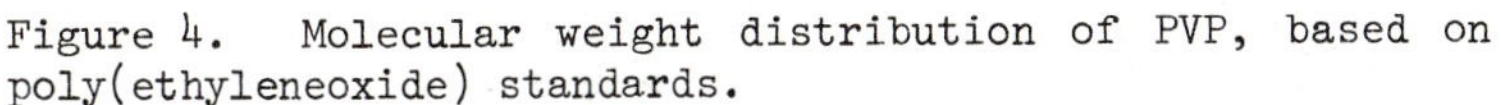
Figure 4. Molecular weight distribution of PVP, based on poly(ethyleneoxide) standards.

amount of two mole percent BPADC, was dispersed in the dry mixture using a Waring blender. It catalyzed the reaction, reducing the reaction temperature by 150°C.

A cure cycle was followed starting at 120°C for 20 minutes, 35 minutes heating from 120°C to 200°C and fifteen minutes at 200°C as a post cure. All reactions were carried out in a hot press at 500 psi using Teflon mold. After the samples were removed they were sectioned. Some of the sections were directly prepared for SEM studies and some were put in a modified Soxhlet extractor using distilled water as the solvent. Etching was continued for 48, 72, and 100 hours then dried at room temperature under vacuum. The resulting systems were characterized using infrared spectroscopy (FTIR), differential scanning calorimetry (DSC), dynamic mechanic thermal analysis (DMTA), and scanning electron mecroscopy (SEM). A Nicolet 60SX FTIR was used with KBr pellets in transmission mode. A Perkin-Elmer Differential Scanning Calorimeter 2, incorporated with a Perkin-Elmer 3600 Data Station, was run under nitrogen. The scanning rates were included on the DSC thermograms. The Dynamic Mechanical Thermal Analysis was acquired on a DuPont Instruments 981 Dynamic Mechanical Analyzer run at 5°C/min under nitrogen between -130°C and 30°C and under air above 30°C. Electron Micrographs were taken with an Amray model 1000A Scanning Electron Microscope using 30 kV. The magnification varies from 50X to 5000X and micron calibration bars were placed on each micrograph.

Results And Discussion

The two systems analyzed were compared and contrasted on many accounts. The infrared spectra of Figures 5 & 6 showed the unreacted materials on top and the reacted species below. The strong absorption band at 2200 cm-1 was attributed to the cyanate group and disappeared after reaction. The new absorption bands at 1370 cm-1 and 1570 cm-1 were due to the triazine ring. The percent conversion was monitored easily in this way. In both cases it was very high, almost 100% for I. Other important information extracted from the spectra was that no significant cross-reaction occurred. It was required that only physical interaction between the separate species should occur if a true SIPN was to be obtained. No chemical interchange should be present which would be more indicative of grafted materials. The finger print region in the spectra showed no shift in absorbances which suggested uniform reactions.

The DSC thermograms were used to follow the reaction scheme. The extent of reaction was also assessed using this method. Running the DSC of the final products showed that system I had no detectable residual reaction, while system II has about 5% post reaction, Figure 7. The role of the catalyst was appreciated when comparing uncatalyzed reactions, Figures 8, 9 & 10, to catalyzed reactions, Figures 11 & 12. No appreciable reaction was seen below 150-160°C without the catalyst. The maximum exotherms appeared about fifty degrees higher than temperatures indicating the onset of reaction. The reaction was not complete until 290°C. It proceeded smoothly without the catalyst, but when two mole percent zinc was added, a drastic reduction in reaction temperature occurred. The mechanism of transition metal coordination had been

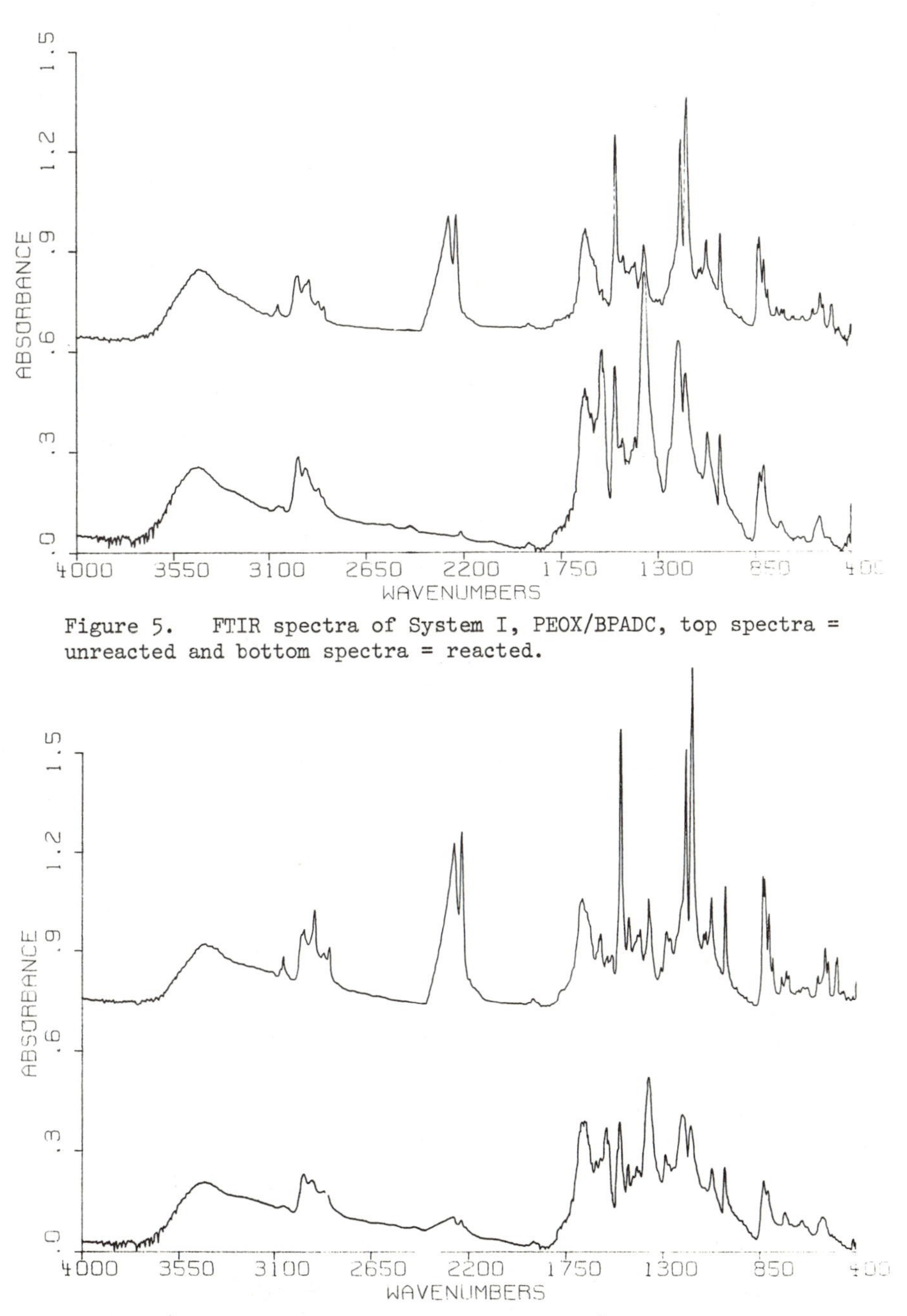

Figure 5. FTIR spectra of System I, PEOX/BPADC, top spectra = unreacted and bottom spectra = reacted.

Figure 6. FTIR spectra of System II, PVP/BPADC, top spectra = unreacted and bottom spectra = reacted.

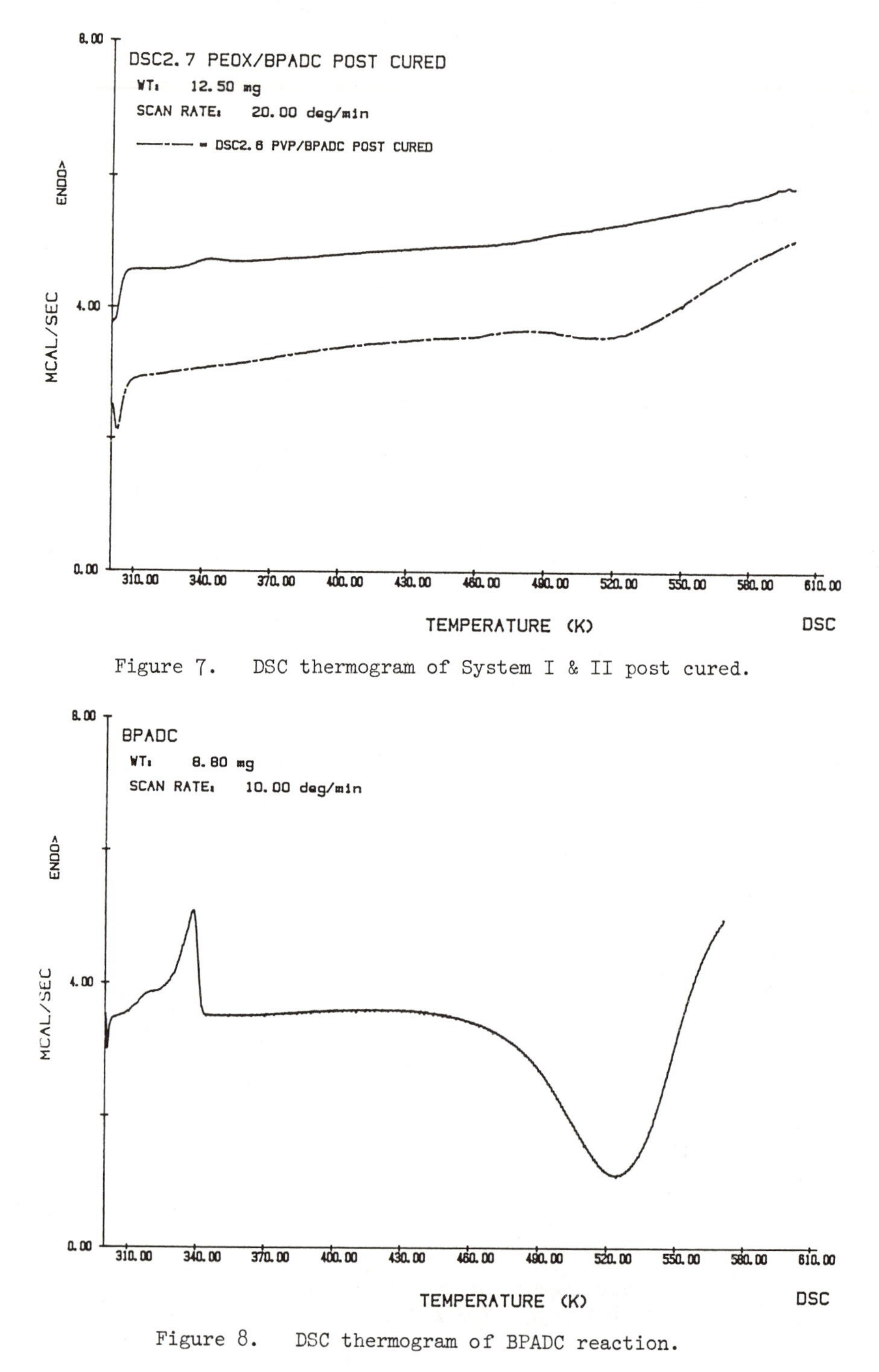

Figure 7. DSC thermogram of System I & II post cured.

Figure 8. DSC thermogram of BPADC reaction.

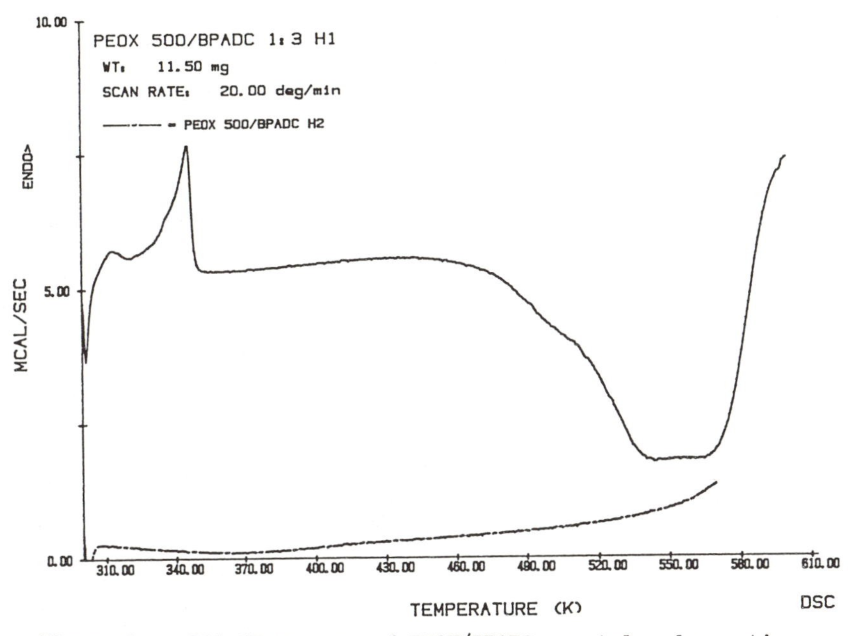

Figure 9. DSC thermogram of PEOX/BPADC uncatalyzed reaction, first and second heating.

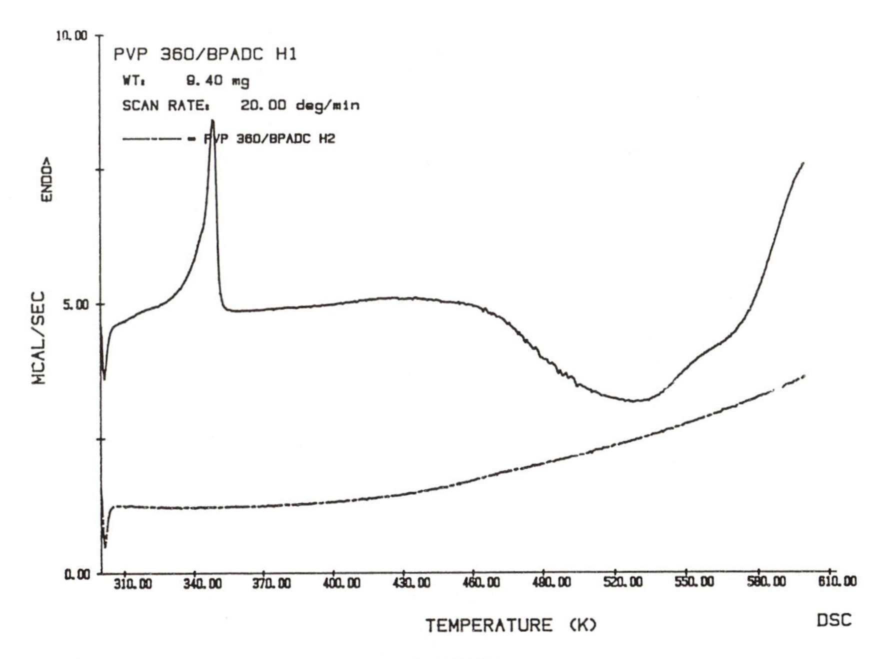

Figure 10. DSC thermogram of PVP/BPADC uncatalyzed reaction, first and second heating.

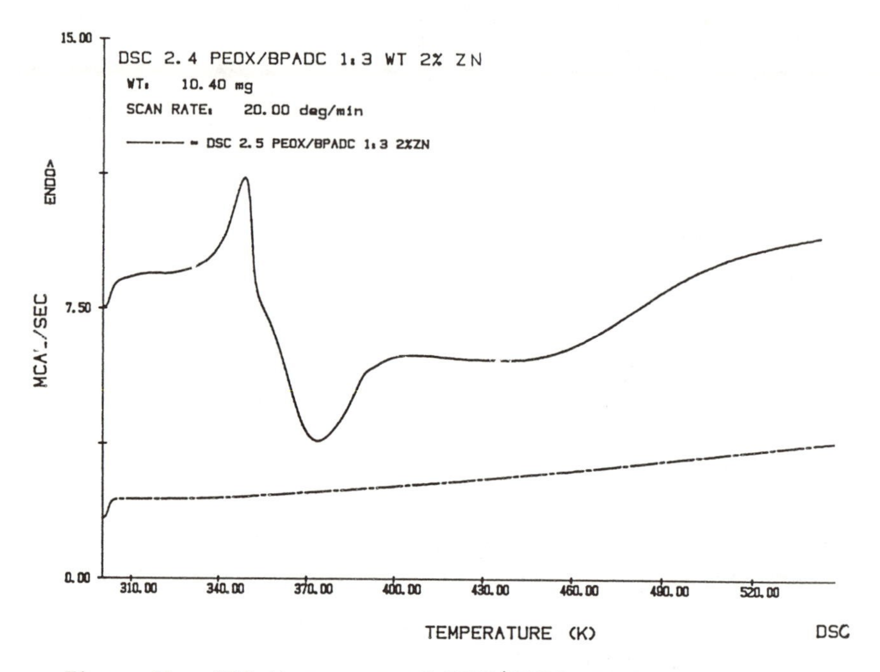

Figure 11. DSC thermogram of PEOX/BPADC catalyzed reaction, first and second heating.

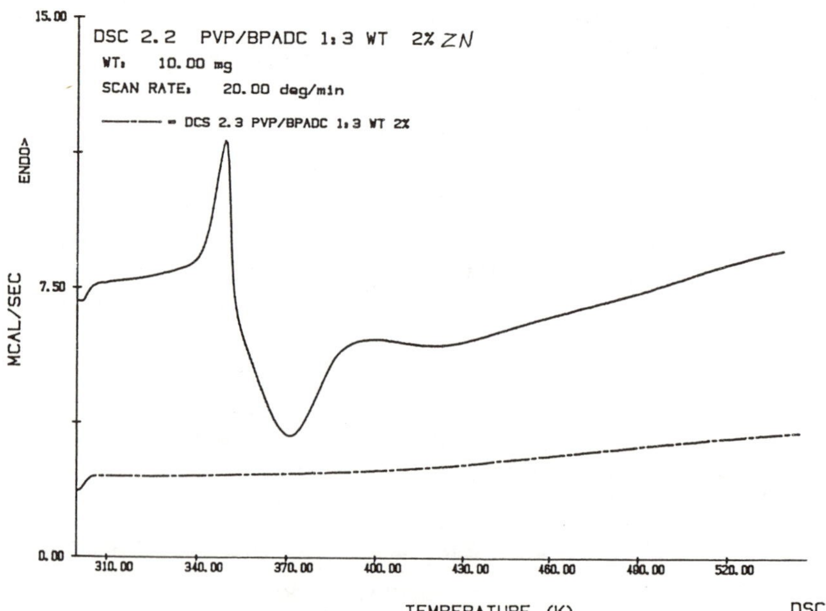

Figure 12. DSC thermogram of PVP/BPADC catalyzed reaction, first and second heating.

discussed previously and there was one important feature emphasized(9,10). The rate limiting step of the reaction was a dimerization of the cyanate groups. The third unit reacted very quickly producing the triazine ring. The metal enhanced the dimerization process thus decreasing the activation energy of the reaction. The DSC thermograms showed the catalyzed reactions beginning immediately after the monomer melted. The maximum exotherm appeared about 150°C lower for both systems. The scheme seemed more complex compared to the uncatalyzed reactions. One thought was that a gel point was quickly reached slowing down the reaction. Another sixty degrees were required to increase the reaction rate again. A second idea to explain the shape of the thermograms was that the heat evolved in the reactions was used to melt the zinc stearate, which melted at 130°C. The melting point of the catalyst may decrease since it was mixed with the other components. This may be the reason why the DSC thermograms were similarly shaped. In both cases, though, the reactions were complete by 200°C

The choice of incorporating a catalyst was made because the stabilities of the thermoplastics at the elevated temperatures was not well defined. The neat triazine thermoset should be post cured at 280-300°C to assure conversions approaching one hundred percent. Thermal deradation of the thermoplastics was a problem. The reaction products in earlier experiments became dark at elevated temperatures, indicating thermal degradation of the reaction media. The decomposition of the thermoplastics was monitored by its texture and color. When put in an oven at 230°C for 20 minutes, they become brown, indicating thermal and oxidative breakdown. Cross reactions were promoted which would drastically influence the chemical and physical properties of the SIPN(3). The resulting properties would be unpredictable and unreproducible. One of the major advantages of using SIPN technology was that the individual components were well characterized. Any decomposition of the thermoplastic during reaction would offset the advantages. A catalyst offered the capability to use lower reaction temperatures while producing uniform results.

One subtle comparison between the thermograms was that in both systems the reaction scheme seemed to be identical. The DSC thermograms showed similar behavior during network formation. The thermoplastic did not seem to alter the process. The reaction mechanism was not dependent on the linear polymer incorporated. This reinforced the notion that a true SIPN was attained in this experiment.

The dynamic mechanical thermal analysis also indicated molecular mixing of the two components. The Tg of the thermoplastic was lower than the composite, as determined from the inflection point of the E' curve seen in Figures 13 & 14. The PEOX glass transition temperature appeared at 62°C, while that for PVP showed up at 145°C. The discrepancy in the Tg of PVP had already been addressed. The Tg extrapolated from the DMTA thermogram was reported since no artifacts were to be introduced. The samples exemplified for all DMTA thermograms were processed in a normalized fashion. Any thermal or hygroscopic influences were eliminated. Some water may be trapped in the neat thermoplastics but this was minimized since they are hot pressed at 150°C and stored under calcium

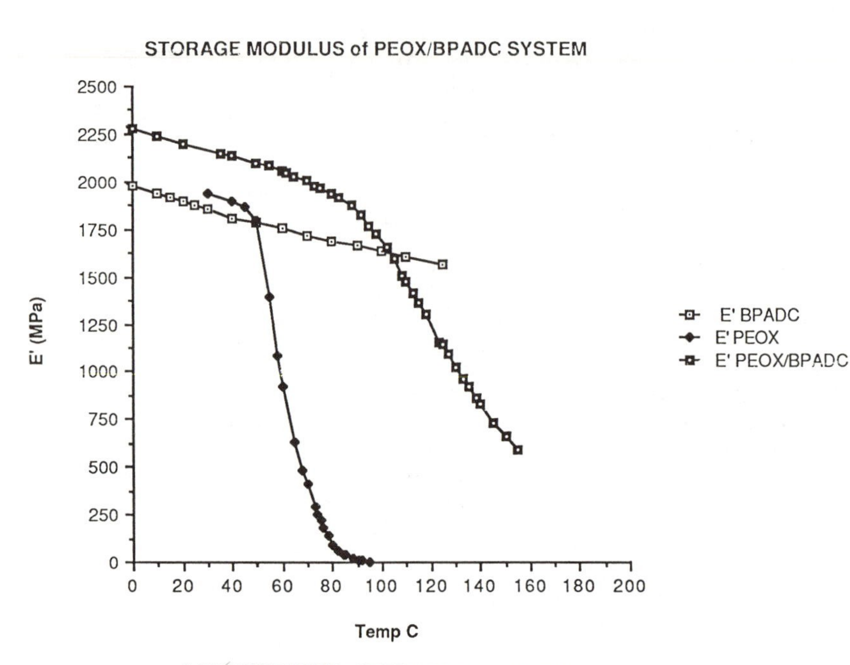

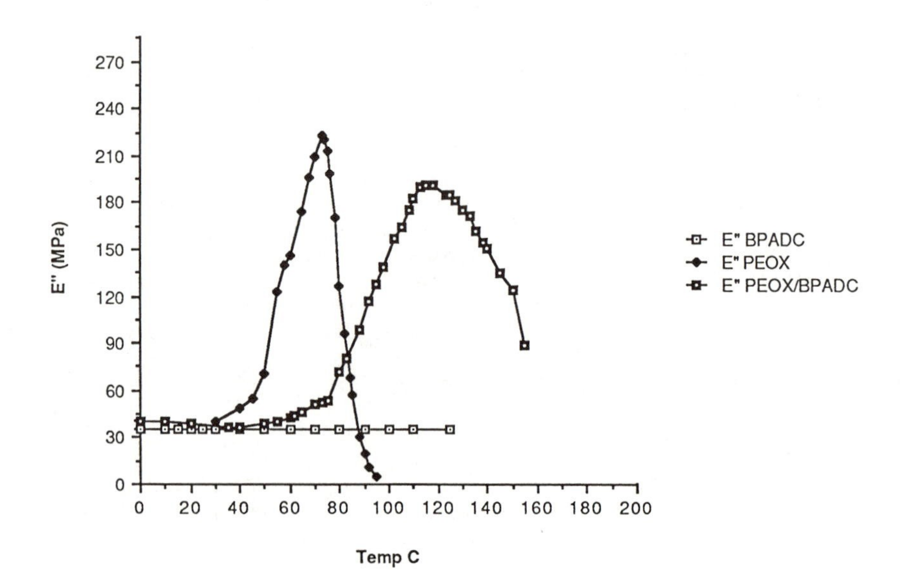

Figure 13. DMTA thermogram of System I run at 5°C/min

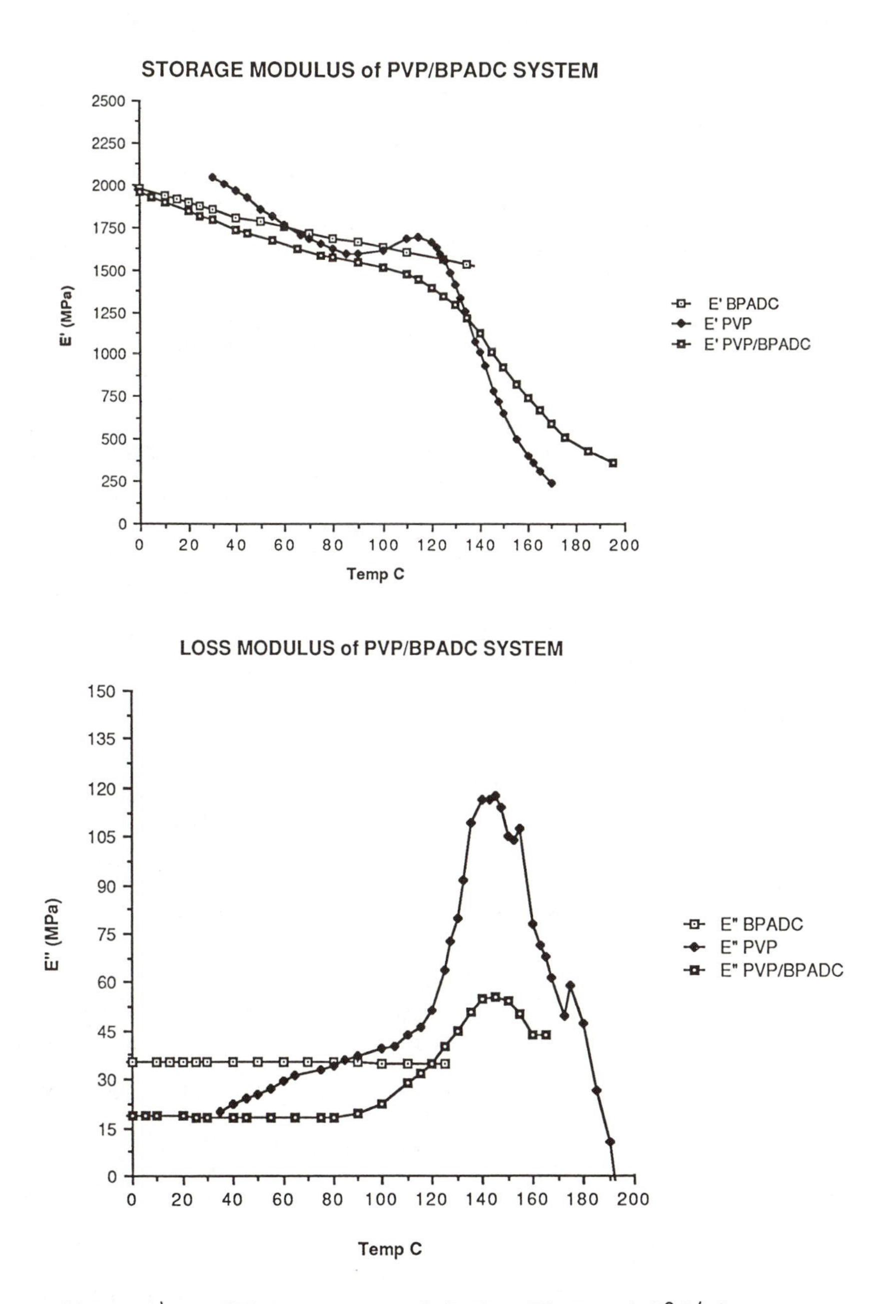

Figure 14. DMTA thermogram of System II run at 5°C/min

sulfate. All samples were stored together. The ambiguity in the Tg of PVP was evidenced by a sample which was thermally treated under vacuum. The Tg approached the limiting value of 180°C. The original sample was included, however, to normalize the results. The Tg of the neat thermoset was undetected. The triazine network showed a small damping peak between -60° to -80°C. This was attributed to a beta transition due to rotation of the bisphenol-A unit. Other workers had reported glass transition points for the cured material at 245°C. The conversions of those reactions were lower than encountered here. These results may indicate mobility of branch points ending with unreacted cyanates.

Usually an indication of molecular mixing can be theoretically calculated based on composite Tg's. When the glass transition temperature of the SIPN was between that of the neat components, molecular interactions were assumed. A definite increase in Tg resulted in these materials. No calculated Tg can be compared since the thermoset did not show a Tg. It was interesting to note that the beta transition of the triazine network was unaffected, but the Tg for the composite was much higher relative to the thermoplastic alone. System II showed a 20°C increase in Tg, while I showed a large increase, changing from 62°C to 125°C.

One would estimate the degree of mixing in I to be greater than in II. Other evidence supporting this idea was observed in the behavior of the original solutions. As described earlier, I was clear while II was slightly turbid. Some difference between the two systems was established on this account but further comparisons observed in the DMTA thermograms indicated that the results were somewhat independent of the thermoplastic. The broadness of the resulting Tg was about the same. The spread as well as the actual shift of the glass transition temperature was critical in depicting molecular interactions(3). The glass transition range for system I was between 73°C and 155°C and for system II was between 102°C and 175°C as extrapolated from the E" curves. The absolute difference was 82°C and 73°C respectively. It seemed that thermoplastic independent processes were present, causing the broadness to be the same for both systems, however unique behavior for each was sustained as exemplified by the change in the glass transition temperature.

The SEM micrographs also indicated the behavior discussed above. The unetched samples, exemplified in Figures 15 & 16, started out relatively flat. Some fluctuation in the topology was seen but this was due to the cure conditions. The mold caused the striations in the surface. Samples etched for 48 and 72 hours show small changes but those etched for 100 hours provided best results. Once the sample was etched for 100 hours many changes in the surface structure arose. Pores had been seen which were occupied by the thermoplastic. System I produced pore sizes, about 0.5 μm in diameter as shown in Figure 17. System II, seen in Figure 18, produced pore sizes on the order of 5 μm in diameter. This too was indicative that the intermolecular mixing was greater in I than in II. The overall development of the SIPN showed great promise in the area of thermoset toughening. The pore size of the thermoplastic was close to the effective size used for rubber toughened epoxies(5). In the SIPN studies, it seemed that a portion of the

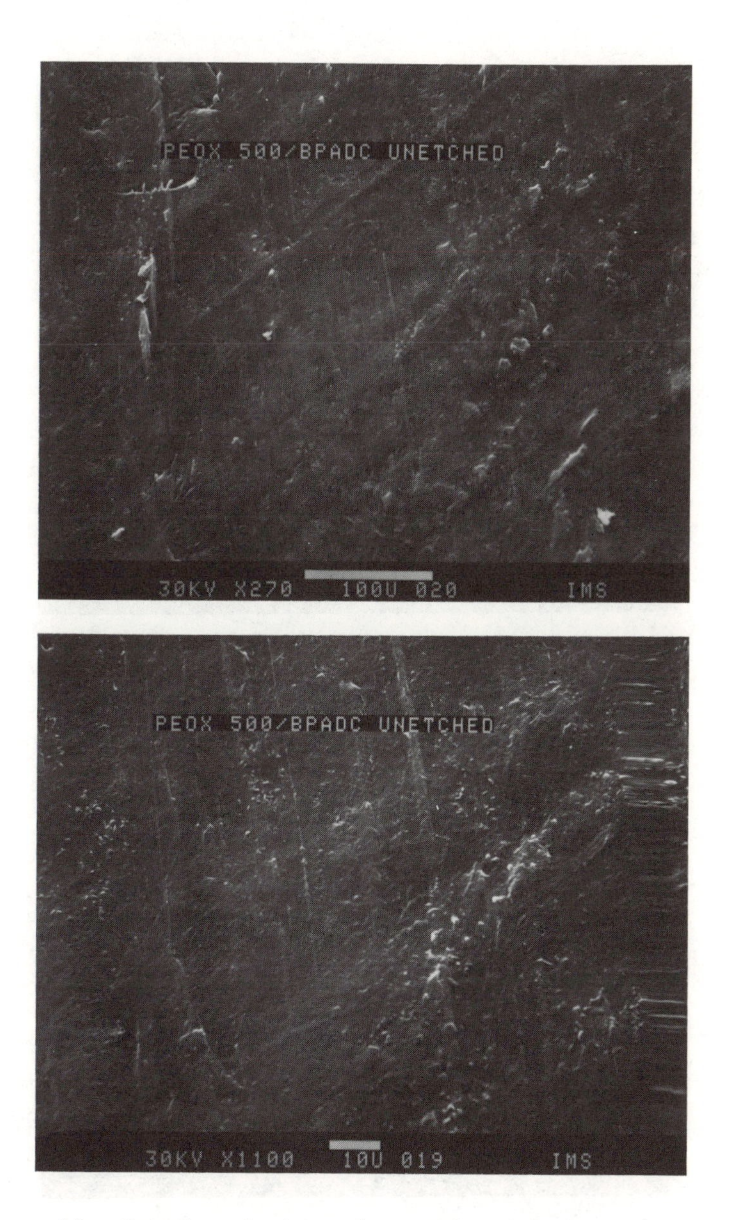

Figure 15. Scanning electron micrographs of System I unetched.

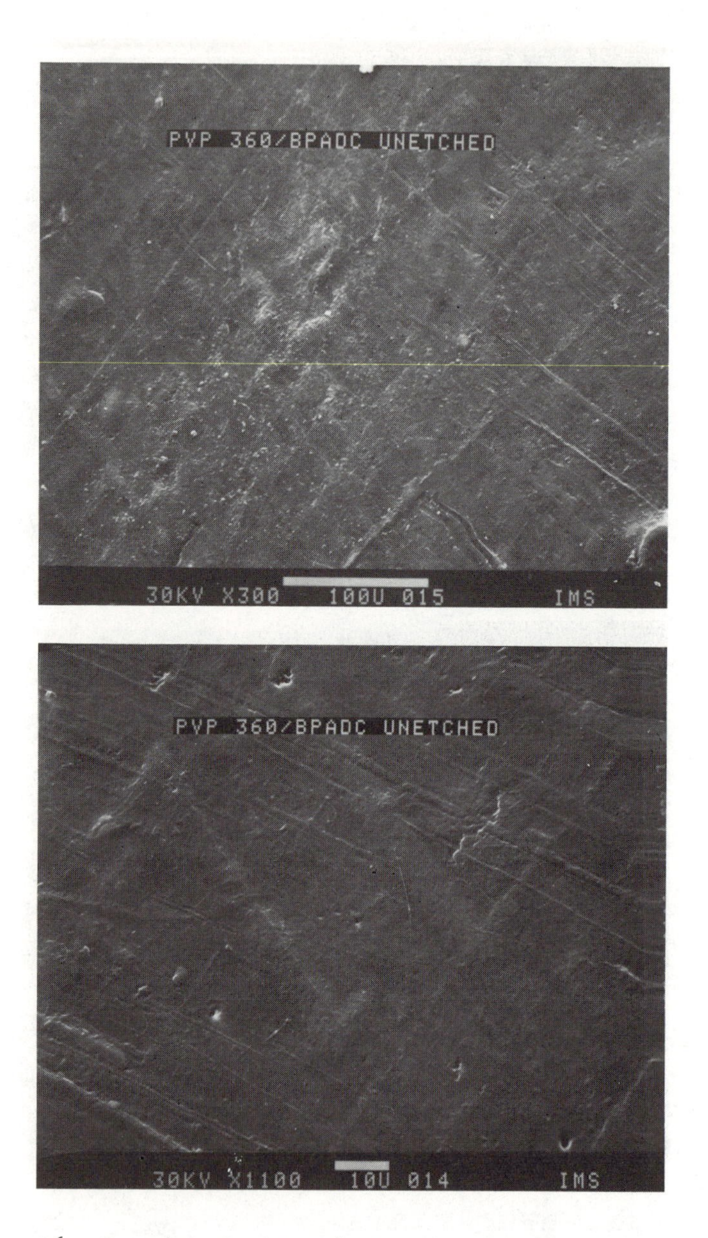

Figure 16. Scanning electron micrographs of System II unetched.

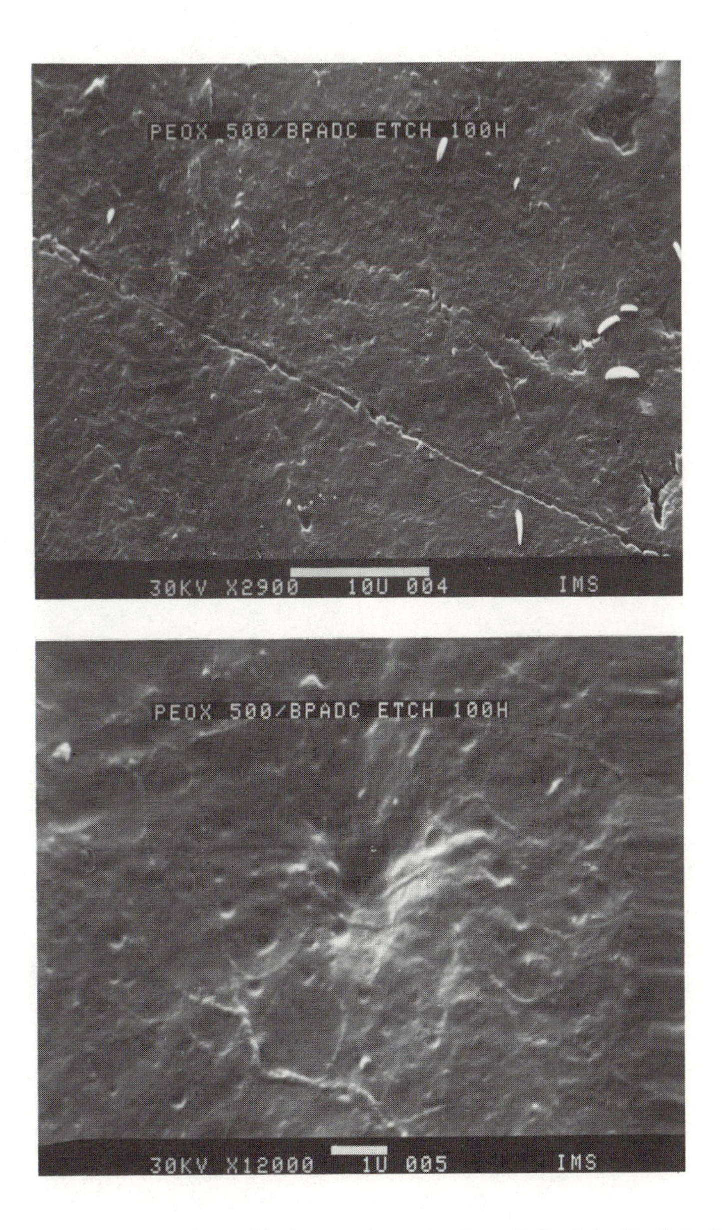

Figure 17. Scanning electron micrographs of System I etched.

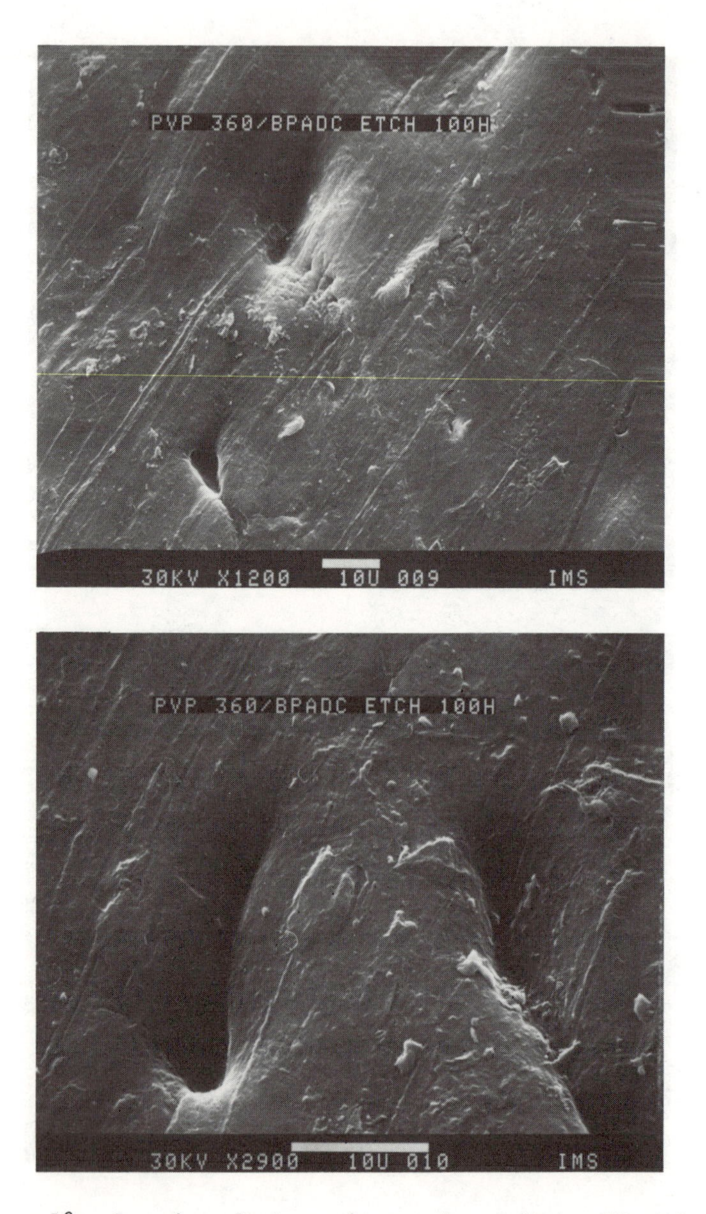

Figure 18. Scanning electron micrographs of System II etched.

thermoplastic was well entangled in the thermoset, but a portion remained as a homogeneous pore. Most of the thermoplastic remained in the triazine network. After etching for one hundred hours no detectable weight reduction was measured.

Biomedical Implications

Enough of the surface is altered to promote the system for biomedical applications. Many thermoplastics used in the body as replacement implants show adhesive failure or slippage between the plastic and the tissue(14,17). Often new surgery is done to rectify the situation by using biomedical glues or replacing the part. This not a desirable situation and it is better if the implant could sustain its properties for a longer time. The bonding strength between the tissue and the synthetic material could be increased either chemically or physically. A new chemical modification of tissue is not preferred and would take a long time to be approved. Many unforeseen side reactions may be expected. If the bond strength could be increased physically, then the tissue would remain unaltered. Plastic prostheses could be formulated from a SIPN as described here. The pore sizes established here are relatively small compared to the size of cellular components in blood (14,15). This surface seems to be on the border between smooth and rough surface domains described earlier. The topology created would not be thought to inhibit blood flow, since the pore sizes are an order of magnitude smaller than the flowing blood cells. Subsequently, the fibrinogen protein layer should deposit since it develops on a molecular level, but the cells would recognize the polymer surface as being approximately smooth. The material is hoped to be competitively thrombosis-resistant. Removing a portion of the thermoplastic leaves sites which can be filled with tissue over a period of time. There is much greater surface area contact between the tissue and the plastic component. As the tissue grows into the network, the physical bonding should be geometrically increased. Less adhesive failure is to be expected. Another advantage is that starting materials can be used which are already accepted for use in the body.

Conclusion

SIPNs can be made from dicyanates and N-alkylamide thermoplastics. The materials should have close solubility parameters or cohesive energy densities to improve compatibility. It seems the production of these systems is not strongly dependent on the thermoplastic or the dicyanate. A broad distribution of interpenetrating networks can be envisioned. The reaction mechanism does not appear to be a function of the dicyanate backbone. Previous work on model compounds and work involving different backbones has revealed the same reaction mechanism. When the thermoplastic is introduced to the thermoset, resulting properties acquire the advantages of the individual components. Various applications require some characteristics of both types. Obtaining the strength and rigidity of thermosets, along with the toughening properties of thermoplastics used at or above Tg, is enticing for several engineering applications. One envisions the ability to tailor specific pairs of materials together, removing one thermoplastic and introducing a second with pressure. Limitless combinations can be assumed. Finally if a very thin film is made using this SIPN technology, the thermoplastic can be removed leaving a thrmally and hydrolytically stable membrane. Further development of semi-interpenetrating networks for engineering and biomedical applications is in progress.

Literature Cited

1. Frisch, K. C. et al., "Topologically Interpenetrating Polymer Networks", Recent Advances in Polymer Blends, Grafts and Blocks, ed. by L. H. Sperling, Plenum Press, N.Y. 1978.
2. Polymer Blends, ed. by D. R. Paul and S. Newman, Academic Press, N.Y., 1978.
3. Sperling, L. H., Interpenetrating Polymer Networks and Related Networks, John Wiley & Sons, N.Y. 1981.
4. Odian, G., Principles of Polymerization, John Wiley & Sons, N.Y. 1981.
5. Bauer, R. S. ed., Epoxy Resin Chemistry II, ACS Symposium Series 221, 1982.
6. Sardelis, K. et al., Polymer, 28, 2, 244 (1987).
7. Vazquez, A. et al., Polymer, 28, 7, 1156 (1987).
8. Wertz, D. H., Plastics Engineering, V. 40, no. 4, 31 (1984).
9. Shimp, D. A., Proc. ACS of PMSE, V. 54, 107, New York, 1986.
10. Cercena, J. L.; Huang, S. J., Polymer Preprints, 25, 2, 320 (1983).
11. Int. Symposium on Adhesives, Sealants and Coatings for Space and Harsh Environments, Proc. ACS of PMSE, V. 56, Denver, 1987.
12. Antoon, M. K. et al., JPS Polym. Phys. ed., V. 19, 1567 (1981).
13. Biomaterials: Interfacial Phenomena and Applications, ed. by S. L. Cooper and N. A. Peppas, ACS Adv. in Chemistry Series 199, Wash. D.C., 1982.
14. Biocompatible Polymers, Metals, and Composites, ed. M. Szycher, Technomic Publ. Co., Lancaster, PA 1983.
15. Cooper, S. L.; Lelah, M. D., Polyurethanes in Medicine, CRC Press, Boca Raton, FLA 1986.
16. Hoffman, A. S., Polymeric Materials and Artificial Organs, p. 13, ed. C. G. Gebelein, ACS Symposium Series 256 Seattle, WA, March 1983.
17. Akeson, W. H., Biomaterials, p. 175, ed. A. L. Bennett Jr., Univ. of Wash. Press, Seattle, WA 1971.
18. Polymers in Medicine, ed. K. Dusek, Springer-Verlag Press, N.Y. 1984.
19. Nojima, Kazuhiko, Polymer, 28, 6, 1017 (1987).
20. Plate, N. A.; Vasil'Ev, A. Y. dd, Polymer Sci. USSR, V. 24 no. 4 743 (1982).
21. Turner, D. T., Polymer, V. 26, 757 (1985).
22. Grigat and Putter, Chem. Ber., 97, 3012 (1964).

RECEIVED November 17, 1987

Chapter 19

Development of Multiphase Morphology in Sequential Interpenetrating Polymer Networks

J. H. An and L. H. Sperling

Polymer Science and Engineering Program, Department of Chemical Engineering, Materials Research Center Number 32, Lehigh University, Bethlehem, PA 18015

Since the start of modern interpenetrating polymer network (IPN) research in the late sixties, the features of their two-phased morphologies, such as the size, shape, and dual phase continuity have been a central subject. Research in the 1970's focused on the effect of chemical and physical properties on the morphology, as well as the development of new synthetic techniques. More recently, studies on the detailed processes of domain formation with the aid of new neutron scattering techniques and phase diagram concepts has attracted much attention. The best evidence points to the development first of domains via a nucleation and growth mechanism, followed by a modified spinodal decomposition mechanism. This paper will review recent morphological studies on IPN's and related materials.

Multicomponent polymer materials are defined as mixtures of two or more structurally different polymeric species, such as polymer blends, blocks, grafts, AB-crosslinked polymers, or interpenetrating polymer networks, IPN's. They can be classified into two groups in terms of type of bonding. Block copolymers, graft copolymers, and AB-crosslinked copolymers contain intermolecular covalent bonds which hold the different species together. On the other hand, polymer blends and IPN's do not have such bonds. However, polymer blends are theoretically separable, but the IPN's, because of interchain crosslinking, are locked together physically.

Most multicomponent systems undergo phase separation because of their positive mixing enthalpies coupled with low entropy of mixing. Morphological features have been central to the study of multicomponent systems, because domain sizes, shapes, and interfacial bonding characteristics determine the mechanical properties. A proper understanding of these features often allow synergistic behavior to be developed.

0097-6156/88/0367-0269$07.75/0

Interpenetrating polymer networks are defined in their broadest sense as an intimate mixture of two or more polymers in network form [1,2]. Ideally, they can be synthesized by either swelling the first crosslinked polymer with the second monomer and crosslinker, followed by in-situ polymerization of the second component (sequential IPN's) or by reacting a pair of monomers and crosslinkers at the same time through different, non-interfering reaction mechanisms, simultaneous interpenetrating networks, SIN's. In fact, many variations of these ideas exist in both the scientific and the patent literature. In any case, at least one of the two components must have a network structure, as an IPN prerequisite.*

In their synthetic route, IPN's are characterized by:

(1) The presence of crosslinks, which, especially in both polymers, limits domain sizes.
(2) Polymerization is accompanied simultaneously by phase separation, requiring diffusion through increasingly viscous media to form their domains.
(3) The final product is often thermoset, forbidding flow or dissolution.

These phenomena cannot be treated independently. Consequently, the morphology of IPN's is often at a quasi-equilibrium state determined by a balance among the several kinetic factors [3]. Therefore, in order to understand the domain formation process in IPN's, we should take into consideration the route taken to the final morphology as well as the chemical and physical properties of each constituent.

In many of the early studies of IPN's, interest centered on the characterization of morphology and thermo-mechanical properties of fully polymerized IPN's [4-7], particularly their interrelationships among composition, crosslinking level, and temperature. Recently, scattering techniques such as small-angle x-ray scattering (SAXS) [8-10] and small-angle neutron scattering (SANS) [11-13] have been applied to the study of the morphological features of IPN's, especially during polymerization [3].

This review will examine the morphological features of sequential IPN's, starting with transmission electron microscopy (TEM) and modulus, and continuing on with SAXS, and SANS. Dual phase continuity in IPN's will be explored, with emphasis placed on spinodal decomposition in IPN systems.

Morphology of Sequential IPN's via Electron Microscopy

Of direct interest to the investigator is the prediction and control of the domain size of phase separated materials. While the first synthesis of IPN's in this laboratory (1969) [14] lacked microscopy, those first experiments virtually cried out for such analysis. The first electron microscopy study of *cross*-polybutadiene-*inter*-*cross*-polystyrene, PB/PS, sequential IPN's was carried out by Curtius et al. [4]. Using samples stained with

*This discussion omits the thermoplastic IPN's, which have physical crosslinks.

osmium tetroxide, TEM studies revealed an irregular cellular structure of a few hundred Angstroms in diameter with the first polymerized component, PB, making up the cell walls. The size of the cells was found to decrease with increasing crosslink density in the PB networks. The impact resistance of those IPN's, Table I, was significantly higher than commercial HIPS of comparable composition, reflecting their finer domain structure and dual phase continuity.

Table I. Comparison of impact resistance of cross-polybutadiene-inter-cross-polystyrene IPN's with HIPS

Material	Thickness of sample, in.	Styrene content	Impact resistance, ft lb/in
PS	0.166	100	0.34
HIPS	0.252	∿ 90	1.29
IPN	0.177	85.6	5.06

SOURCE: Reproduced with permission from ref. 4. Copyright 1972 *Polymer Engineering and Science*.)

More elaborate electron microscopy studies were done on cross-poly(ethyl acrylate)-inter-cross-poly(styrene-stat-methyl methacrylate) by Huelck et al. in 1972 [5,6]. In this experiment, the effect of miscibility between two networks was investigated by changing the ratio of the styrene and MMA systematically in network II. Figure 1 shows typical transmission electron micrographs. Figure 1a shows a characteristic cellular morphology of about 1000 Å simultaneously with a fine structure of the order of 100 Å. The cellular structures decreased in size and were replaced by the fine structures with increasing MMA-mer content, which makes the system closer to semimiscible pair, Figure 1b. At that time, it was thought that the cellular structures were spheroidal in nature. It was not understood that the domains might actually be cylindrical or more likely, worm-like, but rather the cellular structure was interpreted in terms of spherical, dispersed domains.

The influence of the crosslinking level in networks I and II on the domain size was investigated in IPN's based on SBR and polystyrene in 1976 by Donatelli et al., Figure 2 [15,16]. In general, the following conclusions were drawn. Increasing the crosslinking level of polymer I significantly decreased the domain size of polymer II whereas the effect of crosslink density of polymer II was relatively small. If both polymers are crosslinked, there was a strong indication that two continuous phases were evolved. While the electron micrographs superficially suggested spheres embedded in a matrix, closer examination revealed cylinders or worm-like interconnected structures, especially when it was noted that many of the domains

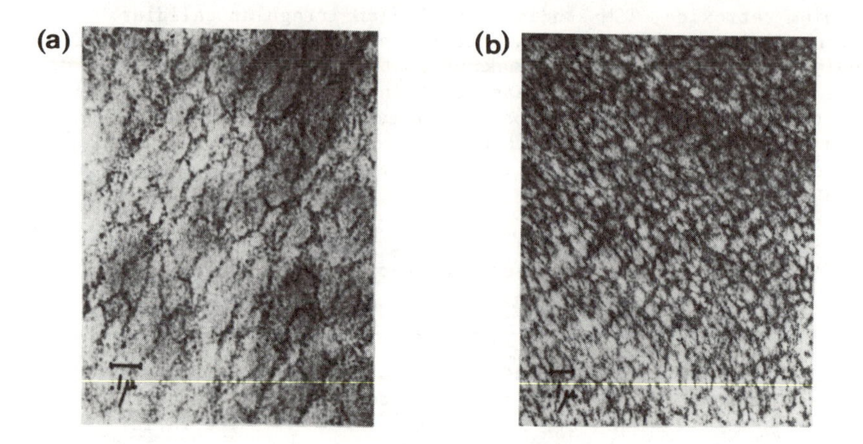

Figure 1. Morphology of sequential IPNs. (a) *Cross*-poly (ethyl acrylate)-*inter-cross*-polystyrene, showing typical cellular structure and a fine structure within the cell walls. (b) *Cross*-poly (ethyl acrylate)-*inter-cross*-polystyrene-*stat*-(methyl methacrylate), showing smaller domain structure. PEA structure stained with OsO_4. (Reproduced from ref. 5. Copyright 1972 American Chemical Society.)

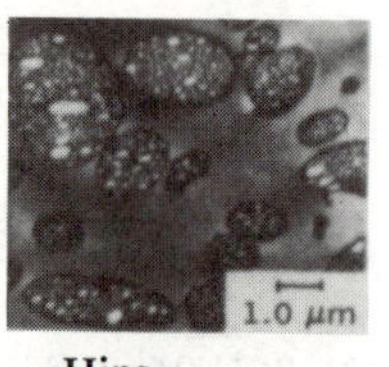

Hips
Phase inverted

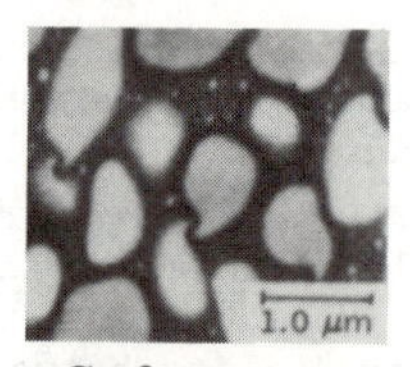

Graft
No phase inversion

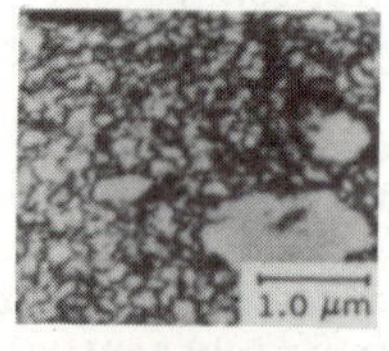

Semi—I

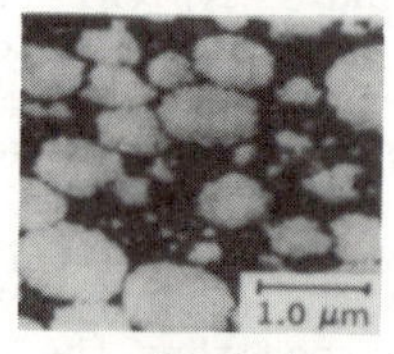

Semi—II

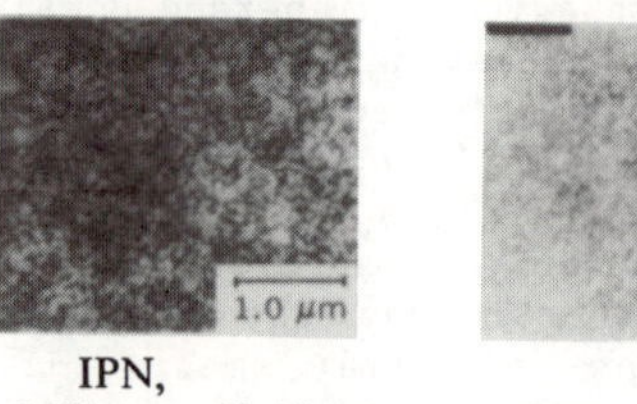

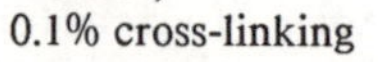

IPN,
0.1% cross-linking

IPN,
0.2% cross-linking

Figure 2. Morphology of various *cross*-polybutadiene-*inter-cross*-polystyrene sequential IPNs and graft copolymers via transmission electron microscopy. The double bonds in the polybutadiene phase are stained dark with osmium tetroxide. (Reproduced from ref. 15. Copyright 1976 American Chemical Society.)

had an ellipsoidal shape probably caused by cutting the cylinders at an angle.

Figure 3 [17] shows a higher magnification version of an IPN crosslinked with 0.1% dicup in the elastomer phase. Lightly stained samples reveal a fine structure on the order of 100 A , with indications of connectivity between them. At that time, the domain formation process was speculated as follows: As polymerization of monomer II inside of the swollen network I proceeds, the free energy of mixing goes from negative to positive, phase separation starts. A cellular structure with polymer I as the cell walls develops. As the polymerization proceeds, a second phase separation occurs, resulting in the 100 A fine structure. This last was based on the observation that the fine structure was observed mainly inside the cell wall. However, it was not possible to confirm when the fine structure exactly starts to form or whether the fine structure has same origin as the larger cellular structures. In practice, electron microscopy analysis of high rubber content materials, corresponding to the early stages of polymerization in IPN's, is difficult due to problems in microtoming soft samples. Also, observation of 100 A level dimensions requires very careful staining techniques to avoid artifacts [17].

The variation of the domain sizes with crosslink density was recognized by Yeo et al. [18], investigating cross-poly(n-butyl acrylate)-inter-cross-polystyrene. Figure 4 shows the morphology of 50/50 compositions as a function of network I crosslinking level. The cellular structures are gradually transformed to finer, and more obviously cylindrical or worm-like shapes with increasing crosslink density.

The search for theoretical relations concerning domain size and shape is ongoing, and several efforts to derive a theoretical formulation have been made. The first attempt was by Donatelli et al. in 1977 [19], who derived an equation especially for semi-IPN's of the first kind (polymer I crosslinked, polymer II linear) and extended to full IPN's by assuming that the molecular weight of polymer II is infinite. Later, this equation was simplified for several limiting cases by Michel et al. [20]. These were pioneer works, but of a semiempirical nature.

A new set of theoretical equations based solely on the properties of networks and their interaction was derived by Yeo et al. [21]. Assuming spherical domains, a crosslink density of ν_1 and ν_2 for networks I and II, respectively, an interfacial tension of δ, and a volume fraction of each phase, ϕ_1 and ϕ_2, the spherical diameter of network II domains, D_2, may be written [21]:

$$D_2 = \frac{4\delta}{RT\ (A\nu_1 + B\nu_2)} \tag{1}$$

$$A = \frac{1}{2}\ \frac{1}{\phi_2}\ (3\phi_1^{1/3} - 3\phi_1^{4/3} - \phi_1 \ln \phi_1) \tag{2}$$

$$B = \frac{1}{2}\ (\ln \phi_2 - 3\phi_2^{2/3} + 3) \tag{3}$$

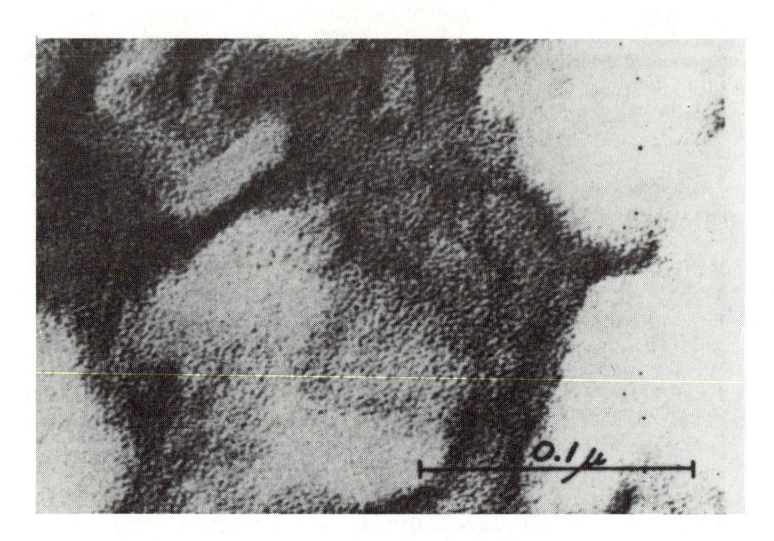

Figure 3. High magnification electron micrograph of a semi-I IPN shown in Figure 2, revealing the fine structure detail. (Reproduced with permission from Ref. 17. Copyright 1974 Plenum Press.)

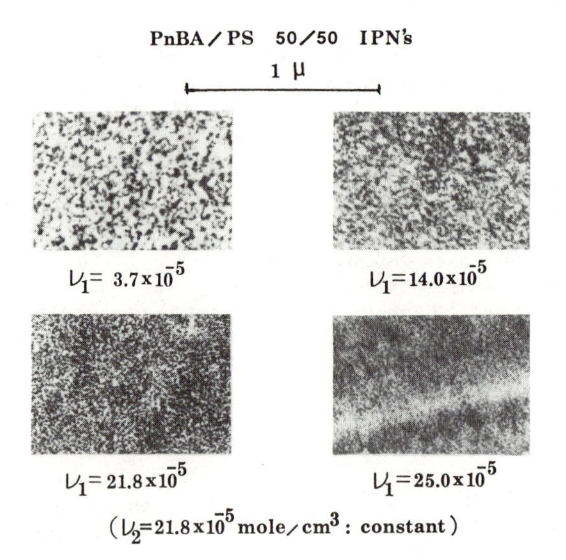

Figure 4. Transmission electron microscopy morphology of 50/50 *cross*-poly(*n*-butyl acrylate)-*inter-cross*-polystyrene IPNs as a function of network I cross-link density. (Reproduced with permission from ref. 18. Copyright 1982 *Polymer Engineering and Science*.)

A comparison of the theoretical and experimental results for the IPN system cross-poly(n-butyl acrylate) inter-cross polystrene is given in Table II [18,21]. The agreement between theory and experiment for this system as well as other systems was better than expected, noting the approximations required to obtain a usable result. It must be pointed out that Yeo et al. had to use spherical shapes for their mathematical treatment, even though it was already recognized that most of the domains were cylindrical.

Dual Phase Continuity

A major interest to sequential IPN's relates to dual phase continuity, defined as a region of space where each of two phases maintain some degree of connectivity. An example is an air filter with the air flowing through it. A Maxwell demon can transverse all space within both the filter phase and the air phase, both phases being continuous.

While several experimental techniques provide information relating to dual phase continuity, the two most important methods involve scanning electron microscopy and dynamic mechanical spectroscopy [16,22-24]. Donatelli, et al [16] performed the first mechanical study on PB/PS IPN's. Figure 5 [16] illustrates the fit provided by the Davies equation [22] and the Budiansky equation [25,26], both of these equations derived on the assumption of dual phase continuity.

While the evidence for dual phase continuity provided by Figure 5 does not indicate directly any mechanism for phase separation, or the shape of the phases, dispersed, spherical polystyrene domains probably would not yield results of this type. By hind sight, the data are consistent with the notion of spinodal decomposition and cylindrical domains.

Yeo, et al. [23,24] went on to make more complete studies of modulus-composition data using cross-poly(n-butyl acrylate)-inter-cross-polystyrene, PnBA/PS, see Figure 6. Both the Davies and the Budiansky models fit reasonably well over wide ranges of composition, especially the Budiansky model. Other models, which in one form or another assume one continuous and one disperse phase, fit much less well.

While significant evidence supporting the notion of dual phase continuity was obtained, particularly from modulus-composition studies, no direct observation had been successful with TEM due to its two-dimensional limitation.

The question of dual phase continuity has been more directly examined by scanning electron microscopy [27-29], where PnBA/PS IPN's were synthesized using a labile crosslinker, acrylic acid anhydride, AAA in polymer I. This allowed decrosslinking and extraction of that component. Scanning electron microscopy of the remaining phase (composed of network II materials), Figure 7 [27], revealed that surprisingly, network II phase domain structure was continuous above 20% of polymer II network, providing evidence that both phases were cocontinuous.

Recently, the effect of dual continuity on a general property, P, of the system was demonstrated by Lipatov et al. [30,31]. They used an equation proposed by Nielsen [32].

Table II. Experimental and theoretical domain sizes for _cross_-poly(n-butyl acrylate)-_inter_-_cross_-polystyrene IPN's

Variables		Domain Diameter (Å)	
$\nu \times 10^5$ (mol cm^{-3}) $M \times 10^{-5}$ (g mol^{-1})	Volume ratio	Experiment	Calculated
ν_1 = 3.7	25/75	800	845
ν_2 = 21.8	40/60	650	644
	50/50	550	572
ν_1 = 21.8	25/75	200	207
ν_2 = 21.8	40/60	170	169
	60/40	150	143
ν_2 = 21.8:const			
ν_1 = 3.7		550	572
ν_1 = 14.0	50/50	260	224
ν_1 = 21.8		195	154
ν_1 = 25.0		120	136

*TEGDM for crosslinker I and DVB for crosslinker II.
**ν_1, ν_2: crosslink density of network I and network II, respectively.

Source: Reproduced with permission from Ref. 2.

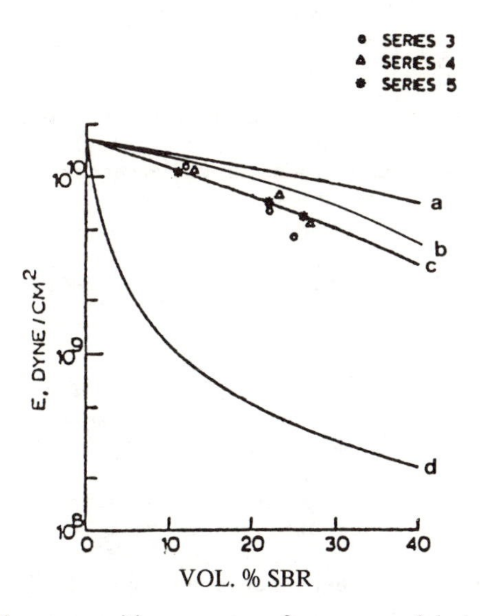

Figure 5. Modulus–composition curves for *cross*-polybutadiene-*inter-cross*-polystyrene semi-I and full IPNs (*16*). (a) Kerner equation (upper bound); (b) Budiansky model; (c) Davies equation; and (d) Kerner equation (lower bound). (Reproduced from ref. 23. Copyright 1981 American Chemical Society.)

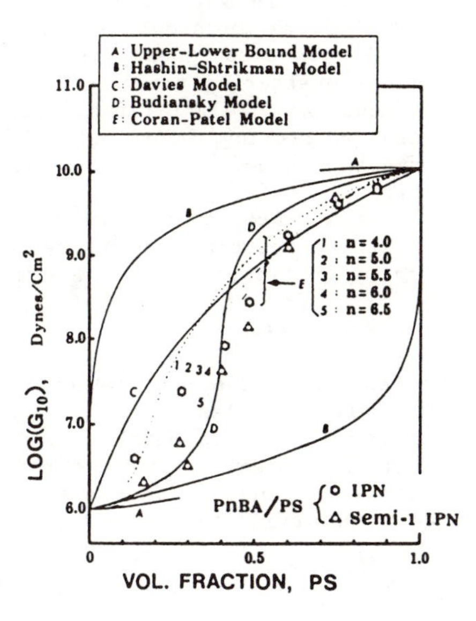

Figure 6. Modulus–composition behavior of *cross*-poly(*n*-butyl acrylate)-*inter-cross*-polystyrene IPNs and semi-I IPNs at 25 °C. (Reproduced with permission from ref. 23. Copyright 1981 *Polymer Engineering and Science*.)

$$P = P_A\phi_{A\parallel} + P_B\phi_{B\parallel} + \frac{P_A P_B(\phi_{A\perp} + \phi_{B\perp})}{P_A\phi_{A\perp} + P_B\phi_{B\perp}} \quad (4)$$

where the symbols have the following meaning: $\phi_{A\parallel}$ is the volume fraction of material A which behaves as though it is a continuous phase. $\phi_{A\perp}$ is the volume fraction which behaves as a dispersed phase to some force F. An analogous meaning applies to $\phi_{B\parallel}$ and $\phi_{B\perp}$. The definitions of connectivity (C_A or C_B) of phases are

$$C_A = \phi_{A\parallel}/\phi_{A\perp} \quad (5)$$

$$C_B = \phi_{B\parallel}/\phi_{B\perp} \quad (6)$$

Figure 8 [30] shows the reduced value of P as a function of the connectivity of component A obtained by computer simulation. The algorithm was based on a matrix of size 100 x 100 elements where each element was assigned by either 1 or 0 to form a interconnected structure obtained by the spinodal phase separation. As phase connectivity, C_A, increases at fixed overall composition, the final properties of the composite, P, is improved. Such trends became more obvious as the difference between properties of two components became larger.

Phase Dimensions via Small-Angle Neutron Scattering, SANS

Even though TEM and SEM played major roles in the study of IPN morphological features, there are various shortcomings, such as staining artifacts, difficulties in sample preparation for very rubbery materials, and the two-dimensional viewing limit for the former. Recently, various scattering techniques have been applied to measure the phase dimensions of IPN's via statistical treatment. The principles of neutron scattering theory as applied to the phase separated materials have been described in a number of papers and review articles [33-36].

For a two-phased system where domains are randomly dispersed, Debye [37,38] showed that the scattered intensity can be expressed in terms of a correlation function, $\gamma(r)$, in a simple exponential form.

$$\gamma(r) = \exp(-r/a) \quad (7)$$

where the quantity a represents the correlation distance which defines the size of the heterogenities in the system.

The specific interfacial surface area, S_{sp}, defined as the ratio of interfacial surface area, A, to the volume, V, is calculated from the value of correlation distance obtained by plotting the scattered intensity in the Debye fashion [12,38].

$$S_{sp} = \frac{A}{V} = \frac{4\phi_1(1-\phi_1)}{a} = \frac{4\phi_1 \cdot \phi_2}{a} \quad (8)$$

where ϕ_1, ϕ_2 represent the volume fraction of each network, respectively.

The transverse length across the domains, the average distance of straight lines drawn across a domain, are given by [39,40]

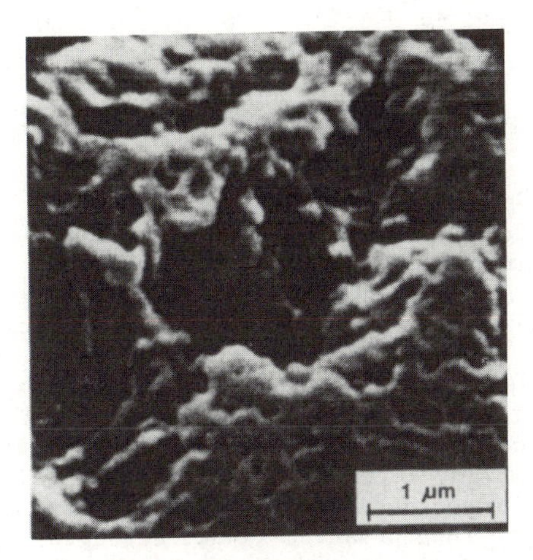

Figure 7. High magnification scanning electron micrograph of decrosslinked and extracted cross-poly(n-butyl acrylate)-inter-cross-polystyrene IPN (80/20). The poly(n-butyl acrylate) phase was extracted. (Reproduced from Ref. 27. Copyright 1982 American Chemical Society.)

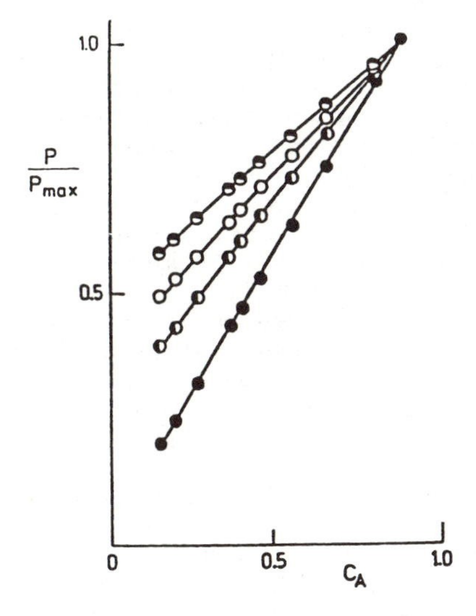

Figure 8. Reduced property, P/P_{max} as a function of the connectivity coefficient, C_A [31]. P_A/P_B: ◒ 2; ○ 5; ◐ 10; ● 100.

$$\ell_1 = \frac{a}{(1-\phi_1)} \tag{9a}$$

$$\ell_2 = \frac{a}{(1-\phi_2)} \tag{9b}$$

It must be pointed out that the above equations do not require any assumption of domain shape in their derivation.

Table III shows the result of SANS analysis on fully polymerized PB/PS IPN's, semi-IPN's, and chemical blends by Fernandez et al. [11]. The specific interfacial surface area was shown to increase with increasing crosslink density, S_{sp} decreasing in the order full-IPN's, semi-I IPN's, semi-II IPN's and chemical blends, as expected from many earlier studies. Its value ranges from 20 to 200 m^2/gm, in the range of true colloids. This result is particularly important because interfacial surface area is closely related to toughness and impact strength.

So far, most of the experimental studies have been limited to fully polymerized samples or samples with a high plastic content. That is because the earlier interest was mainly focused on the effect of properties of the constituents, such as crosslink density and miscibility, ease of TEM studies, etc.

In order to understand the domain formation process, an investigation of the intermediate stages before formation of the final morphology is required. There are several different ways to prepare such intermediate materials [3,42,43], see Figure 9. The characteristic domain dimensions of PB/PS IPN's are compared in Figures 10 and 11 [3,12,41].

In Figure 10, the transverse length of the polystyrene domains increase steadily during the polymerization of monomer II, showing a more rapid increase later in the polymerization. Specific interfacial surface area (Figure 11) does not increase monotonously with PS content. Rather, it shows a maximum at near the midrange of PS content depending on the synthetic detail.

The rapid increase of transverse length and the presence of a maximum in the specific surface area suggests a macroscopic development of dual phase continuity. A similar trend was obtained in another study of IPN's with SANS [13].

In another SANS study, McGarey et al. [13] investigated the size of polystyrene domains in sequential IPN's where poly(dimethyl siloxane) was the first network. Figure 12 shows the radius of the PS domains calculated as a function of composition, via Porod analysis, as well as these calculated from Yeo's equation [21]. The experimental domain sizes qualitatively follow Yeo's equation, notably samples with less than 40% of polystyrene, even though calculated values are about four times larger than the experimental value. It may be speculated that some of the discrepancy might arise from grafting of the poly(dimethyl siloxane) networks to the polystyrene due to the vinyl content of the former, lowering the interfacial tension value from that used in the calculation.

Interestingly, McGarey, et al. also observed a peak in their scattering pattern, suggesting the existence of a preferred distance between domains. The exact cause of such a peak remains uncertain. Comparison of the experimental scattering pattern with

TABLE III. Phase dimensions of cross-polybutadiene-inter-cross-polystyrene IPN's and chemical blends.

Sample	Volume Fractions		Correlation Lengths	Intercept Lengths, a(Å)			Diameters (Å) of PS	
Code	PB	PS	a(Å)	PB	PS	S_{sp} (m^2/g)	D Values[a]	D Range[b]
A	0.193	0.807	106	131	549	58.8	D_s 197	300-2000
B	0.273	0.730	53	73	194	149	D_s 109	150-1200
							D_p 294	350-600
C	0.230	0.770	58	75	252	122	D_s 113	150-1200
							D_p 378	500-1100
D	0.240	0.760	37.5	49	154	196	D_s 74	300-600
							D_p 234	400-800
E	0.116	0.884	161.1	182.2	1388.8	25.5	D_s 273	150-300
							D_p 2080	1000-2250
F	0.116	0.884	182.6	206.6	1574.1	22.5	D_s 310	200-400
							D_p 2360	1200-2500

[a] From SANS
[b] From TEM measurements

*PB: Polybutadiene phase
PS: Polystyrene phase

*Dicumyl peroxide, Di-Cup, for crosslinker I and DVB for crosslinker II
**A, Semi-I IPN, 0.1% Di-Cup; B, Semi-I IPN, 0.2% Di-Cup; C, Full IPN, 0.1% Di-Cup; D, Full IPN, 0.2% Di-Cup; E, Semi-II IPN; and F, Chemical Blend

Source: Reproduced from Ref. 11.

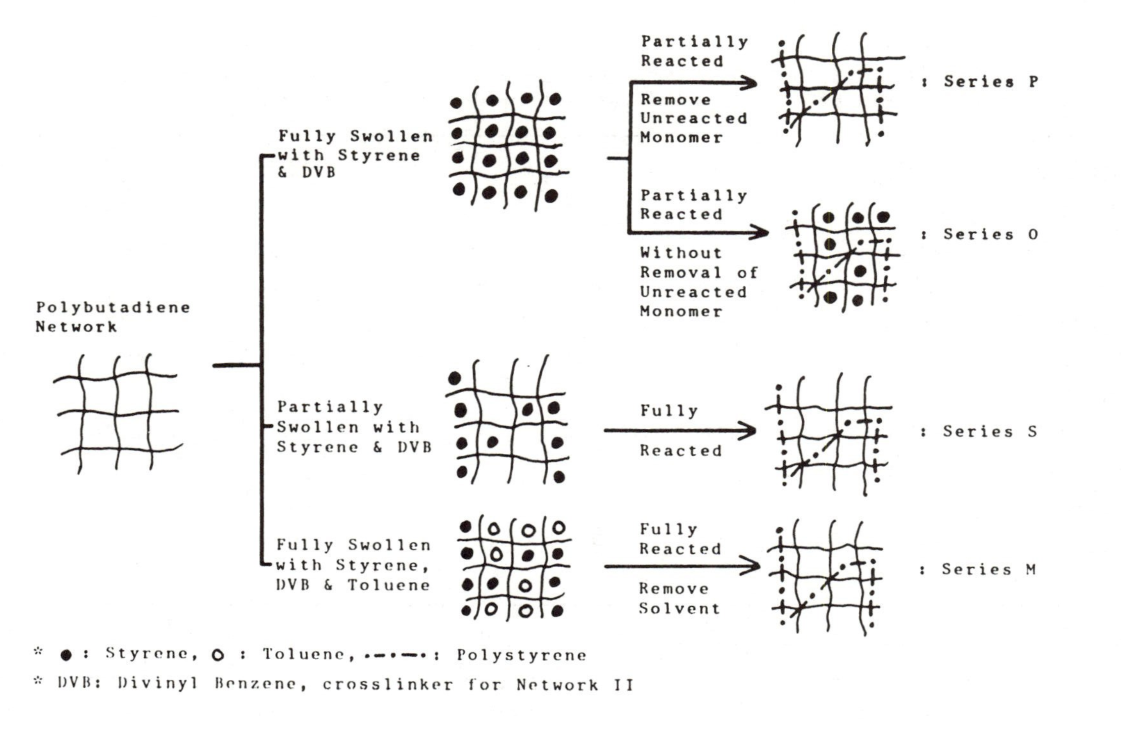

Figure 9. Typical ways preparing partially formed cross-polybutadiene-inter-cross-polystyrene IPN's.

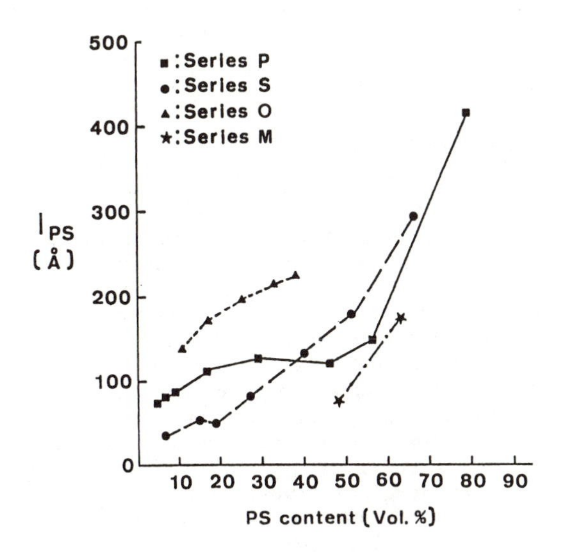

Figure 10. Increase of the polystyrene transverse length with polystyrene content [41].

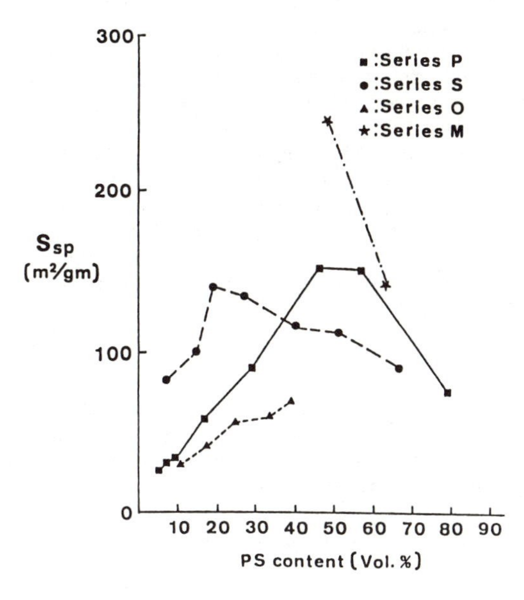

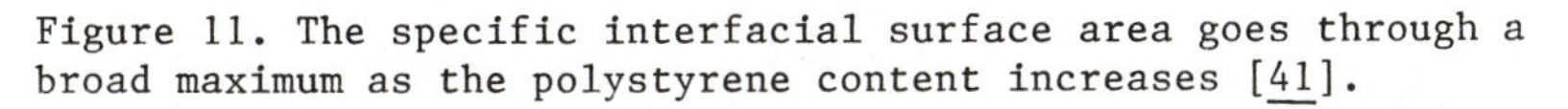
Figure 11. The specific interfacial surface area goes through a broad maximum as the polystyrene content increases [41].

that calculated with assumption of spherical domains did not show good agreement, even allowing for a broad distribution in domain sizes, Figure 13. As discussed below, the domains are more likely cylindrical.

Phase Separation in Sequential IPN's

When Fernandez et al. studied partially polymerized Series P PB/PS systems, SANS results yielded peaks in the scattering patterns, see Figure 14 [43]. Early attempts at understanding these peaks were troublesome due to the multiple causes for such peaks. Later, similar peaks were observed in another set of samples, which were partially swollen, then fully reacted (series S) by An et al. [41], but, in this case, shoulders instead of distinct maximum were observed (Figure 15). In both cases, such maximum or shoulders appeared at mid-range compositions, then were replaced by smoothly decreasing curves on further polymerization of monomer II.

At the same time however, other workers [44-46] were making contributions which would permit a preliminary understanding of such peaks in terms of spinodal decomposition.

There are two known mechanisms of phase separation, nucleation and growth [47], and spinodal decomposition [48], Figure 16. Nucleation and growth is associated with metastability, implying the existence of an energy barrier and the occurrence of large composition fluctuations. Domains of a minimum size, so called "critical nuclei", are a necessary condition. Spinodal decomposition, on the other hand, refers to phase separation under unstable conditions in which the energy barrier is negligible, so even small composition fluctuations grow.

For nucleation and growth, the assumptions are that the polymerization rate of monomer II is under steady state conditions, and that spheres are forming. Since the volume of the spheres increases linearly with time, the third power of the diameter also increases linearly with time. It is well known that for small spheres the scattering power increases as the sixth power of the radius. Thus, the scattering power is expected to increase as the square of the time during the early part of the polymerization of monomer II. The intensity, I, of light scattered at a particular low angle is then given by

$$I = kt^2 \tag{10}$$

where t is the time.

The scattering relationships for spinodal decomposition are more subtle. Following Cahn [48], Nishi, et al. [2] described the rate of growth of the amplitude of the phase domains, $R(\beta)$, as a function of the wave number, β, as

$$R(\beta) = -D\beta^2 \, [1 + 2K\beta^2/f''] \tag{11}$$

where D is the diffusion coefficient, f" is the second derivative of the Gibbs free energy with respect to composition, and K represents the gradient-energy coefficient. If $R(\beta)$ is negative, then the composition modulation will decay with time. However, positive values of $R(\beta)$ result in rapid growth of the amplitude.

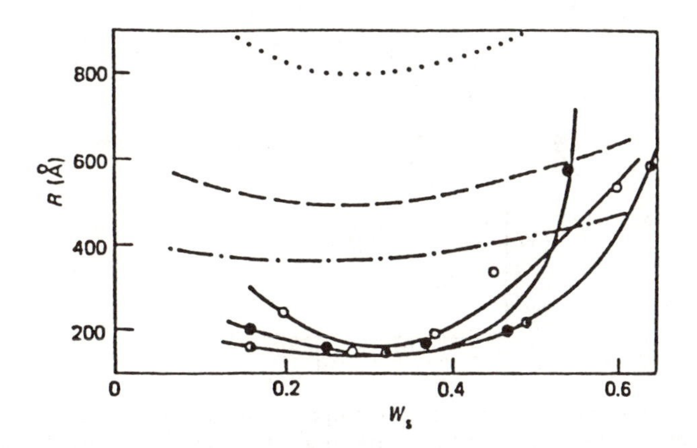

Figure 12. Radius of poly(dimethyl siloxane) phase as a function of weight fraction in cross-poly(dimethyl siloxane)-inter-cross-polystyrene sequential IPN's with three different crosslink densities of network I. Broken lines are theoretical values from Yeo's equation: (O, - - - -) PDMS 3; (◐, -----) PDMS 2; (●, ...) PDMS 1. (Reproduced with permission from Ref. 13. Copyright 1986 Butterworth Scientific Ltd.)

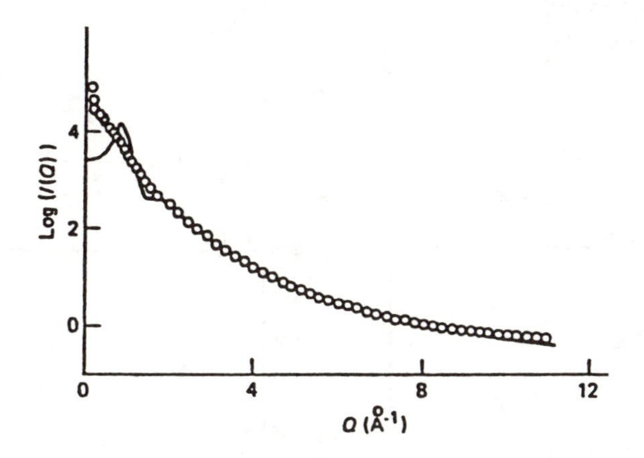

Figure 13. Comparison between experimental scattering (O) and calculated curve (-) assuming spherical domains. (Reproduced with permission from Ref. 13. Copyright 1986 Butterworth Scientific Ltd.)

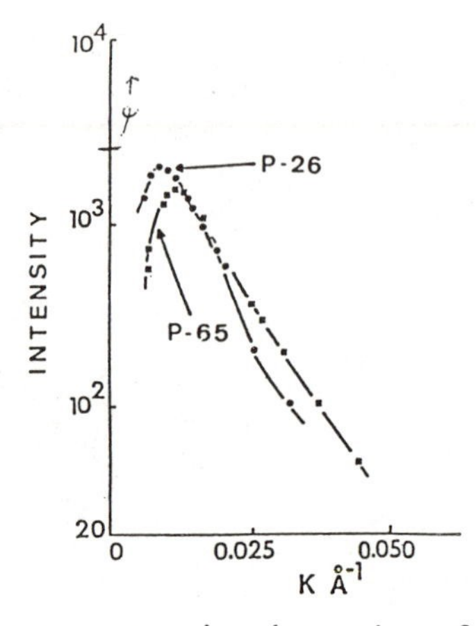

Figure 14. Coherent scattering intensity of series P sample, conversion controlled, showing maxima at around K = 0.01 cm^{-1}. (Reproduced with permission from Ref. 43. Copyright 1984 A. M. Fernandez.)

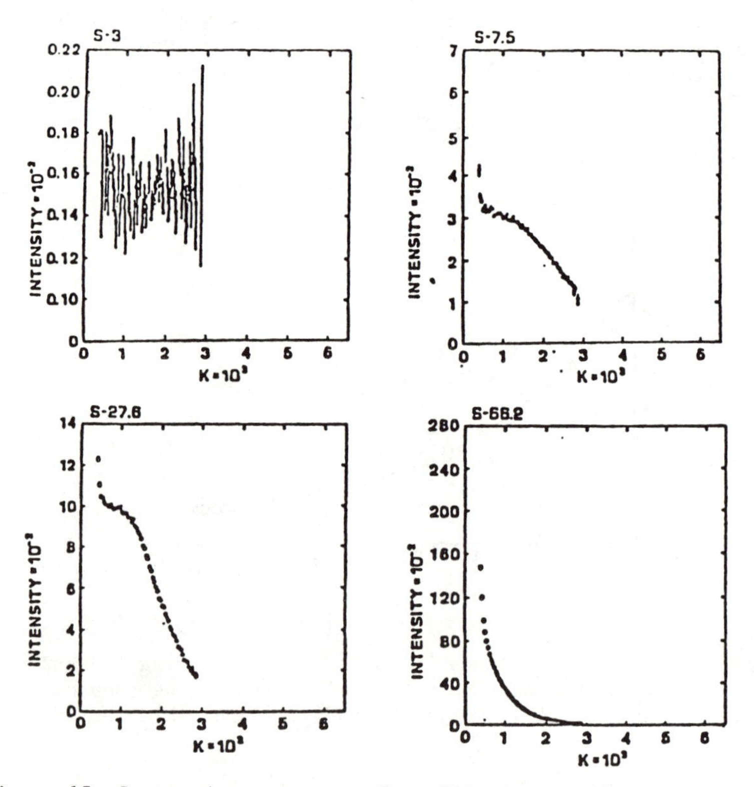

Figure 15. Scattering patterns of swelling controlled, series S, cross-polybutadiene-inter-cross-polystyrene IPN's, showing peaks at intermediate composition [41].

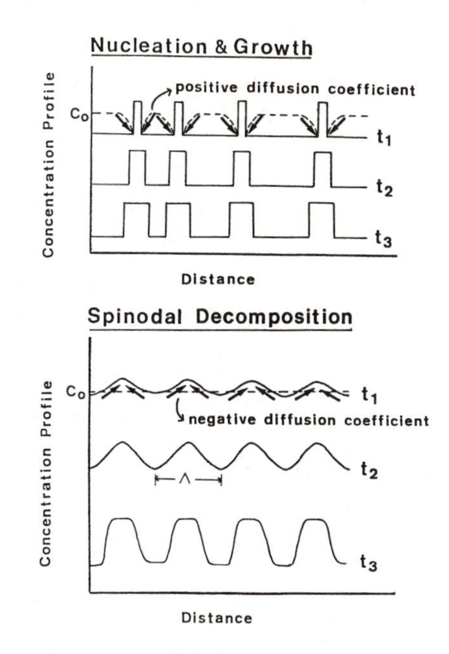

Figure 16. Mechanisms of phase separation.

Therefore, the quantity $R(\beta)$ is the major controlling factor that can be measured. The composition change during the initial stages of phase separation can be written

$$c - c_o = \sum_{\text{all } \beta} A(\beta) e^{R(\beta)t} \cos\beta x \tag{12}$$

where c represents the actual composition at time, t, c_o the average composition before phase separation, x is a space variable, and $A(\beta)$ is a coefficient.

Nishi, et al. [49] then find the value of β at which $R(\beta)$ becomes zero, defined as the critical wave number, β_c:

$$\beta_c = (-f''/2K)^{1/2} \tag{13}$$

They also note that since $R(\beta)$ has a sharp maximum at $\beta_m = \beta_c/2^{1/2}$, the initial composition fluctuation with wave number β_m will have the maximum growth rate.

Then, Nishi, et al. [41] calculated the total change in polymer I concentration in the polymer II-rich phase, Q:

$$Q = \left[\left| \int_{\frac{\pi}{2\beta_m}}^{\frac{3\pi}{2\beta_m}} (c-c_o)\, dx \right| \right]^3 = [2/\beta_m A(\beta_m)]^3 e^{3R\beta_m t} \tag{14}$$

Upon taking the logarithm,

$$Q = 3\ln [2/\beta_m A(\beta_m)] + 3R(\beta_m)t \tag{15}$$

The interesting feature that Lipatov, et al. [19] assumed was that the light-scattering intensity at low angles depends on the square of the difference between c and c_o, which arises from the appearance of $(dn/dc)^2$ in the scattering optical constant H, noting that the size and shape of the domains is constant with time. Hence,

$$\ln I(\beta,t) = \ln I(\beta,0) + 2R(\beta)t \tag{16}$$

Equations (10) and (16) indicate that a plot of the logarithm of the scattering intensity vs. time will behave quite differently for spinodal decomposition than for nucleation and growth, even though both mechanisms undergo an increase in the scattering intensity with time.

In addition,

$$R(\beta) = -D\beta^2 - 2M\gamma\beta^4 \tag{17}$$

where M is the diffusion mobility, γ is $2K/f''$; then, the scattered intensity at an angle over time yields the quantity R in equation (16), and hence a determine of D from equation (17). Thus, the magnitude and sign of the diffusion coefficient can be determined via light-scattering.

The wavelength of spinodal decomposition can be determined from the wavenumber at which the maximum in the scattering occurs. This wavelength, Λ, is given by

$$\Lambda = 2\pi/\beta_m \tag{18}$$

The main features of spinodally decomposed systems can be summarized as follows: (1) The logarithm of the scattered intensity increases linearly with time, (2) the presence of an interference tensity maximum, and (3) the formation of a uniform and highly interconnected structure. The rather constant value of Λ during phase separation, with increasing amplitude of the waves, is also illustrated in Figure 16.

Following Cahn's theory, more extended versions were proposed by Langer et al. [50] and Binder et al. [51]. Recently, de Gennes [52], and Pincus [53] applied spinodal decomposition to polymer mixtures. Many of the recent experimental studies on spinodal decomposition of polymer mixtures deal with measuring characteristic scattering maxima with various scattering techniques [54-60].

Lipatov carried out the first experiments [46] on IPN's with an attempt to separate out the extent of spinodal decomposition from nucleation and growth. He studied a semi-II IPN system of poly(butyl methacrylate) and polystyrene, the latter crosslinked with divinyl benzene. During the initial stages of reaction, the light scattering intensity was found to follow a logarithmic relation based on equation (10), see Figure 17, characteristic of spinodal decomposition. This was especially true at high methacrylic concentrations and temperatures. Other conditions showed apparent deviations, especially after the initial phase separation period. A characteristic wavelength of 1-1.5 microns was found for these material. Most importantly, they found negative diffusion coefficients, which are characteristic of spinodal decomposition.

In the case of PB/PS IPN's studied by SANS [41], a "wavelength" of 600 A for the repeat unit was calculated from the angular position of the peak using equation (12). A preliminary light scattering study (61) on the above PB/PS IPN's was found to obey the logarithmic relationship proposed by Cahn [48] based on equation (10), yielding a wavelength of 1.5 microns and a negative diffusion constant. It must be pointed out that SANS and light scattering might measure different types of regularity. Considering the usual domain dimensions of IPN's, 1000 A for full IPN's and a few thousand Angstroms for semi-II IPN's, such differences might reflect the 2000X difference in wavelength.

Binder and Frisch [44] constructed free energy functions for spinodal decomposition in IPN's. They predict initial domain wavelengths which are both larger and smaller than the typical distances between network crosslinks. In the latter case, they anticipate a coarsening of the structure until the domain size becomes comparable to the distances between crosslinks. This, indeed, is what An, et al. [41] found, because a wavelength of

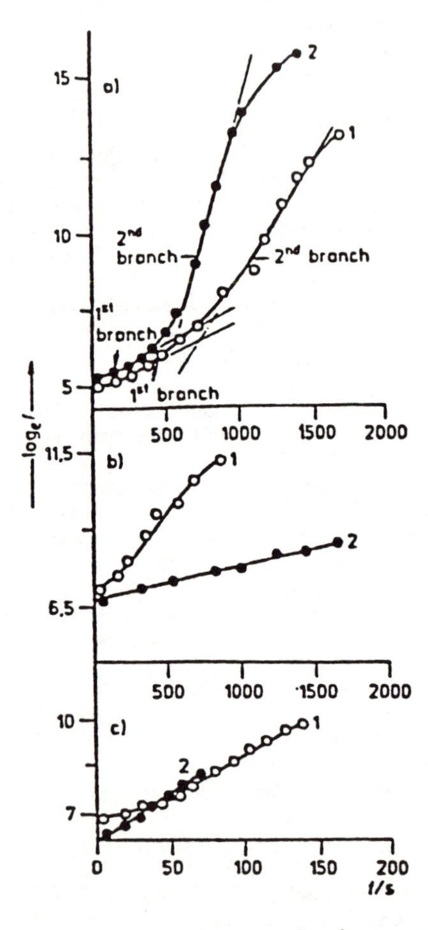

Figure 17. Logarithmic dependence of the scattered light intensity, I, on time, t, for polystyrene-inter-cross-poly(butyl methacrylate) Semi-II IPN's. Registration angle: (●) 10; (O) 20. Temperature: (a) 333 °K; (b) 343 °K; (c) 363 °K. (Reproduced with permission from Ref. 46. Copyright 1985 Huthig & Wepf Verlag.)

600 Å is of the same order of magnitude on the distance between crosslinks in a network containing about 1% crosslinker.

The type of spinodal decomposition encountered in IPN formation differs from the classical temperature quench in the sense that a composition change constitutes the driving force, at a fixed temperature. In this case, the rate of composition change is deeply involved in the phase separation process [62], which severely limits the applicability of current spinodal theory. In fact, Binder and Frisch [44] which assume the polymerization rate is rapid enough to limit the phase separation. On the contrary, in the experimental work by Lipatov et al. [46], the rate of polymerization was kept to a minimum to make the conversion changes during the phase separation minimal.

The phase separation process during polymerization of sequential IPN's bears a resemblance to solvent casting processes of polymer blends. Inoue et al. [63] measured the intensity of scattered light during the casting process of several polymer I-cosolvent-polymer II systems with various rates of solvent evaporation. Among the major findings were an interference maximum in the light-scattering angular profile and a highly interconnected modular structure, which are two major characteristics of spinodal decomposition. The solvent evaporation rate constituted an important factor, faster evaporation rates yielding a shorter wavelength of periodic structure. In some cases, a scattering maxima could not be observed with slower evaporation rates. This might indicate that composition change (evaporation of solvent) acts as a stabilizing factor for the modular structure formed during the early stages of the process as well as a driving force of phase separation. This acts by reducing molecular mobility, presumably by raising either the viscosity or the glass transition temperature.

In a system of significant interest to the present works, Graillard, et al. [62] studied the ternary phase diagrams of the systems polybutadiene-styrene-polystyrene and polybutadiene-*block*-polystrene-styrene-polystyrene. They showed that the presence of block copolymer increased the miscibility of the two polymers, as the styrene component polymerized. Similar effects are probable in the IPN's, as compared with the corresponding blends.

Figure 18 illustrates a model of the three component phase diagram of an IPN, where polymer I, polymer II, and monomer II are chosen for generality in expressing sequential IPN formation. On polymerization of monomer II, first phase separation is initiated, probably by nucleation and growth. However, shortly a modified spinodal decomposition mechanism sets in as the overall composition is driven deeper into the phase separation region.

The collage of electron micrographs shown in Figure 19 illustrates a probable transition from a nucleation and growth phase separation mechanism to a spinodal decomposition. The 7% PS sample shows more or less spherical PS domains, characteristics of nucleation and growth. However, a few domains appear to have some elliptical characteristics. When the PS content increases up to midrange, the shape of the PS domains becomes more obviously elliptical, suggestive of cross-sections of cylinders or other worm-like structures. As the PS content increases further, the

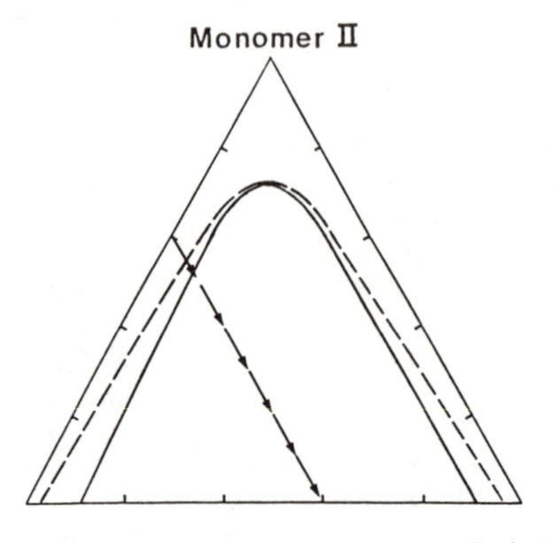

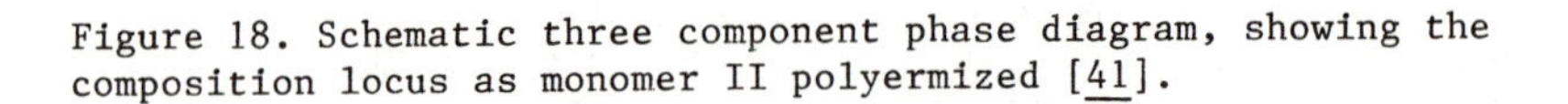
Figure 18. Schematic three component phase diagram, showing the composition locus as monomer II polyermized [41].

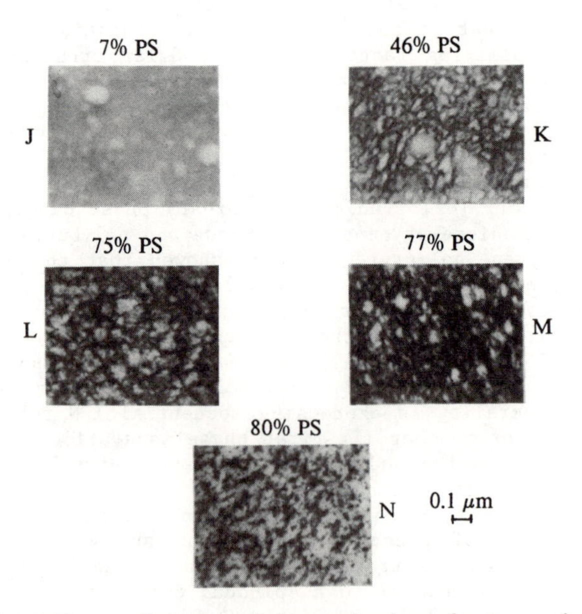

Figure 19. Collage of series P transmission electron micrographs. The morphology changes from spheres to worm-like cylinders as monomer II polymerization proceeds. The morphology is highly suggestive of spinodal decomposition. (Reproduced with permission from Ref. 43. Copyright 1984 A. M. Fernandez.)

phase domains seem to form highly interconnected structures typical of the spinodal decomposition mechanism.

The co-occurrence of nucleation and spinodal decomposition had been observed in the temperature quench experiment of poly(2,6-dimethyl-1,4-phenylene oxide)-toluene-caprolactam system, [64,65], in which the typical morphology formed by nucleation and growth mechanism was observed with electron-microscopy when the quench of temperature is slightly above the spinodal boundary. On the other hand, if the quench temperature is somewhat lower than the spinodal boundary, they observed interconnected structures as well as small droplets.

In the case of IPN's, the presence of crosslinks seems to be more favorable for spinodal decomposition than for linear blends, considering that high molecular weight components generally have a narrower gap between the binodal and spinodal boundaries. The relatively high viscosity of the nascent IPN acts as an unfavorable factor to nucleation and growth, which requires a longer mobility range and is known to be a slow process.

From an experimental point of view, observations of the initial phase separated morphology via electron microscopy is complicated by the removal of unreacted monomer, except for the fully polymerized stage.

In summary, direct evidence for spinodal decomposition in IPN's includes modulus-composition studies showing dual phase continuity, the cylindrical, interconnected form of the phases shown by TEM and SEM, and the maxima or shoulders in the SANS scattering patterns. Figure 20 illustrates a model of interconnected cylinders with occasional spheres, drawn from the sum total of the observation on this system. The interconnected irregular cylinders are a result of the spinodal decomposition, while the more or less spherical domains may be reminescent of the initial nucleation and growth stage, or a coarsening of the structure with breakup of the cylindrical structures during the latter stages of polymerization.

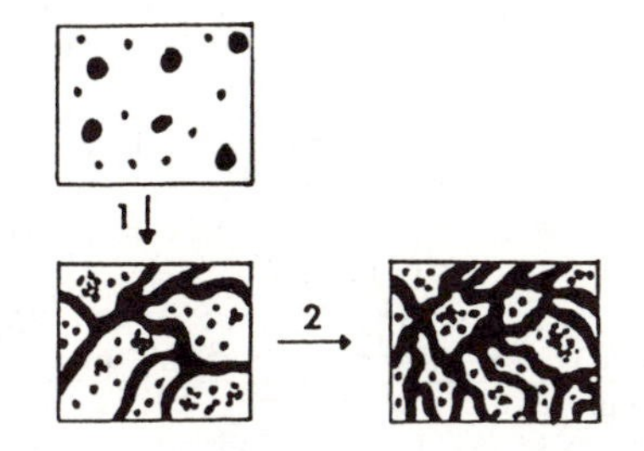

Figure 20. Model of two consecutive growth stages for the polystyrene phase. (1) Early stages: microgels of increasing size. (2) Intermediate to final stages: Domains become wormlike cylinders and then interconnected cylinders that exhibit dual-phase continuity (*41*).

Acknowledgment

The authors wish to thank the National Science Foundation for support under Grant No. DMR-8405053, Polymer Program.

Literature Cited

1. (a) Sperling, L. H. Interpenetrating Polymer Networks and Related Materials; Plenum: New York, 1981; (b) Sperling, L. H. In Multicomponent Polymer Materials; Paul, D. R.; Sperling, L. H.,Eds.; American Chemical Society: Washington, D.C., 1986.
2. Klempner, D.; K. C. Frisch, Eds., Polymer Alloy III, Plenum: New York, 1983.
3. An, J. H.; Fernandez, A. M.; Wignall, G. D.; Sperling, L. H. Polym. Mat. Sci. Eng. 1972, 56, 541.
4. Curtius, A. J.; Covitch, M. J.; Thomas, D. A.; Sperling, L. H. Polym. Eng. Sci. 1972, 12(2), 101.
5. Huelck, V.; Thomas, D. A.; Sperling, L. H. Macromolecules 1972, 5(4), 340.
6. Huelck, V.; Thomas, D. A.; Sperling, L. H. Macromolecules 1972, 5(4), 348.
7. Klempner, D.; Frisch, H. L.; Frisch, K. C. J. Polym. Sci. Part A-2, 1970 8, 921.
8. Shilov, V. V.; Lipatov, Yu. S.; Karbanova, L. V.; Sergeeva, C. M. J. Polym. Sci. Polym. Chem. Ed. 1979, 17, 3083.
9. Lipatov, Yu. S.; Shilov, V. V.; Bogdanovitch, V. A.; Karbanova, L. V.; Sergeeva, L. V. Polym. Sci. USSR 1980, 22, 1492.
10. Lipatov, Yu.; Shilov, S. V. V.; Bogdanovich, V. A.; Karabanova, L. V.; Sergeeva, L. M. J. Polym. Sci. Polym. Phys. Ed., 1987, 25, 43.
11. Fernandez, A. M.; Wignall, G. D.; Sperling, L. H. In Multicomponent Polymer Materials, Paul, D. R.; Sperling, L. H., Eds.; Adv. Chem. Ser. No. 211; American Chemical Society: Washington, D.C., 1986.
12. An, J. H.; Fernandez, A. M.; Sperling, L. H. Macromolecules 1987, 20, 191.
13. McGarey, B.; Richards, R. W. Polymer 1986 27, 1315.
14. Sperling, L. H.; Friedman, D. W. J. Polym. Sci., Part A-2 1969 7, 427.
15. Donatelli, A. A.; Sperling, L. H.; Thomas, D. A. Macromolecules 1976 9(4), 671.
16. Donatelli, A. A.; Sperling, L. H.; Thomas, D. A. Macromolecules 1976 9(4), 676.
17. Donatelli, A. A.; Thomas, D. A.; Sperling, L. H. In Recent Advances in Polymer Blends, Grafts, and Blocks, Sperling, L. H., Ed.; Plenum Press: New York, 1974.
18. Yeo, J. K.; Sperling, L. H.; Thomas, D. A. Polym. Eng. Sci., 1982 22(3), 190.
19. Donatelli, A. A.; Sperling, L. H.; Thomas, D. A. J. Appl. Polym. Sci. 1977, 21, 1189.
20. Michel, J.; Hargest, S. C.; Sperling, L. H. J. Appl. Polym. Sci. 1981, 26, 743.
21. Yeo, J. K.; Sperling, L. H.; Thomas, D. A. Polymer 1983, 24, 307.

22. Davies, W. E. A. J. Phys. D. 1971, 4, 1176.
23. Yeo, J. K.; Sperling, L. H.; Thomas, D. A. Polym. Eng. Sci. 1981 21(11), 696.
24. Yeo, J. K.; Sperling, L. H.; Thomas, D. A. J. Appl. Polym. Sci. 1981 26, 3283.
25. Budiansky, B. J. Mech. Phys. Solids 1965, 13, 223.
26. Budiansky, B. J. Compos. Mater. 1970, 4, 286.
27. Widmaier, J. M.; Sperling, L. H. Macromolecules 1982, 15, 625.
28. Widmaier, J. M.; Yeo, J. K.; Sperling, L. H. Colloid. Polym. Sci. 1982, 260, 678.
29. Sperling, L. H.; Widmaier, J. M. Polym. Eng. Sci. 1983 23(12), 694.
30. Lipatov, Yu. S.; Rosovitskii, V. F.; Maslak, Yu. S.; Zakharenko, S. A. Dokl. Akad. Nauk SSR 1983 278(6), 1408.
31. Lipatov, Yu. S. Pure & Appl. Chem. 1985, 57(11), 1691.
32. Nielsen, L. J. Appl. Polymer Sci. 1977 21, 1578.
33. Wignall, G. D.; Child, H. R.; Samuels, R. J. Polymer 1982, 23, 957.
34. Caulfield, D.; Yao, Y. F.; Ullman, R. X-ray and Electron Methods of Analysis; Plenum: New York; .
35. Higgins, J. S. J. Appl. Cryst. 1978, 11, 346.
36. Sperling, L. H. Polym. Eng. Sci. 1984, 24(1), 1.
37. Debye, P.; Bueche, A. M. J. Appl. Phys. 1949 20, 518.
38. Debye, P.; Anderson, H. R.; Brumberger, R. J. J. Appl. Phys. 1957 28, 649.
39. Kriste, R.; Porod, G. Kolloid-Z. 1962, 184, 1.
40. Miffelbach, P.; Porod, G. Kolloid-Z. 1965, 202, 40.
41. An, J. H.; Fernandez, A. M.; Wignall, G. D.; Sperling, L. H. submitted, Polymer.
42. Adachi, H.; Kotaka, T. Polymer J. 1982, 14(5), 3791.
43. Fernandez, A. M. Ph.D. thesis, Lehigh University, Bethlehem, PA, 1984.
44. Binder, K.; Frisch, H. L. J. Chem. Phys. 1984, 81(4), 2126.
45. Lipatov, Yu.; Shilov, S. V. V.; Gomza, Y. P.; Kovernik, G. P.; Grigor'eva, O. P.; Sergeyeva, L. M. Makromol. Chem. 1984, 185, 347.
46. Lipatov, Yu.; Grigor'eva, S. O. P.; Kovernik, G. P.; Shilov, V. V.; Sergeyeva, L. M. Makromol. Chem. 1985, 186, 1401.
47. Volmer, M.; Weber, A. Z. Phys. Chem. 1925, 119, 277.
48. Cahn, J. W. J. Chem. Phys. 1965, 42(1), 93.
49. Nishi, T.; Wang, T. T.; Kwei, T. K. Macromolecules 1975, 8, 227.
50. Langer, J. S.; Baron, M.; Miller, H. S. Phys. Rev. A. 1975, 11, 1417.
51. Binder, K.; Stauffer, D. Phys. Rev. Lett. 1974, 33, 1006.
52. de Gennes, P. G. J. Chem. Phys. 1980, 72, 4756.
53. Pincus, P. J. Chem. Phys. 1981, 75 1996.
54. Nojima, S.; Ohyama, Y.; Yamaguchi, M.; Nose, T. Polym. J. 1982, 14(11), 907.
55. Hashimoto, T.; Kumaki, J.; Kawai, H. Macromolecules 1983, 16, 641.
56. Snyder, H. L.; Meakin, P.; Reich, S. J. Chem. Phys. 1983, 78(6), 3334.

57. Hill, R. G.; Tomlins, P. E.; Higgins, J. S. Polymer 1985, 26, 1708.
58. Han, C. C.; Okada, M.; Mugora, Y.; McCrackin, F. L.; Bauer, B. J.; Trans-Cong, Q. Polym. Eng. Sci. 1986, 26(1), 3.
59. Baumgartner, A.; Heermann, D. W. Polymer 1986, 27, 1777.
60. Meier, H.; Strohl, G. R. Macromolecules 1987, 20, 649.
61. An, J. H.; Sperling, L. H., to be published.
62. Graillard, P.; Ossenbach-Sauter, M.; Riess, G. In Polymer Compatability and Incompatability, Solc, K., Ed.; MMI Press: New York, 1981.
63. Inoue, T.; Ougizawa, T.; Yasuda, O.; Miyasaka, K. Macromolecules 1985, 18, 57.
64. van Emmerik, P. T.; Smolder, C. A. J. Polym. Sci. Part-C 1972, 38, 73.
65. van Emmerik, P. T.; Smolder, C. A.; Greymaier, W. Europ. Polym. J. 1973, 8, 309.

RECEIVED February 16, 1988

Chapter 20

Polyurethane–Acrylic Coatings

Interpenetrating Polymer Networks

H. X. Xiao, K. C. Frisch, P. I. Kordomenos, and R. A. Ryntz

Polymer Institute, University of Detroit, Detroit, MI 48221

Polyurethane–acrylic coatings with interpenetrating polymer networks (IPNs) were synthesized from a two-component polyurethane (PU) and an unsaturated urethane-modified acrylic copolymer. The two-component PU was prepared from hydroxyethylacrylate–butylmethacrylate copolymer with or without reacting with ϵ-caprolactone and cured with an aliphatic polyisocyanate. The unsaturated acrylic copolymer was made from the same hydroxy-functional acrylic copolymer modified with isocyanatoethyl methacrylate. IPNs were prepared simultaneously from the two-polymer systems at various ratios. The IPNs were characterized by their mechanical properties and glass transition temperatures.

Interpenetrating Polymer Networks (IPNs) are unique types of polymer alloys consisting of two (or more) crosslinking polymers with no covalent bonds or grafts between them (1-3). These intimate mixtures of crosslinked polymers are held together by permanent entanglements, i.e., they are polymeric catenates, produced by homocrosslinking of two or more polymer systems. Formation of IPNs is a way of intimately combining crosslinked polymers with the resulting mixtures exhibiting, at worst, the limited phase separations.

IPNs can be prepared by either the "sequential" or the "simultaneous" technique. IPNs synthesized to date exhibit varying degrees of phase separation depending primarily on the compatibility of the component polymers (4-7).

0097–6156/88/0367–0297$06.00/0

The combination of various chemical types of polymer networks in different compositions, resulting frequently in controlled, different morphologies, has produced IPNs with synergistic behavior. Thus, synergistic properties may be obtained by IPNs such as enhanced tensile and impact strength, improved adhesion and, in some cases, greater sound and shock absorption (4-7).

This paper describes two types of novel urethane-acrylic IPNs for coating applications. The mode of preparation used was the simultaneous or SIN technique. In order to examine the effect of the soft segment on the properties and morphology of IPN coatings, the pendant hydroxy group in the hydroxyethylacrylate-butylmethacrylate copolymer was reacted with caprolactone to increase the chain length of the pendant hydroxy group.

Experimental

Raw Materials. The raw materials used in this study are shown in Table 1. The isocyanatoethyl methacrylate (IEM) was distilled before use. The solvent was dried by a 4A molecular sieve overnight before use. The other materials were used as received.

Synthesis of Urethane Modified Acrylics and Acrylic Modified Urethanes. 1) Synthesis of PBH. Into a reaction kettle equipped with a nitrogen inlet, stirrer, reflux condenser, and an addition funnel charged 500g of xylene. The xylene was heated to reflux and a mixture of 375g butylmethacrylate (BMA), 125g of 2-hydroxyethylacrylate (HEA) and 10g of t-butyl perbenzoate (t-BPB) were added over a period of five hours. After the end of the addition, the solution was kept at 140-150°C for another hour and then 10g of t-BPB was added. The reaction mixture was post-reacted for two more hours and then cooled to room temperature.

2) Synthesis of PBHC. The procedure was the same as in the synthesis of PBH except 350g of BMA and 150g of HEA were used. After the copolymerization was finished, 147g of ε-caprolactone and 0.22g of dibutyltin oxide (DBTO) were added into the reaction product. The ring opening reaction of ε-caprolactone in the presence of DBTO was carried out at 158-160°C for four hours, and then cooled to room temperature.

3) Synthesis of PBHI. Into a reaction kettle, equipped with nitrogen inlet, stirrer, reflux condenser, and thermometer were charged 464g of PBH, 171g of

TABLE 1. Raw Materials

Designation	Description	Supplier
HEA	2-Hydroxethyl acrylate	Rohm and Haas
BMA	Butyl methacrylate	Rohm and Haas
IEM	Isocyanatoethyl methacrylate	Dow Chemical Co.
t-BPB	t-Butyl perbenzoate	Lucidol Corp.
Desmodur N-100	Biuret triisocyanate-adduct of hexamethylene diisocyanate with water	Mobay Chem. Co.
T-12	Dibutyltin dilaurate	Air Products
DMBEHPH	Dimethyl 2,5-bis(ethylhexanoyl peroxy)hexane	Lucidol Corp.
BPO	Benzoyl peroxide	Lucidol Corp
Ca drier	Calcium naphthenate	Mooney Chem. Co.
Pb drier	Lead naphthenate	Mooney Chem. Co.
Co drier	Cobalt naphthenate	Mooney Chem. Co.
Zn drier	Zinc naphthenate	Mooney Chem. Co.
DMBA	N-benzyl-N,N,dimethylamine	Hexcel Chem. Prod.
$CuCl_2$	Copper chloride	Aldrich Chem. Co.
$FeCl_3$	Ferric chloride	Aldrich Chem. Co.
Dabco XDM	N,N-(Dimethylaminoethyl)-morpholine	Air Products
	ε-Caprolactone	Union Carbide Co.
	Hydorquinone	Aldrich Chem. Co.
	Xylene	Aldrich Chem. Co.
Desmodur L-2291A	Adduct of hexamethylene diisocyanate with water	Mobay Chem. Co.

isocyanatoethyl methacrylate (IEM), 1.27g of dibutyltin dilaurate (DBTDL) and 0.17g of hydroquinone. The reaction mixture was kept at 75-80°C for two to three hours until all of the isocyanate was fully reacted as indicated by the IR spectrum. The reaction product was then cooled to room temperature.

4) Synthesis of PBHI. The procedure was the same as in the synthesis of PBHI except 499g of PBHC, 155g of IEM, 1.31g of DBTDL and 0.155g of hydroquinone were used.

Preparation of IPN Coatings. 1) Preparation of P(UA)-1. In 100g of PBHI solution 3% (by weight) of calcium naphthenate, 1% of lead naphthenate, 1% of cobalt naphthenate and 1% of dimethyl 2,5-bis(ethyl hexanoyl peroxy)hexane were added. Films of P(UA)-1 were prepared and cured at 80°C for 30 minutes and post cured at 120°C for 30 minutes.

2) Preparation of P(UA)-2 Coating. In 100g of PBH solution, 45g of Desmodur N-100 and 1g of Dabco XDM catalyst were added. Using a doctor blade, clear films were drawn on glass and steel panels and were cured using the same conditions as in the case of P(UA)-1.

3) Preparation of P(UA)-1' Coating. In 100g of PBHCI solution 5% (by weight) of benzoyl peroxide, 0.03% of 5% $CuCl_2$ in water, 0.03% of 50% $FeCl_3$ in water and 0.2% of n-hexylmercaptane were added. Films of P(UA)-1' were prepared and cured using the same conditons as in the case of P(UA)-1.

4) Preparation of P(UA)-2' Coating. In 100g of PBHC solution, 18.3g of Desmodure L2291A (the same as Desmodure N-100) were added. Films of P(UA)-2' were prepared and cured using the same conditions as in the case of P(UA)-1.

5) Preparation of IPN Coatings. P(UA)-1 nd P(UA)-2 or P(UA)-1' and P(UA)-2' were mixed at different ratios, respectively, and films were prepared and cured using the same conditions as in the case of P(UA)-1.

Results and Discussion

Mechanical Properties of IPN Coatings. The IPN coatings with different compositions exhibited better tensile strength than those of their polymer components as shown in Figures 1 and 2. In particular, the compositions of P(UA)-1/P(UA)-2=60/40 for IPN-I and P(UA)-1'/P(UA)-2'=50/50 for IPN-II exhibited a maximum value, respectively. This is presumably due to the fact that there was a high degree of interpenetration at these compositions. IPN-II exhibited lower tensile strength and hardness but higher elongation than IPN-I due to the introduction of long flexible chain (from ε-

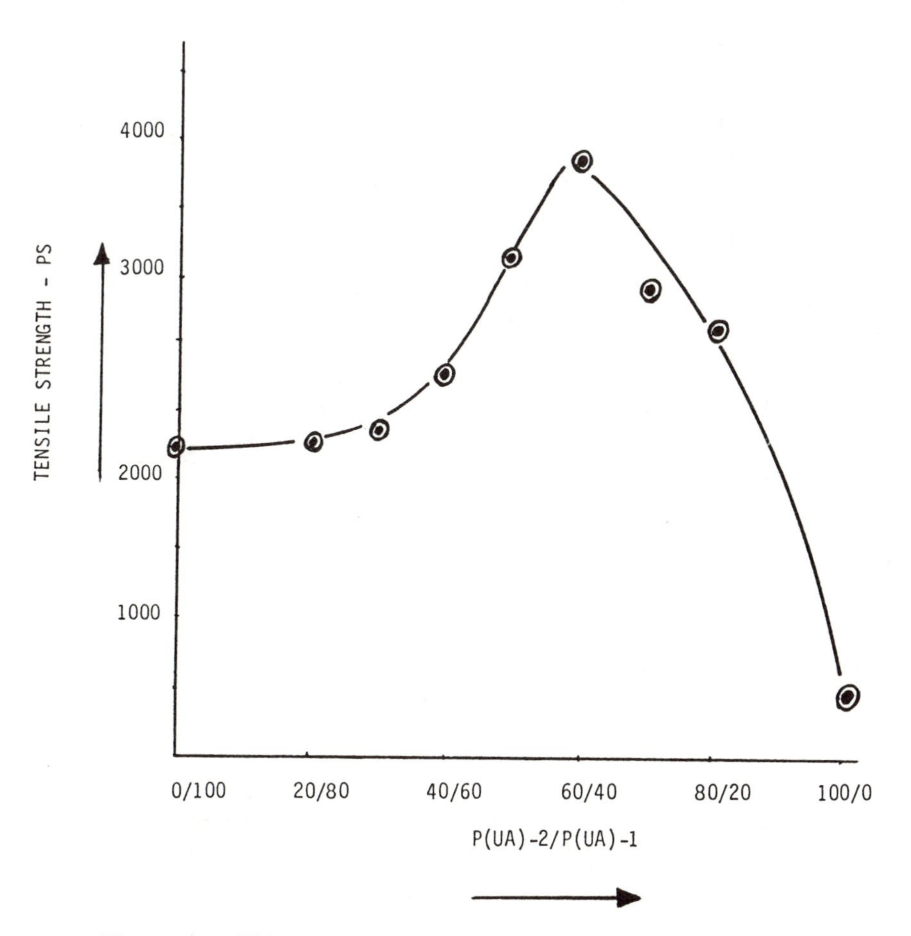

Figure 1. Effect of the composition in the IPN-I on the tensile strength.

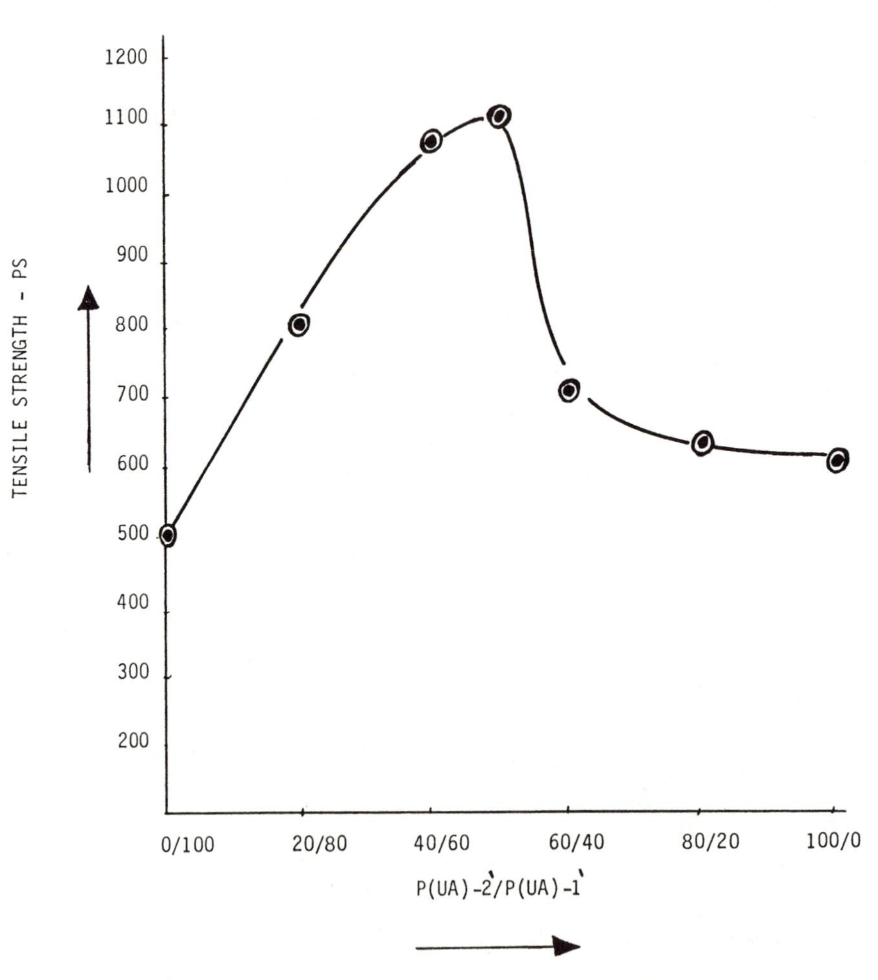

Figure 2. Effect of the composition in the IPN-I on the tensile strength.

caprolactone). As a result the soft segment was increased, hard segment was decreased and the mechanical properties were changed.

Since the hydroxy-containing copolymer (PBH or PBHC) was prepared by solution polymerization at high temperature and in the presence of high concentrations of initiator, the reaction product possessed a rather low molecular weight (MW) and broad molecular weight distribution (MWD). This could improve the miscibility between polymers, because the increasing mixing entropy from low MW and broad MWD could result in a negative free energy of mixing.

The adhesive strengths of IPN coatings to steel panels are shown in Tables 2 and 3. Most IPN coatings exhibited higher lap shear strengths than their original component. The maximum value of lap shear strength was at P(UA)-1/P(UA)-2=60/40 for IPN-I and P(UA)-1'/P(UA)-2'=40/60 for IPN-II. IPN-II exhibited a lower adhesive strength than IPN-I due to the low cohesive energy density from high concentrations of soft segment in IPN-II.

Glass Transitions of IPN Coatings. The glass transition temperatures of IPN-I and IPN-II were measured by thermomechanical analyzer (TMA), using a TMS-2 (Perkin-Elmer), at temperature ranges from -100°C to +100°C and 0.01mm of penetrating range, 80g of penetrating weight, and a heating rate of 10°C/minute.

As shown in Figure 3, all of the IPN coatings with different compositions exhibited a single Tg. The Tg of the IPN decreased with increasing P(UA)-2 or P(UA)-2' due to the increasing flexible structure from P(UA)-2 or P(UA)-2'. Both P(UA)-1, P(UA)-1' and P(UA)-2 and P(UA)-2' possessed low MW, broad MWD and similar structure (urethane) which brought about more hydrogen bonding between the two polymer systems. As a result, the compatibility between the polymer systems was improved, thus, the degree of interpenetration also increased as well as the molecular mixing. Consequently, the single Tg and the improvement of mechanical properties of IPN coatings were obtained. As shown in Figure 3, most of the data for both IPN-I and IPN-II are located in the two straight lines from composition vs. Tg of IPNs except 100% of PU (presumably a possible error in the determination) and the single Tg for each composition of IPN shifted inward compared to the component polymer (P(UA)-1, P(UA)-2, and P(UA)-1' and P(UA)-2'). Presumably it implies that the interpenetration between the polymer networks occurred.

Table 2. Adhesive Strength of IPN-I Coating to Steel Panel

Composition	P(UA)-1/P(UA)-2	Lap Shear (psi)	Type of Failure*
P(UA)-1	100/0	64	Ad
IPN-1	80/20	677	Ad/Co
IPN-2	70/30	557	Ad/Co
IPN-3	60/40	784	Co
IPN-4	50/50	627	Co
IPN-5	40/60	636	Co
IPN-6	30/70	202	Co
IPN-7	20/80	272	Ad/Co
P(UA)-2	0/100	322	Ad/Co

*Ad: Adhesive failure; Co: Cohesive failure

Table 3. Adhesive Strength of IPN-II Coating to Steel Panel

Composition	P(UA)-1'/P(UA)-2'	Lap Shear (psi)	Type of Failure
P(UA)-1'	100/0	120	Ad
IPN-1	80/20	177	Ad
IPN-2	60/40	224	Ad
IPN-3	50/50	224	Ad
IPN-4	40/60	244	Ad
IPN-5	20/80	329	Ad
P(UA)-2'	0/100	150	Ad

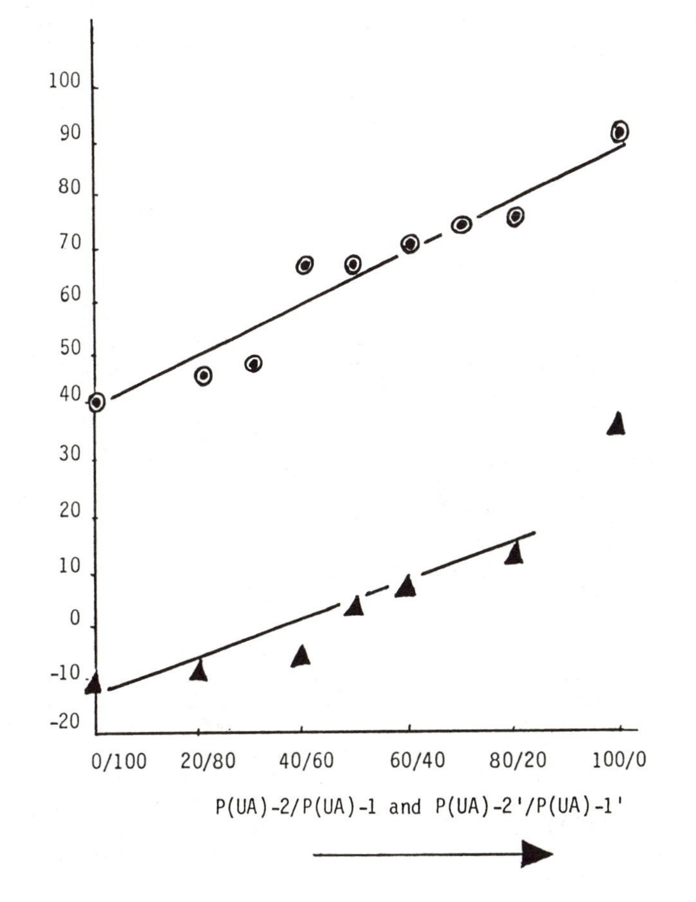

Figure 3. Effect of the composition in the IPN-I and IPN-II on the glass transition temperature.

Morphologies of IPN-I and IPN-II by Scanning Electron Microscopy (SEM). The micrographs from Scanning Electron Microscopy (SEM) for both IPN-I and IPN-II revealed one phase picture, as shown in Figures 4 to 10, which exhibit homogeneous morphologies without phase separation. This could be due to the fact that these IPN coatings possess the molecular mixing due to the introduction of similar structures into both polymer systems and the use of polymers with low molecular weight (MW) and broad molecular weight distribution (MWD).

Conclusion

Two novel IPN coatings were prepared from P(UA)-1 or P(UA)-1' and P(UA)-2 or P(UA)-2'. The former contained both urethane and acrylate structures and were crosslinked by the free radical polymerization of the pendant double bonds, while the latter also contained both urethane and acrylate structures and were crosslinked by reacting the pendant NCO groups with OH groups of Desmodur L-2291A. Both tensile strength and adhesive strength of IPN coatings exhibited higher values than those of their original components (P(UA)-1 or P(UA)-1' and P(UA)-2 or P(UA)-2').

With increasing soft segment content by introducing long chains in the side chain, the tensile strength, hardness and lap shear strength were decreased but elongation was increased. Both IPN coatings exhibited a single glass transition temperature (Tg). The Tg's of the IPN coatings increased with increasing P(UA)-1 or P(UA)-1' due to the rigid structure of P(UA)-1 or P(UA)-1'. The Tg of IPN-II was lower than that of IPN-I because of the presence of soft segment with long chain in IPN-II.

The scanning electron micrographs (SEM) for both IPN-I and IPN-II revealed a one phase morphology which implied that IPN coatings with molecular mixing were obtained by introducing similar structures into both polymer systems and using polymers with low MW and broad MWD at distribution.

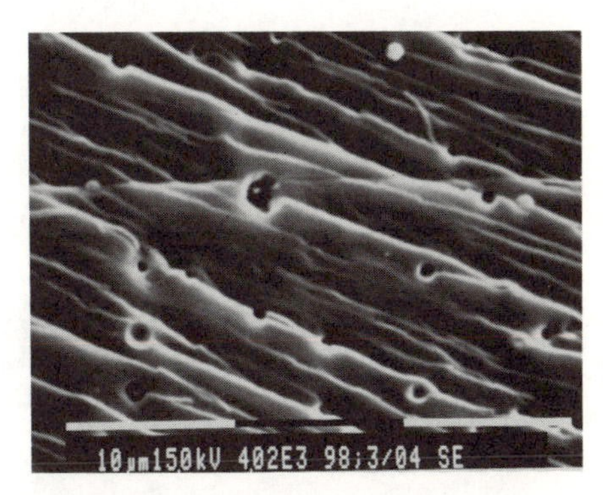

Figure 4. Micrograph of IPN-I with composition of P(UA)-2/P(UA-1) = 100%.

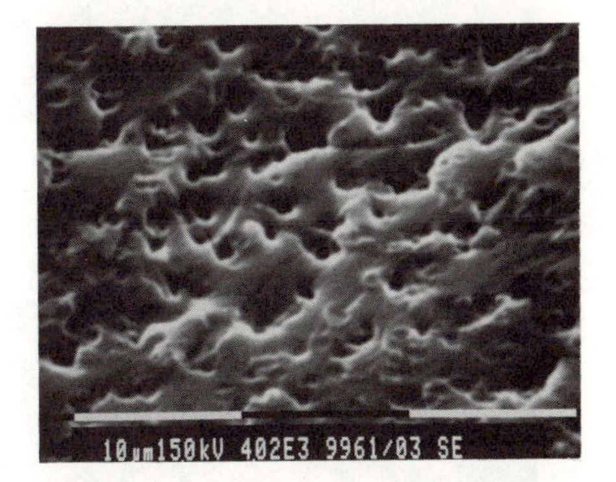

Figure 5. Micrograph of IPN-I with composition of P(UA)-2/P(UA-1) = 80/20.

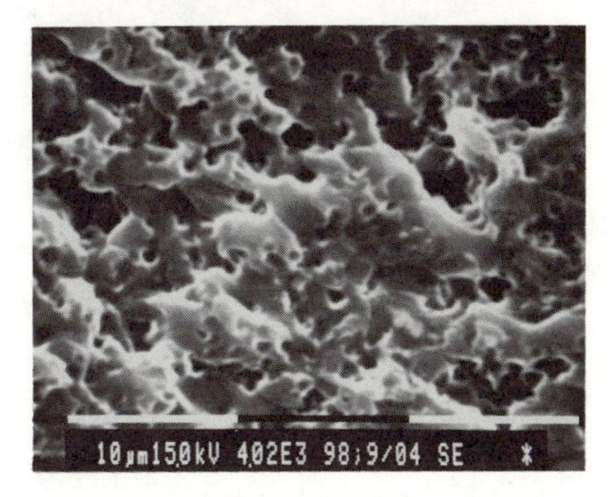

Figure 6. Micrograph of IPN-I with composition of P(UA)-2/P(UA-1) = 70/30.

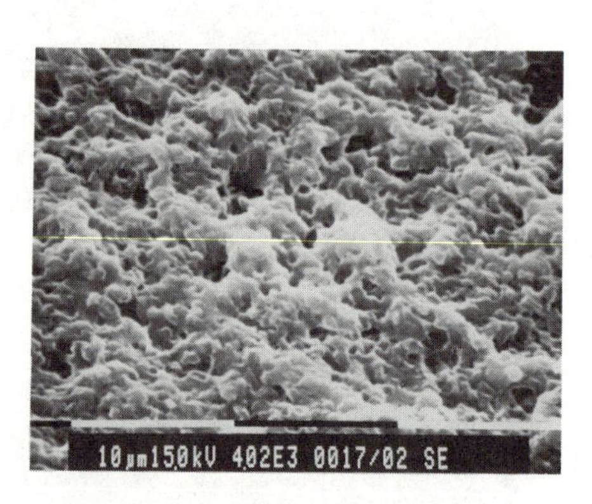

Figure 7. Micrograph of IPN-I with composition of P(UA)-2/P(UA-1) = 60/40.

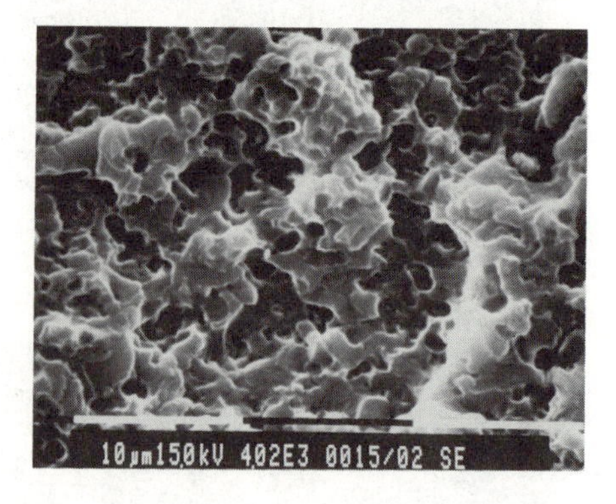

Figure 8. Micrograph of IPN-I with composition of P(UA)-2/P(UA-1) = 50/50.

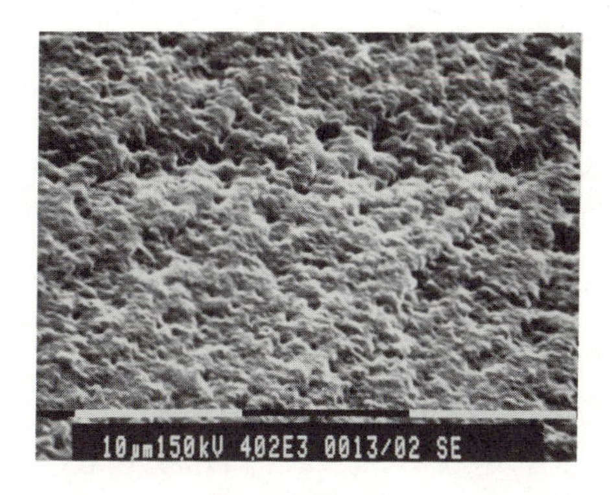

Figure 9. Micrograph of IPN-I with composition of P(UA)-2/P(UA-1) = 40/60.

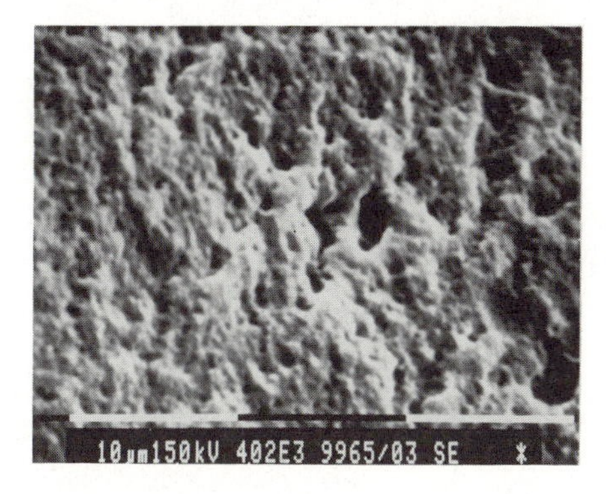

Figure 10. Micrograph of IPN-I with composition of P(UA)-2/P(UA-1) = 30/70.

References

1. Frisch, H. L.; Klempner, D.; Frisch, K. C. *J. Polym. Sci., Polym. Lett.* 1969, *7,* 775; Frisch, H. L.; Klempner, D.; Frisch, K. C. *J. Polym. Sci* 1970, *A-2, 8,* 921.
2. Sperling, L. H.; Friedman, D. W. *J. Polym. Sci.* 1969, *A-2, 7,* 425; Sperling, L. H.; Tayloyr, D. W.; Kirkpatrick, M. L.; George, H. F.; Bardman, D. R. *J. Appl. Polym. Sci.* 1970, *14,* 73; Sperling, L. H.; George, H. F.; Huelck, V.; Thomas, D. A. *J. Appl. Polym. Sci.* 1970, *14,* 2815.
3. Lipatov, Yu; Sergeeva, L. *Independent Interpenetrating Linking Polymer Networks* Naukova Dumka: Kiev, 1979 (in Russian).
4. Sperling, L. H.; Thomas, D. A.; Covitch, M. J.; Curtius, A. J. *Polym. Eng. Sci.* 1972, *12,* 101; Sperling, L. H.; Thomas, D. A.; Huelck, V. *Macromolecules* 1972, *5,* 340.
5. Frisch, K. C.; Frisch, H. L.; Klempner, D.; Mukheigge, S. K. *J. Appl. Polym. Sci.* 1964, *18,* 689; Frisch, K. C.; Klempner, D.; Migdal, S.; Frisch, H. L.; Ghiradella, H. *Polym. Eng. Sci.* 1975, *15,* 339.
6. Xiao, H. X.; Frisch, K. C.; Frisch, H. L. *J. Polym. Sci.* 1984, *22,* 1035; Cassidy, E. F.; Xiao, H. X.; Frisch, K. C.; Frisch, H. L. *J. Polym. Sci.* 1984, 22, 2667; Cassidy, E. F.; Xiao, H. X.; Frisch, K. C.; Frisch, H. L. *J. Polym. Sci.* 1984, *22,* 1851; Cassidy, E. F.; Xiao, H. X. *Science* 1984, *22,* 1839.
7. Frisch, K. C.; Klempner, D.; Frisch, H. L. *Polym. Eng. Sci.* 1982, *22,* 1143.

RECEIVED December 22, 1987

Chapter 21

Ionomer Interpenetrating Polymer Network Coatings from Polyurethane and Vinyl Chloride Copolymer

H. X. Xiao, K. C. Frisch, and S. Al-Khatib

Polymer Institute, University of Detroit, Detroit, MI 48221

Opposite charge group containing ionomer pseudo-interpenetrating polymer network (IPN) coatings were prepared from a carboxyl-containing vinyl chloride copolymer (VMCC) and polyurethanes (PUs) with and without tertiary amine nitrogen in the polymer backbone at various PU–VMCC compositions. The morphology and physical properties of the ionomer pseudo-IPN coatings were compared with those of nonionomer pseudo-IPN coatings. The ionomer pseudo-IPN coatings exhibited improved mechanical properties and higher adhesive strength and did not reveal any phase separation as determined by thermomechanical analysis and scanning electron microscopy.

Interpenetrating polymer networks (IPNs) are a unique blend of crosslinked polymers in which the chains of the two polymers are held together by permanent entanglements formed by homocross-linking of the component polymers. The degree of interpenetration depends upon the miscibility between the constituent polymers. Most polymer pairs are not compatible because, in contrast to low-molecular weight materials, the entropy of mixing two macromolecules containing a large number of segments is relatively small while the enthalpy of mixing is usually positive or near zero, unless specific interactions are present (1-10). To achieve miscibility, the presence of specific interactions is usually required and these include: hydrogen bonding, (11-17) charge-transfer complexes, (18-20) anion-cation interaction (2-9, 21, 22) and ion-dipole interactions. (23-27) These specific interactions generally give rise to the remarkable properties of the resulting polymer blends, alloys, and IPNs.

Pseudo or semi-IPNs are combinations of linear with crosslinked polymers resulting in various degrees of interpenetration. (28-34)

Interest in polyurethane ionomers has increased in the last decade because of their growing uses in water-based coating, adhesives, medical, and semiconductor applications. However, relatively little information has been published regarding the synthesis and

0097–6156/88/0367–0311$06.00/0

properties of ionomer pseudo-IPNs. The purpose of the present study was to study the effects of opposite charge groups in ionomer pseudo-IPNs based on polyurethanes and a vinyl chloride copolymer.

Experimental

A. Raw Materials

The raw materials used in this study are presented in Table 1. The polyether polyol and short chain diols were degassed at 80-90°C under vacuum overnight to remove any moisture. MEK was treated with molecular sieves 4A overnight to remove any moisture prior to use. The other materials were used as received.

B. Procedures

(1) PREPARATION OF POLYURETHANES (PU): A resin kettle under dry nitrogen was charged with H_{12}MDI. A mixture of 1,4-butanediol, or N-methyldiethanolamine, and poly-(oxytetra-methylene) glycol (PTMO) was added at an NCO/OH ratio of 2:1.

The reaction was carried out at 100°C for about two hours until the theoretical isocyanate content, as determined by the di-n-butylamine titration method (27), was reached. The PU prepolymer with or without tertiary amine nitrogen groups was dissolved in dry MEK to obtain a prepolymer solution of 30-40% solids. It was then mixed with a mixture of 1,4-BD/TMP (4:1 by equiv. ratio) at an NCO/OH = 1.05/1.0 ratio in the presence of T-12 catalyst (0.05% based on total weight). The reaction mixture was cast in a metal mold treated with a release agent at ambient temperature. After standing 3-5 hours at room temperature, the mold was placed in an oven and post-cured at 100°C for 16 hours. The samples were then conditioned in a desiccator for one week before testing.

(2) VINYL CHLORIDE COPOLYMER: A copolymer from vinyl chloride, vinyl acetate and maleic acid (VMCC) (1% of maleic acid) was dissolved in MEK to obtain a homogeneous solution of about 30% solids. A stabilizer Thermolite 25, was added to the above solution at 0.5% (by wt.) based on the solids content of VMCC.

(3) PREPARATION OF Pseudo-IPNs: The prepolymer and VMCC solutions were mixed at different PU/VMCC ratios. The catalyst and the crosslinking agent for PU were added at an NCO/OH ratio of 1.05. The mixtures were then cast in metal molds (treated with a release agent) at ambient temperature to obtain film samples of ca. 40-60 mils thickness for determination of the stress-strain properties. The PU cross-linking reaction was carried out at 80°C for 2-3 hours and postcured at 100°C for 16 hours. The film samples were then conditioned in a desiccator for one week before testing. The above solutions were also coated on

TABLE 1 RAW MATERIALS

Designation	Description	Supplier
Terathane 1000(PTMO)	Poly(1,4-oxytetramethylene) glycol MW=1000	Du Pont Co.
$H_{12}MDI$	.Dicyclohexylmethane 4,4' diisocyanate	Mobay Chem Co.
1,4 BD	.1,4-Butanediol	Du Pont Co.
TMP	.Trimethylolpropane	Celanese Chem.Co.
T-12	.Dibutyltin dilaurate	M & T Chem.Co.
N-MDEA	.N-methyldiethanolamine	Penwalt Co.
VMCC	.Poly(vinylchloride-vinyl acetate-maleic acid). MW = 15000, maleic acid = 1%	Union carbide Corp.
Thermolite 25	.Thermostabilizer of PVC	M & T Chem.Co.
MEK	.Methylethyl ketone	Eastman Chem.Co.
-	.Molecular sieves 4A	Eastmen Chem.Co.

aluminum panels for measurement of the Gardner impact and lap shear strength.

Instrumental Techniques and Measurements

The tensile strength, modulus, and elongation at break were measured on an Instron Tensile Tester at a crosshead speed of 20 in./min (ASTM D-412) and the hardness by means of a Shore A Durometer (ASTM D-2240).

The impact resistance of the pseudo-IPN coatings was measured on a Gardner-SPI modified Variable Height Impact Tester using both the direct and the indirect techniques.

A TMS-2 Thermomechanical Analyzer (Perkin-Elmer) was used to determine the glass transition temperatures of the ionomer pseudo-IPNs at temperatures ranging from $-100^{o}C$ to $+100^{o}C$ and 0.01 mm of penetration range, 80g of penetrating weight, and a heating rate of $10^{o}C$/min.

The chemical resistance was measured by placing the coated panels into a 10% NaOH of HCl solution at room temperature for one week.

The reverse side and the edges of the panels were coated with wax to protect the metal surfaces from chemical attack.

To observe the morphology of the pseudo-IPNs with and without opposite charge groups, samples were prepared by freeze-fracturing in liquid nitrogen and applying a gold coating of approximately 200 A thickness. Micrographs were obtained by using a Phillips Scanning Electron Microscope Model SEM 505.

Results and Discussion

A. Effect of Different Compositions of PU/VMCC on the Mechanical Properties of Pseudo-IPN Coatings with and without Opposite Charge Groups

As shown in Figures 1 and 2,the pseudo-IPN coatings with opposite charge groups exhibited improved properties due to neutratization of the carboxyl groups in VMCC by the tertiary amine nitrogen in the PU,as shown in the following schemes:

$$-COO^{\ominus} \cdots\cdots\cdots \overset{\oplus}{H-N}-CH_3$$

As a result, the miscibility between the PU network and the VMCC linear chains was greatly improved and both tensile and lap shear strength were enhanced compared to the materials without opposite charge groups in which only hydrogen bonding between carboxyl and urethane could form instead of ionic bonds.

It has been known that the miscibility between poly(vinyl chloride) and polyesters depends upon the CH_2/C=O ratio, (28) and that the chain length of the

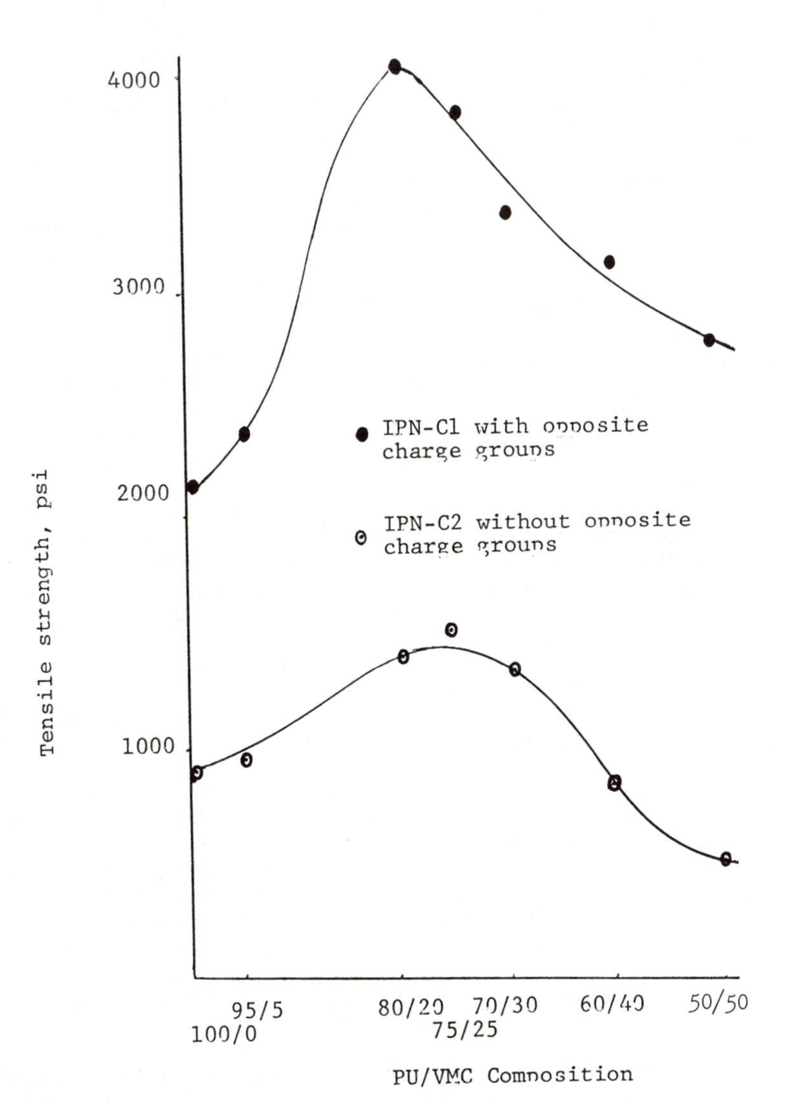

Figure 1. Effect of different compositions of PU/VMCC on the tensile strength of semi-IPN coatings with or without opposite charge groups.

gylcol in the esters plays an important role in the miscibility. (28) Pandyopadhyay and Shaw (29) indicated that the polyurethane based on PTMO exhibited poor miscibility with PVC compared to one based on polycaprolactone glycol due to the fact that the PTMO-based PU physically bonded with PVC could not be produced as readily as with poly-caprolactone glycol-based PU with PVC. (28)

The hydrogen bonding between the carboxyl groups in VMCC and the urethane groups in the PU without tertiary amine nitrogen was not strong enough to improve the miscibility of the two polymer systems presumably due to the very low concentration of carboxyl groups in the VMCC. The pseudo-IPNs without opposite charge groups exhibited low stress-strain properties and low lap shear strength. On the other hand, the pseudo-IPNs with opposite charge groups, even at low concentration of ionic bonds, exhibited relatively high stress-strain properties and lap shear strength.

Increasing the concentration of VMCC resulted in more brittle materials with lower tensile and lap shear strength.

The maximum tensile strength of the pseudo-IPNs appeared at about the 80:20 ratio of PU/VMCC for both pseudo-IPNs with and without opposite charge groups. Presumably the maximum entangle-ment between the VMCC chains and the PU network also occurred at this composition (Figure 1).

The maximum lap shear strength of the pseudo-IPNs, however, appeared at about the 75:25 composition for ionomer pseudo-IPNs with opposite charge groups and 70:30 composition for the pseudo-IPNs without opposite charge groups (Figure 2).

All the pseudo-IPNs without opposite charge groups exhibited very poor elongation due to their poor miscibility. The relatively high tensile strengths of the 75:25 and the 80:20 ratios of PU/VMCC could be due to the higher degree of interpene-tration between the chains of VMCC and the PU networks resulting from improved miscibility because of hydrogen bonding between the carboxyl and urethane groups (Figure 1).

B. Gardner Impact Strength of Pseudo-IPNs with and without Opposite Charge Groups

The ionomer pseudo-IPN coatings with opposite charge groups gave better results than the pseudo-IPNs without opposite charge groups due to their greater flexibility. The impact strength decreased with increasing concentration of VMCC due to its inherent brittleness (Table 2).

C. Solvent and Chemical Resistance of Pseudo-IPN Coatings

As shown in Table 3, both pseudo-IPNs with and without opposite charge groups exhibited poor solvent

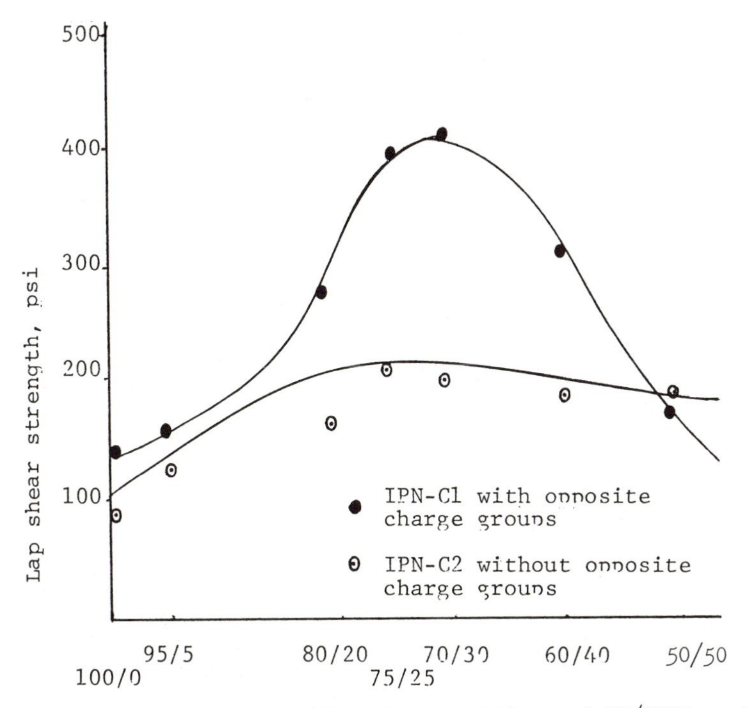

Figure 2. Effect of different compositions of PU/VMCC on the lap shear strength of semi-IPN coatings with or without opposite charge groups.

TABLE 2 IMPACT STRENGTH OF PSEUDO-IPN COATINGS FROM PU AND VMCC WITH AND WITHOUT CHARGE GROUPS

Composition of PU/VMCC	Opposite Charge Groups	Impact Strength, in. lb. Direct	Indirect
100:0	NO	160	160
95:5	YES	150	160
80:20	YES	150	160
75:25	YES	150	150
70:30	YES	140	150
60:40	YES	140	140
50:50	YES	140	120
100:0	NO	160	160
95:5	NO	140	160
80:20	NO	140	130
75:25	NO	130	130
70:30	NO	130	130
60:40	NO	130	120
50:50	NO	120	110

TABLE 3 SOLVENT AND CHEMICAL RESISTANCE OF PSEUDO-IPN COATINGS

Composition of PU/VMCC	Opposite Charge Groups	Solvent Resistance a)		Chemical Resistance b)	
		MEK	Xylene	10% HCI	10% NaOH
100:0	NO	>50	>50	Failure	Failure
95:5	YES	48	45	Failure	Pass
80:20	YES	48	45	Pass	Pass
70:30	YES	41	40	Pass	Pass
75:25	YES	39	38	Pass	Pass
60:40	YES	36	35	Pass	Pass
50:50	YES	34	33	Pass	Pass
100:0	NO	>50	>50	Failure	Failure
95:5	NO	48	48	Failure	Failure
80:20	NO	45	48	Failure	Failure
70:30	NO	40	46	Pass	Pass
75:25	NO	37	46	Pass	Pass
60:40	NO	37	45	Pass	Pass
50:50	NO	34	40	Pass	Pass

(a) Double rub with cotton saturated with solvent the highest rub time without change of the surface was recorded.
(b) At room temperature for one week.

resistance with increasing concentrations of VMCC due to the fact that VMCC is a thermoplastic, linear polymer. In contrast to the solvent resistance, the chemical resistance of pseudo-IPN coatings increased with increasing concentration of VMCC. It is interest-ing to note that the ionomer pseudo-IPN coatings with opposite charge groups exhibited better solvent and chemical resistance than those without opposite charge groups, presumably due to the fact that a high degree of interpenetration occurred between the chains of VMCC and the PU networks, which was brought about by formation of ionic bonds between opposite charge groups.

D. Glass Transition Temperatures of Pseudo-IPN Coatings

The Tg's for both pseudo-IPNs with and without ionic bonds (or opposite charge groups) between the two polymers are shown in Figure 3. Ionomer pseudo-IPNs with ionic bonds only gave single Tg's for each composition. The Tg's increased with an increase of VMCC in PU/VMCC composition due to the introduction of rigid VMCC with a high Tg. In contrast, pseudo-IPNs without ionic bonds gave two Tg's for each composition except at 100% and 95:5 ratios of PU/VMCC. These results imply that some phase separation had occurred in these pseudo-IPN systems due to the fact that the free energy of mixing in the systems may be either positive or zero because of the lack of physical interaction between the two polymers (ionic bonds).

As shown in Figure 3, the glass transition temperatures of both PU and VMCC for pseudo-IPNs without ionic bonds shifted inward with the Tg of PU shifting to higher temperatures and the Tg of VMCC shifting to lower temperatures in the pseudo-IPNs systems without ionic bonds. This implies that some interpenetration between the PU network and the VMCC polymer chains had taken place. Only the 95:5 composition of PU/VMCC exhibited a single Tg, presumably due to the fact that pseudo-IPNs with low concentration of VMCC could be compatible through the hydrogen bonding between the urethane groups and the carboxyl groups of VMCC.

Figure 3 shows the effects of composition on the Tg's for both ionomer and nonionomer pseudo-IPN. They exhibit good linearity for both pseudo-IPNs. Nonionomer pseudo-IPN reveals two lines due to the phase separation. The bottom most one is PU and the top most one is VMCC, but they shifted inward due to the formation of pseudo-IPN.

E. Morphology of Pseudo-IPN Coatings by Scanning Electron Microscopy (SEM)

Micrographs of pseudo-IPN coatings without opposite charge groups are shown in Figure 4 (from A-1 to C-1) at different compositions of PU/VMCC and magnifications. Figure 4 (A-1 and A-2) are micrographs

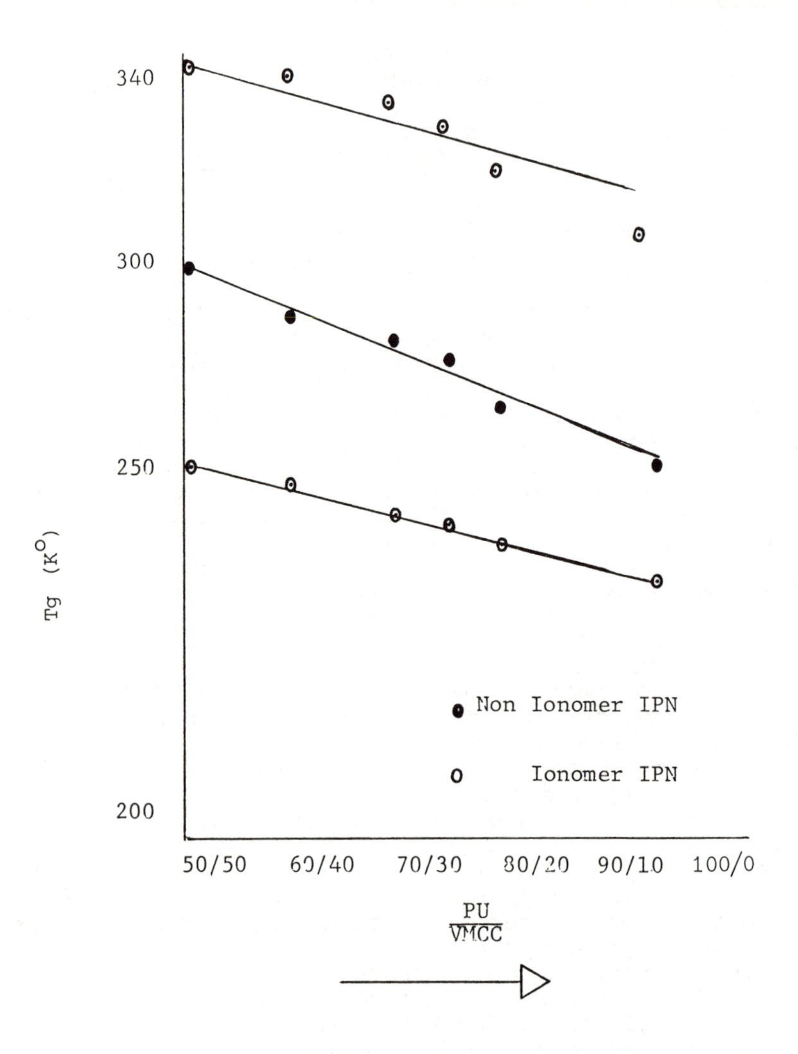

Figure 3. Effect of the composition in the ionomer and nonionomer IPN on the glass transition temperature.

A-1 PU/VMCC=100/0
(no charge groups)

B-1 PU/VMCC=60/40
(no charge groups)

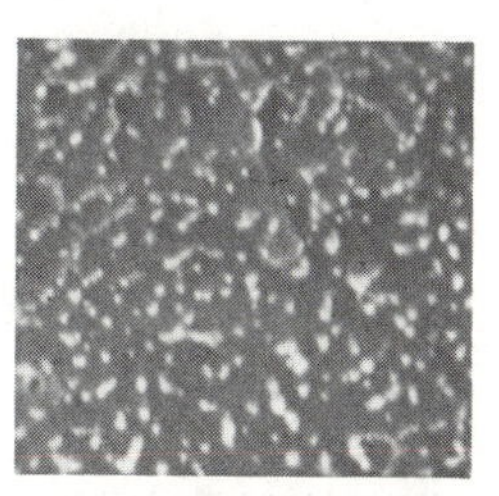

C-1 PU/VMCC=50/50
(no charge groups)

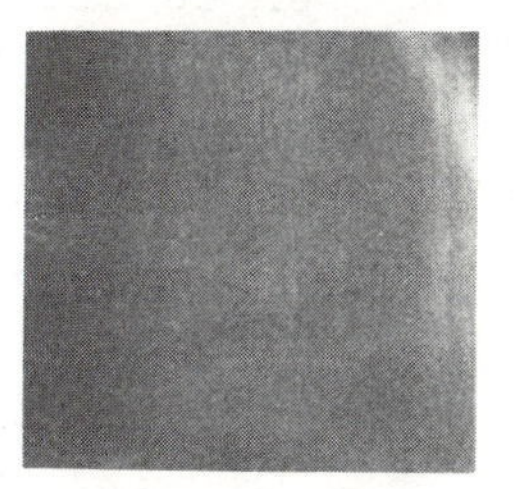

A-2 PU/VMCC=100/0
(with charge groups)

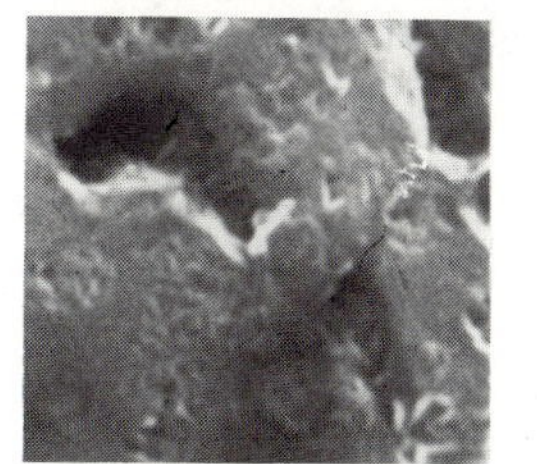

B-2 PU/VMCC=60/40
(with charge groups)

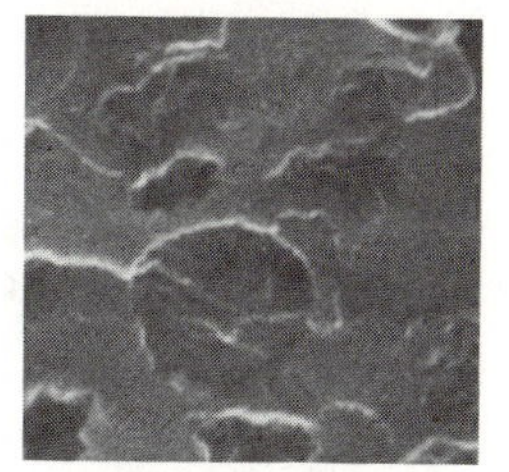

C-2 PU/VMCC=50/50
(with charge groups)

Figure 4. Micrographs of semi-IPN with and without opposite charge groups by scanning electron microscopy (SEM).

of 100% polyurethanes at 5 x 10^3 and 10^4 magnification, respectively. In Figure 4 (B-1 & C-1), the composition of PU/VMCC was changed from 80:20 to 50:50. All of the micrographs exhibited obvious phase separation, in which the white particles of VMCC dispersed into the dark matrix of the PU. Upon increasing the concentration of VMCC, the area of the dispersed phase of VMCC was increased with the VMCC particles in close proximity to each other.

In contrast, the micrographs of the ionomer pseudo-IPN coatings with opposite charge groups, Figure 4 (A-2 to C-2), did not reveal any phase separation. No white particles of the VMCC phase were visible in the dark matrix of the PU phase. Presumably the ionic bonds between the carboxyl and tertiary amine groups provided the best opportunity for interpenetration between the linear chains of VMCC and the networks of PU to prevent any possible phase separation from the ionomer pseudo-IPN microphase. The physical properties of ionomer pseudo-IPNs with opposite charge groups were obviously improved compared to the pseudo-IPNs without opposite charge groups as previously mentioned (see Figures 1 and 2).

Conclusions

The miscibility between poly (vinyl chloride) and poly-urethanes based on poly (oxytetramethylene) glycol can be improved by introducing opposite charge groups to form ionic bonds. The improvement in miscibility from ionic bonds between the two polymer systems provided the best chance for interpene-tration between the linear chains of VMCC and the PU networks in order to obtain good physical properties of the ionomer pseudo-IPN coatings from PU and VMCC.

REFERENCES

(1) Ohno, N. and Kumanotami, J., Polymer J., (1979), 11, 947.
(2) Yoshimura, N. and Fugimoto, K., Rubber Chem. Technol , (1969), 42, 1009.
(3) Eisenberg, A., Smith, P., and Zhou, Z. L., Polymer Eng.Sci., (1982), 22, 1117.
(4) Eisenberg, A., Proceedings of the 28th Macromolecular Symposium, Amherst, MA, p. 877 (1982).
(5) Xiao, H. X., Frisch, K. C., and Frisch, H. L., J. Polymer Sci., Polymer Chem. Ed., (1984), 22, 1035.
(6) Cassidy, E. F., Xiao, H. X. Frisch, K. C., and Frisch, H. L., J. Polymer Sci., Polymer Chem. Ed., (1984), 22,1851.
(7) Cassidy, E. F., Xiao, H. X., Frisch, K. C., and Frisch,H. L., J. Polymer Sci., Polymer Chem. Ed., (1984), 22,2267.
(8) Rutkowska, M. and Eisenberg, A., Macromolecules, (1984), 17, 821.
(9) Rutkowska, M. and Eisenberg, A., J. Appl. Polymer Sci., (1984), 29, 755.
(10) Flory, P., "Principles of Polymer Chemistry," p. 55.
(11) Smith, K. L., Winslow, A. E., and Peterson, D. E.,Ind. Eng. Chem., (1959), 51, 1361.
(12) Djadoun, S., Goldberg, R. N., and Morawetz,H., Macromolecules, (1977), 10, 1015.
(13) Ting, S. P., Bulkins, B. J. , Pearce, E. M. and Kwei,T. K., J. Polymer Sci., Polymer Chem. Ed., (1981), 19,1451.
(14) Otocka, E. P. and Eirich, F. R., J. Polymer Sci., Part A-2, (1968), 6, 895, 913.
(15) Lecourtier, J., Lafuma, F., and Quivoron, C., Macromol.Chem., (1982), 183, 2021.
(16) Ohno, N. Abe. K. Tsuchida E., Macromol. Chem., (1978), 179, 755.

(17) Xiao, H. X., Frisch, K. C., and Frisch, H. L., J. Polymer Sci., Polymer Chem. Ed., (1983), 21, 2547.

(18) Sulzberg, T. and Cotter, R. J., J. Polymer Sci., Part A-1 (1970), 8, 2747.

(19) Ohno, N. and Kumanotami, J., Polymer J., (1979), 11, 947.

(20) Cantow, H. J., Massen, U., Northfleet-Neto, H., Percee, V.,and Schneider, H. A., IUPAC 28th Macromolecular Symposium,1982, Preprint No. 847.

(21) Smith, P. and Eisenberg, A., J. Polymer Sci., Polymer Lett.Ed., (1983) 21, 223.

(22) Zhou, Z. L. and Eisenberg, A., J. Polymer Sci., Polymer Phys. Ed., (1983), 21, 595.

(23) Moacanin, J. and Cuddihy, E. F., J. Polymer Sci., Part C,(1966), 14, 313.

(24) Wetton, R. E., James, D. B., and Whiting, W., J. Polymer Sci., Polymer Lett. Ed., (1976), 14, 577.

(25) Eisenberg, A., Ovans, K., and Yoon, H. N., Adv. Chem Ser.,No., 187, 267 Am. Chem. Soc., (1980).

(26) Harc, M. and Eisenberg, A., Macromolecules, (1984), 17,1335.

(27) David, D. J., "Analytic Chemistry of Polyurethanes," Wiley- Interscience, New York 1969 Vol. XVI, p. 138.

(28) Prud'homme, R. E., Polymer Eng. Sci., (1982), 22, No. 2, 90.

(29) Pandyopadhyay, P. K. and Shaw, M. T., J. Appl. Polymer Sci., (1982), 27, 4323.

(30) Donatelli, A. A., Thomas, D. A., and Sperling, L. H.,Polymer Sci. Technol., (1974), 4, 375.

(31) Lipko, J. D., George, H. F., Thomas, D. A., Hargest, S. C.,and Sperling, L. H., J. Appl. Polymer Sci., (1979), 23, (9), 2739.

(32) Frisch, H. L., Klempner, D., Yoon, H. K., Frisch, K. C.,Macromolecules, (1980), 13, 1016.

(33) Yeo, J. K., Sperling, L. H., and Thomas, D. A., J. Appl.Polymer Sci., 26 (10), 3283 (1981).

(34) Cassidy, E. F., Xiao, H. X., Frisch, K. C., and Frisch,H. L., J. Polymer Sci., Polymer Chem Ed., 22, 1839 (1984).

RECEIVED December 22, 1987

Chapter 22

Liquid Crystalline Oligoester Diols as Thermoset Coatings Binders

Adel F. Dimian and Frank N. Jones

Polymers and Coatings Department, North Dakota State University, Fargo, ND 58105

Liquid crystalline (LC) oligoester diols were synthesized and characterized. They were crosslinked with a hexakis(methoxymethyl)melamine resin (HMMM) to form enamels of high crosslink density. Crosslinking was effected at temperatures between T_m and T_i. Enamels made from LC diols are far harder and tougher than enamels made from non-LC diols. They are hardened without substantially increasing T_g, and they retain the elasticity associated with low T_g. Reasons for the observed property enhancement are unclear, but the effect is substantial.

The mesophases formed by liquid crystalline (LC) polymers are well known to impart strength, toughness, and thermal stability to plastics and fibers (1-5). While LC polymers have been widely studied, their potential utility as coatings binders seems to have been overlooked. Among the very few reports that may describe LC polymers in coatings are patents claiming that p-hydroxybenzoic acid (PHBA), a monomer commonly used in LC polymers, enhances the properties of polyester powder coatings (6-9).

Studies of crosslinked networks of LC polymers have been proposed (10), and prepared without disruption of the mesophase (11-14). Crosslinked elastomers were shown to retain the mesophase up to certain crosslink densities (15,16).

Here we report synthesis, properties, and crosslinking of thermotropic LC oligoester diols, and comparison of the properties of crosslinked enamels to those of enamels made from crystalline and amorphous diols.

LC diols (1a-1g) were synthesized by reacting 4,4'-terephthaloyldioxydibenzoyl chloride (TOBC) with excess aliphatic diols $HO(CH_2)_nOH$, n = 4-12. The semi-rigid, rodlike TOBC segments are well known to impart LC character to high polymers (17).

0097-6156/88/0367-0324$06.00/0

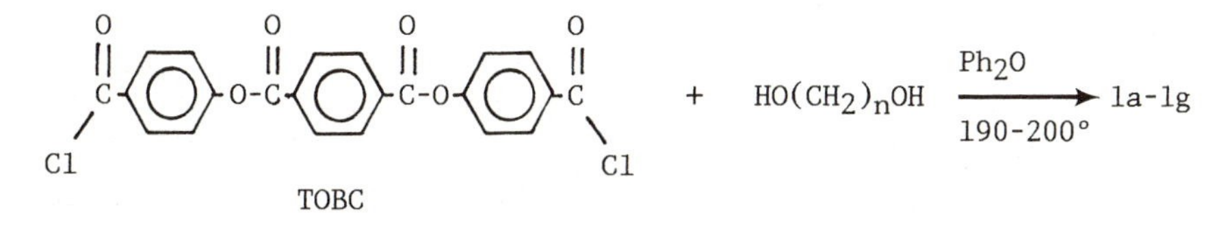

Two types of non-LC oligoester diols were prepared for comparison. In one type, 2a-2g, liquid crystallinity was disrupted by substituting adipoyl chloride for terepthaloyl chloride in TOBC. The structures of 1a-1g and 2a-2g may be represented thus:

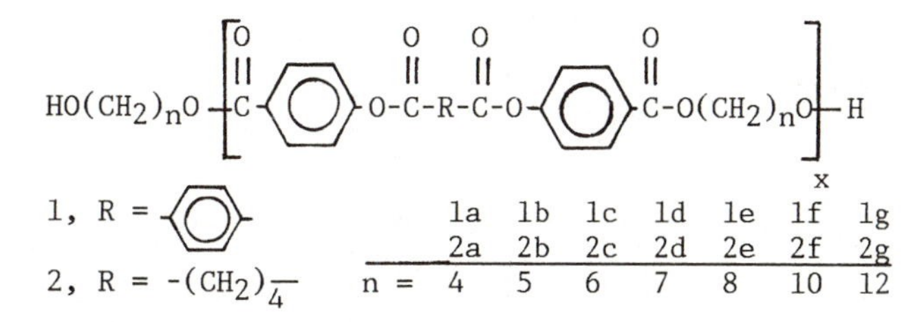

A second series of non-LC diols was amorphous glycol adipates 3a-3g. These were also prepared by Shotten-Baumann oligomerization:

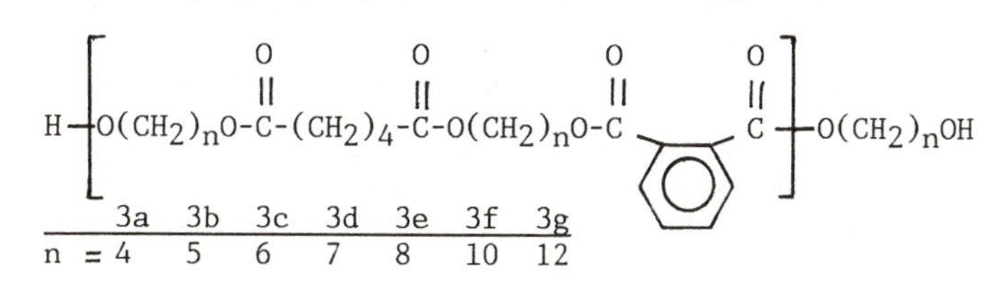

Clear coatings containing oligoester diols 1, 2, and 3 crosslinked with a hexakis(methoxymethyl)melamine (HMMM) resin were prepared, baked and tested.

Experimental Details

Materials. Reagent grade materials were used as received. The hexakis(methoxymethyl)melamine (HMMM) resin was Cymel 303 (American Cyanamid). Bonderite 1000 pretreated cold-rolled 3"x9"x24GA steel panels (Parker) were used for enamel testing. Q-Panel Type DT 3"x6" 0.010" tin-plated steel panels were used for preparation of free films.

Preparation of 1a-1g. TOBC was prepared from terephthaloyl chloride and PHBA as described by Bilibin *et al*. (17). TOBC (0.005 mol), diol (0.025 mol), and diphenyl oxide (10 mL) were placed in 100 mL single-necked round-bottomed flask equipped with a magnetic stirring bar, a distillation adapter, and a septum. The flask was flushed with argon for 15 min, and was stirred and heated in an oil bath at 190-200° under slow argon flow. The reaction mixture became homogeneous after 5 min and the evolution of HCl was observed. The reaction was continued until the evolution of HCl was no longer detectable by moistened litmus paper (4-5 hr). The hot reaction mixture was poured cautiously into 100 mL of toluene and cooled. The oily residue that separated was dissolved in CH_2Cl_2, washed 3x with water, and dried over anhydrous $MgSO_4$. The solution was filtered and concentrated using a rotory evaporator. The residue was precipitated from methanol/water. Yields were 87-92% based on TOBC. ^1H NMR for 1c in $CDCl_3$: 1.4 (br), 3.6 (t), 4.2 (m) 6.8 (d), 8.1(m) ppm. FT-IR for 1c: 3420, 2960, 2938, 1720, 1606, 1512 cm^{-1}. LC diols 1a-1g had similar spectra.

Preparation of 2a-2g. The diacid chloride precursor was prepared by substituting adipoyl chloride for terepthaloyl chloride in Bilibin's procedure (17). Reaction of this precursor with diols was carried out as described for 1a-1g except that the products were not poured into toluene. Diols 2a-2g were resinous solids which solidified on standing.

Preparation of 3a-3g. A two-step Schotten-Baumann synthesis was performed by the method of Bilibin *et al.* (18). Diol (0.04 mol) and adipoyl dichloride (0.01 mol) were placed in 100 mL single-necked round-bottomed flask equiped with magnetic stirring bar and distillation adapter and septum. The system was flushed with argon for 15 minutes and placed in an oil bath at 90-100°. The reaction was continued until the evolution of HCl was no longer detectable by moistened litmus paper (2-3 hr). The reaction mixture was allowed to cool and phthaloyl dichloride (0.01 mol) was added. The system was flushed with argon for 15 min and heated in an oil bath at 160-170° The reaction was continued until the evolution of HCl was no longer detectable by moistened litmus paper (2-3 hr). The product was not further purified. GPC of oligoester diol 3c: M_n 840, M_w 1180, PDI 1.4.

Enamel Preparation. Soluble oligoester diols (1b-1d, 2a-2g, and 3a-3g), HMMM, methyl isobutyl ketone (MIBK) and *p*-toluenesulfonic acid (*p*-TSA) were thoroughly mixed in a 70/30/30/0.3 wt. ratio. The solution was cast on panels and baked at 150° for 30 min. Less soluble LC diols 1e-1g were melted, dispersed in MIBK, mixed with HMMM and *p*-TSA in the above proportions and immediately cast as films. Oligoester diol 1a was too insoluble for enamel formation.

^1H NMR spectra were recorded at 34° on a Varian Associates EM-390 90 MHz NMR spectrometer, using Me_4Si as internal standard. IR spectra were recorded at 25° on a Mattson Cygnus FT-IR using films cast on NaCl plates with polystyrene as standard. A DuPont model 990 thermal analyzer was used for differential scanning calorimetry (DSC) at heating rates of 10°/min. After T_m the temperature was held for 1 min before the scan was resumed. Capillary melting points were used to confirm the thermal data. M_n and M_w were determined by gel-permeation chromatography (GPC) with a Waters model 510 pump equipped with a model R401 refractive index detector, a model M730 data analyzer, and Ultrastragel 100 Å, 500 Å, 10^3 Å, and 10^4 Å columns. GPC was calibrated with polystyrene standards. M_n was also determined by vapor pressure osmometry (VPO) on a Wescan model 233 molecular weight apparatus at a current setting of 50μamp in toluene at 50°. The instrument was calibrated with sucrose octaacetate and solutions of unknowns ranging from 2.4 to 12.0g/L were used to determine the y-intercept. Mass analysis was performed by MicAnal, Tucson AZ. A Leitz Labolux microscope equipped with a polarizing filter was used for optical micrographs at 500x magnification; diols were observed immediately after heating to T_m, enamels were observed at room temperature. Swelling tests were performed by the method of Jones and Lu (19).

Results

Oligoester Diols. The physical properties of LC oligoester diols 1a-1g measured by GPC, VPO, DSC, and polarizing optical microscopy are summarized in Table I.

Table I. Physical Properties of LC Diols 1a-1g

diol	n	M_{Th}[a]	M_n[b]	M_n[c]	M_w[c]	PDI[c]	T_m[d]	T_i[d]	texture
1a	4	550	570	480	720	1.5	110	204	-
1b	5	578	625	530	740	1.4	58	207	smectic
1c	6	606	690	570	810	1.4	75	349	smectic
1d	7	634	720	610	850	1.4	47	300	smectic
1e	8	662	780	650	910	1.4	82	302	smectic
1f	10	718	810	680	950	1.4	80	231	nematic
1g	12	774	910	720	1130	1.4	90	220	smectic

[a]theoretical molecular weight for x=1; [b]determined by VPO; [c]determined by GPC; [d]determined by DSC.

^1H NMR and IR spectra of diols 1a-1g, 2a-2g, and 3a-3g were consistent with the assigned structures. Slightly high H analyses (0.84% higher than theoretical for x=2, 0.11% for x=1) suggested that small amounts of unreacted $HO(CH_2)_nOH$ were present in the products.

The LC nature of diols 1a-1g was demonstrated by DSC (Fig. 1). Two first order transitions were observed. The lower transition temperature appeared to be the crystalline-mesophase transition (T_m), and the higher transition temperature the mesophase-isotropic transition (T_i). Other transitions were not evident in the DSC. The thermal data revealed an odd-even spacer effect for T_m.

The mesophases of LC diols 1a-1g were also observed directly in polarized optical micrographs taken immediately after melting the sample. Textures were identified only by comparison with published micrographs (20), and are therefore tentative. A nematic texture is observed for 1f (Fig. 2), while more highly ordered smectic textures are observed for 1b-1e and 1g (Fig. 3).

In contrast, diols 2a-2g and 3a-3g are not LC materials. Diols 2a-2g were shown to be crystalline by microscopy and by the existence of single first order transitions in the DSC. Diols 3a-3g appeared amorphous in the micrographs and had no first order transitions in the DSC.

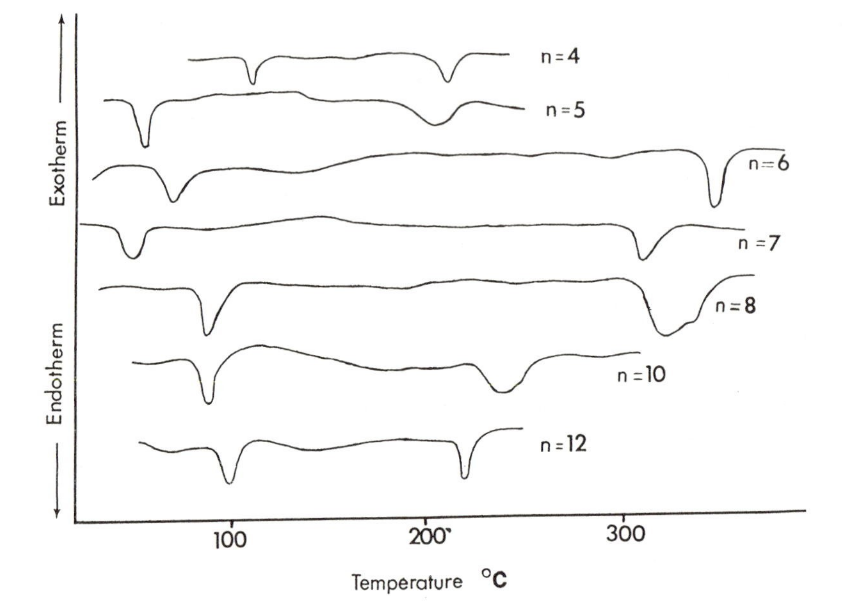

Figure 1. DSC thermograms of LC diols 1a–1g, heating rate 10 °C/min.

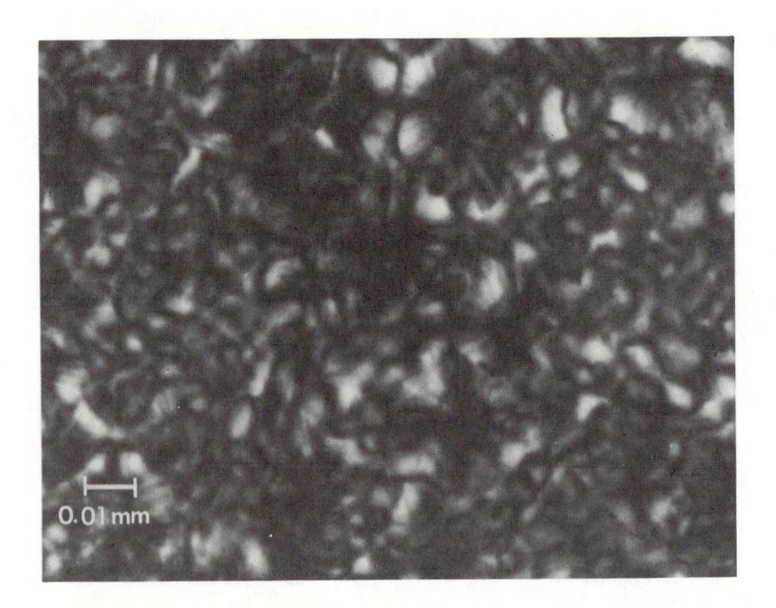

Figure 2. Polarizing optical micrograph of 1g, 400 X magnification.

Crosslinked Enamels--70:30 ratio. All enamels described in this section were obtained using a 70:30 wt. ratio of diol:HMMM. Diols 1b-1g, 2a-2g, and 3a-3g were crosslinked with HMMM at 150°. Cure temperature was within the mesophase temperature range for 1b-1g. Enamel formation of 1a was not feasible because of poor miscibility. The properties of the enamels are summarized in Table II.

Table II. Properties of Enamels Prepared from 1b-1g and 2a-2g. Diol:HMMM:p-TSA 70:30:0.3 by wt., cure cycle 150°/30 min

	Enamels from LC diols						non-LC	
	1b	1c	1d	1e	1f	1g	2a-g	3a-g
spacer length (n)	5	6	7	8	10	12	4-12	4-12
reverse impact[a]	>80	50	>80	50	50	55	8-15	30-45
direct impact[a]	>80	50	>80	50	50	50	10-15	30-45
pencil hardness[b]	6H	6H	5H	6H	5H-6H	6H	H-2H	HB-H
adhesion[c]	5B	5B	5B	5B	5B	5B	5B	5B
acetone rubs[d]	200	200	200	200	200	200	200	200
flexibility[e]	100%	100%	100%	100%	100%	100%	100%	100%
dry film thick.[f]	0.5	0.5	0.5	0.5	0.5	015	0.5	0.5
T_g[g]	17°	35°	23°	16°	15°	22°	17-28°	16-25°
appearance	--------------transparent, glossy-----------------							

[a]ASTM-D 2794, units are in-lb, impact tester limit was 80 in.lb.; [b]ASTM-D 3363; [c]ASTM-D 3359,5B is 100% cross-hatch adhesion; double rubs; [e]ASTM-D 522; [f]ASTM-D 1400, General Electric thickness gauge type B, units are 1/1000 in.; [g]onset of transition, determined by DSC.

All enamels had excellent adhesion, solvent resistance, and flexibility. Enamels made from LC diols 1b-1g were far superior to those made from control diols 2a-2g and 3a-3g in both hardness (5H-6H vs. H-2H; 5H corresponded to about 20 KHN on a Tukon tester) and impact resistance (50 to >80 in-lb vs. 8 to 15 in-lb). Odd spacers 1b and 1d afforded the best properties in these formulations. Spacer variations did not measurably affect enamel properties in the control oligoesters.

DSC thermograms of the crosslinked enamels revealed onset of glass transitions (T_g) ranging from 15° to 35° for all three types of enamels. Attempts to detect first order transitions in the DSC corresponding to T_m or T_i were unsuccessful due to large exotherms starting at about 90°. An odd-even pattern was not observed.

Polarized optical micrographs revealed birefringent heterogeneous regions in the crosslinked enamels of 1b-1g (Fig. 4). Enamels of 2a-2g and 3a-3g appeared amorphous. IR spectra of the baked LC and amorphous enamels had bands attributable to unreacted -OH groups at 3420 cm^{-1} (-OH stretch).

Crosslinked Enamels--Variable Wt. Ratio. All enamels described in this section were prepared from LC diol 1c with the diol:HMMM ratio varied from 90:10 to 60:40. Results are summarized in Table III. The ratio of the maximum absorbances of -OH to -C=O in the FT-IR spectra and swelling tests provide rough estimates of the extent of the crosslinking reactions.

Figure 3. Polarizing optical micrograph of 1c, 400 X magnification.

Figure 4. Polarizing optical micrograph of enamel made from 1c, 400 X magnification.

FT-IR data indicate that at 80:20 and 70:30 ratios a significant fraction of the -OH groups are unreacted even though more than a stoichiometric ratio of HMMM is present. The stoichiometric ratio is estimated at about 84:16.

Swelling data indicate that crosslink density in the continuous phase of the 70:30 and 60:40 networks is high. Crosslink densities were estimated from the data in Table III by the method of Hill and Kozlowski (21). Results were: for 80:20, $\nu_e = 10^{-3}$ moles of elastically effective network chains/cm^3; for 70:30, $\nu_e = 2.5 \times 10^{-3}$ chains/cm^3; for 60:40, $\nu_e = 4.3 \times 10^{-3}$ chains/cm^3. These estimates suggest that the crosslink densities are within the range reported for conventional, highly crosslinked acrylic:HMMM and polyester:HMMM enamels (19,20).

Film properties of this series of enamels is shown in Table III. Enamels made with a 90:10 lc:HMMM ratio were tacky. Microscopic birefringent heterophases were visible immediately after baking, but they disappeared within a week. Enamels with higher proportions of HMMM also had microscopic heterophases, and they remained in the films indefinitely. The best overall properties were attained with the 80:20 ratio. This ratio also had the highest population of visible heterophases.

Table III. Enamel Properties with Varying HMMM Concentrations Prepared from LC Diol 1c

diol:HMMM	IR	film properties				
	OH/C=O	%swell	pencil	impact res.	flex[a]	heterophases[b]
90:10	0.745	-	-	-	-	none
80:20	0.464	35.1%	3H-4H	>80in.lb	100%	most
70:30	0.250	14.2%	5H	25in.lb.	100%	few
60:40	0.099	7.5%	7H	10in.lb.	0%	least

[a]flexibility, ASTM-D 522; [b]from optical micrographs taken after 1 week.

Discussion

Oligoesters Diols. The synthetic method used to make oligoester diols 1a-1g was adapted from Bilibin's method (17) for making main chain LC high polymers. A five-fold excess of $HO(CH_2)_nOH$ was used to suppress molar mass. Spectral, chromatographic and analytical evidence indicated that the expected products were obtained.

GPC, VPO and analytical data suggested that the structures with x=2 and x=1 predominate; smaller amounts of structures with x>2 and of $HO(CH_2)_nOH$ are probably present in 1a-1g and 2a-2g. The discrepancy between GPC and VPO data is not surprising in view of the large structural differences between the polystyrene GPC standards and the oligomers studied.

The thermal behavior of 1a-1g observed by DSC (Fig. 1) confirms the presence of mesophases and is typical of low molecular weight thermotropic LC materials (20). The lower T_ms for 1b and 1d are consistent with the higher entropy of activation for crystallization of odd-n spacers, demonstrated in several main chain LC polymers (23). The apparent absence of nematic-smectic transitions in the DSC suggests the observed morphology is stable throughout the mesophase temperature region. The nematic texture of oligomeric LC diol 1f is the same as reported for the homologous LC high polymer (24).

The mesophases of the uncrosslinked LC diols can be observed below T_m after melting because formation of fully ordered crystals is slow. The mesophases were not observable after one week at room temperature.

Crosslinked Enamels. Properties of enamels made from LC diols optimized with slight variations in HMMM concentration gave 5H hardness (20 KHN Tukon tester) and >80 in-lb. impact resistance. The best properties obtainable for enamels made from non-LC diols were 2H hardness and 45 in-lb. impact resistance.

The greatly enhanced hardness and impact resistance of enamels made from LC diols is not simply explainable by the monomer raising the T_g of the coating. In fact, T_gs of the crosslinked enamels of 1b-1g are abnormally low for hard coatings and are similar to the much softer control enamels (Table II).

Direct evidence for the nature of the microscopic birefringent domains in the cured enamels is lacking. They may be crystalline, liquid crystalline, or mixed. The substantial fraction of unreacted -OH groups in the 80:20 and 70:30 lc:HMMM enamels suggests that the -OH groups may be buried in heterogeneous domains inaccessible to the HMMM. The bulk properties of such domains are probably similar to those of the original oligoester diols. If so, they are crystalline at room temperature and liquid crystalline at elevated temperatures. In the interphase between domains and the crosslinked continuous phase there may be regions that are immobilized by the network and are liquid crystalline at room temperature.

It is disappointing that DSC exotherms, presumably caused by further crosslinking reactions, interfered with the detection of LC phase transitions in the cured enamels. These exotherms were broad, and recurred when the samples were cooled and reheated. This behavior may be attributable to slow consumption of -OH groups in the domains or to slow self-condensation of HMMM (21,22). If these reactions occur to any appreciable extent, their exotherms are likely to obscure the relatively weak endotherms produced by the low concentration of the LC phase.

It is interesting to speculate about what happens during crosslinking. At 25° the 70:30 diol:HMMM mixture appears to be a heterogeneous dispersion of LC diol in liquid HMMM. As this mixture is heated to 150° two processes presumably occur: [1] partial or complete dissolution of the dispersed LC diol in HMMM and [2] reaction of the -OH groups with HMMM to form a covalently crosslinked network (22). As crosslinking proceeds a stage will be reached, probably near the gel point, at which the material is immobilized. Dissolution will virtually stop, but the crosslinking reaction can continue. The morphology of the cured film will probably be fixed at the stage where dissolution of LC particles stops. After cooling, undissolved LC particles would probably crystallize. Properties will be affected by the relative rates of the dissolution and crosslinking processes. Relative rates will be governed by factors such as diol:HMMM ratio, size of the dispersed particles, catalyst levels and heating rates.

Alternatively, it is possible that the LC diols completely dissolve during crosslinking and that microscopic birefringent heterophases form when the crosslinked network cools. This possibility seems unlikely since mobility within the highly cross-linked network is restricted, prohibiting association of large numbers of mesogenic units. Finkelmann and Rehage have stated that LC domains cannot form in highly crosslinked networks (15).

The reason for the observed property enhancement remains a matter of conjecture. The observation that presence of microscopic heterophases in the enamels is associated with enhanced properties suggests a cause and effect relationship, but this correlation may be incidental. Property enhancements could also be attributed to submicroscopic heterogeneous regions or to the inherent stiffness of the oligomer structure. An interesting possibility is that properties may be enhanced by mesogenic units that dissolve during crosslinking and then associate within the network into very small aggregates, perhaps containing as few as two or three mesogenic units.

Conclusions

LC oligoester diols can be crosslinked with HMMM by carrying out the crosslinking reaction at temperatures between T_m and T_i. Cross-link density is high. Certain film properties of enamels made from LC diols are far superior to those of enamels made from non-LC diols. Films are hardened without substantially increasing T_g, and they retain the elasticity associated with low T_g. The mechanism of this property enhancement is uncertain, but the effect is substantial. Inclusion of LC diols in enamels offers vast possibilities for manipulation and improvement of film properties.

Acknowledgment

Financial support by the U.S. Environmental Protection Agency (Grant No. R-811217-02-0) is gratefully acknowledged.

Literature Cited

1. Flory, P.J. In Advances in Polymer Science, Liquid Crystal Polymers I; Gordon, M., Ed.; Springer-Verlag: New York, 1984; Vol. 59.
2. Schwarz, J. Macromol. Chem. Rapid Commun. 1986, 7, 21.
3. Shibaev, V.P.; Plate, N.A. Polym. Sci. U.S.S.R. 1977, 19, 1065.
4. Kwolek, S.L.; Morgan, P.W.; Schaefgen, J.R.; Gulrich, LW. Macromolecules 1977, 10, 1390.
5. Dobb, M.G.; McIntyre, J.E. In Advances in Polymer Science, Liquid Crystal Polymers II/III; Gordon, M., Ed.; Springer-Verlag: New York, 1984; Vol. 60/61.
6. Maruyama, K. et al., Japan. Kokai 75 40,629, 1975; Chem. Abstr. 1975, 83, 133572y.
7. Nakamura, K. et al., Japan. Kokai 76 56,839, 1976; Chem. Abstr. 1976, 85, 110175y.
8. Nogami, S. et al., Japan. Kokai 76 44,130, 1976; Chem. Abstr. 1976, 85, 79835n.

9. Nogami, S. et al., Japan. Kokai 77 73,929, 1977; Chem. Abstr, 1978, 88, 8624u.
10. De Gennes, P.G. Phys. Lett. A 1969, 28(11), 725-726.
11. Bhadani, S.N.; Gray, D. G. Mol. Cryst. Liq. Cryst. 1984, 102, 255-260.
12. Tsutsui, T.; Tanaka, R.; Tanaka, T. J. Polym. Sci., Polym. Lett. Ed. 1979, 17(8), 511-520.
13. Tsutsui, T.; Tanaka, R. Polymer 1981, 22(1), 117-123.
14. Aviram, A. J. Polym. Sci., Polym. Lett. Ed. 1976, 14(12), 757-760
15. Finkelmann, H.; Rehage, G. In Advances in Polymer Science, Liquid Crystal Polymers II/III; Gordon, M., Ed.; Springer-Verlag: New York 1984; Vol. 60/61.
16. Zentel, R.; Rechert, G. Makromol. Chem. 1986, 187 1915-1926.
17. Bilibin, A.Y.; Tenkovtsev, A.V.; Piraner, O.N.; Skorokhodov, S.S. Polym. Sci. U.S.S.R. 1984, 26, 2882.
18. Bilibin, A.Y.; Pashkovsky, E.E.; Tenkovtsev, A.V. Macromol. Chem. Rapid Commun. 1985, 6, 545-550.
19. Jones, F.N.; Lu, D.L. J. Coat. Technol. 1987, 59(751), 73-79.
20. Noel, C. In Polymeric Liquid Crystals; Blumstein, A., Ed.; Plenum Press: New York, 1984.
21. Hill, L.; Kozlowski J. Coat. Technol. 1987, 59(751), 63-71.
22. Jones, F.N. Polym. Mat. Sci. Eng. 1987, 55, 222-228.
23. Ober, C.K.; Jin, J.; Lenz, R.W. In Advances in Polymer Science, Liquid Crystal Polymers I: Gordon, M., Ed.; Springer-Verlag: New York, 1984; Vol. 59.
24. Lenz, R.W. J. Polym. Sci., Polym. Sym. 1985, 72, 1-8.

RECEIVED October 7, 1987

Chapter 23

Synthesis of Cross-Linkable Heterogeneous Oligoester Diols by Direct Esterification with *p*-Hydroxybenzoic Acid

Use in Coatings Binders

Daozhang Wang and Frank N. Jones

Polymers and Coatings Department, North Dakota State University, Fargo, ND 58105

A direct esterification procedure by which a linear polyester diol can be modified with p-hydroxybenzoic acid (PHBA) was demonstrated. The procedure appears adaptable to large scale production. The products are polydisperse oligomers in which phenolic end-groups appear to predominate. Oligomers containing 30 wt% or more of PHBA are probably liquid-crystalline. Baked enamels made by crosslinking these oligomers with a methylated melamine-formaldehyde resin were much harder and tougher than enamels made by crosslinking amorphous polyols and modified polyols derived from m-hydroxybenzoic acid (MHBA).

Liquid-crystal (LC) polymers are the subject of worldwide research and development. (1-5). Commercial films and plastics such as Kevlar (du Pont), Xydar (Dartco) and Ekonol (Sumitomo) utilize the remarkable ability of LC behavior to enhance physical properties. In the field of coatings, however, crystal formation in the film is usually avoided; few reports (6) of LC polymers in coatings binders are known to the authors.

The preceding chapter (7) described synthesis of model LC oligomeric diols and their incorporation into coatings binders of the baking enamel type. The enamels had far better hardness and impact resistance than control enamels made from amorphous or crystalline oligomeric diols. The model LC oligomers were synthesized by a Schotten-Bauman method that would be costly for large-scale production.

We have been seeking a method for synthesizing oligomeric LC polyols that could be adapted to economical production. Reactions of p-hydroxybenzoic acid (PHBA) were investigated. PHBA is commonly used as a monomer in LC polymers (3-4), and its use in binders for powder coatings has been described. (8-11).

0097-6156/88/0367-0335$06.00/0

This chapter is a preliminary report of our results. We found a procedure by which PHBA can be reacted with pre-formed amorphous oligoester diols at 230°C with minimal formation of phenol, an undesirable by-product. The products of such reactions are oligomers containing amorphous and ordered phases. They appear liquid crystalline. While the oligomer structures formed in the procedure described here differ from those of the previous paper, they form baked enamel films of exceptional hardness, toughness, and adhesion.

Experimental Details

Materials. Phthalic anhydride (PA), adipic acid (AA), neopentyl glycol (NPG), p-hydroxybenzoic acid (PHBA), salycilic acid, m-hydroxybenzoic acid (MHBA), xylene, and methyl isobutyl ketone (MIBK) were purchased from Aldrich. p-Toluenesulfonic acid (p-TSA) was purchased from Matheson. "Aromatic 150" solvent and "Resimene 746", a hexakis(methyloxymethyl)melamine (HMMM) resin were supplied by Exxon and Monsanto, respectively. "Bonderite 1000" pretreated cold-rolled steel panels 3"x9"x24GA were purchased from Parker.

Synthetic procedure. A linear oligoester diol was prepared by heating a mixture of PA, AA and NPG in a 1:1:3 mol ratio under N_2 at 230° with removal of H_2O until the acid number was 5 to 10 mg KOH/g.

A mixture of the above diol, PHBA, p-TSA and "Aromatic 150" was heated under N_2 in a 3-neck flask equipped with stirrer, Dean-Stark trap, condenser and thermometer. The PHBA/diol wt. ratio varied from 20/80 to 60/40; 0.2 wt. % of p-TSA was used. About 10 wt. % of "Aromatic 150" was used; the amount was adjusted to maintain the temperature at 230 +/- 3°. Distillate (cloudy H_2O) was collected in the Dean-Stark trap during 9 to 11 hr. The reaction mass was cooled to 115°, and MIBK was added to yield a solution (20/80 PHBA/diol ratio) or dispersion (other PHBA/diol ratios) of the crude polyol.

The linear oligoester diol was heated with salycilic acid and with MHBA using a similar procedure to yield modified polyols. Only 60% to 80% of theoretical distillate was obtained.

Purification. The crude polyols made from 20/80 and 30/70 PHBA/diol ratios were concentrated and dissolved in CH_2Cl_2. The solution was washed 5x with H_2O, dried with Na_2SO_4, and concentrated on a rotary evaporator. The residues were heated at 120° to constant weight. The crude polyols made from 40/60 to 60/40 PHBA/diol ratios were dispersed in CH_2Cl_2 and were purified similarly but were not washed with water. They were heated at about 80° under vacuum on a rotary evaporator to remove small amounts of volatile, crystalline material.

Synthesis of a PHBA-benzoic acid adduct. PHBA and benzoic acid in a 1:1.5 mol ratio were heated at 230+/-3° in the presence of 0.2 wt% of p-TSA and "Aromatic 150" as described above. The product mixture was cooled and diluted with an equal volume of a 1:1 mixture of "Aromatic 150" and cyclohexanone. The resulting paste was filtered, dried in an oven, pulverized, washed 5 times with hot water, and dried at 100°. The material was insoluble in hot MIBK.

Enamel preparation. Solutions or mixtures of polyol, HMMM and p-TSA in a 75/25/0.25 wt. ratio were cast on steel panels and were baked at 175° for the specified time. Dry film thicknesses were 20 to 25 µm.

Characterization and testing. FT-IR spectra were recorded using a Mattson Instruments, Inc. CYGNUS 25 spectrophotometer. Differential scanning calorimetry (DSC), gel permeation chromatography (GPC) and optical microscopy were effected essentially as described in the preceding chapter(7) except for preparation of samples for microscopy: LC polyols were cast on glass slides and were dried and observed at 25°; enamels were baked at 175° for 20 min on glass slides. Impact resistance, pencil hardness and Knoop-Tukon indentation hardness were tested according to ASTM-D 2794, ASTM-D 3363 and ASTM-D 1474 respectively. Solvent resistance was tested by spotting films with methyl ethyl ketone.

Results

Synthesis and characterization of polyols. The procedure described above yields PHBA-modified oligomers, apparently with minimal side reactions. The odor of phenol was barely detectable in the products, indicating that little phenol had been formed. p-TSA catalyst plays a crucial role. When p-TSA was not used in the 30/70 PHBA/diol reaction only 75% of theoretical distillate was collected, and the product smelled strongly of phenol. Solvent plays an important role by helping control temperature and by facilitating removal of water. If desired, the products can be purified as described to remove small amounts of unreacted PHBA and phenol.

Modification of the PA/AA/NPG diol with salicylic and with m-hydroxybenzoic acids apparently did not proceed to completion. The products appeared amorphous by polarizing microscopy.

The FT-IR spectrum of unmodified polyester is shown in Fig. 1A. A band at 3535 cm^{-1} (Fig. 1B) suggesting that the OH groups present are predominately phenolic. Because the average number of PHBA units per molecule in the 20/80 composition is only about 2.3 (see below), the spectrum suggests that a large population of structures having single PHBA units at chain ends is present. The spectrum of 30/70 PHBA/diol polyol has a peak at a slight shoulder at 3535 cm^{-1}. The spectra of 30/70 to 60/40 PHBA/diol polyols are similar, each having a strong peak at about 3365 cm^{-1} (Figs. 2A-2D).

Substraction spectra were obtained by setting the aliphatic C-H stretching band at 2969 cm^{-1} at about zero. The spectrum comparing 40/60 PHBA/diol polyol and unmodified polyester (Fig. 3) is typical. The aliphatic OH band at about 3535 cm^{-1} has been largely replaced by a phenolic OH band at about 3365 cm^{-1}. Subtraction spectra of the other oligomers (Fig. 4) yield similar conclusions.

Incorporation of aromatic groups into the oligomer was confirmed by presence of strong absorption in the 1580-1620 cm^{-1} region. The starting diol had much smaller peaks in this region.

Molecular weights determined by GPC are provided in Table 1. Also provided are rough estimates of the average number of PHBA units per number average molecule. These estimates were obtained by multiplying product M_n by the wt. fraction of PHBA charged and dividing the result by 120, the molar mass of PHBA minus water.

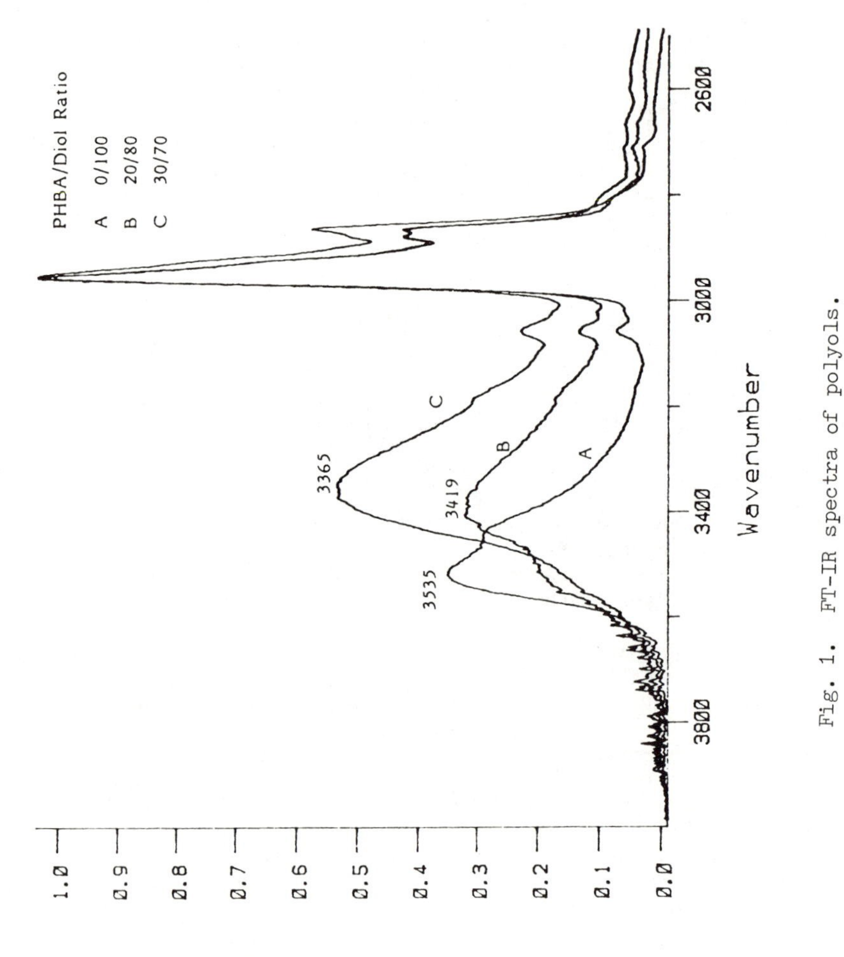

Fig. 1. FT-IR spectra of polyols.

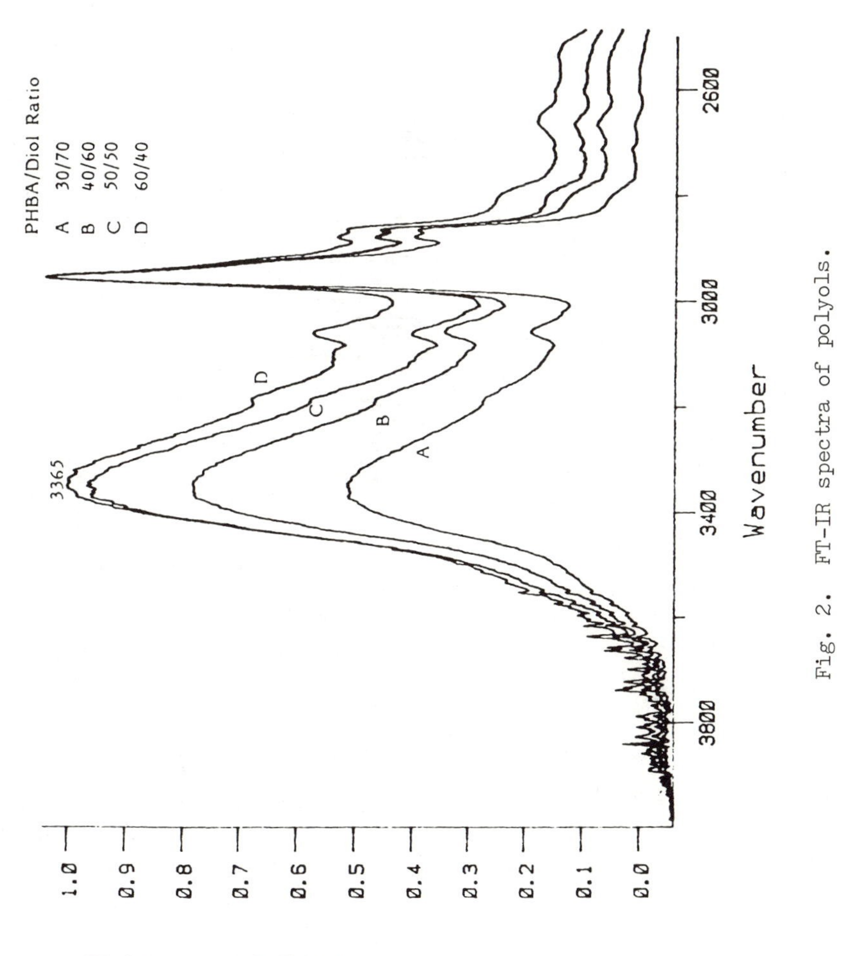

Fig. 2. FT-IR spectra of polyols.

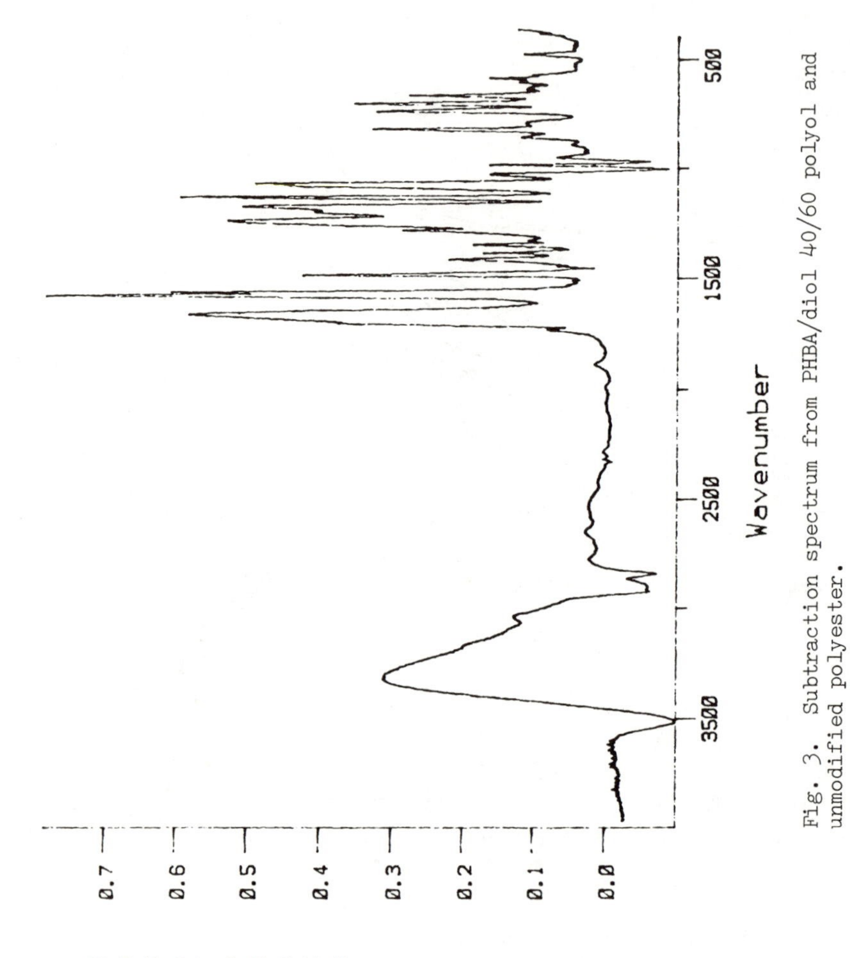

Fig. 3. Subtraction spectrum from PHBA/diol 40/60 polyol and unmodified polyester.

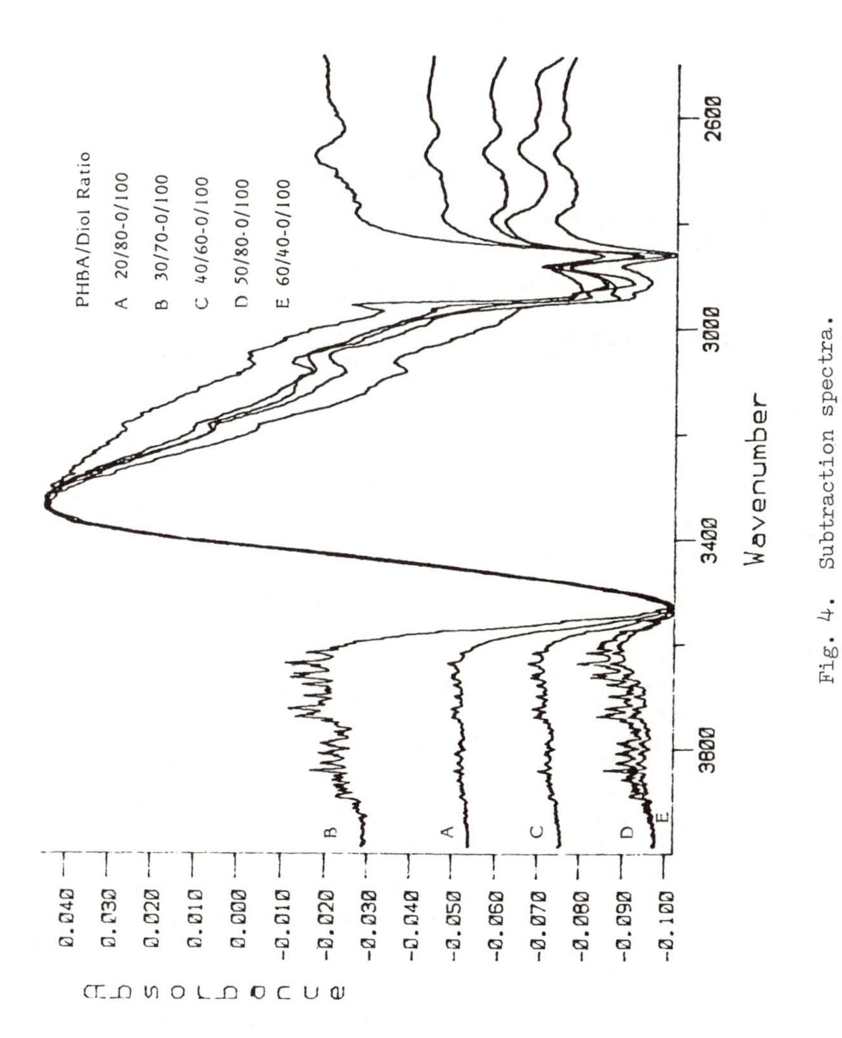

Fig. 4. Subtraction spectra.

Table 1. Gel Permeation Chromatography of Polyols

PHBA/diol wt	ratio mol	M_n	M_w	M_w/M_n	Est. avg. PHBA units/molecule
0/100	---	1200	2000	1.7	---
20/80	2.1/1	1400	2400	1.7	2.3
30/70	3.6/1	1100	1900	1.7	2.8
40/60	5.8/1	970	1600	1.6	3.2
50/50	8.8/1	870	1400	1.7	3.6
60/40*	13/1	830	1400	1.7	4.1

*Filtered to remove a small fraction of THF-insoluble material.

Like most synthetic oligomers, the products of the PHBA-diol reactions are mixtures of individual molecules having different structures and chain lengths. The above evidence strongly suggests that the predominant structures are phenol-tipped oligomers containing an average of two to four PHBA residues per molecule.

Morphology of oligomers. PHBA-containing oligomers were studied by DSC and by polarizing microscopy with the results indicated in Table 2. The polarizing micrograph of the 40/60 PHBA/diol ratio LC polyol (Fig. 5) is typical.

Table 2. Differential Scanning Calorimetry and Polarizing Microscopy of Polyols

PHBA/diol ratio	0/100	20/80	30/70	40/60	50/50	60/40
T_g(°C)	-10	7	14	19	27	14
Appearance, 500x	clear	a few spots	---microscopic birefringement phases---			

DSC indicates a gradual increase in Tg with increasing PHBA/diol ratios except for the 60/40 ratio. Otherwise DSC scans of these polyols are complex and are strongly affected by the thermal history of the materials. The 40/60 and 50/50 PHBA/diol ratio materials often have two endothermic peaks, one near 10° and the other near 70°. The relative size of these peaks vary depending on sample history. Small peaks at about 100° are also sometimes observed.

The 20/80 PHBA/diol ratio polyol is transparent and is soluble in polar solvents. The 30/70 and 40/60 PHBA/diol ratio polyols are transparent at temperatures above 130° but are translucent below 70°. The 50/50 and 60/40 PHBA/diol ratio polyols are translucent throughout the range 25° to 230°.

The 20/80 PHBA/diol ratio oligomer was soluble in MIBK. The 30/70, 60/40, 50/50 and 40/60 oligomers form stable, milky dispersions in MIBK. For example, a roughly 12% "solution" of the 40/60 diol was slightly milky, yet it readily passed through fine filter paper (Whatman 42). A 70% dispersion of this oligomer in MIBK retained its milky appearance for more than one year. No precipitate separated.

In contrast, the benzoic acid-PHBA adduct is insoluble in MIBK. Oligomers of PHBA are also extremely insoluble (13), and physical blends of oligo-PHBA with polymer solutions display no tendency to form stable dispersions (14).

Enamel Properties. Clear coatings were formed by crosslinking the PHBA-modified oligomers with a standard melamine resin. Baking at 175° was necessary to obtain optimal properties. The cured films were glossy and nearly transparent except for films made from 60/40 PHBA ratio polyol. Adhesion was excellent.

The outstanding feature of enamels made from 40/60 to 50/50 PHBA/diol ratio L-C polyols is that they are both very hard and very impact resistant as shown in Table 3 and 4.

Table 3. Reverse Impact Resistance of Baked Enamels

PHBA/diol ratio Baking Time (min) at 175°	0/100	20/80	30/70	40/60	50/50	60/40
20	*	p	p	p	p	f
40	*	p	p	p	p	f
60	*	f	p	p	f	f

p: Passes 80 in-lb reverse impact test. f: Fails this test.
*: Passes this test when baked at 149°; when baked at 175° enamels appear to pass but crack after several days.

Table 4. Knoop-Tukon Indentation Hardness, KHN, and (Pencil Hardness) of Baked Enamels

PHBA/diol ratio Baking Time (min) at 175°C	0/100	20/80	30/70	40/60	50/50	60/40
20	4(HB)	7(H)	13(H)	21(3H)	20(4H)	22(5H)
40	4(HB)	6(H)	12(H)	23(3H)	22(4H)	29(5H)
60	3(HB)	6(H)	17(2H)	23(4H)	26(5H)	36(6H)

In interpreting the information in Tables 3 and 4 it is useful to recall that a Knoop hardness of 12 KHN is adequate for acrylic auto enamels. Thus the 40/60 and 50/50 enamels (21-26 KHN) are much harder than auto enamels. It is quite probable that they are also more flexible, although direct comparisons are not possible from the data at hand.

The enamels with Knoop hardness above 20 KHN and hardness of 3H or higher had excellent solvent (methyl ethyl ketone) resistance.

The salycilic acid modified oligomers did not cure at 175°. The m-hydroxybenzoic acid modified oligomers cured at 175° to give hard but brittle films. All failed the 80 in-lb impact resistance test.

Polarizing micrographs at 25° showed microscopic birefringent domains in the enamel films. A typical one made from the 40/60 PHBA/diol ratio polyol is shown in Fig. 6. Such domains were not visible in cured films made from the PA/AA/NPG polyol or from the MHBA-modified enamels.

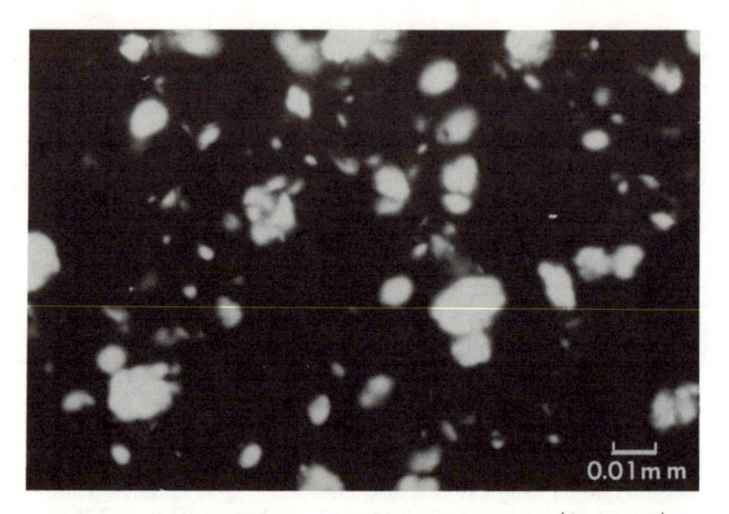

Fig. 5. The polarizing micrograph of the 40/60 PHBA/diol ratio LC polyol.

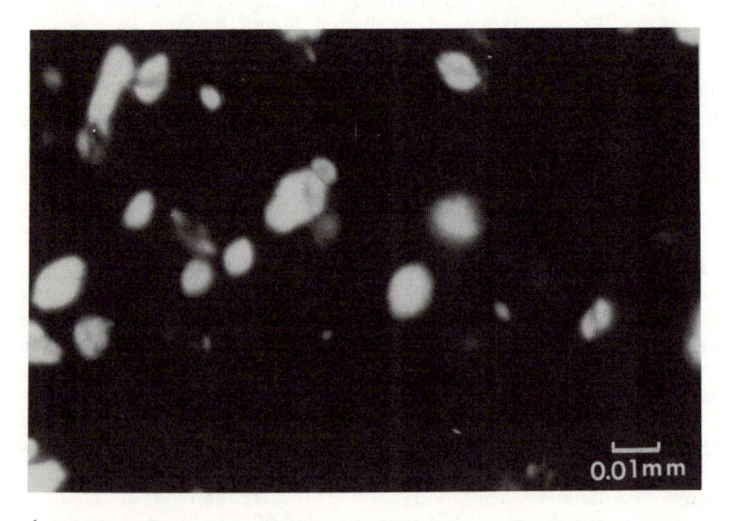

Fig. 6. The polarizing micrograph of the enamel film made from the 40/60 PHBA/diol ratio polyol.

Discussion

Direct esterification of PHBA monomer is problematic because the rate of decarboxylation is often comparable to the rate of (co)polymerization (15,16). The problem seems largely overcome by two process refinements introduced here: [1] Use of p-TSA apparently catalyzes esterification and more than it catalyzes formation of the undesirable by-product, phenol. [2] "Aromatic 150" solvent helps maintain constant reaction temperature and facilitates water removal. Refluxing solvent also prevents accumulation of sublimed PHBA in the upper parts of the reactor.

Based on the limited data available it can be tentatively proposed that the reaction of PHBA with the PA/AA/NPG diol occurs in two stages (see Scheme I):

[1] The predominant reaction in the early stages is esterification of aliphatic OH groups of the oligomeric diol with the COOH group of PHBA-tipped oligomers. Evidence for this reaction course is found by examination of the 20/80 PHBA/diol product: It has a higher M_n than the original diol, its FT-IR indicates that phenolic OH groups predominate, and its amorphous character indicates that relatively few PHBA-PHBA blocks are present.

[2] A second stage becomes significant as aliphatic -OH groups on oligomer ends are consumed. In the present case this stage is reached when the PHBA/diol ratio is above 20/80. During this stage two reactions, apparently slower than the reaction of stage one, occur. They are transesterification and esterification of the terminal phenolic groups. Evidence is the gradual reduction of M_n and a gradual increase in PHBA units per number average molecule as PHBA/diol ratios increase (see Table 1). Apparently these two reactions occur at comparable rates. Transesterification leads to oligomer chain scission, while esterification of phenolic groups leads to formation of blocks of oligo-PHBA at chain ends. The latter reaction is thought to be responsible for the unusual character of the products.

The above description of the process is tentative because it is based on limited data. If it is correct, the predominate structures in the PHBA-modified products have amorphous PA/AA/NPG center sections end-capped with single units or short blocks of oligomeric PHBA. Random distribution of the PHBA cannot be ruled out, but the heterogeneiety of the products suggests that a substantial fraction of PHBA is incorporated into short blocks. The FT-IR and GPC data are consistent with the proposal that short, phenolic-tipped oligomers are the predominant structurs present. The possibility that the materials are physical mixtures of oligo-PHBA and amorphous diols can be virtually ruled out on the basis of the extreme insolubility of oligo-PHBA (13) and of the model PHBA-benzoic acid adduct synthesized in this study. These materials separate readily from solutions and dispersions of PHBA copolymers.

The 30/70, 40/60, and 50/50 PHBA/diol oligomers are clearly heterogeneous. The fact that they form stable dispersions in MIBK, their appearance, and polarizing microscopy indicate that they are

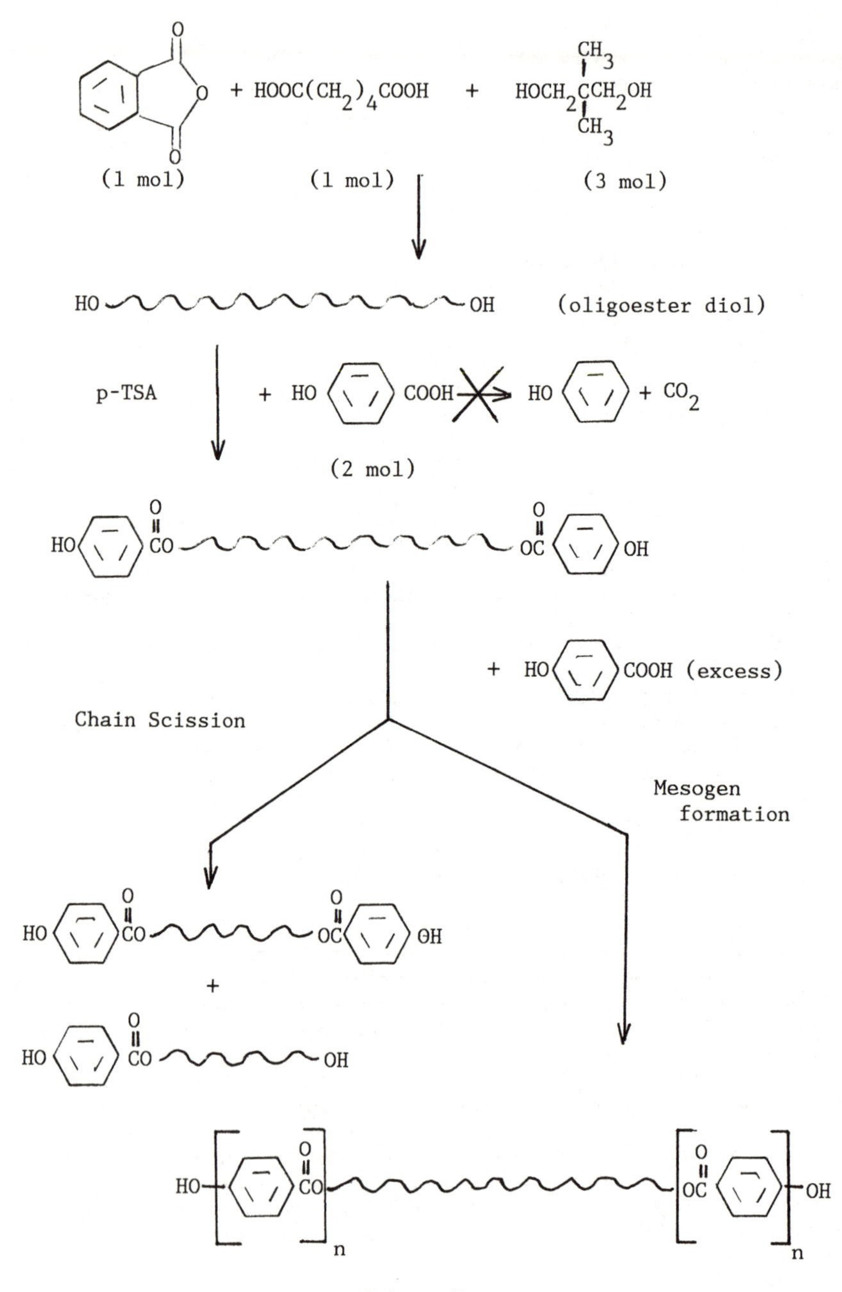
+ HOOC(CH2)4COOH
+
CH3
HOCH2CCH2OH
CH3
(1 mol)
(1 mol)
(3 mol)
HO
OH
(oligoester diol)
p-TSA
+ HO
COOH
HO
+ CO2
(2 mol)
HO
CO
OC
OH
+ HO
COOH (excess)
Chain Scission
Mesogen formation
HO
CO
OC
OH
+
HO
CO
OH
HO
CO
OC
OH
n
n

Scheme I.

liquid crystalline. Further investigation of their morphology is warranted.

The structures of the polyols prepared in this study are quite different from those of the model LC polyols described in the preceding chapter (7). Here the mesogenic units are apparently at the ends of the oligomers and amorphous segments are in the middle. In the preceding chapter it is the other way around. While they are different, the two types of polyols share two characteristics, presence of mesogenic units and presence of amorphous segments that can act as flexible spacers. Flexible spacers are known to substantially affect the properties of liquid crystalline polymers (4,5). The present study indicates that oligomers of both types can impart exceptional hardness and toughness to coatings and presumably to other highly crosslinked materials.

Conclusions

A direct esterification procedure by which a linear polyester diol can be modified with p-hydroxybenzoic acid (PHBA) was demonstrated. The products are oligomers in which phenolic end-groups appear to predominate. They are heterogeneous and are probably liquid-crystalline when 30 wt% or more of PHBA is incorporated. The procedure has the advantage that appears adaptable to large scale production but the disadvantages that the polyols are predominately phenolic and are contaminated with small amounts of phenol and unreacted PHBA.

Baked enamels made by crosslinking these polyols with a melamine resin contained birefringent domains in the films. These enamels were both harder and tougher than enamels made by crosslinking amorphous polyols and modified polyols derived from m-hydroxybenzoic acid (MHBA). Hardness and toughness of enamels made from these oligomers were similar to the exceptional hardness and toughness attained with enamels made from model liquid crystalline oligomers described in the preceding chapter (7).

Acknowledgment

Financial support by the U. S. Environmental Protection Agency (Grant No. R-811217-02-0) is gratefully acknowledged.

Literature Cited

1. Blumstein, A. "Polymeric Liquid Crystals", Proceedings of the Second Symposium on Polymeric Liquid Crystals, Division of Polymer Chemistry of the American Chemical Society Meeting: Washington, D.C., August 28-31, 1983; Plenum: New York. 1985.
2. Gordon, M.; Plate, N.A., Ed.; "Liquid Crystal Polymers 1"; Advances in Polymer Science 59; Springer-Verlag: Berlin, Heidelberg, New York, Tokyo, 1984.
3. Shibaev, V.P. and Plate, N.A. Polym. Sci. USSR, 1977, 19, 1065-1122.

4. Ober, C.K.; Jin, J.; Zhou, Q.; Lenz, R.W. In "Liquid Crystal Polymers 1. Advances in Polymer Science 59"; Gordon, M.; Plate, N.A., Ed.; Springer-Verlag: Berlin, Heidelberg, New York, Tokyo, 1984.
5. Lenz, R.W. J. Polym. Sci., Polym. Symp. 1985, 72, 1-8.
6. Chen, D. S.; Jones, F. N. Polym. Mater. Sci. Eng. 1987, 57, 217-221.
7. Dimian, A. F.; Jones, F. N. Preceding Chapter.
8. Maruyama, K.; Sholi, A.; Honma, M. Japan. Kokai 75 40,629, 1975; Chem. Abstr. 1975, 83, 133572.
9. Nogami, S.; Nakamura, K.; Waki, K.; Suzuki, Y.; Nishigaki, Y. Japan. Kokai 76 44,130, 1976; Chem. Abstr. 1976, 85, 79835.
10. Nakamura, K.; Sasaguri, K.; Matsumoto, Y.; Matsuo, S.; Sato, M.; Hayashi, Y.; Uda, B. Japan. Kokai 76 56,839, 1976; Chem. Abstr. 1976, 85, 110175.
11. Nogami, S.; Nishgaki, Y.; Waki, K.; Suzuki, Y. Japan. Kokai 77 73,929, 1977; Chem. Abstr. 1978, 88,8624.
12. Demarest, B. O.; Harper, L. E. J. Coat. Technol. 1983, 55, (701), 65-77.
13. Chen, D. S.; Jones, F. N. J. Polym. Sci. [A], Polym. Chem. Ed. 1987, 25, 1109-1125.
14. Chen, D. S.; Jones, F. N. Unpublished observations.
15. Linden, G. L. U.S. Patent 4,331,782, 1982.
16. Berry, D. A. Paint Manuf. 1965, 4, 45-50.

RECEIVED December 18, 1987

Chapter 24

Benzocyclobutene in Polymer Synthesis

Novel Cross-Linking Systems Derived from a Bisbenzocyclobutene and Bismaleimides, Dicyanates, and Bisphenylacetylenes

Loon Seng Tan[1], Edward J. Soloski[1], and Fred E. Arnold[2]

[1]University of Dayton Research Institute, 300 College Park Avenue, Dayton, OH 45469-0001

[2]Air Force Wright Aeronautical Laboratories, Wright-Patterson Air Force Base, OH 45433

A number of blends were prepared from a bisbenzocyclobutene monomer (BCB) and various bisdienophiles (Kerimid 353; dicyanates such as bisphenol A dicyanate, tetramethylbisphenol F dicyanate and thiodiphenol dicyanate; and bisphenyl-acetylenes: 1,3-PPPO and PhATI) in 1:1 molar ratio. Nonstoichiometric mixtures were also concocted from BCB and K-353. All were completely compatible, as evidenced by a single initial Tg, except the BCB/dicyanate blends, which showed small exothermic transitions attributable to crystallization phenomenon. The results from the thermal analyses of these blend systems, in particular the K353/BCB system, lend credence to our belief that curing via Diels-Alder cycloaddition may predominate. As such, the blend systems were more stable toward thermo-oxidative degradation than their pure bisdienophile components.

It has been established that under appropriate thermal conditions, the strained 4-membered ring of benzocyclobutene undergoes electrocyclic ring opening to generate in-situ an extremely reactive diene [1]. The temperature at which such a valence isomerization process occurs is influenced by the nature of the substituents at the alicyclic carbons. Aromatic substitution does not appear to have any significant effect. The electron-yielding groups exert the opposite effect. Unsubstituted benzocyclobutene usually begins its transformation at about 200°C in solution [2].

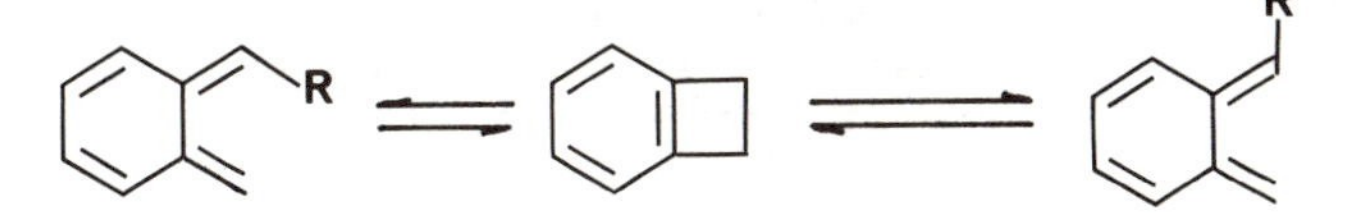

0097-6156/88/0367-0349$06.00/0

In the presence of a suitable dienophile, the more reactive form, i.e. o-xylylene (or o-quinodimethane) will be engaged in a Diels-Alder reaction, otherwise, it will react with itself to form dimers and polymeric materials. Both processes are known to be quite facile in solution, and with proper synthetic design, they can be adapted and made amenable to melt polymerization. In fact, we have recently prepared a number of bis(benzocyclobutene)-terminated monomers and oligomers [3], as well as AB aromatic benzocyclobutene-alkyne [4] and benzocyclobutene-maleimide monomers [5], and found that the thermosetting systems derived from them possess excellent thermal properties. In addition, a certain bisbenzocyclobutene (BCB) and a commercially available bismaleimide formed compatible mixtures in various molar ratios, which were cured to afford surprisingly thermo-oxidatively stable materials with high Tg's [6]. In this paper, we describe our work on some AABB thermosetting systems involving bis(benzocyclobutene), BCB, and various bisdienophiles whose idealized Diels-Alder reactions and their products are shown in Figure 1.

Experimental

The preparation of 2,2-bis[4-(N-4-benzocyclobutenyl phthalimido)]-hexafluoropropane (BCB) has been described elsewhere [6]. Although a two-step synthesis of 1,1'-(1,3-phenylene)bis-(3-phenyl-2-propyn-1-one) (1,3-PPPO) has been reported [8], we found that 1,3-PPPO could also be prepared from a one-step route, starting from iso-phthaloyl chloride and phenylacetylene, employing the $PdCl_2(PPh_3)_2$/CuI catalytic system and triethylamine as the solvent and HCl-acceptor [9] in a 40-50 percent yield. 2,2-Bis[4-(N-3-phenylethynyl)phthalimido]hexafluoropropane (PhATI) was prepared from 3-aminotolan and 2,2-bis(4-phthalimido)hexafluoropropane in N,N-dimethylacetamide, using acetic anhydride/pyridine as the cyclodehydrating agent. 3-Aminotolan was obtained from sodium hydrosulfite reduction of 3-nitrotolan (custom synthesis sample), which was prepared from the coupling reaction of copper(I) phenylacetylide with 3-bromonitrobenzene. (B.A. Reinhardt, Air Force Wright Aeronautical Laboratories, Wright-Patterson Air Force Base, personal communication) A commercially available bismaleimide mixture, Kerimid-353, was a gift from Hexcel Corporation. The dicyanate monomers, bisphenol A dicyanate (BADCy), tetramethyl bisphenol F dicyanate (METHYLCy) and thiodiphenol dicyanate (THIOCy) were kindly supplied by Dr. Richard B. Graver of Interez Inc., formerly Celanese Specialty Resins Division, Celanese Corporation. Both K-353 and the dicyanate monomers were used as received. All the blends were prepared by dissolving completely the components in methylene chloride, and the solvent was subsequently removed by slow evaporation with mild heating. The solid blends were then dried *in vacuo* at 65-75°C overnight before being subjected to thermal analysis.

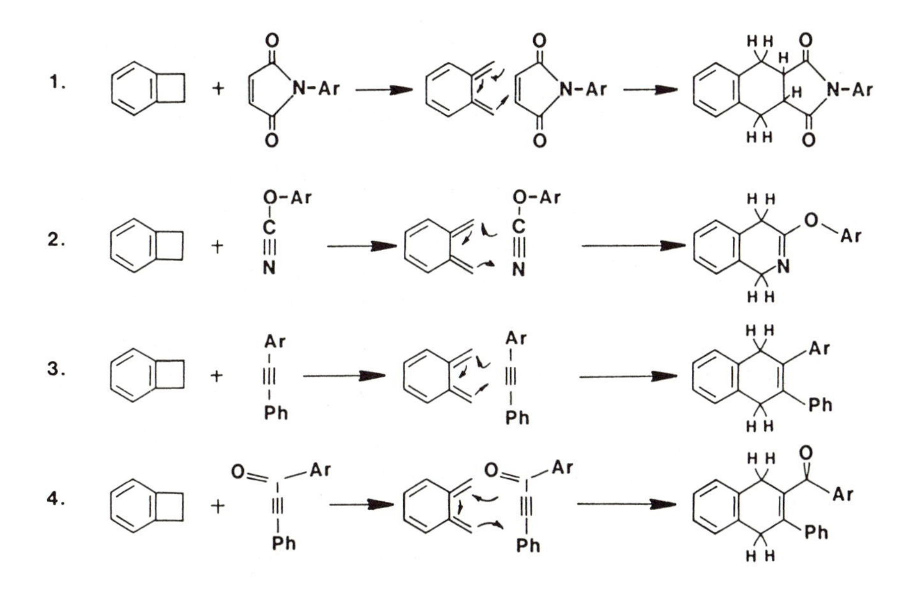

Figure 1. Idealized Diels-Alder reaction between benzocyclobutene and various dienophiles.

Results and Discussion

BCB/K-353. Four compatible mixtures of a bis(benzocyclobutene) terminated monomer (BCB) and K-353 were mixed in accordance with the following molar ratios (BCB:K-353): 1:1; 1:1.5; 1:2; 1:3. The molecular structures of BCB and the components of K-353 are shown in Figure 2. K-353 is actually a eutectic mixture of three bismaleimides, A, B, and C, with the composition of approximately 50:40:10 (by weight), respectively. The samples were examined by differential scanning calorimetry (DSC) and thermogravimetric analysis (TGA), and their thermal characteristics are tabulated in Table I. A representative DSC thermogram is displayed in Figure 3, indicating complete compatibility as evidenced by a single initial glass transition temperature (Tg).

In brevity, all blends exhibit relatively low initial Tg's (41-58°C), and the trend is consistent with the relative amounts of K-353 present in the mixtures. The component C of K-353 is a flexible aliphatic bismaleimide which can, therefore, exert the depressing effect on the glass transitions of the other components. The polymerization exotherms are very well defined (onset: 222-226°C; max.: 261-262°C) and characteristic of the benzocyclobutene-based systems. Consequently, the BCB/K-353 blend systems possess excellent processing windows (161-181°C), which are defined as the temperature difference between the onset of polymerization and the initial Tg. The samples of mixtures and pure BCB as well as pure K-353 were cured at about 250°C under a nitrogen atmosphere for eight hours. Cured samples were subsequently subjected to isothermal gravimetric analysis (ITGA) (see Figure 4). The ITGA results indicate that after 200 hours at 650°F (343°C) in circulating air, the cured K-353 sample was almost completed degraded, whereas the BCB/K-353 mixtures with molar ratios of 1:1 and 1:1.5 exhibited thermo-oxidative stabilities similar to BCB (12-14 percent weight loss), and those with molar ratios of 1:2 and 1:3 displayed lower, yet respectable, thermo-oxidative tolerance (22-24 percent weight loss). It was expected that the cure site structure of the mixtures, being composed of three fused rings (ladder-like), should be more thermoxidatively stable than that of BM1. Such observation, therefore, lends support to the conviction that Diels-Alder polymerization is the predominant, if not the only, curing process in the benzocyclobutene/maleimide systems. Furthermore, the remarkable thermo-oxidative resistance of the nonstoichiometric mixtures (1:1.5, 1:2, and 1:3) in comparison with the pure K-353 does imply that a synergistic effect can exist in the BCB/K-353 combination. Presumably the polymeric structure, generated from the Diels-Alder cycloaddition of benzocyclobutene and maleimide groups, which in theory should be linear, can impart thermoplastic character, i.e. toughness, to the networks formed by the homopolymerization of the excess K-353. This is in agreement with the fact that a cured sample of 1:1 BCB/K-353 distintegrated without dissolution in methylene chloride in less than an hour, whereas a cured sample of K-353 remained intact in the same solvent even after a month of immersion. Whether such a synergistic

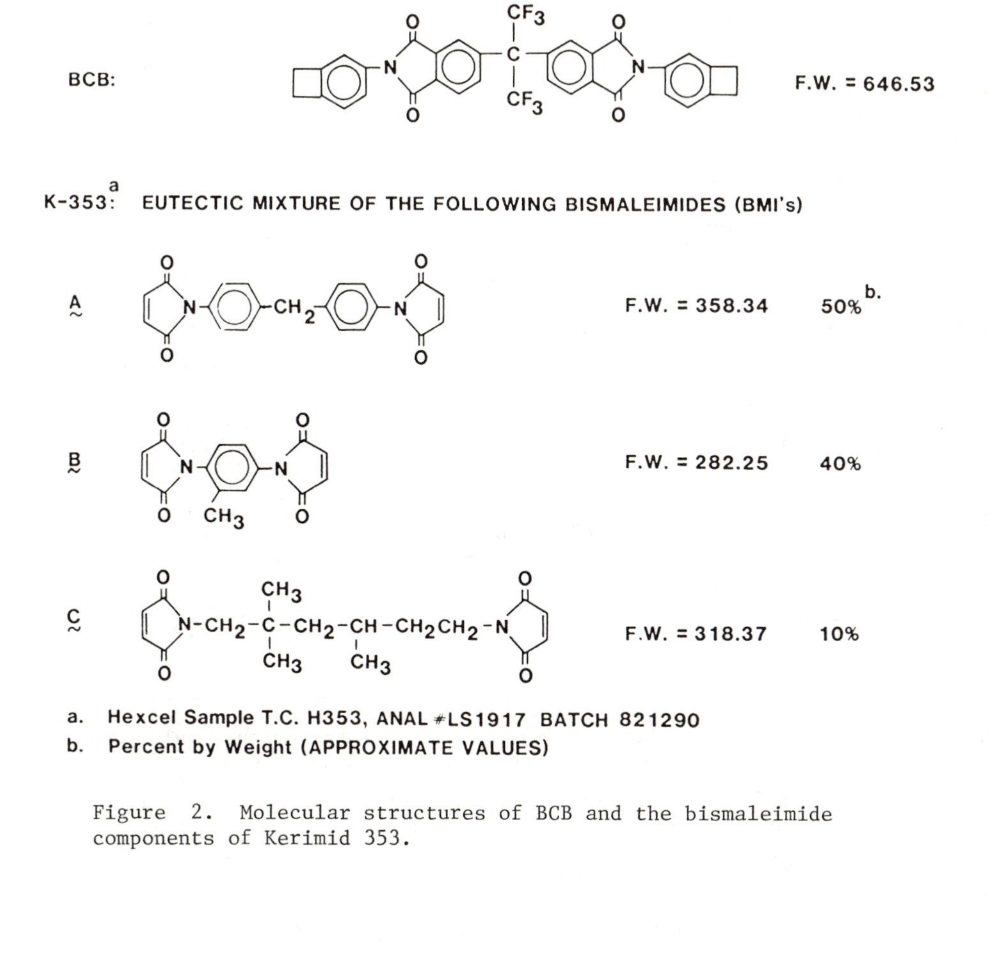

Figure 2. Molecular structures of BCB and the bismaleimide components of Kerimid 353.

TABLE I. THERMAL PROPERTIES OF COMPATIBLE MIXTURES OF BCB/K-353 (BMI)[a]

MOLAR RATIO BCB:K-353	T_g	T_{poly} ONSET	T_{poly} MAX.	T_g (CURE)[e]	CURED SAMPLE[b] T_g (CURE) DSC	CURED SAMPLE[b] T_g (CURE) TMA
0 : 1	15°	206°	234°	--[c]	--[c,g]	--[c]
1 : 1	58°	226°	262°	285°	280°[d] (283°)[e]	265°
1 : 1.5	51°	225°	262°	272°	265°[d] (267°)[e]	260°
1 : 2	47°	224°	261°	259°	235°[d] (257°)[e]	--[f]
1 : 3	41°	222°	261°	243°	230°[d] (255°)[e]	256°

NOTES: (a) K-353 obtained from Hexcel; all temperature values in °C; (b) samples were cured at 250°-253°C/8 hrs./N_2; (c) not observed; (d) small exotherms were observed: for 1:1 sample, 294°C (Max.); for 1:1.5 sample, 348°C (Max.); for 1:2 sample 336°C (Max.); for 1:3 sample, 338°C (Max.); (e) values obtained from rescanning the samples after previously heating to 450°C; (f) sample was too brittle; (g) a reaction exotherm was observed (300°C Max.).

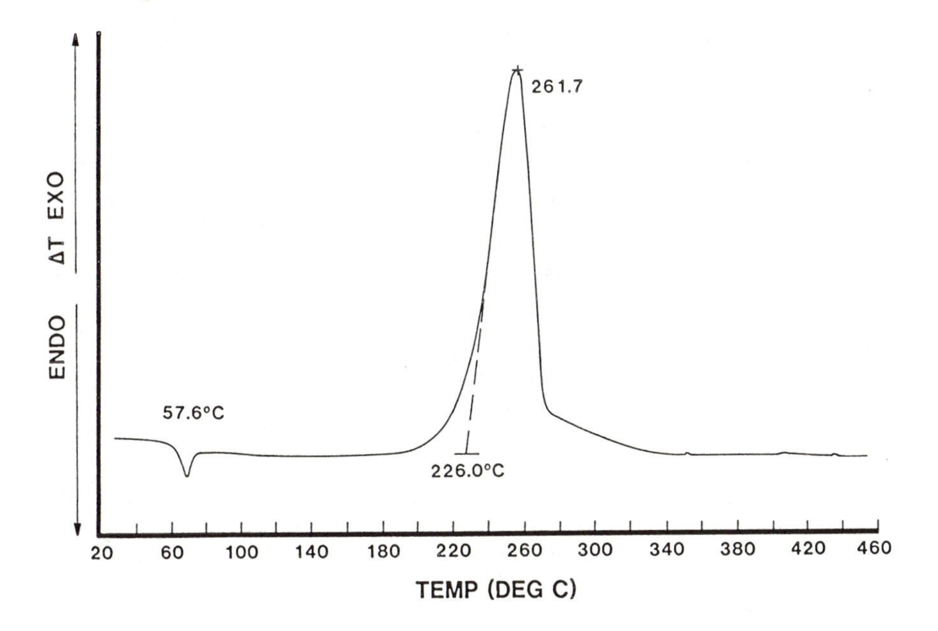

Figure 3. DSC thermogram of 1:1 molar mixture of BCB and K-353.

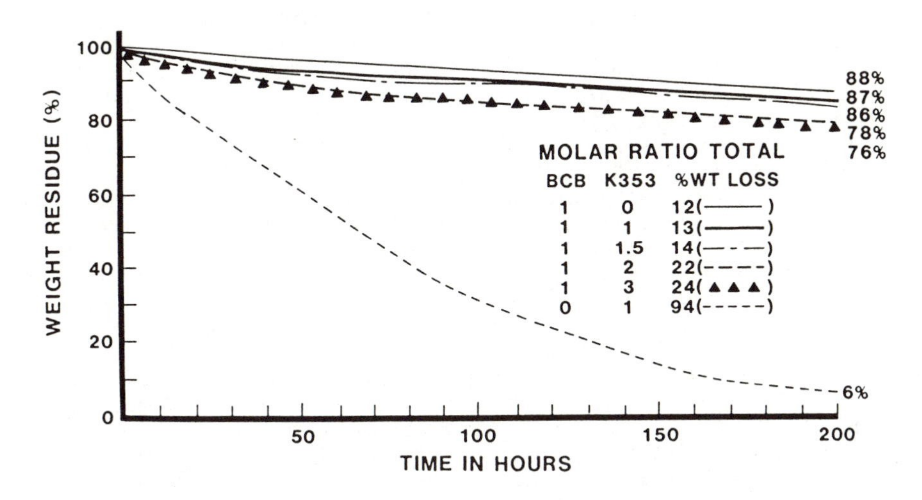

Figure 4. Isothermal aging studies of the thermosetting resins derived from the mixtures of BCB and K-353, in comparison with those derived from pure BCB and K-353.

effect also extends to the mechanical properties of BCB/K-353 systems remains to be seen. Efforts are being made to confirm or disprove the aforementioned deductions.

BCB/Dicyanate. With the encouraging results from the BCB/K-353 systems, we proceeded to study the BCB/dicyanate blend systems. See Figure 5 for molecular structures of the dicyanate monomers. The dicyanates were selected primarily because they are commercially available, and the cyanate group with its carbon-nitrogen triple bond can serve as a dienophile. Although monomeric dicyanates can be cured thermally after B-staging, the curing time is much shorter when a catalyst is used [10]. During cure, the monomeric dicyanates undergo cyclotrimerization to form a network structure with triazine (or cyanurate) rings at the crosslinking junctures. Recently, a new class of high-performance, high-temperature plastics, which are actually semi-IPN's derived from dicyanatebisphenol A, and various commercial engineering thermoplastics has been reported [11].

As a preliminary study, three 1:1 molar mixtures, BCB/THIOCy, BCB/METHYLCy, and BCB/BADCy, were prepared. The suggested catalytic system [10] was not included in the blends since we wanted to minimize the competitive cyclotrimerization of the dicyanate monomers. The samples, along with pure dicyanate samples, were then subjected to thermal analysis by DSC and TGA. The results are summarized in Table II. A representative DSC thermogram of the 1:1 mixtures is shown in Figure 6.

Under the DSC conditions (N_2, scanning rate = 10°C/min), it is apparent that the decomposition processes are occurring at a much faster rate at or near the temperature at which cure is taking place in all the pure dicyanate samples. Both BADCy and THIOCy showed small exotherms (onset at 277°C and 226°C and peak at 308°C and 289°C, respectively). Their major decompositions began about 251°C and 246°C, respectively, as observed by TGA. On the contrary, all the 1:1 BCB/dicyanate blends displayed the expected thermal transitions. Besides initial Tg's (20-28°C) and Tm's (171-183°C), all samples showed small exotherms in their DSC scans with maxima at 147-151°C. This is attributable to the thermally-induced crystallization in the mixtures, which also led to some initial phase separation. The polymerization exotherms are consistent with the typical temperature ranges for the known benzocyclobutene-based systems (onset: 229-233°C; max.: 259-266°C).

For comparative isothermal aging studies, all the samples of pure BCB and dicyanate monomers, as well as their 1:1 molar mixtures, were cured in a single batch at 200-220°C for 40 hours under nitrogen atmosphere. The cured samples of BADCy, METHYLCy, and THIOCy were all transparent and yellow/amber, and their blends with BCB were also transparent but dark red in color. The cured sample of BCB was translucent and yellow. The Tg's (cure) of the thermosets derived from the dicyanate monomers are relatively high, 224°-261°C as determined by thermomechanical analysis (TMA). There is an increase of 10-31°C in Tg (cure) values in their blends with BCB. The ITGA results of the cured samples of BCB,

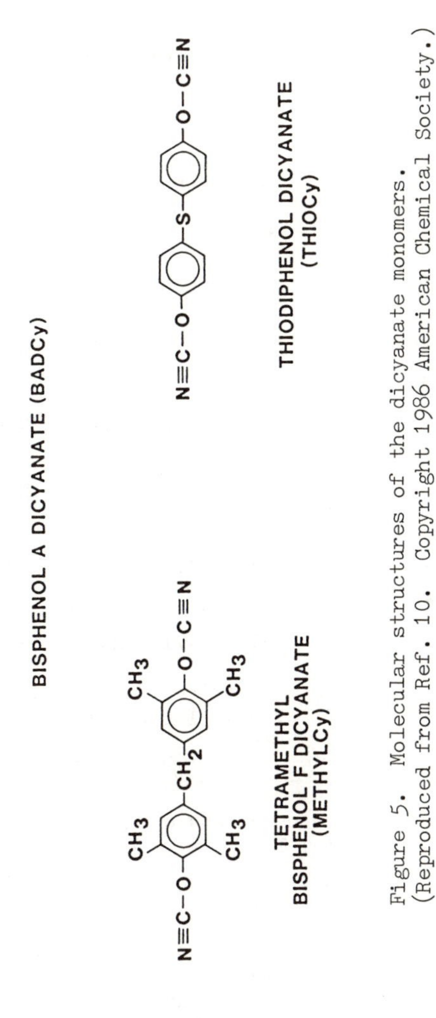

Figure 5. Molecular structures of the dicyanate monomers. (Reproduced from Ref. 10. Copyright 1986 American Chemical Society.)

TABLE II. THERMAL CHARACTERISTICS OF BCB, DICYANATES AND THEIR BLENDS (1:1)[a]

COMPOSITION	T_g(ini.)	T_{cryst}[b]	T_m	T_{poly} ONSET	T_{poly} MAX.	T_g(cure) DSC	T_g(cure) TMA	T_d[c]
METHYLCy	--	--	110	--[e]	--[e]	--	224[f]	255
BADCy	--	--	84	277	308[d]	--	245[f]	251
THIOCy	--	--	98	226	289[d]		261[f]	246
BCB	161	--	219	230	258	281	--	470
BCB/METHYLCy	27	151	183	232	268	215	255[f]	427
BCB/BADCy	28	147	171	233	266	--	269[f]	404
BCB/THIOCy	20	148	181	229	259	232	271[f]	392

NOTES: (a) All values are expressed in °C. Blends are in 1:1 molar ratio. (b) T_{cryst} = temperature (max.) at which crystallization from amorphous phase (possile phase separation occurs. (c) T_d = extrapolated temperature at which major decomposition occurs as observed by TGA. (d) decomposition (major exothermic process) also occurs. (e) only decomposition/volatilization were observed, beginning at 234°C. (f) T_g(cure)'s of the samples which were previously cured at 200-220°C/N_2/40h; determined by thermomechanical analysis (TMA).

dicyanate monomers and BDB/dicyanate blends are depicted graphically in Figure 7. After 200 hours of isothermal aging at 650°F (353°C) in an atmosphere of circulating air, the dicyanate-based thermosets had retained only 4-10% of their original weight. Under the same conditions, the BCB/dicyanate mixtures lost 24-28% of their original weights as opposed to 12% weight loss observed for pure BCB. Although the results were not as good as those with BCB/K-353, it is quite clear there was definitely some interaction, most likely in the Diels-Alder fashion, between BCB and the dicyanate monomers.

BCB/Diacetylenes. Our previous studies demonstrated that a 1:1 stoichiometric mixture of BCB and a bis(phenylacetylene) terminated monomer (PhATI) [12], when subjected to DSC study, showed a residual polymerization exotherm (about 367°C) attributable to the homopolymerization of the phenylacetylene group [6]. A similar characteristic polymerization exotherm was also observed for the AB aromatic benzocyclobutene/alkyne monomers with phenylacetylene being deactivated by terminal electron-donating phenoxy substituents [4]. It is thought that a strong electron-demanding group such as carbonyl, in conjugation with the carbon-carbon triple bond, should greatly enhance the dienophilicity of the phenylacetylene group (see Figure 8). Therefore, a 1:1 molar mixture of BCB and 1,3-PPPO was prepared. Figure 9 depicts the DSC thermogram of the BCB/1,3-PPPO mixture. A single, low initial Tg (38°C) was observed indicating the compatibility of components, followed by the polymerization exotherm with onset at ca. 217°C and maximum at ca. 264°C. A small overlapping exotherm peaking at 308°C was also observed. It is probably due to the simultaneous occurrence of the homopolymerization and decomposition of the residual 1,3-PPPO. A final Tg at 233°C was achieved from the DSC rescan of the BCB/1,3-PPPO sample after it had been heated to 450°C. Together with the comparative TGA results (see Figure 10), it is ostensibly that the BCB/1,3-PPPO mixture is more thermo-oxidatively stable than 1,3-PPPO alone when cured under the same conditions. However, both 1,3-PPPO and 1:1 BCB/1,3-PPPO blend lost CO after they were cured at 263-265°C for 40 hours under N_2, as evidenced by the formation of relatively large voids in the samples, which also showed disappearance of ν(CO) at 1646 cm^{-1} in their IR spectra. Unfortunately, it is not clear when the evolution of CO from the samples occurred. Therefore, it is inconclusive as to whether there is an enhanced dienophilicity in 1,3,-PPPO due to the carbonyl group conjugated to the carbon-carbon triple bond. The thermal properties of BCB, PhATI, 1,3-PPPO, and their 1:1 molar mixtures are summarized in Table III.

Based on the assumption that there is no chemical interaction or any synergistic effect caused by physical changes in the blends, theoretical weight losses of the blends can be calculated, using the isothermal aging data of the individual components as well as their weight percents in the blends. The results are compared with the experimental data in Table IV (BCB/K-353) and Table V (BCB/Dicyanate).

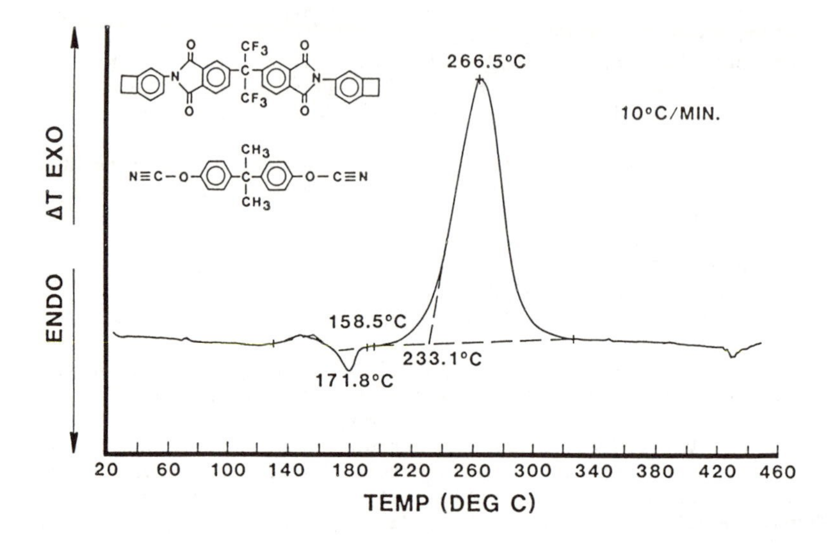

Figure 6. DSC thermogram of 1:1 molar mixture of BCB and BADCy.

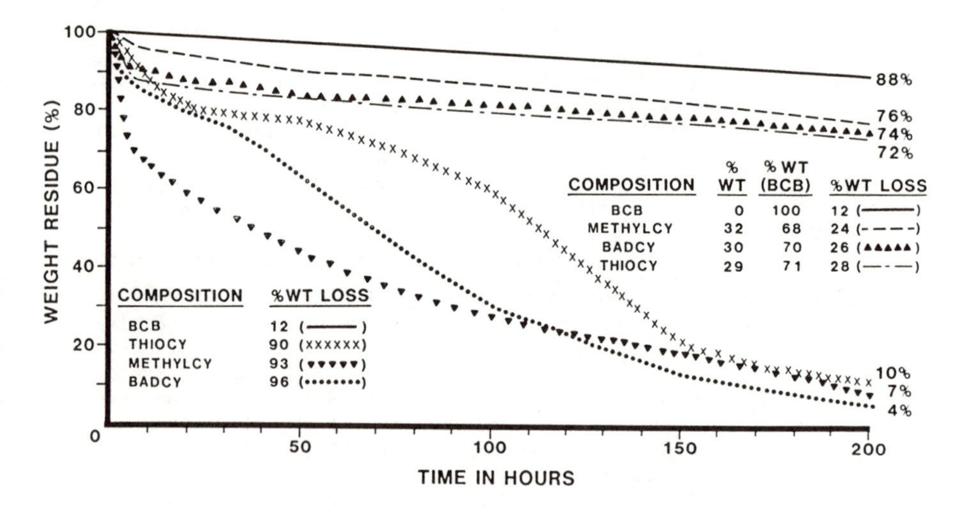

Figure 7. Isothermal aging studies of the thermosetting resins derived from the 1:1 molar mixture of BCB and the dicyanate monomers, in comparison with those derived from pure BCB and the dicyanate monomers.

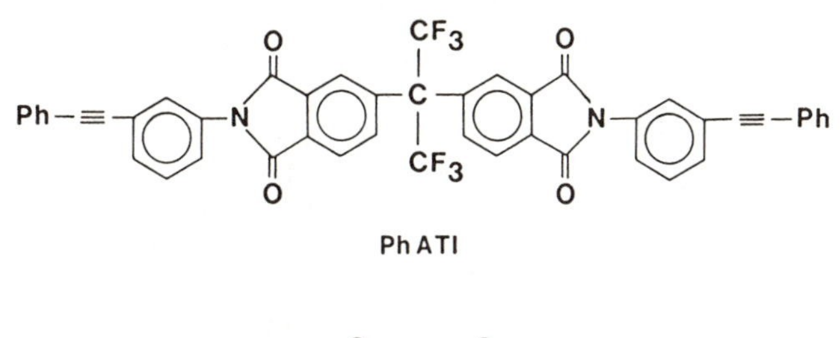

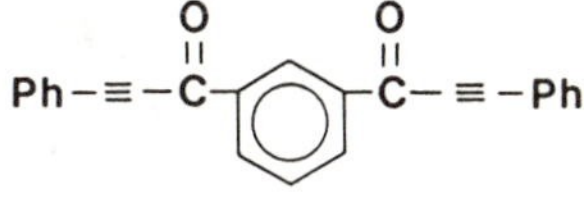

Figure 8. Molecular structures of PhATI and 1,3-PPPO.

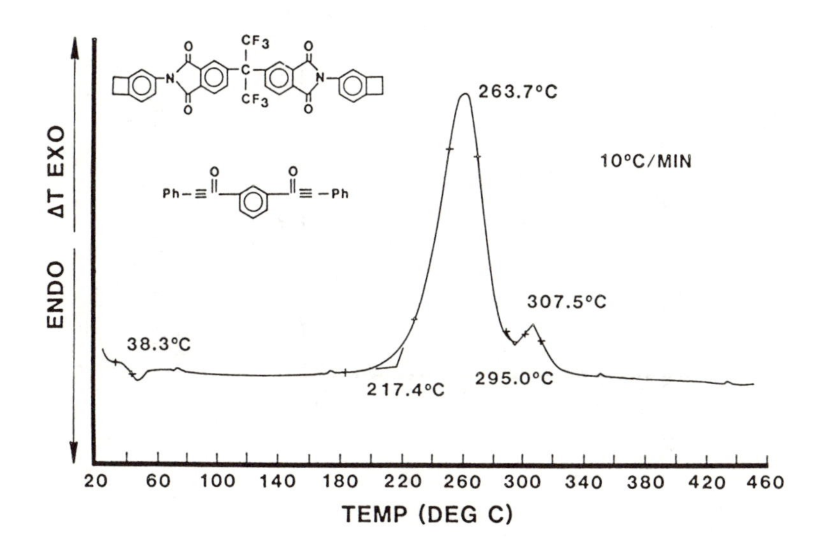

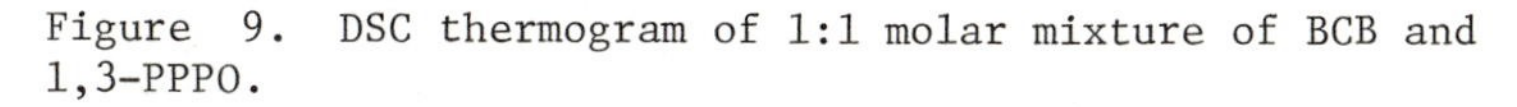
Figure 9. DSC thermogram of 1:1 molar mixture of BCB and 1,3-PPPO.

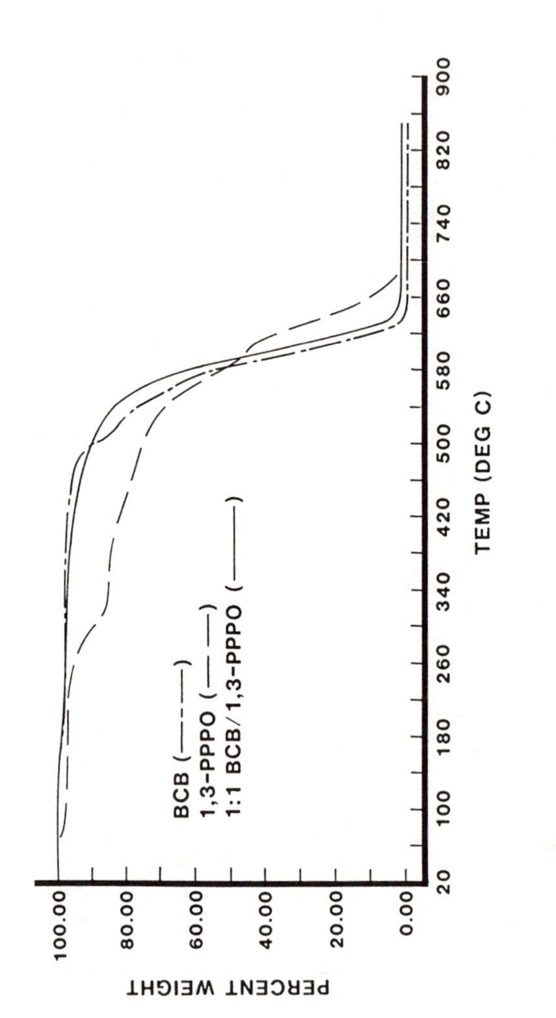

Figure 10. Composite TGA thermograms of BCB, 1,3-PPPO, and (1:1) BCB/1,3-PPPO.

TABLE III. THERMAL PROPERTIES OF BCB, PhATI, 1,3-PPPO AND THEIR BLENDS (1:1)[a]

COMPOSITION	T_g(ini.)	T_m	T_{poly} ONSET	T_{poly} MAX.	T_g(cure)	TGA T_d[b]	TGA $T_{10\%}$[b]	ITGA[d] % Wt. Loss
BCB	161	219	230	259	281[c]	470	500	12
PhATI	111	-	305	370	358[c] (202)[e,f]	460	540	16
1,3-PPPO	-	118	266	312	270[e]	264	298	84
BCB/PhATI	96	-	230 321	268 367	250 (264)[e,f]	477	525	9
BCB/1,3-PPPO	38	-	217	264 308	233 (226)[e]	400	508	50

NOTES: (a) All values are expressed in °C. (b) T_d = extrapolated temperature at which major decomposition started. $T_{10\%}$ = temperature at which 10% weight loss was observed. (c) Samples were previously cured for 8 hours under N_2. BCB (250°C), PhATI (350°C). (d) 650°F (343°C), 200 hours, in circulating air. (e) Samples were previously cured at 263-265°C for 40 hours under N_2. (f) DSC showed residual exotherms with onset at 312°C for PhATI and at 340°C for 1:1 PhATI/BCB.

TABLE IV. COMPARISON OF THEORETICAL AND EXPERIMENTAL THERMO-OXIDATIVE STABILITIES OF BCB/K-353 SYSTEMS (650°F/AIR/200h)

MOLAR RATIO BCB:K-353	% BY WT. BCB:K-353	% WT. LOSS THEO	EXPTL	Δ (DIFFERENCE)
1 : 1	67 : 33	39	13	26
1 : 1.5	57 : 43	47	14	33
1: 2	50 : 50	53	22	31
1 : 3	40 : 60	61	24	37

TABLE V. COMPARISON OF THEORETICAL AND EXPERIMENTAL THERMO-OXIDATIVE STABILITIES OF BCB/ DICYANATE SYSTEMS (650°F/AIR/200h)

COMPOSITION % BY WT.	% WT. LOSS THEO	EXPTL	Δ (DIFFERENCE)
BCB : METHYLCy 68 : 32	38	24	14
BCB : BADCy 70: 30	37	26	11
BCB : THIOCy 71 : 29	35	28	7

The fact that the differences (Δ-values) in the theoretical and experimental weight loss are all positive indicates that there exists some synergistic interaction between the components of the blend. Furthermore, the general trend of the Δ-values is consistent with that of dienophilicity of the bisdienophiles, suggesting that Diels-Alder polymerization plays an important role in the BCB/bisdienophile systems.

Conclusion

Our studies show that the bisdiene generated in-situ from bisbenzocyclobutene is so reactive that it can react with a variety of bisdienophiles. Excellent thermal and thermo-oxidative stabilities can be realized with the combination of BCB and strong bisdienophiles such as bismaleimides. Furthermore, it is expected the toughness of the blend systems will increase considerably, in comparison to the thermosetting systems based on the individual components, if Diels-Alder polymerization can be maximized. As a consequence, promising high-temperature thermosetting resin systems can emerge from the numerous combinations of bisbenzocyclobutene and bisdienophiles, providing a range of selections to suit each individual application.

Acknowledgments

We are grateful to the generosity of Hexcel and Interez Corporations for sending us the free samples of K-353 and dicyanate monomers, respectively. We also thank Marlene Houtz and Dr. Ivan J. Goldfarb for their contributions in the thermal analysis, and Prof. R. G. Bass for suggesting bispropynones.

Literature Cited

1. For leading references, see Tan, L.S.; Arnold, F.E. Am. Chem. Polym. Div. Polym. Preprints 1985, 26(2), 176.
2. Oppolzer, W. Synthesis 1978, 793.
3. See Ref. 1. Also a number of other bis(benzocyclobutene) terminated monomers have been independently synthesized by others: Kirchhoff, R.A.; Baker, C.E.; Gilpin, J.A.; Hahn, S.F.; Schrock, A.K. Proceedings 18th International SAMPE Technical Conference 1986, 18, 478.
4. Tan, L.S.; Arnold, F.E. Am. Chem. Soc. Polym. Div. Polym. Preprints 1985, 26(2), 178.
5. Tan, L.S.; Soloski, E.J.; Arnold, F.E. Am. Chem. Soc. Polym. Div. Polym. Preprints 1986, 27(2), 240.
6. Tan, L.S.; Soloski, E.J.; Arnold, F.E. Am. Chem. Soc. Polym. Div. Polym. Preprints 1986, 27(1), 453.
7. Harris, F. W.; Beltz, M.W. Am. Chem. Soc. Polym. Div. Polym. Preprints 1986, 27(1), 114.
8. (a) Bass, R.G.; Cooper, E.; Connell, J. W.; Hergenrother, P.M. Am. Chem. Soc. Polym. Div. Polym. Preprints 1985, 27(1), 313; (b) Sinsky, M.S.; Bass, R.G.; Connell, J. W.; Hergenrother, P.M. J. Poly. Sci., Part A: Poly. Chem. 1986, 24, 2279.
9. Tohda, Y.; Sonogashira, K.; Nagihra, N. Synthesis 1977, 777.
10. Shimp, D.A. Proceedings Am. Chem. Soc. Polym. Matls. Sci. Engr. 1986, 54, 107.
11. Wertz, D.H.; Prevorsek, D.C. Polym. Engr. and Sci. 1985, 25, 804.
12. Unroe, M.R.; Reinhardt, B.; Arnold, F.E. Am. Chem. Soc. Polym. Preprints 1985, 26(1), 136.

RECEIVED February 16, 1988

Chapter 25

Characterization of Bisbenzocyclobutene High-Temperature Resin and Bisbenzocyclobutene Blended with a Compatible Bismaleimide Resin

Lisa R. Denny, Ivan J. Goldfarb, and Michael P. Farr[1]

Air Force Wright Aeronautical Laboratories, Materials Laboratory (AFWAL/MLBP), Wright-Patterson Air Force Base, OH 45433

Cured Bisbenzocyclobutene (BCB) terminated resin systems exhibit good mechanical properties with 70% to 85% retention of properties at 260°C and high thermal stability. The Materials Laboratory has studied these materials for use as high temperature structural matrix resins in composites. They are well suited for this use since they do not require the use of catalysts and cure without the evolution of volatiles.

The samples studied were a BCB terminated aromatic imide oligomer and a BCB terminated imide monomer blended with a compatible bismaleimide (BMI) resin. Neat properties studied on both resin systems included the thermal and rheological properties of the uncured specimens subsequently used to determine appropriate cure conditions. Thermal and mechanical properties of the cured materials are also discussed.

This laboratory has been involved in the synthesis and characterization of new high temperature resin systems for use as structural matrix materials in composites at use temperatures exceeding the capabilities of currently available resins. There are a number of criteria which these resins must meet to be suitable for this use, including good mechanical properties and thermal stability in the cured specimens as well as low moisture absorption. The resin itself must exhibit a reasonable shelf life and relative ease in processing. A series of new bisbenzocyclobutene (BCB) terminated resins currently being studied by this laboratory appear to be promising for this use. These resins are stable at room temperature and polymerize upon heating without the use of catalysts and cure via an addition reaction without the evolution of volatiles. The properties exhibited by the cured neat resins are indicative of those necessary for a composite matrix.

[1]Current address: Materials Science Department, Pennsylvania State University, University Park, PA 16802

A variety of BCB terminated resins have been synthesized in this laboratory, a review of which is also being given in this symposium [1]. The characterization of two of those resin systems will be presented here. The first resin to be discussed is a BCB terminated imide oligomer (BCB oligomer) with the structure shown below.

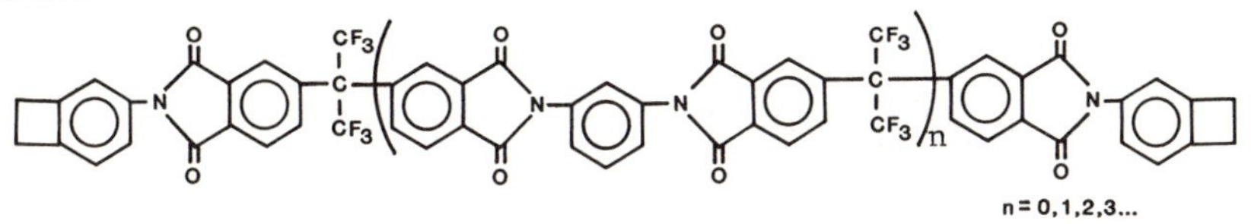

BCB OLIGOMER

The synthesis and proposed cure mechanisms of this resin are described in reference 2. While the cure mechanism of the BCB terminated resin is not yet known, it is speculated that it reacts via one of two different routes. Initially the strained four member ring of the benzocyclobutene undergoes a thermally induced ring opening. The opened rings then react with one another by a linear type addition to form a network type of structure or by cycloaddition to form linear polymer chains. An illustration of the proposed polymerization mechanism of benzocyclobutene (BCB) terminated resins is shown below.

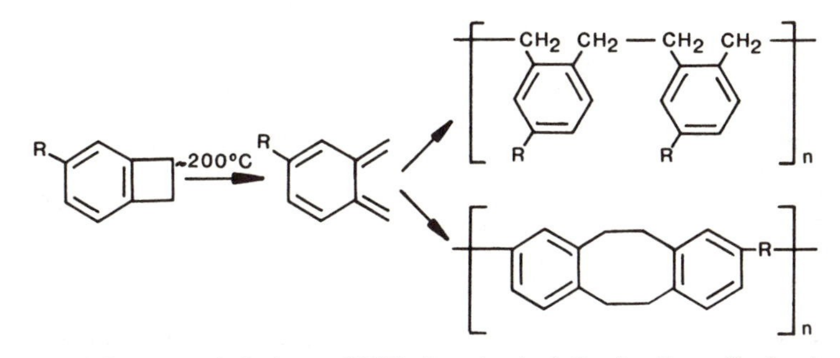

Proposed Benzocyclobutene (BCB) Terminated Resin Cure Mechanism

The structure formed through a linear addition of the molecules is shown at the top of the illustration. The benzocyclobutene groups on either end of the BCB terminated resin molecules react with two other BCB groups as shown to form a poly(o-xylyene) strands. With BCB reactive groups present at both ends of the resin molecules, a double strand of poly(o-xylyene) is formed which is bridged by the aromatic imide chains.

The second proposed polymerization mechanism shown at the bottom of the illustration involves a cycloaddition reaction. The strained four member benzocyclobutene ring undergoes a thermally induced ring opening as described previously, the opened ring then reacts with one other opened ring to form an eight member ring between the two molecules. As shown this mechanism results in the linear polymer molecules with no branch or network formation. The BCB terminated resins may actually undergo a combination of both reactions during the cure.

The other BCB resin system to be discussed is a blend of a BCB terminated resin synthesized in this laboratory [3] and a bismaleimide (BMI) resin commercially available from Aldrich Chemical Co. The two components were blended in a 1:1 molar mixture and the structure of each is shown below.

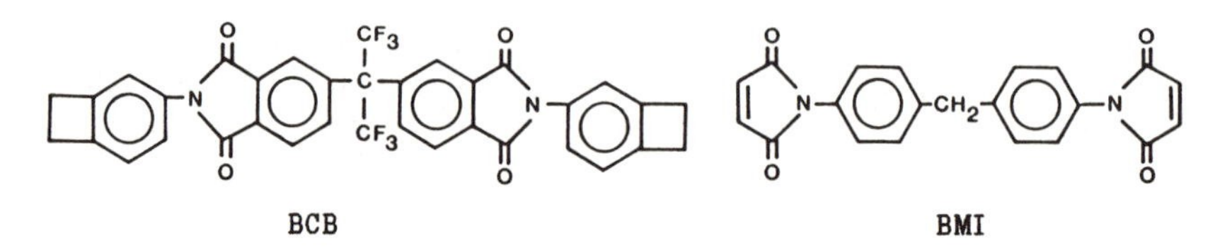

The cure mechanism of the BCB/BMI blend is believed to differ from that of the BCB oligomer. Following the thermally induced ring opening of the strained four member ring of the benzocyclobutene it can then react with the dienophile introduced to the system by the BMI resin via a Diels-Alder reaction [3]. An illustration of that reaction is shown below.

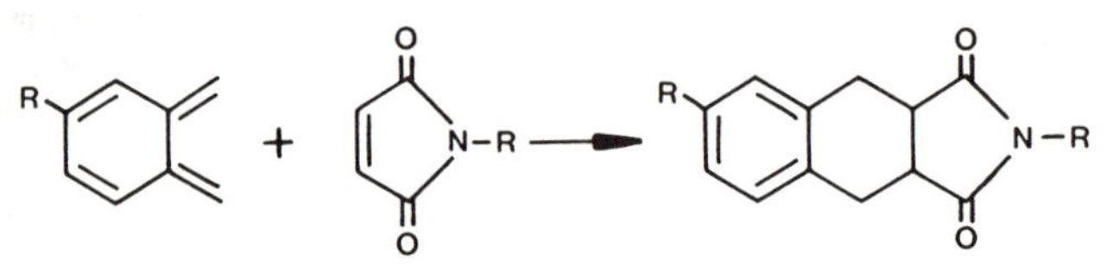

BCB/BMI Resin Polymerization Mechanism

The Diels-Alder reaction mechanism is discussed in greater detail in reference 3. As seen above the reaction between a benzocyclobutene terminated resin and a BMI results in the formation of linear polymer chains.

EXPERIMENTAL

With the diversity of resins being studied in this laboratory a set of standardized testing techniques have been adopted. Generally only limited quantities of each resin are available for preliminary evaluation. Therefore, tests have been developed to yield as much useful data as possible from fifty grams of sample and are described in more detail elsewhere [4].

The criteria by which these resins are evaluated include: thermal analysis, cure kinetics and rheological studies of the uncured resin. Mechanical properties including hot/wet sample testing and thermal analysis are then obtained from cured neat resin specimens. The results from these tests run on the neat resin will give some indication of the suitability of that resin for use as a composite matrix material and future studies to be conducted.

Resin Analysis. Thermal analysis of the BCB resin systems included Differential Scanning Calorimetry (DSC) using a DuPont 910 cell and Thermal Gravimetric Analysis (TGA) run on a DuPont 951 TGA, the entire thermal analysis system run using an Omnitherm model 35053 controller. Thermal analysis scans were run at 10°C/minute, DSC was run in a nitrogen atmosphere and TGA scans were run in air.

Dynamic mechanical properties of the uncured resin were obtained using Torsion Impregnated Cloth Analysis (TICA) [5]. Using this technique the resin is evenly distributed throughout a small rectangle of fiberglass cloth. The impregnated cloth is then folded in the dimensions of a torsion bar (2.5" x 0.5") and run with slight tension applied on a Rheometrics (RDS-7700) Spectrometer to obtain cure rheology data. Temperature scans were run at 2°C/minute in nitrogen with a frequency of 1.6Hz. Resin viscosity was determined by parallel plate rheology using the Rheometrics Spectrometer. Scans were run on 1" diameter parallel plates in a nitrogen atmosphere at 2°C/minute with a frequency of 1.6Hz.

An appropriate cure cycle was established based on the results obtained from the thermal analysis and cure rheology studies of the resin and cured BCB bar and dogbone shaped samples were fabricated for testing. Bar shaped specimens had the dimensions of 3.5" x 0.5" x 0.125" and were used to make compact tension specimens for fracture toughness studies and for dynamic mechanical analysis of a torsion bar. Dogbone shaped specimens for tensile tests had a gauge area of 1" x 0.15" and were approximately 0.040" thick.

Mechanical Properties. Mechanical properties obtained on the cured resins included tensile strength and fracture toughness. Tensile tests were run on an Instron model 1122 Universal Tester with a crosshead speed of 0.02"/minute. Tests were run on dry and saturated samples in air. Fracture toughness (K_Q) [6] values have been obtained using a MTS 810 Materials Testing System at 0.02"/minute at ambient and elevated temperatures in air. The compact tensile specimens tested were 0.5" x 0.5" x 0.125" in dimension. Mechanical properties data are based on the results from four or more tests run at each condition.

Thermal analysis, moisture uptake and dynamic mechanical analysis was also accomplished on cured specimens. Thermal analysis parameters used to study cured specimens are the same as those described earlier to test resins. The moisture uptake in cured specimens was monitored by immersing dogbone shaped specimens in 71°C distilled water until no further weight gain is observed. A dynamic mechanical scan of a torsion bar of cured resin was obtained using the Rheometrics spectrometer with a temperature scan rate of 2°C/minute in nitrogen at a frequency of 1.6Hz. The following sections describe the results obtained from tests run on the two different BCB resin systems. Unless otherwise noted all tests have been run as specified above.

RESULTS AND DISCUSSION

Uncured BCB Oligomer Resin Properties. The uncured BCB oligomer is a golden brown powder at room temperature. Liquid chromatograpy (LC) run on the uncured resin using a UV detector at 254nm indicates the following composition: 22% n=0 (see structure in introduction), 20% monomer, 17% dimer, 12% trimer, and higher molecular weight for the remaining material. As seen in Figure 1, an initial T_g of 153°C was observed by DSC in the uncured resin, the polymerization exotherm maximum is at 257°C, with the onset of that exotherm at approximately 220°C. TGA of the BCB oligomer resin in air indicates that no weight loss occurs in the resin until the onset of degradation at 430°C. The cure kinetics of the BCB oligomer was studied by DSC using a technique developed in this laboratory. Temperature scans were run in nitrogen at various heating rates ranging from 5°C/minute to 80°C/minute and correlated via the method outlined by Goldfarb and Adams [7]. Using this technique an average activation energy for the curing of the resin was 31.7 kcal/mole, and an average heat of reaction of 31.3±5.7 cal/gram was obtained. This kinetic analysis was utilized to produce reaction window plots to aid in choosing cure cycles.

An initial T_g of 190°C (G''_{max}) is observed by TICA with the cure maximum occurring at 247°C (G''_{max}) as illustrated in Figure 2. A cured T_g of 399°C is observed by TICA when scanning the uncured resin up in temperature from 0°C to 400°C. When the same sample is then cooled down from 400°C at 2°C/minute the T_g of the material decreases 20° to 379°C, indicating that some degradation has taken place as the sample was heated up to 400°C. Parallel plate rheology studies of this material indicate that the minimum viscosity is greater than 10^3 poise. The high minimum viscosity exhibited by the uncured resin prohibited further viscosity studies of this material due to instrument limitations. Using the data obtained from the preliminary analysis of the uncured resin the cure cycle described in the following section was developed to fabricate specimens.

Cure Cycle. The following cure cycle was developed to obtain cured dogbone and bar shaped specimens of BCB oligomer. Specimens were fabricated by compression molding the material under controlled temperatures and pressures in a set of matched metal molds using a Wabash 30 ton microprocessor-controlled hot press. The material was degassed at 100°C for 1 hour prior to molding to remove any residual volatiles or water. The resin powder was precompacted into a small bar for ease of handling, the bar was then placed in a preheated 185°C mold.

The mold and sample were placed in the press also preheated to 185°C and the cure cycle started immediately. Press temperature and pressure were microprocessor-controlled for consistent cure cycles. The cure cycle consisted of an initial 30 minute hold at 185°C with 1300 psi of pressure followed by 45 minutes at 221°C with 850 psi pressure and finally 15 minutes at 260°C with 300 psi pressure. The mold was removed from the press immediately following the completion of the final step and the sample removed from the mold while still hot.

Cured specimens were a transparent golden brown color. DSC indicated less than 10% residual exotherm. Following the cure cycle described above the specimens were postcured for 1 hour at 300°C in nitrogen. The extent of cure was determined by DSC to be greater than 95% following postcure. The cure cycle and postcure described above have been used in the fabrication of all cured BCB specimens.

Cured Oligomer Properties. The T_g of the cured specimen was not obvious by DSC, however the T_g was observed at 383°C (G''_{max}) by dynamic mechanical testing of a torsion bar. The T_g obtained by this method has been driven up in temperature by the slow heating rate of 2°C/minute used in the test, the actual T_g is somewhat lower

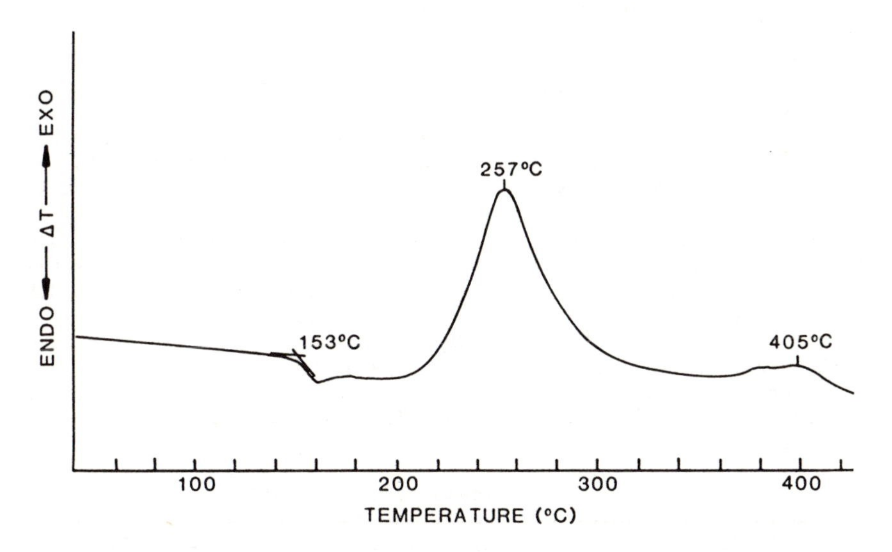

Figure 1. DSC trace of uncured BCB oligomer resin, the initial T_g and the polymerization exotherm are indicated.

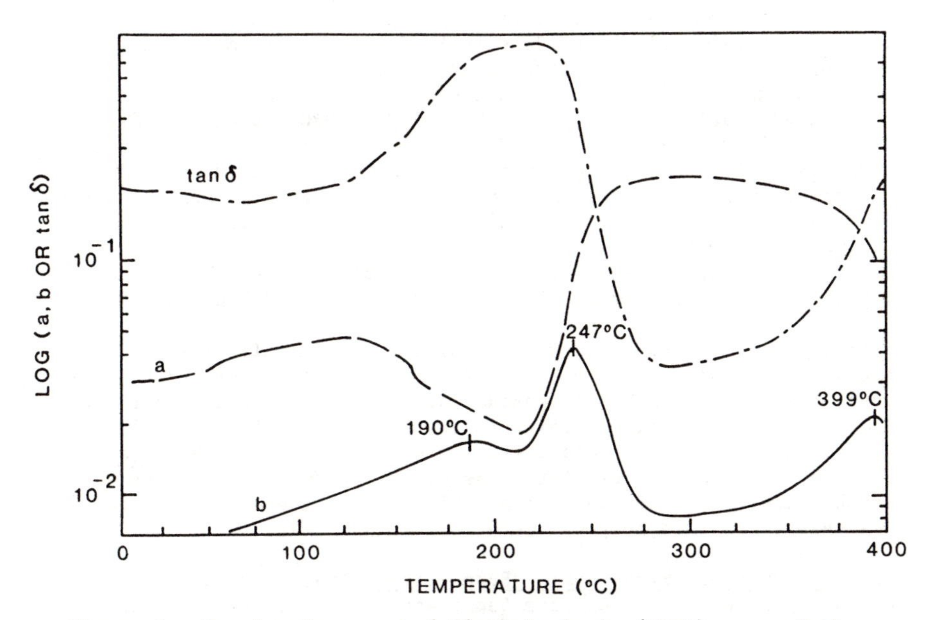

Figure 2. Torsion Impregnated Cloth Analysis (TICA) scan of the uncured BCB oligomer. Tan δ, the in-phase (a) response and the out-of-phase (b) response are plotted as a function of temperature.

and will be determined in future studies. Cured BCB oligomer dogbone shaped specimens exhibited a 3.5% water uptake after being saturated over 600 hours, relatively low when compared with moisture uptake exhibited by some currently used state-of-the-art epoxy systems.

The TGA in air of BCB oligomer cured as previously described indicated good thermal stability with no weight loss prior to the onset of degradation at 430°C, the weight loss occurred in a single step with the extrapolated onset of degradation observed at 552°C. Isothermal aging in an air atmosphere of the cured material was done by holding samples at 343°C (650°F) and periodically weighing the sample to determine the weight loss at that temperature. BCB oligomer cured as described in the previous section exhibited a weight loss of only 9% after 200 hours at 343°C, indicative of the high thermal stability of this resin system.

Tensile tests on cured BCB oligomer were done at ambient temperature on dry and saturated dogbone shaped specimens. The results are shown in Table I.

TABLE I. BCB Oligomer Tensile Results

CONDITION	STRESS(KSI)	STRAIN(%)	MODULUS(MSI)
Dry	10.2±1.5	2.3±0.6	0.46±0.02
Saturated	9.7±1.8	2.2±0.4	0.46±0.03

The moisture uptake in the cured samples appears to have no effect on the room temperature tensile properties as demonstrated in Table I. A critical study to be done is to determine the moisture effect on properties at elevated temperatures.

Fracture analysis of cured compact tensile specimens gave the following K_Q values: RT= 0.80±0.05MPam$^{1/2}$, 100°C = 0.83±0.05MPam$^{1/2}$, 177°C = 0.61±0.06MPam$^{1/2}$, 260°C = 0.55±0.05MPam$^{1/2}$. As a point of comparison Narmco 5208 epoxy was tested by a similar method and exhibited a K_Q of 0.48MPam$^{1/2}$ at room temperature, and Acetylene Terminated Bisphenol-A previously studied by this laboratory had a RT fracture toughness (K_Q) of 0.44MPam$^{1/2}$. In addition to the good mechanical properties, the BCB oligomer also exhibits retention of 70% of those properties at elevated temperatures as high as 260°C.

Discussion. The BCB Oligomer, partially as a result of some linear propagation, and partially as a result of the length of the oligomer used, has a considerable toughness for a high temperature thermoset. The cure reaction also appears to result in a quite stable linkage as illustrated by the isothermal aging results as well as the excellent retention of toughness at elevated temperatures. Use temperatures of greater than 250°C are therefore considered reasonable for this polymer.

Uncured BCB/BMI Blend Resin Properties. The uncured resin is a bright yellow powder at room temperature. LC of each of the two components indicated that the BMI was greater than 98% monomer and the BCB was greater than 90% monomer prior to blending. To make the resin blend, a 1:1 molar mixture of BCB and BMI was completely codissolved in methylene chloride, and the solvent was then stripped off using a rotary evaporator. Any residual solvent remaining in the blend was removed in the degassing step prior to processing.

Preliminary analysis of the uncured resin by DSC (Figure 3) showed a single initial T_g of 61°C, indicating the two resins were thoroughly blended. The polymerization exotherm onset of the BCB/BMI resin blend was 224°C with the maximum at 259°C. Although a different cure mechanism is involved with the BCB/BMI resin blend than with the BCB oligomer the cure occurs in the same temperature range. An early weight loss is observed by TGA in the BCB/BMI uncured resin, approximately 2% weight loss occurring from 70° to 185°C which may be due to residual solvent from the synthesis or blending of the two monomers. The early weight loss is eliminated by the deaerating step prior to processing. The onset of degradation in the resin occurs at 440°C.

DSC study of the cure kinetics revealed an average activation energy of 38.2 kcal/mole and an average heat of reaction of 83.4$\pm$4.5 cal/gram. The average activation energies of the BCB oligomer and the BCB/BMI resin are in the same range but a substantial increase in the heat of reaction is observed in the BCB/BMI as compared to that of the BCB oligomer due to the different reaction mechanisms taking place.

In the TICA temperature scan shown in Figure 4, the uncured resin showed two initial T_gs at 50°C (G''_{max}) and 92°C (G''_{max}). This may be an indication that the technique by which the cloth was impregnated with the resin resulted in the phase separation of the two components, or possibly an indication of an impurity present in the mixture. The polymerization maximum of the resin is observed at 252°C (G''_{max}) and a small peak is seen over the range from 140°C to 200°C with the maximum at 174°C (G''_{max}). The origin of the 174°C peak is unknown at this time. It may be due to the phase separation initiated during sample preparation. In this TICA scan the uncured resin sample was heated up from 0°C to 400°C and then scanned down in temperature at 2°C/minute. During the scan up in temperature a single cured T_g was observed at 372°C (G''_{max}), and when the sample was then scanned down from 400°C (the sample was held at 400°C approximately 10 minutes prior to scanning down) the single cured T_g peak had dropped to 358°C (G''_{max}) due to degradation occurring at the elevated temperatures.

A parallel plate rheology temperature scan of BCB/BMI resin degassed for 0.5 hour at 145°C shown in Figure 5 illustrates the minimum viscosity of the resin blend is 400 poise at 185°C. BCB/BMI resin which has not been degassed exhibits the same viscosity minimum however, an anomaly is observed as the viscosity decreases with heating. The resin begins to drop in viscosity at 120°C, but at 140°C the resin exhibits a short increase in viscosity for approximately 20° before the viscosity drops once again until the onset of polymerization at 193°C. The resins unusual viscosity profile may be related to the unexplained peak observed in TICA at 174°C and could be attributed to the release of residual solvent or some sort of anomalous reaction taking place over that temperature range. The increase in viscosity upon heating is eliminated when the sample is deaerated for 0.5 hour at 145°C prior to running, consistent with residual solvent being the cause.

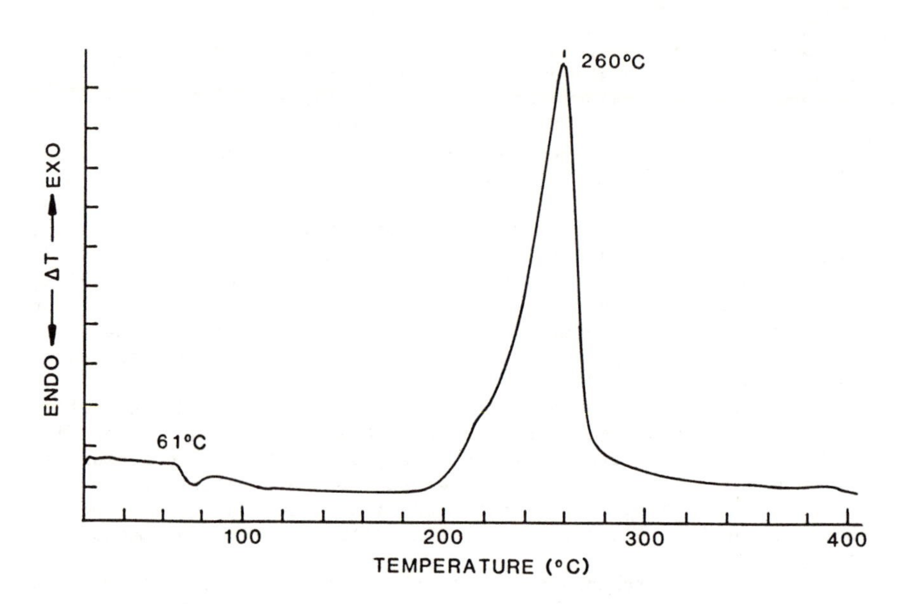

Figure 3. DSC of BCB/BMI resin blend, the initial T_g and the polymerization exotherm are indicated.

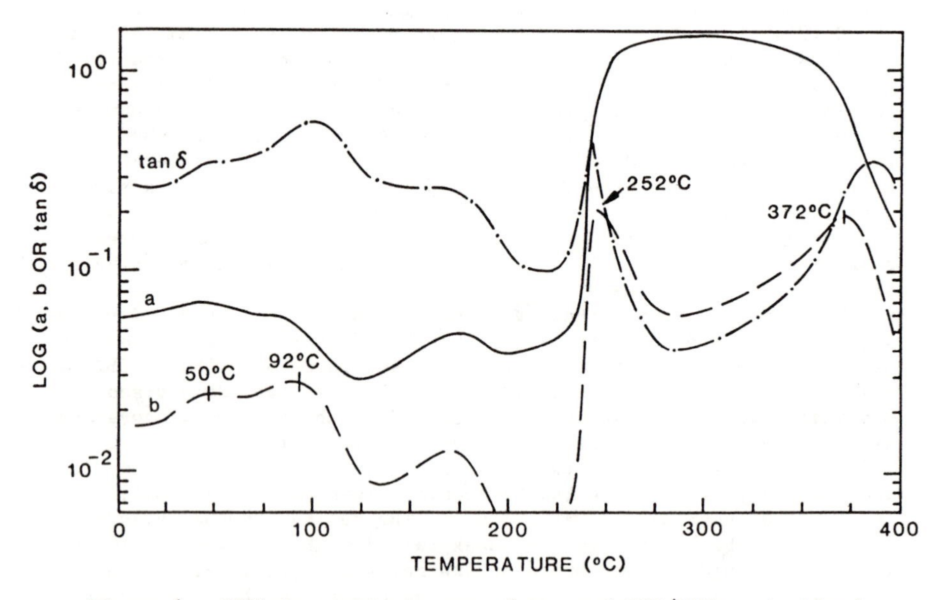

Figure 4. TICA temperature scan of uncured BCB/BMI resin blend. Tan δ, the in-phase (a) response and the out-of-phase (b) response are given as a function of temperature.

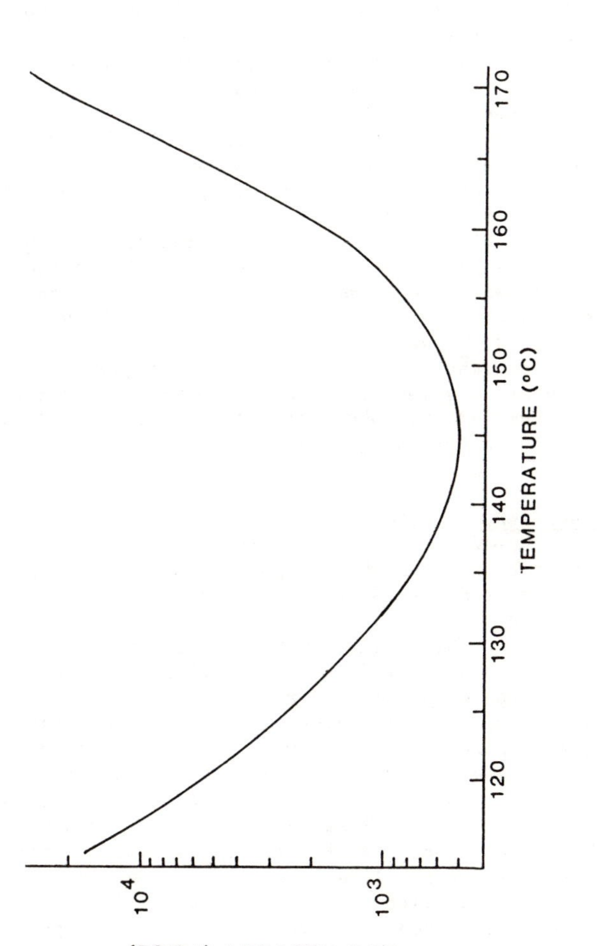

Figure 5. Parallel plate rheology scan showing viscosity of degassed BCB/BMI resin as a function of temperature.

Cure Cycle. BCB/BMI specimens were cured using a cure cycle similar to that used for the BCB oligomer since the cures of the two resins occurred over the same temperature range and similar cure profiles are exhibited. Some allowances were made for the lower minimum viscosity exhibited by BCB/BMI. Lower pressures were used throughout the cure cycle and the material was allowed to precure slightly prior to the application of any pressure. The sample was deaerated at 145°C for 0.5 hour before processing. The resin powder was then placed in an ambient temperature matched metal compression mold. The mold was placed in a 185°C preheated press and allowed to heat to 185°C for 10 minutes. The sample and mold were then held at 185°C for 12 minutes to allow the sample to precure prior to processing. Following the precure, 600 psi of pressure were applied while the temperature remained at 185°C for 20 minutes. The temperature was increased to 221°C while the pressure was decreased to 450 psi and held there to 45 minutes. The temperature was then increased to 260°C and the pressure decreased to 300 psi and held there for 15 minutes prior to removing the sample from the mold.

Cured BCB/BMI specimens are a transparent golden brown in color, similar to what was seen for the BCB oligomer. Once the specimens have been removed from the mold they are postcured for 1 hour at 300°C in nitrogen resulting in an extent-of-cure of greater than 95%.

Cured BCB/BMI Properties. Dynamic mechanical testing of a cured BCB/BMI torsion bar showed a glass transition temperature of 378°C (G''_{max}), this value being somewhat high due to the 2°C/minute heating rate that was used. 4.1% water uptake was seen in cured BCB/BMI dogbone shaped specimens immersed in 71°C water until saturation (600 hours). This value represents a 0.6% increase in the water uptake over that exhibited by the BCB oligomer, due to the presence of the BMI group in the blended resin.

TGA in air of cured BCB/BMI indicated good thermal stability with no weight loss observed in the sample prior to the onset of degradation at approximately 430°C. The weight loss occurs in two steps, the first step is observed to occur in the temperature range from the initial onset of weight loss seen at 430°C to 510°C and represents a 15% weight loss, with the extrapolated onset of degradation for the first weight loss step observed at 486°C. The extrapolated onset of degradation for the second weight loss step is observed at 566°C. The second weight loss step corresponds to the weight loss observed in the BCB oligomer, the first weight loss step is due to the presence of the BMI in the resin blend.

Isothermal Aging of BCB/BMI samples cured as described in the previous section was done in an air atmosphere at 343°C (650°F). The samples were weighed regularly to determine weight loss over time at that temperature. After 200 hours BCB/BMI exhibited a weight loss of only 15%. This is slightly higher than the 9% weight loss of the BCB oligomer tested under the same conditions, the increased weight loss seen in Isothermal Aging with the presence of the BMI agrees with the results obtained by TGA.

Tensile tests run on dry and saturated dogbone shaped specimens at room temperature resulted in the values shown in Table II.

TABLE II. BCB/BMI Tensile Results

CONDITION	STRESS(KSI)	STRAIN(%)	MODULUS(MSI)
Dry	10.9 ± 1.4	2.4 ± 0.3	0.44 ± 0.02
Saturated	9.8 ± 1.0	1.9 -	0.48 ± 0.02

The tensile properties of BCB/BMI are not affected by the absorption of water as shown in Table II. Future studies will be done to determine the effect of water uptake on the mechanical properties at elevated temperatures.

Fracture toughness (K_Q) values were obtained by fracture analysis of compact tensile specimens at the following temperatures: RT = $0.80\pm0.04 MPam^{1/2}$, 100°C = $0.85\pm0.06 MPam^{1/2}$, 177°C = $0.67\pm0.01 MPam^{1/2}$, and 260°C = $0.68\pm0.05 MPam^{1/2}$. BCB/BMI exhibits excellent retention of fracture toughness at elevated temperatures with 85% of it's original toughness at 260°C.

Discussion. The combination of BCB with BMI significantly improves the properties of the BMI, apparently by tying up the maleimide in a Diels-Alder reaction with the transient diene formed from the opening of the cyclobutene ring. Both the thermal and the thermo-mechanical properties are improved almost to the level of the BCB oligomer itself.

SUMMARY

BCB and BCB/BMI resin systems cure through different types of addition reactions. These materials cure via addition type reactions without the use of catalysts or evolution of volatiles and have been compression molded using moderate pressures and temperatures to produce void-free cured specimens.

The properties exhibited by the two BCB terminated resin systems studied indicate that these resins may be viable materials for use as structural matrices in composites. Cured samples of both the BCB oligomer and the BCB/BMI blended resin system exhibit good mechanical properties with excellent thermal stability and toughness. Isothermal aging of cured samples at 343°C (650°C) in air exhibited weight losses of 9%-15% after 200 hours. Samples tested at temperatures as high as 260°C retained 70% to 85% of the fracture toughness observed in room temperature tests.

Previous BCB/BMI resin studies have demonstrated that the addition of 40% to 60% BCB to a BMI results in a resin with thermal stability dramatically improved over that of the BMI itself [1]. The 1:1 molar ratio BCB/BMI system used in this study has exhibited the retention of thermal mechanical properties at elevated temperatures as well.

Further studies are continuing to determine the mechanical properties at elevated temperatures of cured BCB resin systems when saturated, and to determine the properties of BCB resins when used as matrices in composites. Work is currently being conducted to determine the neat resin properties of the BCB monomer (without BMI) and of the BCB oligomer blended with a BMI resin.

ACKNOWLEDGMENTS

The BCB samples were synthesized in this laboratory by Dr. Fred Arnold and Dr. Loon-Seng Tan, a description of which is also presented in this symposium (1). We would like to thank Charles Benner of the University of Dayton Research Institute (UDRI) for the liquid chromatography results, and Ed Soloski and Marlene Houtz of UDRI for their work in the thermal analysis of these materials.

A portion of this work was performed by Universal Energy Systems, Inc., under Air Force Contract F49620-85-C-0013.

REFERENCES

1. Loon-Seng Tan, Edward J. Soloski and Fred E. Arnold, Chemistry, Properties, and Applications of Crosslinking Systems, ACS Symposium Series this volume.
2. Loon-Seng Tan and Fred E. Arnold, ACS Polymer Preprint, 1985, 26(2), 176.
3. Loon-Seng Tan, Edward J. Soloski and Fred E. Arnold, ACS Polymer Preprint, 1986, 27(1), 453.
4. L. R. Denny, C. Y-C Lee and I. J. Goldfarb, ACS Polymer Preprint, 1985, 26(1), 138.
5. C. Y-C Lee and I. J. Goldfarb, Polymer Engineering Science, 1981, 21, 787.
6. C. Y-C Lee and W. B. Jones, ACS Organic Coatings and Plastics Preprint, 1982, 47, 109.
7. I. J. Goldfarb and W. W. Adams, ACS Organic Coating and Plastics Preprint, 1981, 45, 133.

RECEIVED October 7, 1987

Chapter 26

Carbon-13 NMR Investigation of the Oligomerization of Bismaleimidodiphenyl Methane with Diallyl Bisphenol A

Keith R. Carduner and Mohinder S. Chattha

Ford Motor Company, Dearborn, MI 48121

The structure of a soluble prepolymer, resulting from the addition reaction of bismaleimidodiphenyl methane to diallyl bisphenol A at approximately 150 °C, is confirmed by a detailed high resolution ^{13}C NMR investigation. The NMR results also provide the rate of reaction, which requires about 5 hrs for 30% of the reactants to be consumed. Curing of this system at 240 to 250 °C leads to the formation of a highly cross-linked polymeric material with a high glass transition temperature. Thermal mechanical analysis shows no significant softening of this material up to at least 320 °C. Solid state ^{13}C NMR experiments indicate a high cross-linking density for the high temperature product.

Early research on high temperature polymers concentrated primarily on thermal stability and paid little attention to their processability and cost. However, for a polymer to be successful as a commercially viable structural matrix, it must exhibit a favorable combination of processability, performance characteristics, and price. In particular, a desirable high temperature polymeric system for coatings, composites, and adhesives applications must exhibit adaptability to conventional processing techniques at low temperature and pressure, should exhibit good mechanical properties, acceptable repairability, weatherability, and cost effectiveness.

Recently, researchers at Ciba-Geigy Corporation have described reactions of bismaleimides and alkenyl phenols that produce high temperature thermosets. (1-3) Unfortunately, many bismaleimides are not suitable for coatings application because they are insoluble in common organic solvents. We have obtained soluble oligomers by the reaction of bismaleimidodiphenyl methane, 1, (see below) and diallyl bisphenol A, 2. When heated to 240 to 250 °C, these oligomers form extremely hard coatings with many desirable mechanical and thermal properties. In this paper, we describe a ^{13}C NMR study of the condensation mechanism of bismaleimide 1 with diallyphenol 2 and the structure of the soluble prepolymers. In part, the experiment was

0097-6156/88/0367-0379$06.00/0

undertaken to confirm a reaction mechanism proposed by workers at Ciba-Geigy Corporation. (1-3) Some preliminary solid-state ^{13}C NMR data is also presented on the structure of the rigid material formed from this prepolymer by curing at 250 °C.

Experimental

Materials. Bismaleimidodiphenyl methane, 1, was obtained from Ciba-Geigy Corporation. It is a commercial grade yellow powder containing greater than 85% of theoretical maleimide double bond content.

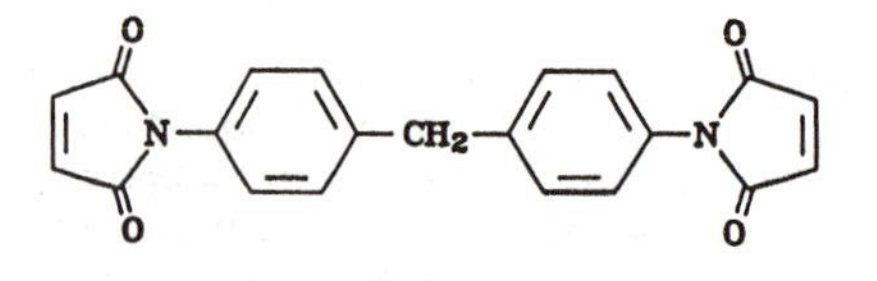

1

Diallyl bisphenol A, 2, also supplied by Ciba-Geigy, is a commercial grade amber colored viscous liquid at room temperature. Typical hydroxyl content is greater than 0.62eq/100g (0.66eq/100g theoretical).

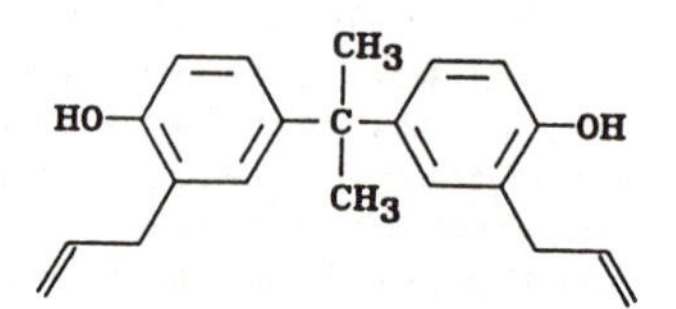

2

Preparation of Copolymer. 23.0 g of 1 and 17.5 g of 2 were mixed in a round bottom flask and heated by an oil bath at 150 °C for 15 min. Continuous stirring led to a homogenous melt.

A series of samples were prepared for liquid-state ^{13}C NMR experiments for the purpose of studying the condensation reaction as a function of time. The melt was returned to 145 °C and six samples were drawn off at hourly time intervals between 1 and 6 hr. When the individual samples were dissolved in a DMSO(5%)/$CDCl_3$ solvent system, it was observed that for longer heating times, an increasing larger insoluble fraction was created. This fraction was filtered out of the samples before data acquisition. Locking and shimming of the NMR static magnetic field was done on the $CDCl_3$ ^{2}H resonance.

High Resolution ^{13}C NMR spectroscopy was performed at 75.4 MHz using a Bruker MSL300. To acquire liquid-state spectra, approximately 0.1 g of solute was dissolved in the mixed solvent in 10 mm NMR sample tubes. Spectra were acquired under conditions of broad band (BB) decoupling to remove the scalar proton carbon interaction. (4) Normally, about 500 transients were averaged using a 30 deg read pulse and a 1 sec repetition rate. (5) TMS was added to the sample to reference the chemical shifts. Some qualitative Nuclear Overhauser Effect (NOE) enhanced, off-resonance decoupling experiments were performed to verify the assignments made on the basis of the chemical shifts. (4)

Magic Angle Spinning (MAS) ^{13}C NMR spectra were also acquired for this study with a Bruker CPMAS probe. For these experiments, ca. 300 mg of cryoground polymer was packed into 7 mm Lowe-type double-beam rotors. The rotor speed was typically 4 KHZ. Carbon signal was created by proton cross-polarization for 1 msec followed by proton decoupling (field strength in excess of 80 KHz) during the acquisition of the carbon transient precession. (6) The pulse sequence for this experiment, CPMAS, is illustrated in Figure 1-A. Sideband interference is eliminated by the total suppression of spinning sidebands technique (TOSS). (7) The pulse sequence for CPMAS with TOSS is shown in Figure 1-B. Protonated carbon suppression by delayed decoupling was also combined with TOSS to verify assignments made on the basis of chemical shifts. (8)

Results

Results from Differential Scanning Calorimetry (DSC) studies of this system display low and high temperature exotherms occurring at approximately 130 °C and 255 °C respectively. (2,3) The DSC data from reference 2 is reproduced here as Figure 2. On the basis of the NMR results explained below, the following addition reaction of 1 to 2 producing the soluble prepolymer is expected to correspond to the low temperature exotherm (2,3):

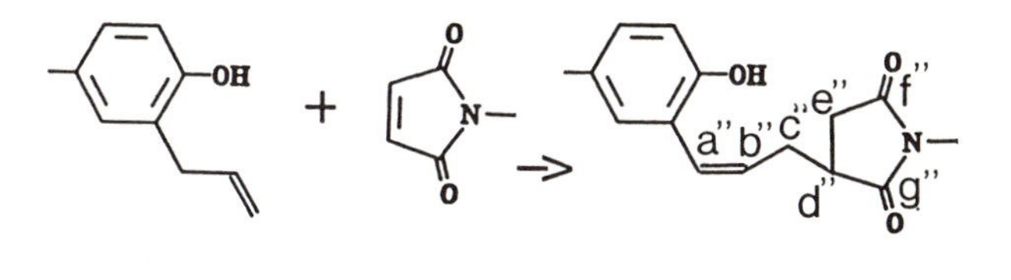

The high temperature exotherm at 252 °C corresponds to the reaction that produces the rigid coating material. Reaction mixtures held at 150 °C for 15 mins and then brought to 250 °C for approximately 2 hrs produce a hard, brownish polymer that is totally resistant to the DMSO/$CDCl_3$ solvent. Thermal Mechanical Analysis (TMA) performed on a duPont 943 at a heating rate of 10 °C/min shows no observable softening of this polymer to at least 320 °C. Results of the TMA are shown in Figure 3. Likewise, DMTA results, illustrated in Figure 4, show no reduction in the modulus until at least 320 °C, or 70 °C

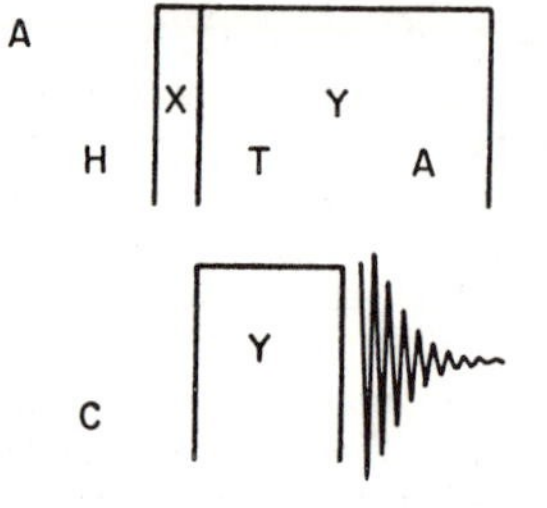

Figure 1
Pulse sequence for ^{13}C MAS NMR. A) CPMAS. B) CPMAS with TOSS. T is magnetization transfer; A, acquisition; X,Y, -X are pulse phases; and * represents a delay.

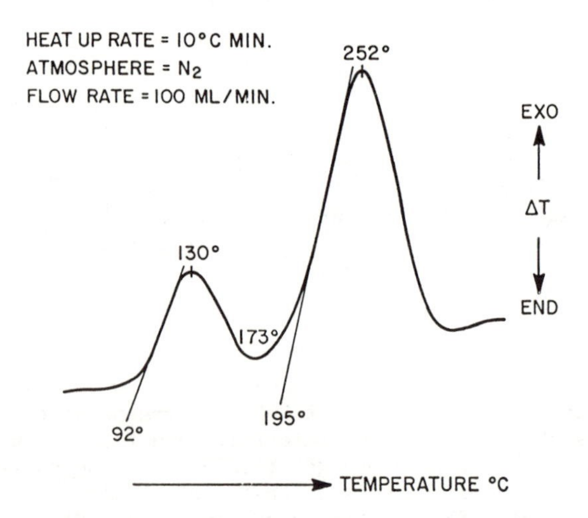

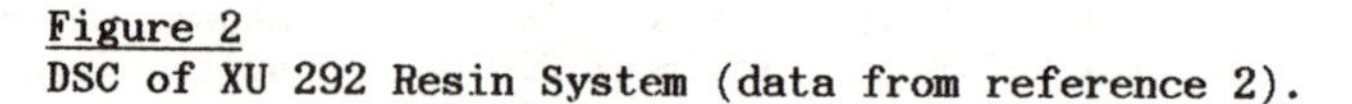

Figure 2
DSC of XU 292 Resin System (data from reference 2).

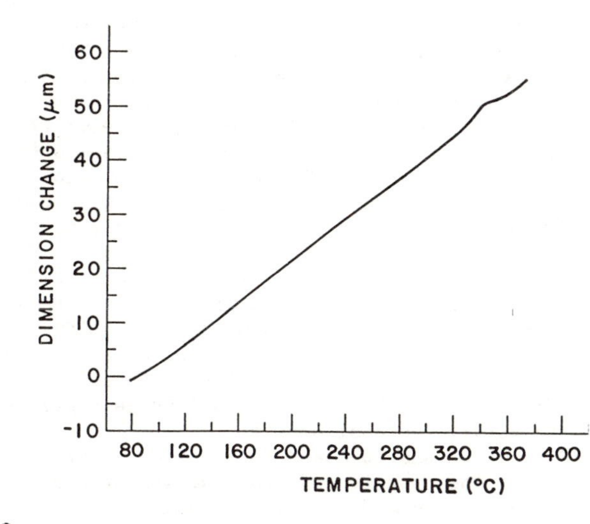

Figure 3
Thermal Mechanical Thermogram of material cured at 250 °C for 3 hr. Thermogram acquired using 5 mm of sample under argon at a rate of 10 °C/min.

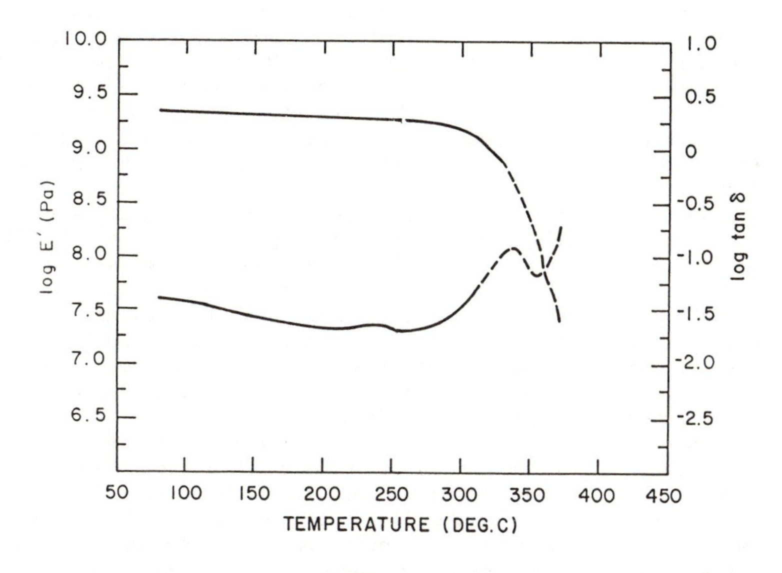

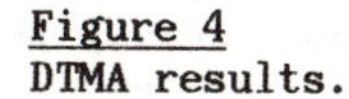
Figure 4
DTMA results.

higher that the reaction temperature. The plot of Figure 5 shows the TMA softening point as a function of cure time. This data indicate that substantial solid state reaction occurs within the first 1 hr at 250 °C. Other types of mechanical measurements (2,3) demonstrate that this thermoset is an extremely hard and durable material characterized by a highly cross-linked network. At present, no specific information is available on the details of the 250 °C cross-linking reaction. Preliminary ^{13}C MAS NMR results that provide some structural detail of this material are given below. It should be noted that these results are not inconsistent with the mechanism proposed in reference 2 and 3 for the 250 °C curing reaction.

To begin the NMR experiments, high resolution ^{13}C spectra were acquired of the two reactants 1 and 2. These spectra are reproduced as Figures 6 and 7, respectively. The assignments were made on the basis of the general principles of carbon chemical shifts as well as by comparison with reference spectra. The spectra agree exactly with the structures of compounds 1 and 2 given above. Off-resonance decoupling experiments, which reveal J coupling multiplicities, were also performed to verify the assignments.

The 150 °C reaction was followed by studying samples heated for as long as 6 hrs. The spectra of this series is characterized by intense lines from unconsumed reactants and gradually increasing intensity of new lines, presumably associated with the product(s). Figure 8-A presents the ^{13}C spectrum for the sample heated for 6 hrs. Figures 8-B and 8-C are expansions of different spectral regions of 8-A.

Comparison of the peaks listed in Table I with the spectra of Figure 8 verifies the assertion made above that the most intense lines in the 6 hr spectrum are from the reactants. The sub-spectrum of smaller lines are the new features, and so it seems that even after 6 hrs, some reactant still remains. More reaction had probably occurred than is indicated by the spectra since some of the product was lost to the NMR in the insoluble fraction. Concentrating on the sub-spectrum of 20 new peaks, at least five spectral features are evident that support the mechanism for the condensation reaction given above.

1) There are two new singlets (no attached protons as verified by off-resonance decoupling) in the carbonyl region at 178.66 and 175.75 ppm. The slight downfield shift and the appearance of two lines are expected for the carbonyls in the 150 °C product (carbons f" and g" in 3) since they are not equivalent. Comparison of the intensity of these two lines to the carbonyl line at 169.56 ppm, which comes from carbon a of compound 1, in the 5 hr spectrum allows one to conclude that the rate of reaction approximately corresponds to 30% consumption of reactants in 5 hrs at 150 °C.

2) There is an absence of product resonances in the vicinity of reactant resonances c' and c, as may be seen in Figure 8-B. This is expected since these carbons are absent in the product.

3) In the spectral range 114-116 ppm of Figure 8-B, where there are two reactant peaks g' and h', only one product peak is present. This peak can be assigned to the carbon corresponding to the h' resonance (an aromatic ring carbon, see compound 2). The product proposed above does not contain a carbon corresponding to carbon g'.

4) It appears that the remaining reactant lines in the spectral range 112 to 154 ppm (labeled in figure 8-B) can be brought into

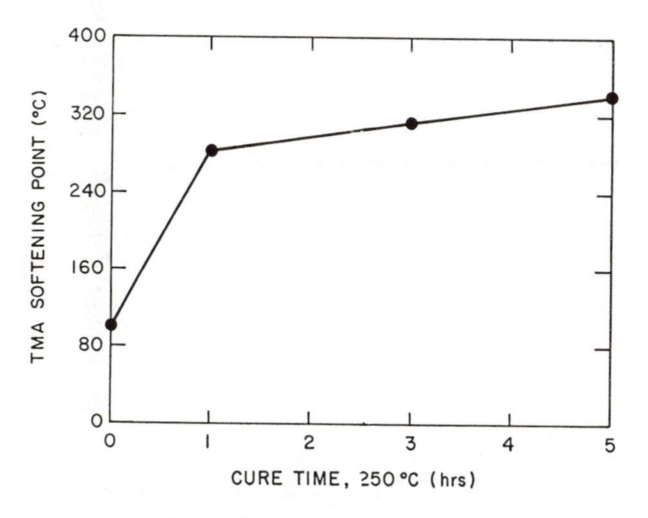

Figure 5
TMA softening point as a function of cure time at 250 °C. B/D is the molar ratio of Bismal to Diallyl.

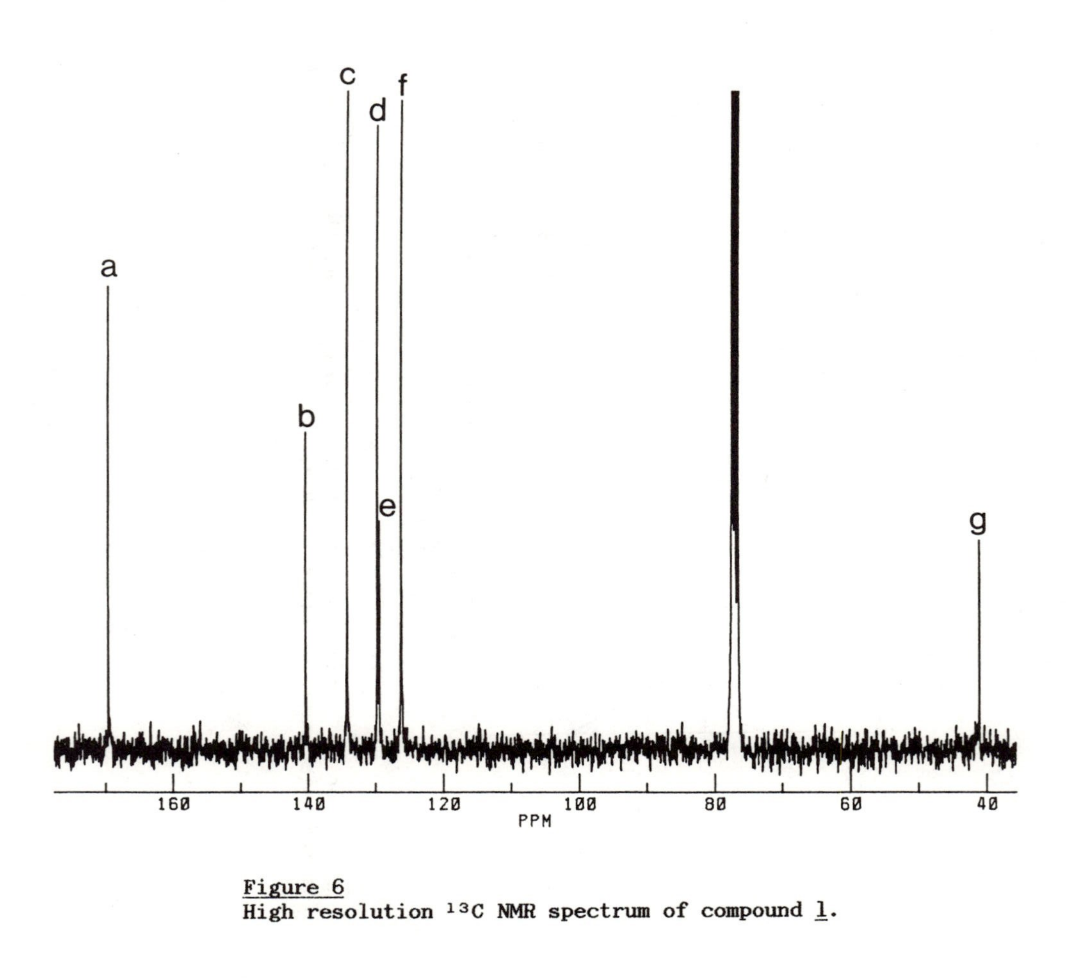

Figure 6
High resolution ^{13}C NMR spectrum of compound 1.

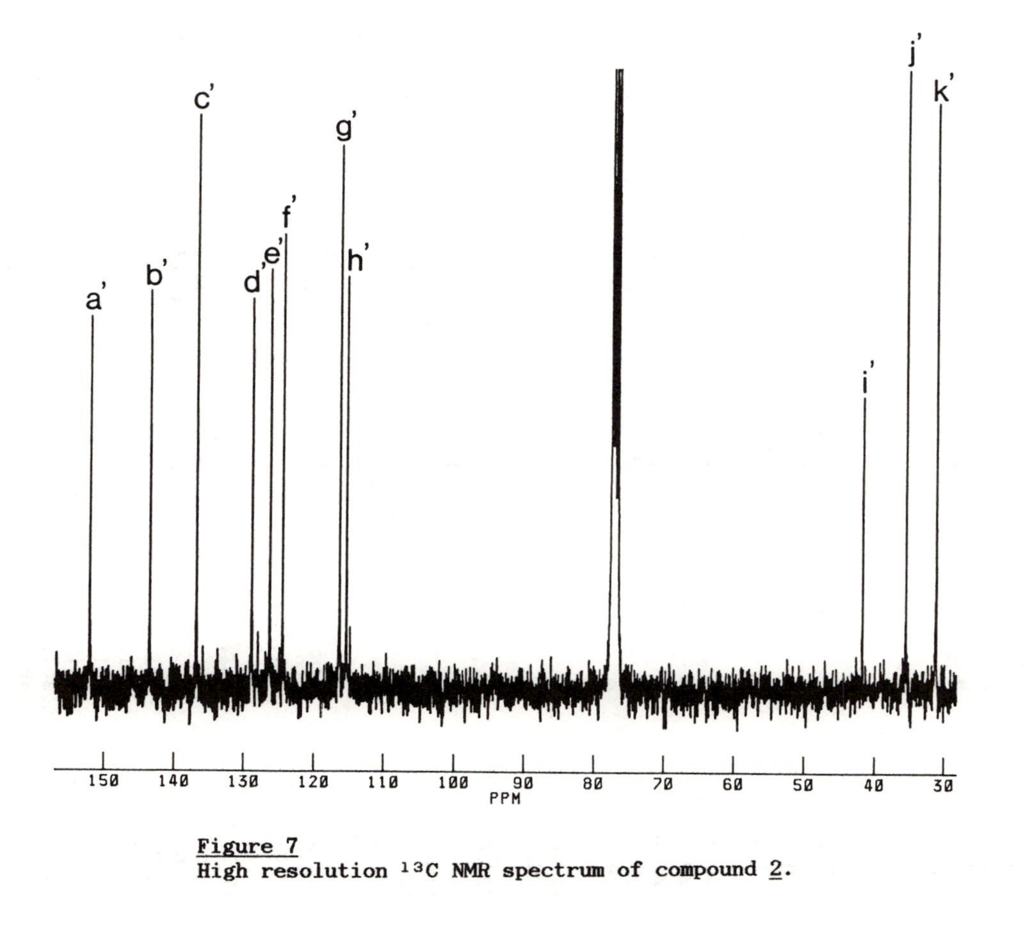

Figure 7
High resolution ^{13}C NMR spectrum of compound 2.

Table I. Spectral Lines of Figure 8

shift(ppm)[a]	intensity[b]	identification[c]	corresponding reactant resonance
178.66	w	p	
175.75	w	p	
169.56	s	a	169.51
152.37	w	p	
152.31	s	a'	151.93
152.21	w	p	
142.43	s	b'	143.41
142.24	w	p	
142.17	w	p	
140.95	w	p	
140.37	s	b	140.34
137.26	s	c'	136.69
134.22	s	c	134.17
130.23	w	p	
129.66	s	d	129.69
129.43	s	e	129.38
128.43	s	d'	128.79
128.32	w	p	
127.66	w	p	
127.33	w	p	
126.71	w	p	
126.22	s	e',f	126.22,126.15
125.69	w	p	
125.22	w	p	
125.14	s	f'	124.41
124.05	w	p	
122.74	w	p	
115.58	w	p	
115.21	s	g'	116.24
114.79	s	h'	115.25
77.61	sol.[d]		
77.18	sol.		
77.76	sol.		
41.57	s	i'	41.72
41.07	s	g	41.05
40.81	sol.		
40.54	sol.		
40.26	sol.		
39.98	sol.		
39.70	sol.		
39.42	sol.		
39.14	sol.		
35.11	w	p	
34.77	s	j'	35.48
33.65	w	p	
31.22	s	k'	31.16
31.18	w	p	

[a]wrt TMS. [b]w, weak; s, strong. [c]p, product.

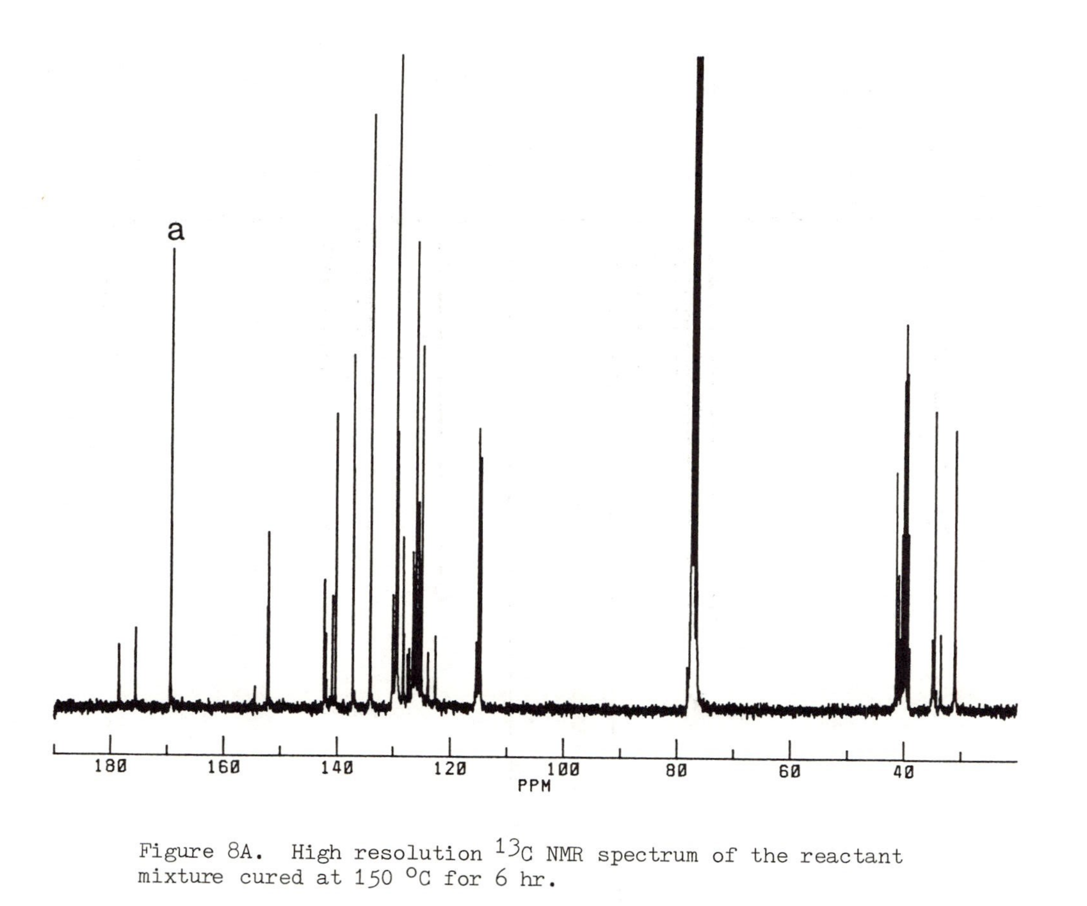

Figure 8A. High resolution ^{13}C NMR spectrum of the reactant mixture cured at 150 °C for 6 hr.

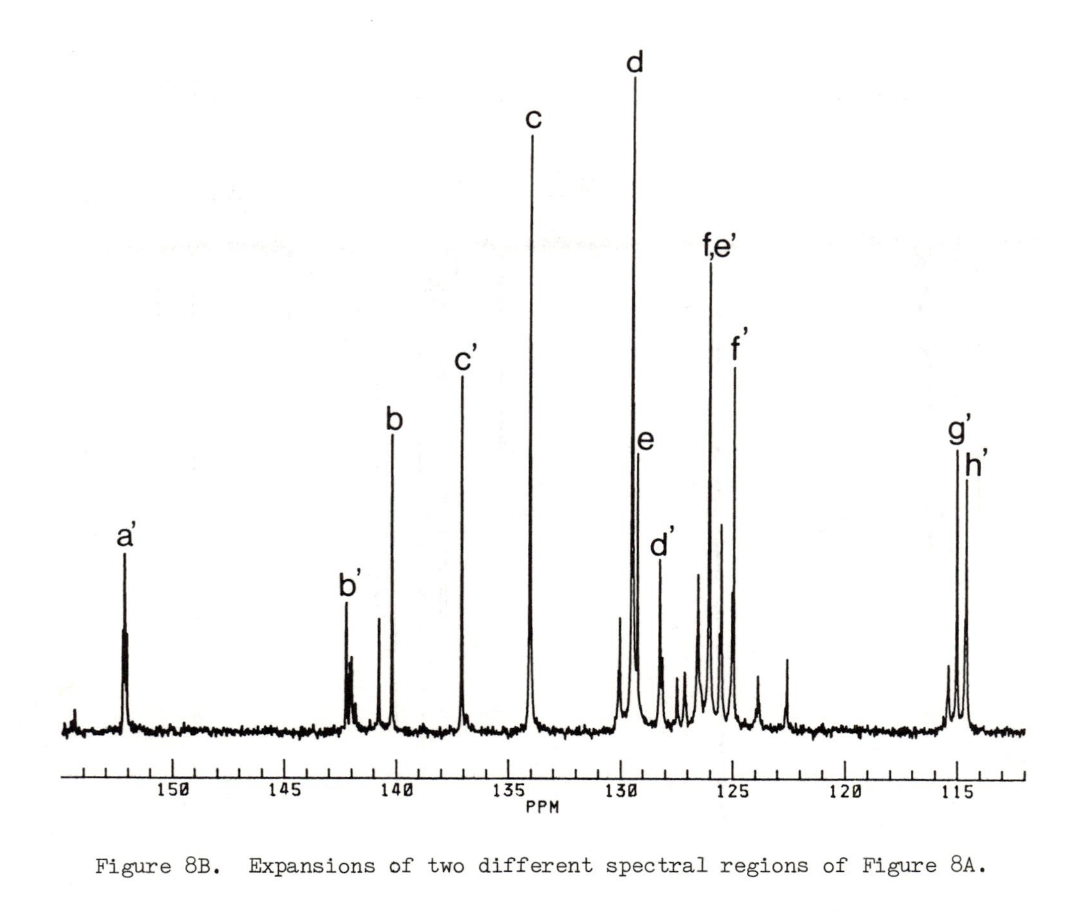

Figure 8B. Expansions of two different spectral regions of Figure 8A.

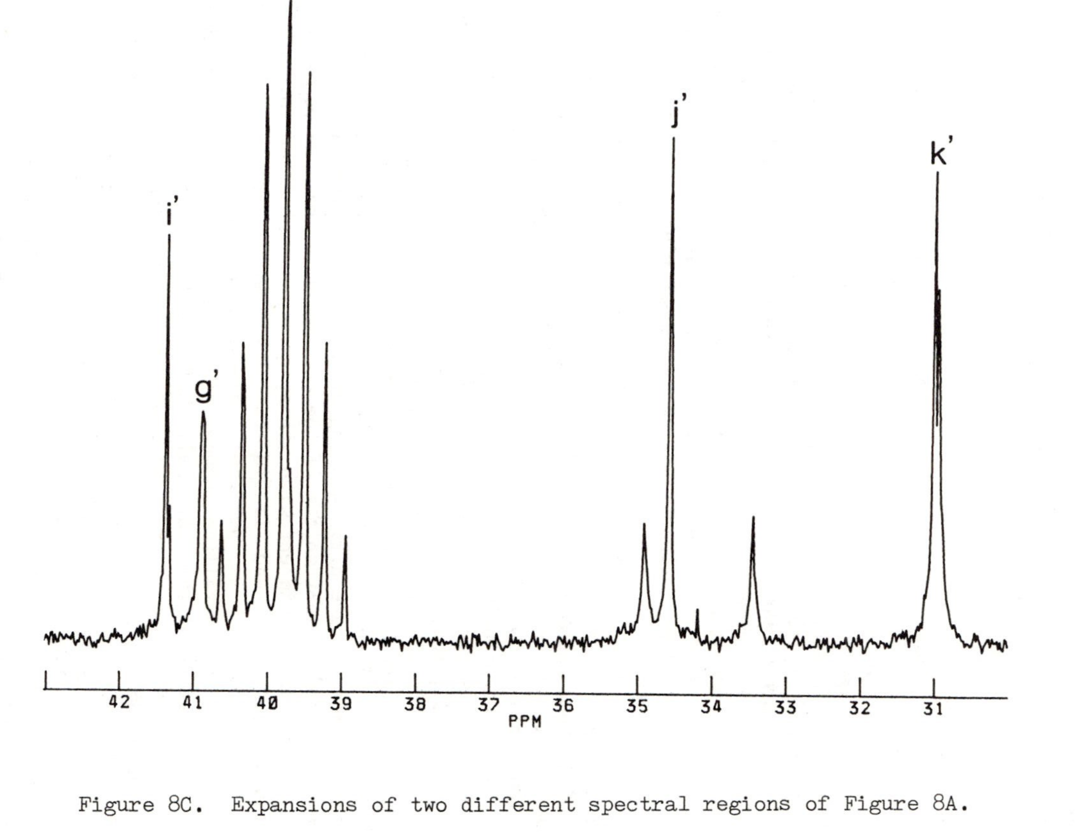

Figure 8C. Expansions of two different spectral regions of Figure 8A.

correspondence with nearby product resonances. This suggests that the product is structurally similar to the reactants. The slight changes in chemical shifts between corresponding reactant and product carbons are to be expected to result from the "ene" coupling reaction proposed above. Some of the product lines are split into two or more peaks (notably at 152 and 142 ppm). This probably indicates the formation of more than one type of closely related products. One possibility is multiple "ene" reactions on a single reactant molecule. The peaks at 124.05 and 122.74 ppm are assignable to the a" and b" carbons of the product.

5) In the spectral range 30-42 ppm, one can identify three new resonances at 35.11 (triplet or two attached protons), 33.65 (triplet), and 31.18 ppm (quartet or three attached protons), as illustrated in Figure 8-C. The two new triplets may be identified with the c" and e" carbons, which are two of the three new methylene groups found in the product. The resonance of carbon d", the third methylene carbon, would also be found in this region, although it is probably lost beneath the DMSO band. The lines from the other product carbons to be found in this region are probably overlapped by the corresponding reactant lines. For this reason, one does not see the pattern of less intense lines that characterized the other parts of the spectrum. The exact overlap is a consequence of the fact that the "ene" reaction should have the least amount of effect on the carbons in this region. A definite exception to this is the new line at 31.18 ppm, which is the product resonance corresponding to the reactant k' resonance. It is also possible that the small peak just to low field of the i' line is the corresponding product peak.

When the prepolymer was cured at 250 °C, it became insoluble in DMSO. This suggests, in agreement with the previously reported DSC results (2,3), that a second reaction occurs. A sample of material cured for 15 mins at 250 °C was powdered in a cryogrinder and packed into a rotor for a ^{13}C MAS spectrum. The results are illustrated in Figure 9-A. Figure 9-B shows the high resolution spectrum of the still soluble mixture after heating at only 150 °C for 15 min. The broadening and slight shifts of the corresponding resonances between 9-A and 9-B are consistent with those generally observed in comparisons between solid and liquid state spectra. (9) The lines in the MAS spectrum are, in addition, somewhat broader than would be observed for a crystalline powder. (9) This indicates the highly amorphous nature of this material. Unfortunately, the combination of amorphousness with a fair number of inequivalent carbons produces a MAS spectrum with poorly separated resonances. This greatly complicates the problem of extracting information about changes in carbon functionality associated with the 250 °C reaction. Further solid state NMR experiments, including relaxation experiments, are presently in progress to gain insight into the 250 °C curing reaction and the structure of the high temperature material.

Conclusion

A soluble prepolymer results from the addition of bismaleimidodiphenyl methane 1 to diallyl bisphenol A 2 at approximately 150 °C. In the present study, a detailed 13C NMR investigation has provided strong evidence for the proposed structure of the soluble oligomer. The NMR has also provided some idea of the rate of

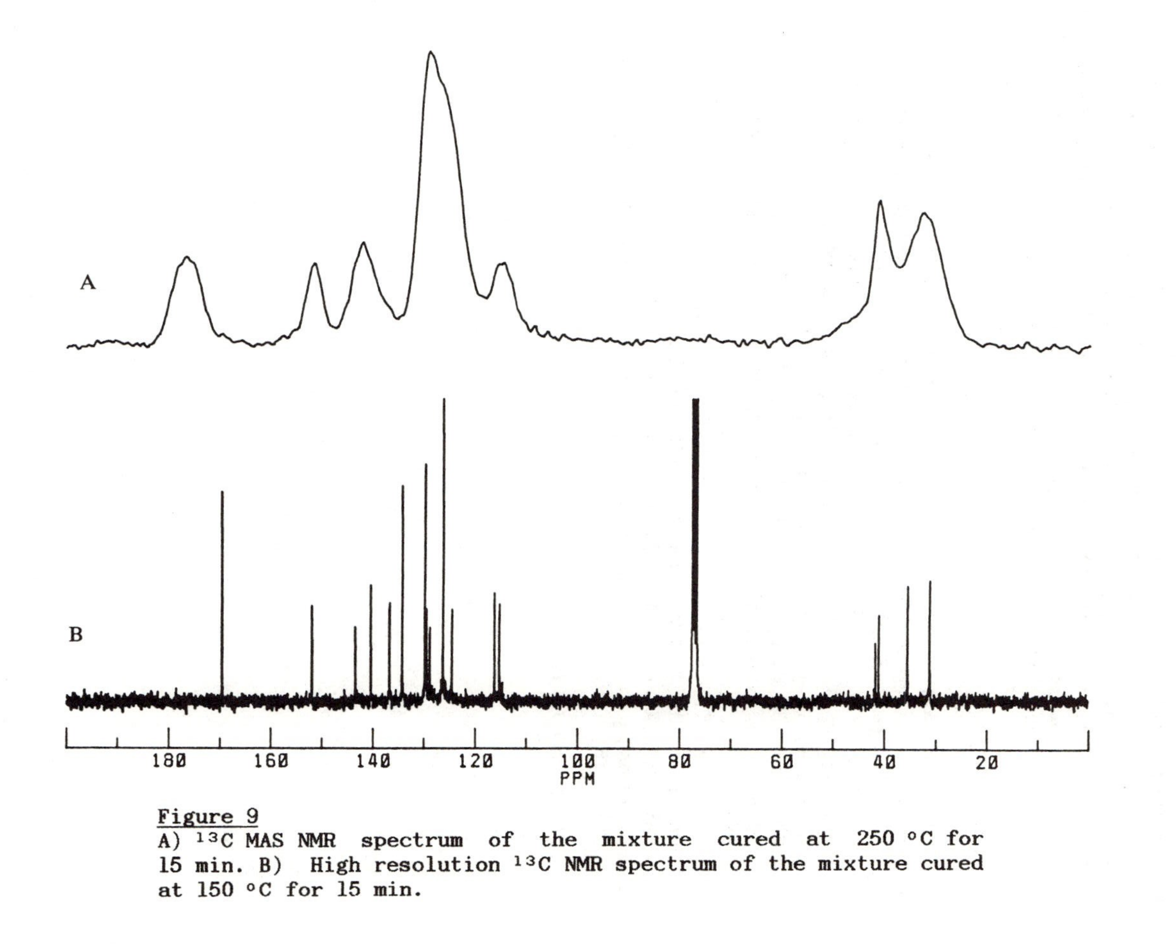

Figure 9
A) ^{13}C MAS NMR spectrum of the mixture cured at 250 °C for 15 min. B) High resolution ^{13}C NMR spectrum of the mixture cured at 150 °C for 15 min.

reaction, which corresponds to 30% consumption of reactants in 5 hrs. A further reaction occurs at 250 °C. The product of this reaction is an extremely hard polymeric material with a high glass transition temperature. TMA shows no significant softening of this material occurs up to at least 320 °C, approximately 70 °C above the reaction temperature. The material has a highly amorphous nature, as determined by the solid state NMR linewidths. The linewidths are consistent with the notion that the high temperature thermoset is densely cross-linked and that little crystallization has occurred.

Literature Cited

1. Zahir, S. A.; Renner, A. U. S. Patent No. 4 100 140, 1978.
2. King, J. J.; Chaudhari, M. A.; Zahir, S. A. Proceedings of the 29th National SAMPE Conference, 1984; pp. 392-408.
3. Chaudhari, M. A.; Galvin, T. J.; King, J. J. Proceedings of the 30th National SAMPE Conference, 1985; pp. 735-746.
4. Ernst, R. R. J. Chem. Phys. 1966, 45, 3845.
5. Shoolery, J. N. Prog. in NMR Spec. 1977, 11, 79.
6. Pines, A.; Gibby, M. G.; Waugh, J. S. J. Chem. Phys. 1973, 59, 569.
7. Dixon, W. T.; Schaefer, J.; Sefick, M. D.; Stejskal, E. O.; Mckay, R. A. J. Magn. Reson. 1982, 49, 345.
8. Carduner, K. R. J. Magn Reson. 1987, 72, 173.
9. Garroway, A. N.; Ritchey, W. M.; Moniz, W. B. Macromolecules 1982, 15, 1051.

RECEIVED October 7, 1987

Chapter 27

Characterization of Cobalt(II) Chloride-Modified Condensation Polyimide Films

Properties Before and After Solvent Extraction

J. D. Rancourt and L. T. Taylor

Department of Chemistry, Virginia Polytechnic Institute and State University, Blacksburg, VA 24061-0699

Cobalt(II) chloride was dissolved in poly(amide acid)/N,N-dimethylacetamide solutions. Solvent cast films were prepared and subsequently dried and cured in static air, forced air or inert gas ovens with controlled humidity. The resulting structures contain a near surface gradient of cobalt oxide and also residual cobalt(II) chloride dispersed throughout the bulk of the film. Two properties of these films, surface resistivity and bulk thermal stability, are substantially reduced compared with the nonmodified condensation polyimide films. In an attempt to recover the high thermal stability characteristic of polyimide films but retain the decreased surface resistivity solvent extraction of the thermally imidized films has been pursued.

Metal ion modified polyimide films have been prepared to obtain materials having mechanical, electrical, optical, adhesive, and surface chemical properties different from nonmodified polyimide films. For example, the tensile modulus of metal ion modified polyimide films was increased (both at room temperature and 200°C) whereas elongation was reduced compared with the nonmodified polyimide (1). Although certain polyimides are known to be excellent adhesives (2) lap shear strength (between titanium adherends) at elevated temperature (275°C) was increased by incorporation of tris(acetylacetonato)aluminum(III) (3). Highly conductive, reflective polyimide films containing a palladium metal surface were prepared and characterized (4). The thermal stability of these films was reduced about 200°C, but they were useful as novel metal-filled electrodes (5).

Much of the recent work pertaining to metal ion modified polyimides has been aimed at increasing the surface electrical conductivity of condensation polyimides (6,7,8). This has been accomplished to a significant extent (surface resistivity reduced up to 11 orders of magnitude (8,9)) but in most systems the bulk

0097-6156/88/0367-0395$06.00/0

thermal stability of the polymer is also reduced. The purpose of this work was to determine the effect solvent extraction has on the properties of cobalt chloride modified polyimide films. The goal was to obtain enhanced bulk thermal stability and retain the low surface electrical resistivity characteristic of these systems.

The data indicate that the properties of the lower glass transition temperature metal ion modified polyimides are altered more than the properties of the higher glass transition temperature metal ion modified polyimides. Extraction removes both cobalt and chlorine from the films and slightly increases bulk thermal stability and both surface resistivity and bulk electrical resistivity. Details pertaining to the structure, analysis and properties of these novel gradient composites are discussed.

Experimental

Chemicals. 3,4,3',4'-Benzophenonetetracarboxylic dianhydride, BTDA (Aldrich Chemical Co.) was vacuum dried at 100°C for two hours. 4,4'-Oxydianiline, ODA (Aldrich Chemical Co.) was sublimed at 185°C and less than 1-torr pressure. 4,4'-Bis(3,4-dicarboxyphenoxy)-diphenylsulfide dianhydride, BDSDA, was used as received from NASA, Langley Research Center, Hampton, VA. Reagent grade N,N-dimethyl-acetamide, DMAc, distilled in glass, was stored over molecular sieves under a nitrogen atmosphere and sparged with dry nitrogen prior to use. Anhydrous cobalt(II) chloride was prepared by heating the hexahydrate for three hours at 120°C under vacuum.

Polymer Synthesis and Modification. The condensation reaction between either BTDA or BDSDA and ODA was performed in DMAc at room temperature under a nitrogen atmosphere. ODA (0.004 mole) was added to a nitrogen-purged glass septum bottle with 7 ml DMAc. One of the dianhydrides (0.004 mole) was then added to the diamine solution with an additional milliliter of DMAc resulting in 15-25 wt% solids depending upon the monomer combination. The resulting solution was stirred for 20-24 hours to form the poly(amide acid), a polyimide precursor. For the modified polyimides, anhydrous cobalt(II) chloride (0.001 mole) was added as a solid within one-half hour after the dianhydride.

Film Preparation. Poly(amide acid) solutions were centrifuged at ca. 1700 rpm, poured onto clean, dust-free soda-lime glass plates, and spread with a doctor blade with a 16 mil or 8 mil blade gap to obtain films having a final cured thickness of approximately 3.4 or 1.5 mil, respectively. The films were cured at 80°C for 20-30 minutes with subsequent drying and thermal imidization in a forced air, static air, or an inert gas oven with controlled humidity at 100°, 200° and 300°C each for one hour. After cooling to room temperature, some of the polyimides were removed from the glass plate by soaking in distilled water. The surface of the film in contact with the soda-lime glass plate during imidization is referred to as the glass-side; while, that in contact with the cure atmosphere of the curing oven is referred to as the air-side.

Extraction of Films. Cobalt modified polyimide films were extracted by one of several techniques: (1) films on the casting plate were soaked in a tray of distilled water at room temperature, (2) films were soxhlett extracted with distilled water, (3) films were soaked in DMAc at room temperature, and (4) films were soxhlett extracted with DMAc.

Transmission Electron Microscopy: Transmission electron microscopy data were obtained by personnel in the Ultrastructure Laboratory at the Virginia-Maryland College of Veterinary Medicine using a JEOL 100CX-II transmission electron microscope. Samples were imbedded in Poly-bed 812 epoxy resin and cured at 50-60°C for 2-3 days. Samples were then sectioned to between 800 and 1000 Å on either a Sorval MT2B or an LKB IV Ultramicrotome using glass knives and were placed on 200 mesh copper grids.

Spectroscopic Methods. X-ray photoelectron spectra were recorded with a Perkin-Elmer Phi Model 5300 ESCA system. Auger Electron spectroscopy and depth profiling via argon ion etching were recorded with a Perkin-Elmer Phi Model 610 Scanning Auger Microprobe System. Energy dispersive X-ray analysis was obtained by personnel in the Ultrastructure Laboratory at the Virginia-Maryland College of Veterinary Medicine using a Tracor Northern 5500 X-ray microanalysis system. Sections were prepared in a manner similar to that used for TEM evaluation except that the samples were sectioned to between 900 and 1200 Å and were placed on 190 mesh carbon coated nylon grids.

Thermal Methods. Thermogravimetric analysis of a single disk (1/4" diameter) of each film was obtained with a Perkin-Elmer Model TGS-2 thermogravimetric system at 20°C/min heating rate in dynamic air purge. Differential scanning calorimetric analysis of two or more disks of each film sealed aluminum sample pans was obtained with a Perkin-Elmer Model DSC-4 Differential Scanning Calorimeter at 10°C/min heating rate in in a dynamic nitrogen purge.

Electrical Methods. Room temperature surface and volume direct current electrical resistivities of 85 mm diameter polymer films were determined using a Keithley high voltage source (Model 240A), a Keithley electrometer (Model 610C), and a Keithley three electrode assembly (Model 6105 Resistivity Adapter). Variable temperature electrical resistivity determinations were obtained with a computer controlled instrument developed in our laboratory (10). The system controls sample temperature, atmosphere and measurement mode in addition to automatically storing, printing, and plotting the data with the associated peripherals. The electrode geometry for the variable temperature test cell is the same as the Keithley 6105 Resistivity Adapter.

Results and Discussion

General Properties. It has been shown (6) that the incorporation of cobalt chloride into a poly(amide acid) solution and processing the solvent cast film to temperatures up to 300°C in an appropriate atmosphere results in a highly anisotropic polymeric structure. X-ray photoelectron spectroscopy was used to determine the final chemical state of the dopant. Typically, the cobalt chloride forms cobalt oxide near the air-side of the polymer film as shown by Auger electron spectroscopy with depth profiling via argon ion sputtering. The electrical properties of the polymers are also anisotropic. The air-side behaves as a conductor, whereas, the bulk behaves as a dielectric with both dielectric loss and d.c. conduction. Variable temperature electrical resistivity determinations suggest that the near-surface region of the polymer contains discrete metal oxide domains because the surface resistivity thermoreversibly increases as the sample is heated through the glass transition temperature of the polymer (independently determined by differential scanning calorimetry). An additional feature of these gradient composites is that, although some of the dopant is converted to cobalt oxide, the bulk resistivity is still many orders of magnitude greater than it should be for the amount of incorporated cobalt chloride. This leads to the speculation that the dielectric constant of the polymer is not high enough to support dissociation of an appreciable amount of the cobalt salt. The structural features envisioned based on the analytical techniques employed for the cobalt chloride modified BTDA-ODA polyimide films cured in a moist air atmosphere were verified using transmission electron microscopic evaluations of ultramicrotomed of cross sections (Figure 1, parts A, B and E). EDAX verified that the particles dispersed throughout the film were composed primarily of cobalt and chlorine. It was attempted, on the basis of these data, to use extraction techniques to modify the polymer structure and properties.

Thermal Properties. Each of the polyimide film samples was evaluated by differential scanning calorimetry to determine the glass transition temperature (Table I). The general observation is that the BTDA-ODA polyimide films have a higher glass transition temperature than the BDSDA-ODA polyimide films whether they are nonmodified or are modified with cobalt chloride. This is in agreement with the work of Frye (11) in which the dianhydride moiety, not the diamine, was found to control the polyimide glass transition temperature.

More specifically, the glass transition temperature of the BDSDA-ODA polyimide film is slightly higher for the modified than for the nonmodified sample. During the first heating cycle there is an endotherm which is absent on subsequent reheating that has been attributed to a stress relaxation phenomenon in the glass transition region for each BDSDA-ODA sample. The glass transition is also increased slightly after the first heat-cool cycle due most likely to annealing (Figure 2). DSC analysis to only 250°C (50° lower than the process temperature, 300°C) of a sample also results in loss of the stress relaxation endotherm and an increase in the glass transition temperature after this first heat-cool cycle. We take this as further support that the higher glass transition temperature observed during the second heating cycle is due to annealing and not to further reaction (imidization) or to further solvent (DMAc) loss.

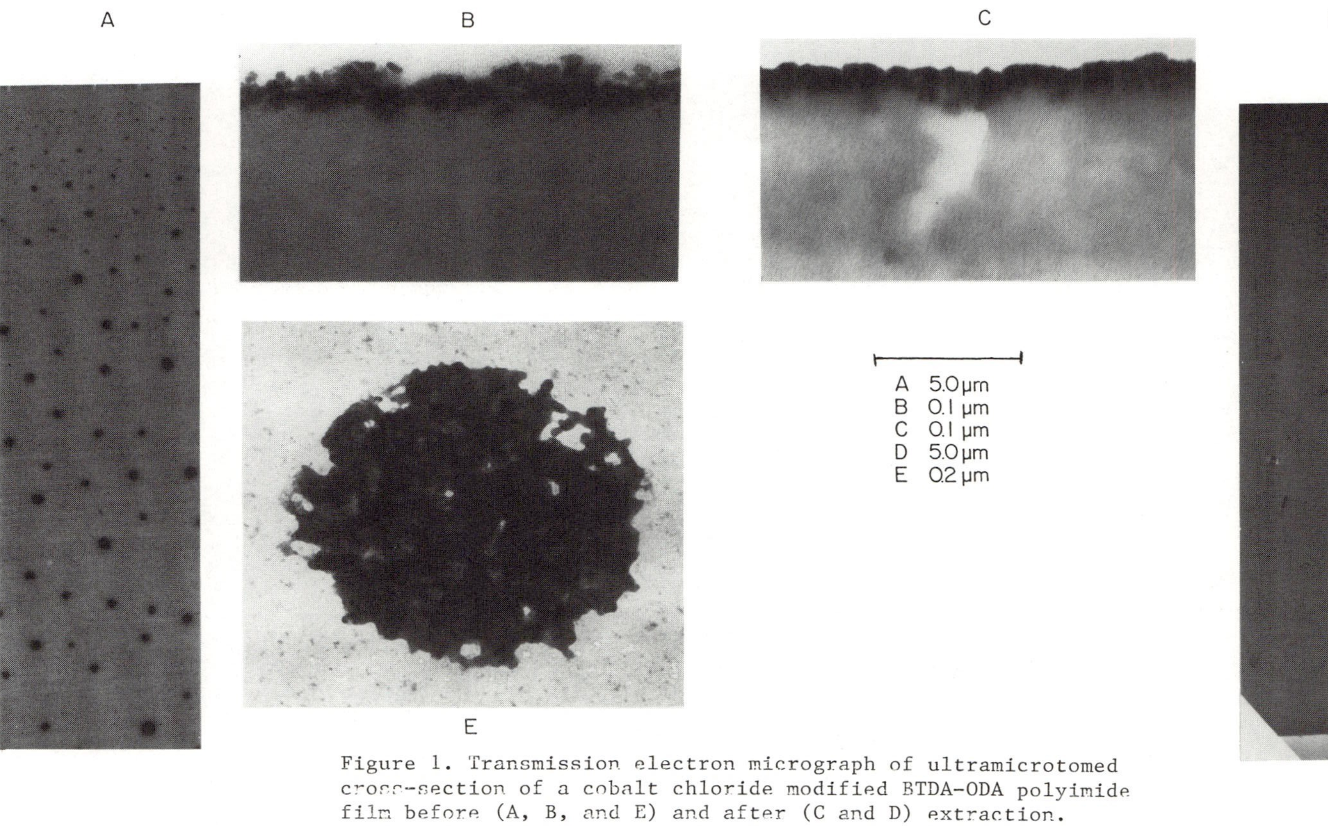

Figure 1. Transmission electron micrograph of ultramicrotomed cross-section of a cobalt chloride modified BTDA-ODA polyimide film before (A, B, and E) and after (C and D) extraction.

Table I. Thermal Properties of Dry Air Cured Cobalt Chloride Modified Condensation Polyimide Films

Film No.	Polyimide	Doped	Treatment	Mass Lost (%)[a]	PDT (°C)[b]	AGT (°C)[c] 1st Heat	AGT (°C)[c] 2nd Heat
1	BTDA/ODA	No	Nonextracted	1.36	553	275	284
2	BTDA/ODA	Yes	Nonextracted	2.07	503	272,345	302
3	BTDA/ODA	Yes	#2, 24 hr water soak	1.83	505	279,360	302
4	BTDA/ODA	Yes	#3, 24 hr water soxhlett	2.30	505	273,357	300
5	BTDA/ODA	Yes	Nonextracted	2.60	504	269	306
6	BTDA/ODA	Yes	#5, 24 hr DMAc soak	2.00	505	276,336	300
7	BTDA/ODA	Yes	#6, 24 hr DMAc soxhlett	15.00	385	---	300
8	BDSDA/ODA	No	Nonextracted	0.41	580	212	212
9	BDSDA/ODA	Yes	Nonextracted	1.24	512	215	217
10	BDSDA/ODA	Yes	#9, 24 hr water soak	1.21	532	215	217

[a] 200°C
[b] Temperature corresponding to 10% weight lost in dynamic air
[c] AGT = Apparent glass transition temperature

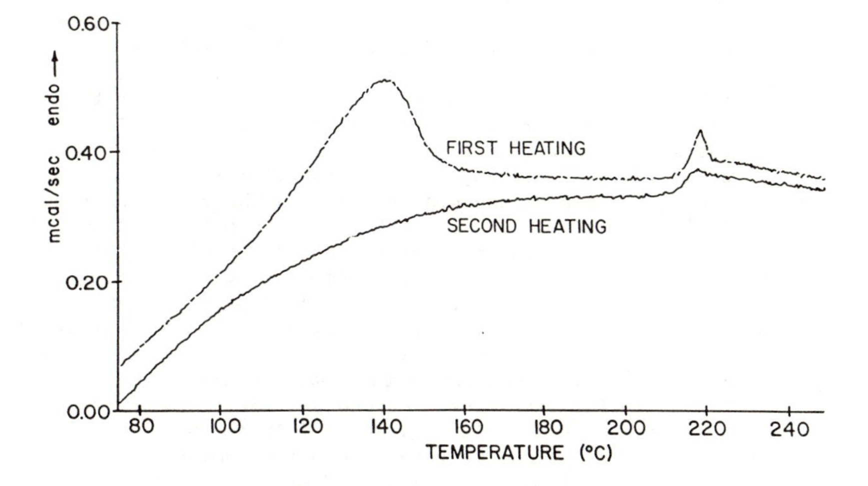

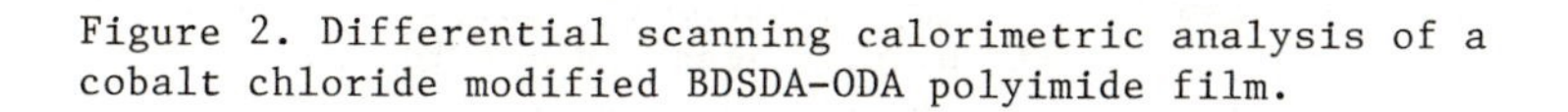
Figure 2. Differential scanning calorimetric analysis of a cobalt chloride modified BDSDA-ODA polyimide film.

DSC analysis of the cobalt chloride modified BTDA-ODA polyimide films is more complex than the BDSDA-ODA case. During the first heating cycle two changes in heat capacity occur with the BTDA-ODA polyimide films which directly suggest the presence of two glass transition temperatures and thereby two types of polyimide phases. This is particularly interesting in view of the TEM cross-section (Figure 1, part A) that shows a depletion zone, approximately 5% of the sample thickness, and a zone which shows discrete particles dispersed throughout the film. The higher glass transition temperature, which correlates well with thermomechanical analysis data (Figure 3, part C), may be due to a filler-effect. The lower glass transition temperature may reflect the polyimide in the depleted zone. The magnitude of the higher temperature glass-transition of 350°C is greater than the magnitude of the glass-transition of 280°C which suggests more of the sample is responsible for the higher glass transition. This in combination with the TEM data supports the assignment of the 350°C Tg to the particle-containing portion of the sample. Heating to 400°C during DSC analysis may allow partial polymer relaxation around the cobalt chloride particles. This may be the reason for the intermediate glass transition temperature observed during the second heating cycle: 302°C instead of ~280 and ~350°C. DSC analysis of the films after extraction revealed that the extraction solvents (DMAc or water) and the extraction conditions (soak at room temperature or soxhlett extraction) do not seem to have an appreciable influence on the glass transition temperature of either the cobalt chloride modified BTDA-ODA or BDSDA-ODA polyimide films.

Each sample was evaluated by thermogravimetry to determine if the thermal stability could be enhanced by removing some residual cobalt chloride. The BTDA-ODA polyimide film thermal stability is reduced about 50°C due to the cobalt chloride dopant. Soaking or extraction with water has no positive effect on the thermal stability whereas soxhlett extraction with DMAc severely degrades the polymer stability. For the BDSDA-ODA polyimide films the incorporation of cobalt chloride also reduces the bulk polymer thermal stability. Soaking this film in water for 24 hours, however, increased the bulk thermal stability slightly from 512° to 532°C.

The volatile content of the nonmodified and cobalt chloride modified polyimide films was estimated by determining the mass loss that occurred up to 200°C during the thermogravimetric evaluation. In general, compared with the nonmodified controls, ion incorporation does not significantly alter the volatile content of the cobalt chloride modified polyimides. Also, the volatile content is comparable to that observed for DuPont Kapton polyimide film (12) (i.e. 1.3 wt% at 50% R.H., 2.9 wt% at 100% R.H.). In no case did soaking or extracting the polyimide films significantly alter the volatile content of the film except for the DMAc extracted BTDA-ODA samples which contained about 15% volatiles.

Electrical Characteristics. The d.c. electrical resistivity of the polyimide films was determined as a function of temperature in vacuum. Insofar as the nonmodified polyimides are concerned the BTDA-ODA polyimide film has a slightly higher volume and surface

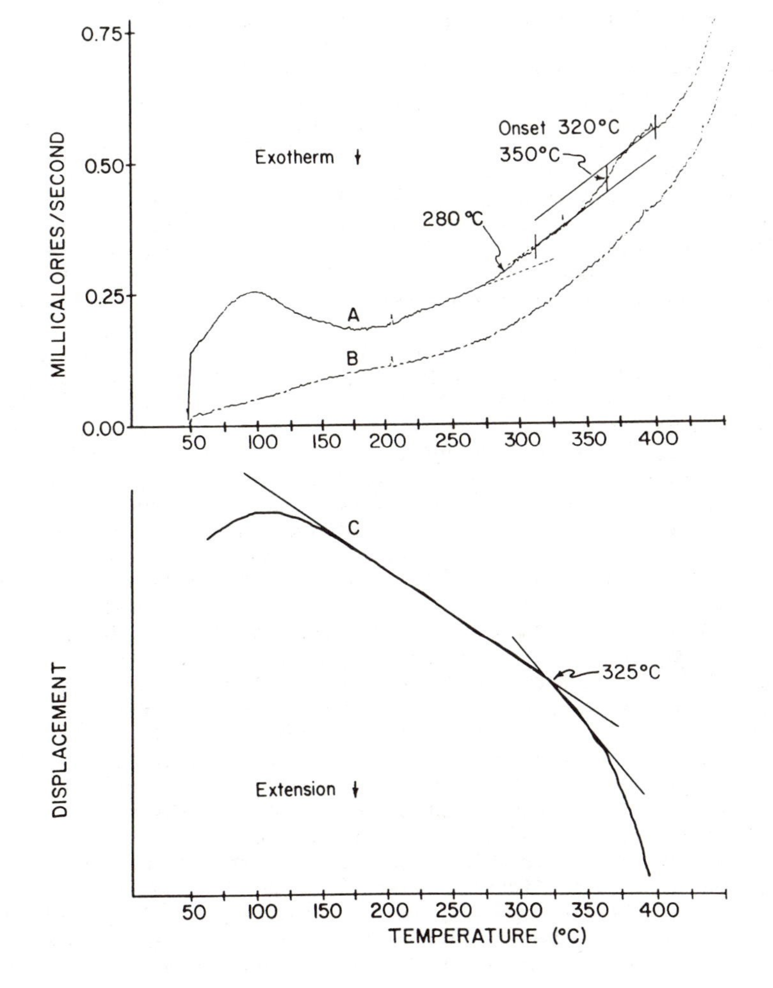

Figure 3. Comparison of differential scanning calorimetric data (parts A and B) with thermomechanical analysis data (part C) for a cobalt chloride modified BTDA-ODA polyimide film.

resistivity than the BDSDA-ODA polyimide as expected based on both the higher glass transition temperature (lower ionic mobility) and lower dielectric constant (lower free ion content) of the former. For both systems incorporation of cobalt chloride reduces the bulk and surface resistivity and, as expected, more so for the BDSDA-ODA polyimide films (Table II). The use of extraction techniques to increase the bulk resistivity of the films was effective in all cases except for the 24 hour water soak of the cobalt chloride modified BTDA-ODA polyimide film. Although the volume resistivity, in general, was increased, so too was the air-side resistivity. The air-side resistivity of the cobalt chloride modified polyimide film was increased to the value observed with a nonmodified BTDA-ODA polyimide film; while, the cobalt chloride modified BDSDA-ODA polyimide film had an increase in surface resistivity of only about three orders of magnitude after soaking this film in water. The variable temperature air-side surface resistivity profiles for the cobalt chloride modified BDSDA-ODA polyimide film before and after a water soak are shown in Figure 4.

Metal Ion Content. The metal ion content of cobalt chloride modified condensation polyimide films was assessed using elemental analysis. The data (Table III) indicate that the weight percentage of both cobalt and chlorine are usually reduced following soaking or extraction. However, except for the polyimide film that was Soxhlett extracted with DMAc the metal ion content is not reduced very much. It is believed that the polyimide film, even with ion incorporation, is an efficient barrier against extraction. Proof that if the polyimide were more permeable then greater extraction efficiency would be attained was obtained by the following experiment. The same TEM grid, used to obtain the micrograph shown in Figure 1, part A was soaked in distilled water. Prior to soaking, EDAX was used to verify that the particles were primarily composed of cobalt and chlorine and should therefore be soluble in water. After soaking the grid in water for 24 hours the transmission electron micrograph shown in Figure 1, parts C and D was obtained. All the spherical particles are absent but most of the near-surface deposit is present. The spherical particles are removed because the ultramicrotomed cross-section thickness, comparable to the diameter of these particles, makes the particles accessible to the water which can dissolve them. Thus, had the polyimide allowed sufficient solvent into the film the desired increase in volume resistivity and bulk thermal stability would have been realized. The surface resistivity was increased after soaking because the continuity of the near-surface metal oxide was lessened.

Additional details regarding the samples were obtained by inferring from the elemental analysis the amount of cobalt present as oxide and chloride. The only source of chlorine in these samples is the cobalt chloride dopant. Thus, based on the chlorine concentration we can estimate the amount of cobalt that must be present as cobalt chloride. For example, film #5 is calculated to contain 78.7% of the cobalt as chloride and by difference about 21% as cobalt oxide. Film #7 (which is film #5 after both a 24 hour soak and a 24 hour extraction with DMAc) contains 20% of the cobalt

Table II. Electrical Properties of Dry Air Cured Cobalt Chloride Modified Condensation Polyimide Films

Film No.	Polyimide	Doped	Treatment	Volume Resist.[a] (ohm cm)	Volume Dielectric Constant[b]	Air-Side Resist.[a] (ohm)
1	BTDA/ODA	No	Nonextracted	16.82	3.5	15.77
2	BTDA/ODA	Yes	Nonextracted	15.56	5.23	14.14
3	BTDA/ODA	Yes	#2, 24 hr water soak	14.97	4.03	14.53
4	BTDA/ODA	Yes	#3, 24 hr water soxhlett	16.04	2.76	15.33
5	BTDA/ODA	Yes	Nonextracted	14.76	5.69	14.88
6	BTDA/ODA	Yes	#5, 24 hr DMAc soak	16.07	4.52	15.53
7	BTDA/ODA	Yes	#6, 24 hr DMAc soxhlett	c	c	c
8	BDSDA/ODA	No	Nonextracted	16.22	---	15.48
9	BDSDA/ODA	Yes	Nonextracted	13.48	---	9.65
10	BDSDA/ODA	Yes	#9, 24 hr water soak	16.22	---	12.80

a Log_{10} of average of first heating and first cooling cycle at 200°C in vacuum with +100 VDC

b Room temperature at 1000 Hz in ambient

c Film was destroyed

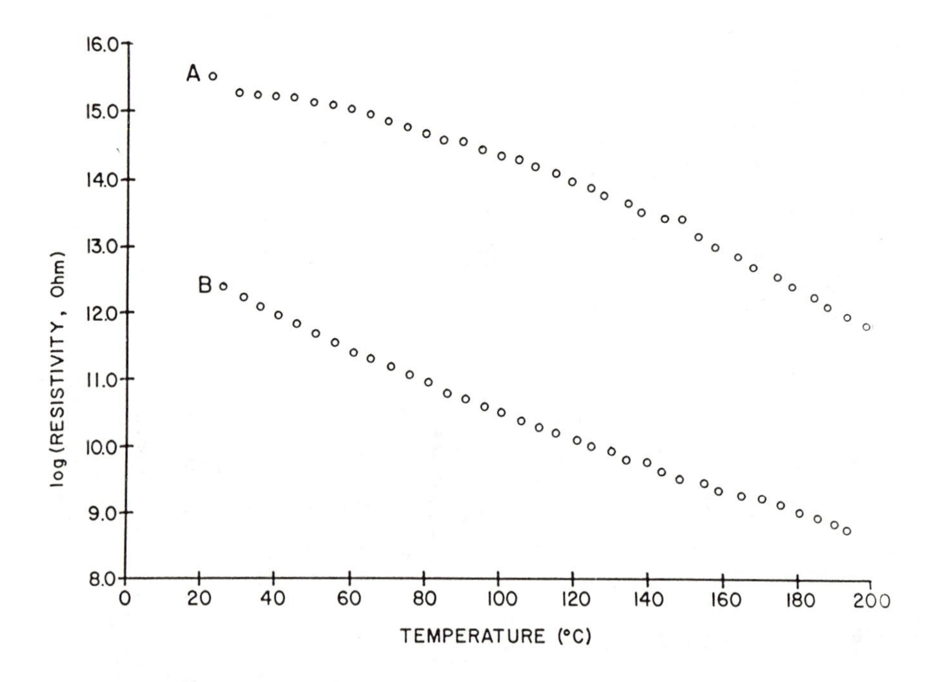

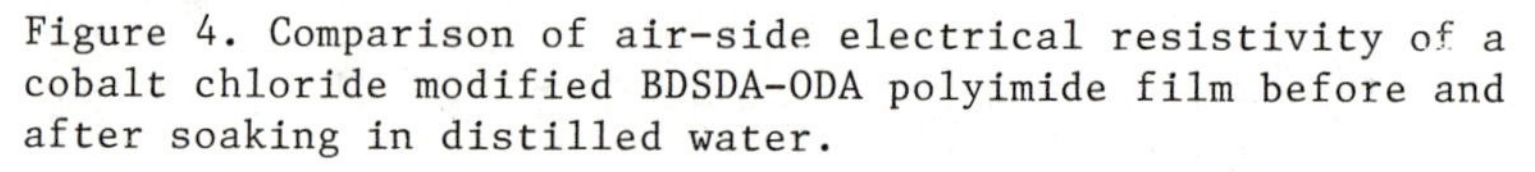
Figure 4. Comparison of air-side electrical resistivity of a cobalt chloride modified BDSDA-ODA polyimide film before and after soaking in distilled water.

Table III. Elemental Composition of Dry Air Cured Cobalt Chloride Modified Condensation Polyimide Films

Film No.	Polyimide	Doped	Treatment	Chlorine, wt% Found	Chlorine, wt% Expect.	Cobalt, wt% Found	Cobalt, wt% Expect.
1	BTDA-ODA	No	Nonextracted	0.04	0.00	---	0.00
2	BTDA-ODA	Yes	Nonextracted	3.20	3.30	2.37	2.74
2	BTDA-ODA	Yes	Nonextracted	---	3.30	2.63	2.74
3	BTDA-ODA	Yes	#2,24 hr water soak	2.52	3.30	2.30	2.74
4	BTDA-ODA	Yes	#3,24 hr water extr.	2.17	3.30	2.00	2.74
5	BTDA-ODA	Yes	Nonextracted	2.29	3.30	2.42	2.74
5	BTDA-ODA	Yes	Nonextracted			2.47	2.74
6	BTDA-ODA	Yes	#5,24 hr DMAc soak	2.62	3.30	2.32	2.74
7	BTDA-ODA	Yes	#6,24 hr DMAc extr.	0.16	3.30	0.66	2.74
8	BDSDA-ODA	No	Nonextracted	0.01	0.00	---	0.00
9	BDSDA-ODA	Yes	Nonextracted	2.02	2.44	1.66	2.03
10	BDSDA-ODA	Yes	#9,24 hr water soak	1.99	2.44	1.47	2.03

as cobalt chloride and 80% of the cobalt as oxide. As a cross-check, the data for film #5 implies that complete extraction of all the cobalt chloride but no extraction of cobalt oxide should result in 0.53 wt% cobalt (as cobalt oxide) for film #7. This is in good agreement with the experimental data (0.66 wt%) for film #7 suggesting mostly cobalt chloride, not cobalt oxide, is extracted from the film by this harsh technique.

Acknowledgment

The financial assistance of the National Aeronautics and Space Administration, Langley Research Center is gratefully appreciated. The services of L. Horning in performing many of the extraction experiments is appreciated.

Literature Cited

1. Taylor, L. T. and St. Clair, A. K. J. Appl. Polym. Sci., 1983, 28, 2393.
2. Progar, D. J. and St. Clair, T. L. 7th National Sampe Technical Conference, Albuquerque, NM, Oct. 1975.
3. Taylor, L. T.; St. Clair, A. K. NASA Langley Research Center, "Aluminum Ion Containing Polyimide Adhesives", U. S. Patent 284,461 (1981).
4. Wohlford, T. L.; Schaff, J.; Taylor, L. T.; St. Clair, A. K.; Furtsch, T. A.; Khor, E. Conductive Polymers; Seymour, R. B., Ed.; Plenum Publishing Corporation: New York, 1981, p 7.
5. Furtsch, T. A.; Finklea, H. O.; Taylor, L. T. Polyimides: Synthesis, Characterization and Applications, Vol. 2; Mittall, K. L., Ed.; Plenum Press: New York, 1984, p 1157.
6. Rancourt, J. D.; Boggess, R. K.; Horning, L. S.; Taylor, L. T. J. Electrochem. Soc. 1987, 134, 85.
7. Ezzell, S. A.; Furtsch, T. A.; Taylor, L. T. J. Polym. Sci. Polym. Chem. Ed., 1983, 21, 865.
8. Ezzell, S. A. and Taylor, L. T. Macromolecules, 1984, 17, 1627.
9. Rancourt, J. D.; Porta, G. M.; Taylor, L. T. Thin Films, submitted for publication, 1987.
10. Rancourt, J. D.; Swartzentruber, J. L.; Taylor, L. T. Am.Lab. March 1986, 68.
11. Frye, M. Polyimides: Synthesis, Characterization and Applications; Mittal, K. L., Ed.; Plenum Press: New York, 1984, p 377.
12. DuPont Kapton Polyimide Film, DuPont DeNemours Literature.

RECEIVED October 7, 1987

Chapter 28

Chain Cross-Linking Photopolymerization of Tetraethyleneglycol Diacrylate

Thermal and Mechanical Analysis

J. G. Kloosterboer and G. F. C. M. Lijten

Philips Research Laboratories, P.O. Box 80000, 5600 JA Eindhoven, Netherlands

Network formation by photopolymerization has been studied for tetraethyleneglycol diacrylate (TEGDA) using isothermal calorimetry (DSC), isothermal shrinkage measurement and dynamic mechanical thermal analysis (DMTA). Due to vitrification the polymerization does not go to completion at room temperature. The ultimate conversion as measured by DSC seems to depend on light intensity. This can be explained by the observed delay of shrinkage with respect to conversion.

However, mechanical measurements show that especially at low intensities, the photopolymerization process continues for a considerable time at a rate which cannot be detected by DSC. At equal doses the temperatures of maximum mechanical loss, T(tan δ_{max}), were observed to be the same.

DMTA of partially polymerized samples of TEGDA reveals an increase of Young's modulus due to thermal aftercure near 120°C. Parallel DSC-extraction experiments show that this aftercure requires the presence of free monomer. Near the end of the polymerization the free monomer is exhausted and only crosslinks are formed. T(tan δ_{max}) then increases markedly with double bond conversion.

Photopolymerization and photocrosslinking processes have been in use for many years in the electronics industry, for example in the making of printed circuit boards and in the fixation of color dots in TV tubes. More recent applications of light-induced chain crosslinking polymerization processes are the replication of optical discs (1,2) of aspherical lenses (3,4) and the in-line coating of optical fibers (5,6).

In many applications densely crosslinked, glassy polymers are desired. Such polymers can be made by photopolymerization of suitable diacrylates. Important parameters are the rate of polymerization, the maximum extent of reaction and the

0097–6156/88/0367–0409$06.00/0

presence of unreacted monomer, as well as dimensional stability and adhesive properties. Shrinkage and refractive index are of particular importance for the lens replication while Young's modulus, thermal expansion coefficient and glass transition temperature are critical parameters with the optical fiber coatings. All these parameters will depend on the duration and the intensity of the UV exposure and on the temperature of reaction. In the development of the production processes mentioned above we were prompted to study some of these relationships by investigating the photopolymerization of simple model monomers.

In this contribution we present results obtained with tetra-ethyleneglycol diacrylate (TEGDA). This compound was chosen since its polymer shows an easily discernible maximum in the mechanical losses as represented by tan δ or loss modulus E'' versus temperature when it is prepared as a thin film on a metallic substrate. When photopolymerized at room temperature it forms a densely crosslinked, glassy polymer, just as required in several applications. Isothermal vitrification implies that the ultimate conversion of the reactive double bonds is restricted by the diffusion-limited character of the polymerization in the final stage of the reaction. Therefore, the ultimate conversion depends strongly on the temperature of the reaction and so does the glass transition.

With a somewhat stiffer monomer, 1,6-hexanediol diacrylate, (HDDA) we have previously observed that the ultimate conversion as measured with differential scanning calorimetry (DSC) also depends on light intensity. This has been attributed to the experimentally observed delay of shrinkage with respect to chemical conversion (7). In principle, such a dependence of conversion on intensity should show up in the mechanical properties as well. However, these are difficult to measure with thin samples of HDDA.

Since the polymerization of TEGDA can easily be studied with DSC as well as with dynamic mechanical thermal analysis (DMTA) we have repeated our study with this monomer in order to see whether or not mechanical properties depend on the intensity rather than on the dose of UV irradiation. DMTA also reveals whether or not postcuring occurs during thermal after-treatment, similar to what has been observed with other thermosetting materials (8).

Combination of the results with solvent extraction/liquid chromatography data may elucidate the role of free, unreacted monomer in the post-curing process. The main conclusions have already appeared elsewhere (9). Here we report in much more detail on the experimental techniques as well as on new results on the delay of shrinkage with respect to chemical conversion.

EXPERIMENTAL

Chemicals

TEGDA (Polysciences (Warrington, Pa., USA)) was purified by washing several times with an aqueous NaOH solution (10%) followed by washing twice with an aqueous solution of $CaCl_2$. After centrifugation the extracted monomer was dried over anhydrous $CaCl_2$ and filtrated.

The photoinitiator α,α-dimethoxy-α-phenylacetophenone (DMPA) (Ciba Geigy, Basle, Switzerland) was used without further purification.

Calorimetry

Isothermal DSC measurements were made with a Perkin Elmer DSC-2C apparatus, modified for UV irradiation (Figure 1). The aluminum sample holder enclosure cover contains two windows, one for the sample and one for the reference compartment. The windows consist of cylindrical quartz cuvettes which have been evacuated in order to prevent moisture condensation. The windows were mounted by using a thermally cured epoxy adhesive.

The polyurethane foam filled outer draft shield contains one large evacuated cuvette. Here, the evacuation also assists in maintaining sufficient thermal insulation. On top of the draft shield an electrically driven shutter (Prontor E/64, 24 V) is mounted. The shutter also contains a filter holder to accommodate neutral density filters. On top of the shutter a small, 4 W fluorescent lamp (Philips, TL08, length 15 cm) is mounted. The lamp emits at 350 nm, the bandwidth at half intensity is about 45 nm. With this set-up the maximum intensity at the sample position is 0.2 $mW.cm^{-2}$. Intensities were measured using an International Light IL 745A UV curing radiometer.

A tenfold increase of the intensity may be obtained at the cost of a somewhat reduced thermal stability by omitting the draft shield and mounting the shutter and lamp assembly directly on top of the sample holder enclosure block. Since the heat production of the fluorescent lamp is very small the whole irradiation equipment could be mounted within the standard glove-box.

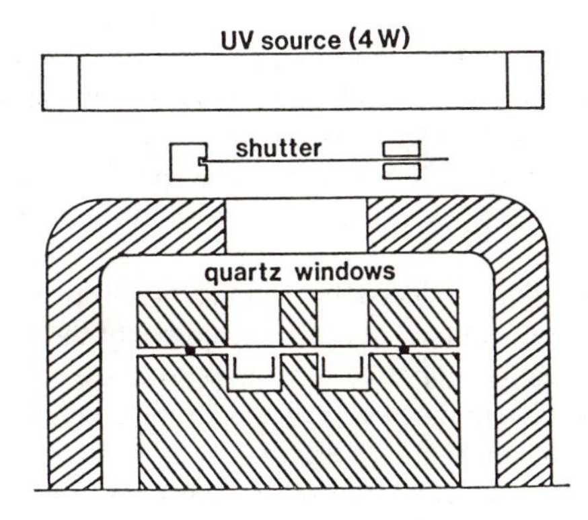

Figure 1. **Schematic view of the modified DSC apparatus.** See text for explanation.

The oxygen level of the purge gas (nitrogen) could be reduced to a level as low as 2 ppm, measured at the exit in the outflowing gas using a Teledyne Model 311 oxygen meter. This was achieved by replacing the originally present teflon gas tubing with copper tubing. Reduction of the oxygen concentration down to this level is important since a concentration of only 60 ppm may already have a significant influence on the rate of photopolymerization of a diacrylate (10).

Rate vs. time curves are are presented as per cent C=C conversion per second by using the experimentally determined heat of polymerization of acrylate groups of 78 $kJ.Mol^{-1}$ (11). Extents of reaction were reproducible to within 0.7%, the accuracy depends on the accuracy of the heat of reaction. The distortion of DSC curves of fast reactions is discussed below in the section on shrinkage and conversion.

Shrinkage

Shrinkage measurements were made by monitoring the decrease of thickness of a sandwich composed of two glass plates with a thin layer (50 μm) of sample in between. A sample, consisting of monomer with dissolved photoinitiator will polymerize when irradiated through one of the glass plates. Polymerization shrinkage then causes a decrease of the thickness of the sample layer and therefore of the sandwich. Provided that the polymerizing sample adheres well to the glass plates most of the volume shrinkage will occur as a decrease of thickness, with only a minor lateral contribution. The construction of the sandwich is shown in cross section in Figure 2a and in perspective in Figure 2b. The lower glass plate consists of a quartz glass block with dimensions of 2 x 8 x 18 mm. In its top face a rectangular recess with dimensions of 8 x 10 mm and a depth of 50 μm has been machined. This recess is filled with monomer with dissolved photoinitiator and covered with the upper glass plate. The latter is made from a microscope cover glass with dimensions of 8 x 14 mm. It has a thickness of 0.14 mm. The cover glass is coated with an aluminum reflective layer. In this way UV light, entering through the lower plate will pass the sample twice, just as in the DSC samples where light is reflected at the bottom of the aluminum sample pan.

During polymerization, induced by irradiation through the bottom plate, the polymerization shrinkage will tend to cause a decrease of the sample thickness. Since the thin upper glass plate is supported at two opposite edges and since the adhesion of the polymerizing system to the glass plates is strong enough, the thinner of the two will be curved by the shrinkage process.

The displacement of the center part of the cover glass is continuously monitored by a displacement transducer. The actual measurements were performed using a modified thermo-mechanical system (TMS), a Perkin Elmer TMS-2 apparatus.

The oven of the TMS apparatus has been removed and the open ended quartz tube of the TMS is surrounded by a closed quartz tube fitted with a flat quartz window at the bottom and a screw cap on its side to allow sample introduction (Figure 2c).

The sample sandwich described above is placed in the TMS, the screw cap is closed and the closed tube is flushed with nitrogen via small inlet and outlet ports. Next the transducer shaft is lowered till it touches the upper plate. A thin, flat ended TMS penetration probe was used as the transducer shaft throughout the experiments.

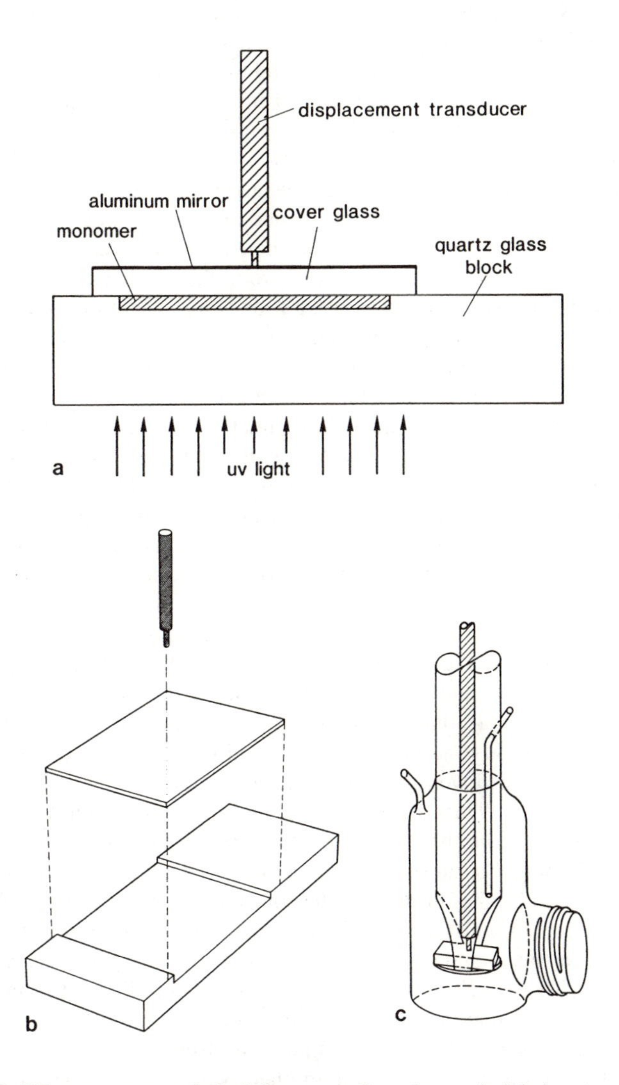

Figure 2. **Measurement of shrinkage.** a. Sample sandwich in cross section. b. Disassembled sandwich. c. Modified TMS apparatus.

After equilibration of the sample solution with the surrounding atmosphere the irradiation is started. Irradiation is performed with a similar lamp and shutter assembly as used with DSC. The lamp is mounted in a closed box, installed below the TMS apparatus. In this way, almost identical conditions during shrinkage and conversion measurements could be obtained. It is important to stipulate that special measures are required to attain an acceptable oxygen level in the sample compartment. Stainless steel tubing is used where possible. The concentration of oxygen in the outflowing gas is monitored as with the DSC. The oxygen concentration was about 3 ppm, slightly higher than with the DSC, since some leakage occurs along the driving shaft of the displacement transducer. As a consequence of the unfavorable geometry of the TMS samples (large sandwich with a small monomer/gas interface) the removal of dissolved oxygen from these samples takes several hours whereas a few minutes are sufficient with the open sample pans used with DSC. Therefore, the former were flushed overnight. Minimum flushing times were determined from rate measurements: when prolonged flushing did not result in an increase of the maximum rate anymore, the process was considered to be complete.

Rates are obtained by graphical differentiation of the displacement vs. time curve, since the derivative curve, generated by the instrument, proved to be seriously distorted. In order to allow a proper comparison with the DSC curves, it is desirable to plot the rate of shrinkage also as per cent of total shrinkage per second. This requires an estimate of the theoretical amount of total shrinkage at 100 per cent C=C conversion. However, due to the occurrence of vitrification this situation cannot be reached. Moreover, in the glassy state the volume shrinkage is per se not in equilibrium. The higher the glass transition temperature with respect to the temperature of observation, the larger the excess free volume. Therefore we have just simply assumed that the samples are in equilibrium at the end of the reaction, when both the rate of polymerization and the rate of shrinkage have decreased till below the limit of detection. So if at the end of the reaction a C=C bond conversion of 90 per cent has been reached we assume that the observed shrinkage is also 90 per cent of the theoretical value. Then the transformation to percentage shrinkage per second can easily be made. Since the highest temperature of maximum mechanical loss at 1 Hz was 49°C, the glass transition temperature will be at a somewhat lower value, presumably near 30°C. This means that the contribution of excess free volume, equal to $(\alpha_l - \alpha_g)$ $(T_g - T)$, must be very small indeed. (The difference between volume expansion coefficients α_l and α_g of liquid and glassy polymers, respectvely, is usually of the order of 3×10^{-4} K^{-1}, both for linear and for crosslinked polymers (12,13).

It should be noted that the change of thickness of the sample layer can be determined much more accurately than the thickness itself: the cover glass is always somewhat tilted with respect to the bottom plate, variations of total thickness of the sandwich can be as large as 5 μm across the width of the sandwich. Therefore the estimated inaccuracy of the relative change of thickness of the sample layer is ± 10%. Changes of thickness of 5 μm can be determined with an inaccuracy of less than 0.05 μm, better than 1%. Rate of shrinkage measurements are obviously also subject to errors of up to 10% at the maxima. Towards the end of the reaction the errors may become even larger. An alternative method of obtaining the rate of shrinkage, for example digitization, curve fitting and analytical differentiation could decrease the scatter of the points as obtained from graphical differentiation but it would not reduce the large absolute error.

During photopolymerization of TEGDA at a light intensity of 0.2 $mW.cm^{-2}$ the relative change of the sample thickness was 11 ± 1%. Density measurements of larger samples made under the same conditions gave a value of 10.5 ± 0.2%. So it can be concluded that within the (large) experimental error total shrinkage shows up as a decrease of thickness, due to the large area to thickness ratio the lateral contribution can be neglected.

Dynamic mechanical thermal analysis

DMTA measurements were made with a Polymer Labs instrument. Samples were clamped in the single cantilever mode in a frame of 22 mm using 6 mm clamps with 0.5 mm faces. The sample length between the clamps was 8 mm. Measurements were performed at a frequency of 1 Hz, a strain amplitude of 0.063 mm and a heating rate of 5 $K.min^{-1}$. Clamping was checked by monitoring the strain amplitude on an oscilloscope. The measurements were carried out in air. Values of the temperature of maximum mechanical loss, T (tan δ_{max}), were reproducible to ± 2 K.

Sample preparation for DSC, TMS and DMTA

A solution of 3.36 w% initiator in TEGDA was used for DSC, TMS and DMTA measurements. DSC sample weights were about 1 mg, corresponding to a thickness of about 60 μm when aluminum lids of standard sample pans are used.

After polymerization in the DSC the samples were extracted with 2-propanol for one week and the extracts were analyzed for unreacted monomer using liquid chromatography as described in the next subsection.

TMS samples were made by depositing 3 - 5 μl of the sample solution in the recess of the lower quartz plate. Special care was taken to avoid inclusion of air bubbles when mounting the cover glass.

DMTA samples were made by spreading the liquid sample on a CrNi steel substrate with a thickness of 200 μm and an area of 7x25 mm. These samples were placed in a vessel provided with a clear quartz window. Before and during irradiation the vessel was flushed with pure nitrogen (< 2 ppm O_2).

Irradiation was performed with a similar lamp as used with DSC. Prior to irradiation the samples were flushed with nitrogen for 15 minutes. After the irradiation the samples were kept in the vessel overnight to allow for volume relaxation and decay of trapped radicals. Next a similar layer was prepared on the back of the substrate in order to obtain a symmetrical sample.

Solvent extraction of DSC samples

The method of solvent extraction/liquid chromatography analysis of relatively large and thin samples (10 − 15 μm layer on a glass disc with a diameter of 40 mm, sample weight 15 − 20 mg) has been described in a previous publication (14).

In this work a slightly modified procedure was used: the 1 mg DSC samples were extracted with 400 μl 2-propanol in 6 mm bore reagent tubes, flattened at their lower part in order to keep the sample pans in an upright position. With non-flattened tubes, samples could stay upside down at the bottom of the tube, thereby preventing rapid extraction. Samples were kept in a shaking machine for at least one week. Tetraethyleneglycol dibenzoate (TEGDB) was added as a probe for monitoring the extraction process. Complete recovery of the probe was considered to be indicative of complete extraction of the unreacted monomer.

In a previous study of HDDA we used the corresponding dipropionate instead of the dibenzoate since this compound is most similar to the polymerized acrylate (15). However, the propionate has a very weak UV absorption at the wavelength of detection (210 nm) which means that about 3% of the probe had to be added, even when a very sensitive Kratos Spectroflow 773 UV detector was used. Since we had observed that the extraction behavior of hexanediol dipropionate is indeed very similar to that of hexanediol dibenzoate (16), we used the dibenzoate also with TEGDA as well. With TEGDB only 0.3 w% of the probe had to be added. HPLC analyses were performed on a Hewlett Packard 1090A chromatograph using a ternary solvent (water, acetonitrile and methanol) and a filter photometric detector. HPLC determinations were reproducible to ± 1% of the original monomer content. However, variations in irradiation dose may cause sample to sample variations of up to 5%.

RESULTS AND DISCUSSION

Calorimetry

Fig. 3 shows the maximum extents of double bond conversion x, obtained at various light intensities for polymerizations of TEGDA at 20 and 80°C, respectively. The increase of ultimate conversion with light intensity is observed at both temperatures. This effect is not caused by self-heating of the polymerizing samples (9).

When the samples polymerized at 20°C were heated in the dark to 80°C, additional reaction occurred: 3, 2 and 1% conversion at 0.002, 0.02 and 0.2 $mW.cm^{-2}$, respectively. This indicates the presence of trapped radicals which are remobilized by heating (9,15). After standing in air no aftercuring was observed.

Shrinkage and conversion

In Figure 4 the rate of polymerization, as measured with DSC, is compared with the rate of shrinkage, as measured with TMS. Contrary to our previous observations with HDDA (7) it seems that the shrinkage process is faster than the polymerization. The relative rate of shrinkage not only exceeds the relative rate of polymerization, it also reaches its maximum value at an earlier time. Since TEGDA polymerizes much faster than HDDA, this effect may well have been caused by the inertia of the DSC. As a check we measured the response times of the two systems separately.

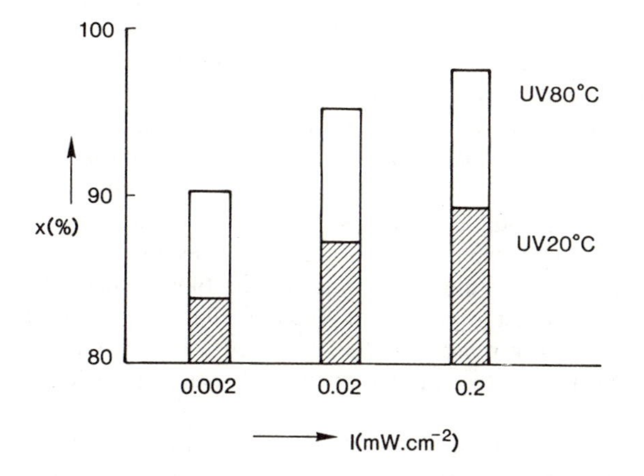

Figure 3. **Maximum extent of reaction** x measured with DSC at 20°C and at 80°C for various light intensities. TEGDA with 3.4 w% DMPA.

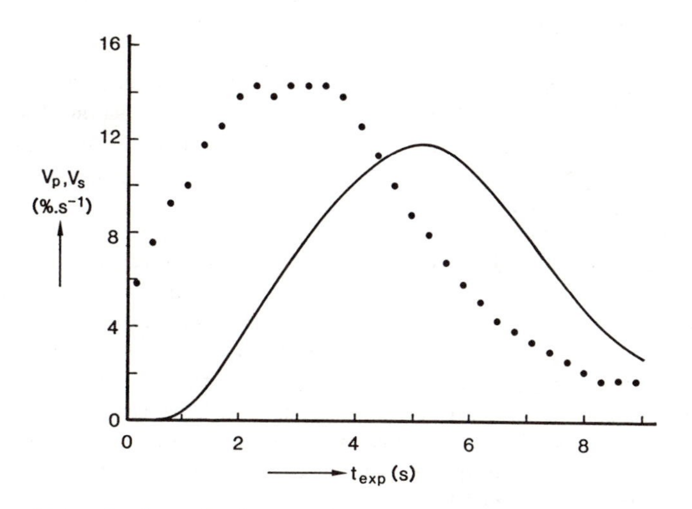

Figure 4. **Rate of polymerization** (continuous curve) **and rate of shrinkage** (dotted curve) for TEGDA containing 3.4 w% DMPA and 0.3 w% TEGDB. Temperature: 20°C. Light intensity: 0.2 mW.cm^{-2}.

The TMS was tested by mounting an empty cell, putting a weight of 1 g on the tray of the TMS (which causes bending of the cover glass) and measuring the response after the sudden addition or removal of an additional weight of 2 g. The DSC was tested by focusing the beam of a 15 W halogen lamp on the empty sample holder and actuating the shutter. The results are shown in Figure 5. It can be seen that the TMS reacts considerably faster than the DSC. The difference increases when a flattened sample lid, such as used for small samples instead of the standard sample pan, is placed in the sample compartment. Then it makes no difference whether or not the lid is covered with a 60 μm layer of poly (TEGDA). Unflattened lids gave the same results. Obviously, in Figure 4 the DSC curve has been shifted towards longer times such that the delay observed with HDDA is completely masked.

Next the time to reach the maximum signal and the relative "peak area" were determined as a function of the duration of the shutter opening (Figure 6). Relative peak area is almost constant after 1 s, whereas the time to reach the maximum amplitude becomes constant beyond 6 s. These findings are consistent with the presence of barriers to heat transfer in the DSC itself and between sample pan and the thermistor, which delay the transfer of the heat rather than changing the total amount of heat detected.

Reduction of the light intensity slows the reaction down and at the same time the curves come closer together (Figure 7). At the end there appears to be a small delay of shrinkage with respect to conversion. Initially the opposite seems to occur, so obviously the DSC curve is still distorted and by shifting it to the left the delay of shrinkage towards the end of the reaction would be more pronounced. However, the rate of polymerization reaches a distinctly higher maximum value than the rate of shrinkage. Upon a further reduction of the light intensity to 0.002 $mW.cm^{-2}$ (Figure 8) the initial parts of both curves coincide, so the unrealistic delay of conversion with respect to shrinkage now has disappeared. The lower shrinkage maximum is accompanied by an enhanced rate towards the end of the reaction. The Figure shows only a part of the curves, the reaction was followed for about 30 min. As a consequence of the representation of the rates as percent change per second and the assumption of equilibrium at the end of the reaction (see experimental part) the areas under both curves must be equal.

After having established the delay of shrinkage with respect to conversion we now can understand the data of Figure 3: the delay of shrinkage with respect to chemical conversion causes a temporary excess of free volume. During this period pendent double bonds and free monomer molecules have a relatively high mobility and therefore they can easily react up to a high conversion, provided that enough radicals are available. Later on, the propensity to react will decrease due to additional shrinkage. So if the photons are generated at a lower rate a larger part of the reaction will have to occur in a more relaxed state, that is at a much lower rate which will soon become undectable with DSC.

DMTA measurements

Figure 9 shows the DMTA curves neasured after various irradiation times. On reducing the exposure time from 600 to 0 seconds the maximum in tan δ shifts progressively towards lower temperatures. For exposure times of 5 s and less, thermally induced

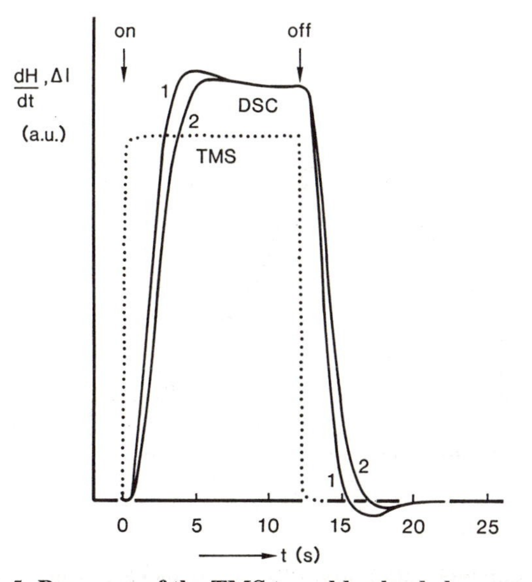

Figure 5. **Response of the TMS to sudden load changes and of the DSC to a block-shaped irradiation pulse.** Dotted curve: TMS. Continuous curves: DSC. (1) in the absence and (2) in the presence of a flattened lid of a standard sample pan.

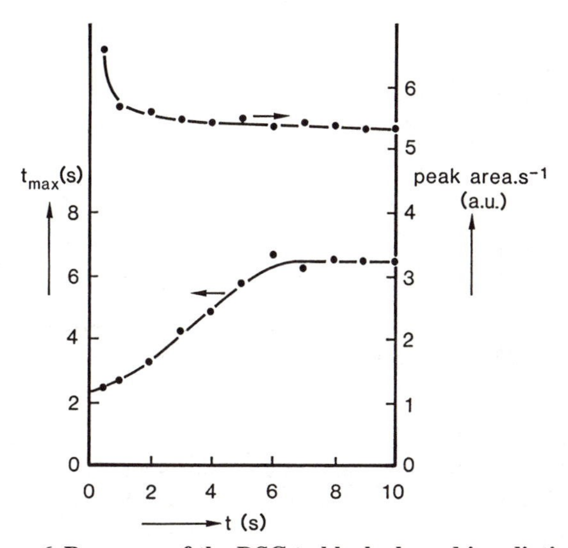

Figure 6. **Response of the DSC to block-shaped irradiation pulses.** Time to reach the maximum amplitude (lower curve) and peak area per unit of time (upper curve) as a function of the pulse duration.

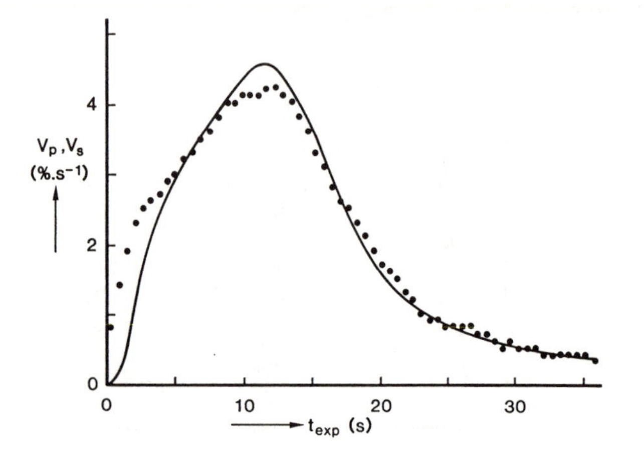

Figure 7. **Rate of polymerization and rate of shrinkage.** Same experiments as in Figure 4 but with a light intensity of 0.02 $mW.cm^{-2}$.

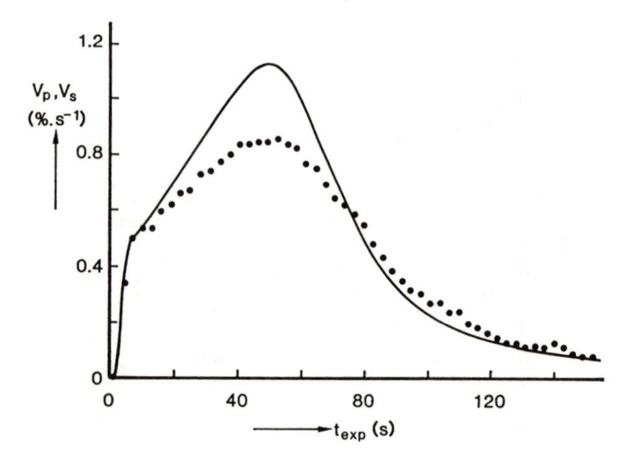

Figure 8. **Rate of polymerization and rate of shrinkage.** Same experiments as in Figure 4 but with a light intensity of 0.002 $mW.cm^{-2}$.

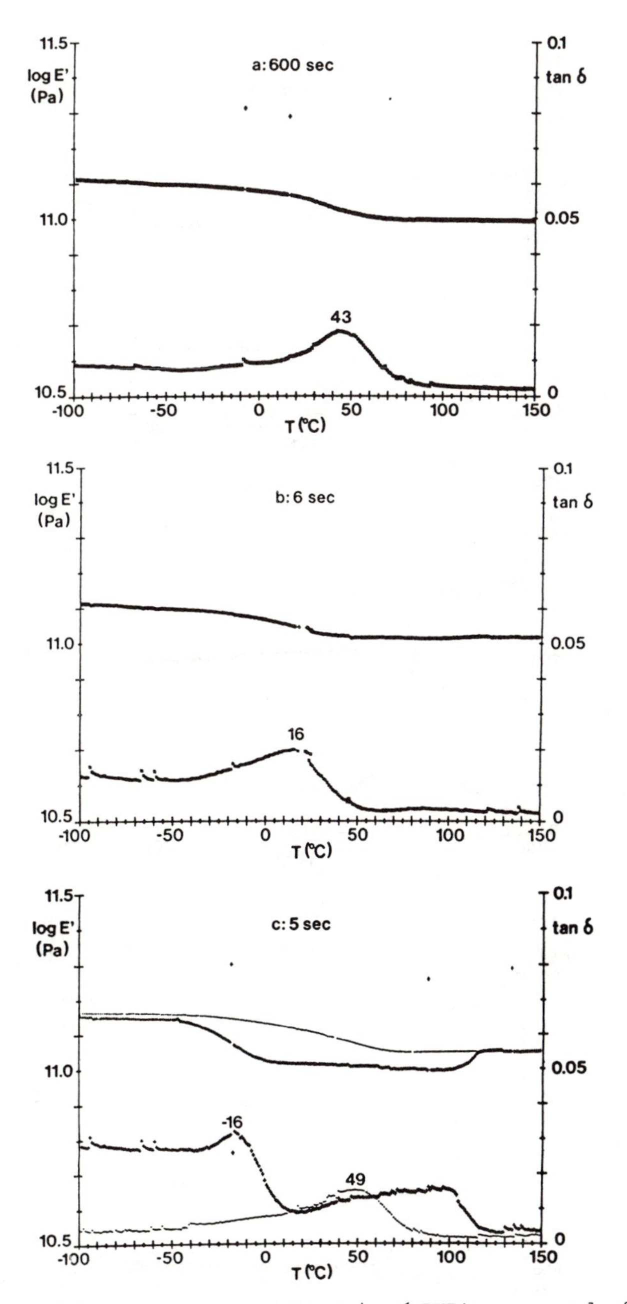

Figure 9. DMTA curves of TEGDA + 3.4 w% DMPA, measured after various exposure times. Intensity: 0.2 $mW.cm^{-2}$. Upper curves: log E'. Lower curves: tan δ. Frequency: 1Hz. (Thin line for t = 5s: repeat.) (Reproduced with permission from Ref. 9. Copyright 1987 Polymer.) (Continued on next page.)

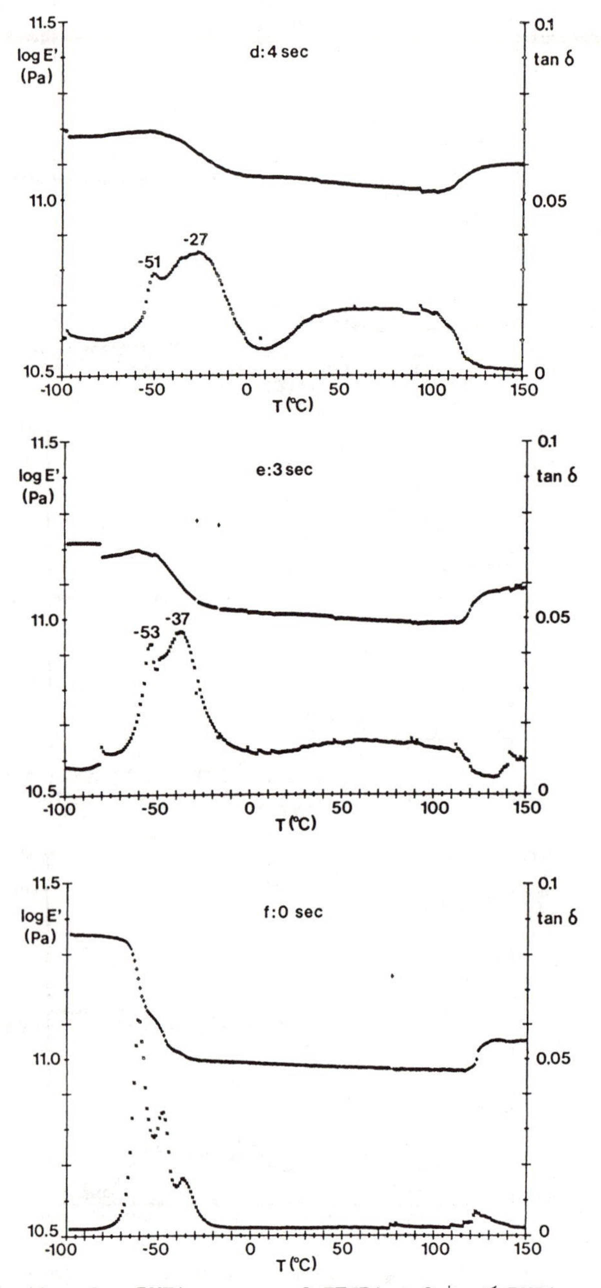

Figure 9 Continued. DMTA curves of TEGDA + 3.4 w% DMPA, measured after various exposure times. Intensity: 0.2 mW.cm^{-2}. Upper curves: log E'. Lower curves: tan δ. Frequency: 1 Hz. (Thin line for t = 5s: repeat.) (Reproduced with permission from Ref. 9. Copyright 1987 Polymer.)

polymerization occurs around 120°C, together with an increase in E′. (Since all samples were coated on a metallic substrate and since it is difficult to control the thickness very accurately the changes in E′ are more significant than their absolute values.)

The repeated scan in Figure 9c shows that the maximum in tan δ shifts towards an even higher temperature than in Figure 9a. This can be readily explained since before the DMTA experiment the sample in Figure 9a had not been subject to as high a temperature as the one in Figure 9c. Repetition of the scan of Figure 9a caused a shift of the maximum to 48°C. The thermally induced polymerization is even more marked in Figures 9d and e, as can be seen from the jump in E′. Moreover the low-temperature maximum in tan δ splits up into two peaks, one at a rather constant position around − 50°C, the other at a steadily decreasing value. It is tempting to attribute the low-temperature peak to unreacted monomer (see below).

In Figure 9f, finally, our attempt to measure unreacted monomer is depicted. This is very difficult to accomplish, however, since upon melting, the sample starts flowing off the substrate. The thermal reaction shown in the Figure is caused by the unquantified remainder of liquid sample on the substrate. From the low-temperature maxima it can at best be concluded that melting occurs in the vicinity of − 50°C. No satisfactory explanation for the splitting can be offered at the present stage, although delamination of the sample might well contribute to the complicated picture. With DSC a glass transition was observed at − 70°C. Since the DSC experiment is carried out on a much longer time scale this may well be the same transition as was observed with DMTA near − 60°C.

Combination of calorimetric and mechanical results

We now turn to a correlation of the DMTA results with DSC measurements. In Figure 10 the extent of C=C double bond conversion, as measured with DSC, is plotted versus exposure time. Also plotted is the amount of monomer extracted afterwards from the DSC samples. It can be seen that for exposure times longer than 6 s the rate of polymerization decreases suddenly to a much lower value, but not to zero. At the same time the free monomer is exhausted so further reaction necessarily means further crosslinking by reaction of pendent double bonds. According to our mechanical measurements thermal aftercuring also ceases to have an observable effect on E′ (Figure 9a and b).

From this we conclude that significant thermal aftercuring requires the presence of unreacted monomer. In its absence only a change in T(tan δ_{max}) was observed. This change may be either due to volume relaxation or to some additional polymerization that is insufficient to cause an observable increase of Young's modulus. Alternatively, the change could also have been caused by evaporation of unreacted photoinitiator. During prolonged exposure only little conversion takes place, but since mainly crosslinks are formed the temperature of maximum loss continues to change considerably (Figure 11). Beyond 80% conversion (t_{exp} > 6 s) a very steep increase in the position of the maximum is observed.

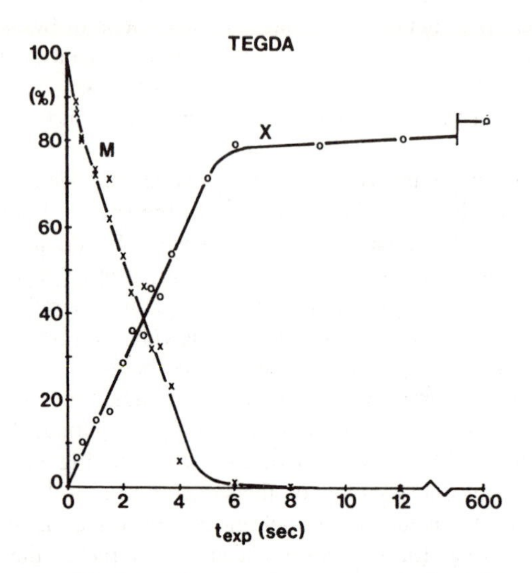

Figure 10. Extent of C=C bond conversion measured with DSC (o) and fraction of extractable monomer (x) vs. exposure time. TEGDA + 3.4 wt% DMPA. 0.2 mW.cm^{-2}. Reproduced with permission from ref. 9. Copyright 1987 *Polymer*.

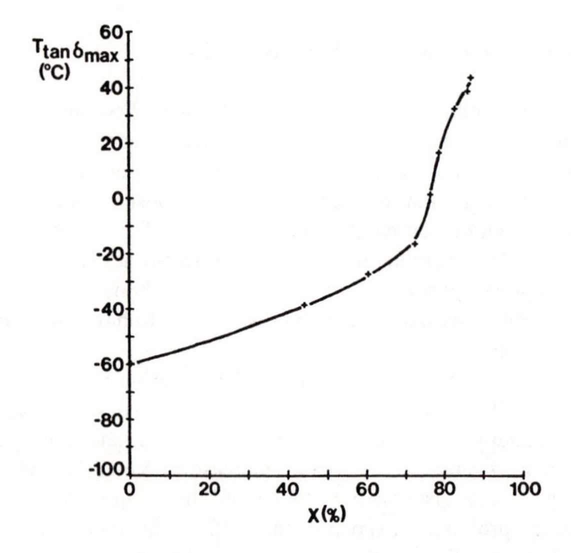

Figure 11. T(tan δ_{max}) vs. extent of C=C bond conversion x. TEGDA + 3.4 wt% DMPA. 0.2 mW.cm^{-2}. Reproduced with permission from ref. 9. Copyright 1987 *Polymer*.

Next we turn back to the original question,i.e. whether or not variations in light intensity show up in the mechanical properties of photopolymers. Table 1 shows the values of T (tan δ_{max}) that were measured on samples polymerized at different light intensities.

TABLE 1. Values of T (tan δ_{max}) for samples polymerized at various light intensities

Intensity ($mW.cm^{-2}$)	Irrad.time (min)	T (tan δ_{max}) (°C)	Irrad.time (min)	T (tan δ_{max}) (°C)
0.002	12	23	2400	48
0.02	6	32	240	43
0.2	4	38	24	42
2.0	2.5	45	2.4	44

Values reported in the third column were obtained from samples polymerized under the same conditions and during the same time as in the DSC experiments. The latter were stopped at the time the rate of polymerization fell below the limit of detection of the DSC apparatus (about 0.2% of the maximum rate observed during an experiment). There is a clear increase of T (tan δ_{max}) with light intensity and, therefore, with conversion.

However, the applied doses also increase with light intensity. When the doses are kept equal (last column) the difference between the samples vanishes within experimental error. This shows again that, especially at low light intensities, reaction continues at an extremely low rate when the vitreous state has been reached. This rate cannot be measured with isothermal DSC. The results illustrate the self-decelerating character of the reaction during isothermal vitrification: the mobility of unreacted groups determines the rate of reaction but additional conversion reduces the mobility, which reduces the rate, and so on. The closed-loop dependence of mobility, rate and extent of reaction strongly resembles physical aging (17). This similarity has been discussed in some detail elsewhere (18).

CONCLUSIONS

The rate of photopolymerization of TEGDA shows a sudden decrease when free monomer is exhausted and only crosslinking can still occur.

The maximum extent of double bond conversion in TEGDA as measured with DSC increases not only with temperature but also with light intensity. Mechanical measurements show, however, that the intensity dependence vanishes when equal doses are applied. This means that at low intensities the polymerization continues for a considerable time at a rate which is imperceptible with DSC.

With decreasing UV dose T(tan δ_{max}) also decreases. When free monomer is still present additional polymerization causes a stepwise increase of E′ and T(tan δ_{max}) during a thermal scan. In the presence of sufficient monomer the tan δ peak splits up into two peaks: one at a constant and the other at a dose-dependent position, representing monomer and network, respectively.

ACKNOWLEDGMENT

The contributions of L.A.J. Swerts and J. Boven, who performed the chromatographic measurements and extraction probe synthesis, respectively, are gratefully acknowledged.

LITERATURE CITED

1. Van den Broek, A. J. M., Haverkorn van Rijsewijk, H. C., Legierse, P. E. J., Lippits, G. J. M. and Thomas, G. E., J. Rad. Cur. 1984, **11** (1), 1.
2. Kloosterboer, J. G. and Lippits, G. J. M., J. Rad. Cur. 1984, **11** (1), 10.
3. Zwiers, R. J. M. and Dortant, G. C. M., Appl. Optics 1985, **24**, 4483.
4. Zwiers, R. J. M. and Dortant, G. C. M., in: *"Integration of Fundamental Polymer Science and Technology"*, p. 673. L. A. Kleintjens and P. J. Lemstra (Eds.). Appl. Science Publ., London 1986.
5. Broer, D. J. and Mol, G. N., J. Lightwave Techn. 1986, **LT-4**, 938.
6. Blyler, L. L., Eichenbaum, B. R. and Schonhorn, H., *"Optical Fiber Telecommunications"*, Ch. 10. S. E. Miller and A. G. Chynoweth, Eds. Academic Press, New York 1979.
7. Kloosterboer, J. G., Van de Hei, G. M. M., Gossink, R. G. and Dortant, G. C. M., Polym. Commun. 1984, **25**, 322.
8. Lee, C. Y-C. and Goldfarb, I. J., Polym. Eng. Sci. 1981, **21**, 390.
9. Kloosterboer, J. G. and Lijten, G. F. C. M. Polymer, 1987, **28**, 1149.
10. Kloosterboer, J. G. and Lijten, G. F. C. M., Polymer Commun. 1987, **28**, 2.
11. Brandrup, J. and Immergut, E. H. (Eds.), *"Polymer Handbook"*, 2nd. ed., pp. II-421. Wiley, New York 1975.
12. Haward, R. N. (Ed.), *"The Physics of Glassy Polymers"*, p. 600. Applied Science, London 1973.
13. Oleinik, E. F., Adv. Polym. Sci. 1986, **80**, 49.
14. Kloosterboer, J. G., Van Genuchten, H. P. M., Van de Hei, G. M. M., Lippits, G. J.M. and Melis, G. P., Org. Coat. Plast. Chem. 1983, **48**, 445.
15. Boots, H. M. J., Kloosterboer, J. G., Van de Hei, G. M. M. and Pandey, R. B., Brit. Polym. J. 1985, **17**, 219.
16. Swerts, L. A. J. and Kloosterboer, J. G., unpublished results.
17. Struik, L. C. E., *"Physical Aging in Amorphous Polymers and Other Materials"*, p.8. Elsevier, New York 1978.
18. Kloosterboer J. G. and Lijten G. F. C. M., Proc. Networks '86: *"Biological and Synthetic Polymer Networks"*, O. Kramer, Ed. Elsevier Applied Science, London 1987.

RECEIVED February 16, 1988

Chapter 29

Structure and Properties of Polydimethacrylates

Dental Applications

D. T. Turner, Z. U. Haque, S. Kalachandra, and Thomas W. Wilson

Department of Operative Dentistry and Dental Research Center, University of North Carolina, Chapel Hill, NC 27514

Dimethacrylate monomers were polymerized by free radical chain reactions to yield crosslinked networks which have dental applications. These networks may resemble ones formed by stepwise polymerization reactions, in having a microstructure in which crosslinked particles are embedded in a much more lightly crosslinked matrix. Consistently, polydimethacrylates were found to have very low values of Tg by reference to changes in modulus of elasticity determined by dynamic mechanical analysis. Also, mechanical data on the influence of low volume fractions (0.03-0.05) of rigid filler particles provide evidence of a localized plastic deformation which would not seem understandable by reference to a uniformly crosslinked network. A non-uniformly crosslinked matrix might also be invoked to account for insensitivity of the rate of diffusion of water on the apparent degree of crosslinking. However, an observed increase in the uptake of water with apparent degree of crosslinking remains unexplained.

Crosslinked polymers are widely used as dental materials (1-3). Perhaps the most challenging application is in the restoration of teeth (4). The monomers must be non-toxic and capable of rapid polymerization in the presence of oxygen and water. The products should have properties comparable to tooth enamel and dentin and a service life of more than a few years. In current restorative materials such properties are sought using so-called "dental composites" which contain high volume fractions of particulate inorganic fillers (5-7). However in the present article attention is concentrated on one commonly used crosslinked polymeric component, and on the way in which some of its properties are influenced by low volume fractions of fillers.

Up to the present time, use has been made almost entirely of dimethacrylates which are polymerized by free radical mechanisms to yield crosslinked products (8). Polymerization is initiated by redox systems, such as benzoyl peroxide/aromatic amine, and by

0097-6156/88/0367-0427$06.00/0

photopolymerization with either ultra-violet or longer wavelength light. A considerable fraction of the double bonds remains unreacted because of immobilization, due to vitrification or steric isolation (9-14). Even so, the shrinkage which accompanies the conversion of double to single bonds is believed to be the cause of failures in service (15,16). One approach to this problem is to reduce the concentration of double bonds in reactants by using dimethacrylates of high molecular weight such as urethane dimethacrylates (17,18) and BIS-GMA i.e. the adduct of bisphenol-A and glycidyl methacrylate (19). A more ambitious approach is to avoid shrinkage by recourse to ring opening polymerization reactions, but this inevitably requires a long term effort because of the difficulties of developing any new polymer to the stage of application (20,21).

The main deficiency of polydimethacrylates for the restoration of teeth is their poor wear resistance (4-7). This can be improved by inclusion of particulate fillers which are harder than the polymeric matrix. An ambitious goal would be to match the remarkable properties of dental enamel, which contains more than 95 vol-% of hydroxyapatite crystallites tightly packed into an intricate microstructure. In comparison, the current composite restorative materials have a crude microstructure with no more than 65 vol-% inorganic filler. This is not for want of application of current knowledge of the technology of particulate polymer composites. A wide range of fillers of varying shapes and sizes, ranging from colloidal dimensions to tens of microns, is being used in varying combinations. Silane coupling agents are being used, in extension of Bowen's pioneering work (19), to bond these particles to the polymeric matrix. Yet, the current composite materials have much lower wear resistance than the silver amalgams which they are designed to replace. Despite this deficiency there are diverse cogent reasons, such as esthetics and avoidance of mercury pollution of the environment, which spur on their further development.

Crosslinked polymers are also used to bond composite restorations to dentin (22) and thereby minimize the occurrence of marginal gaps which could result in bacterial invasion (23). In this application, there is a carry-over of the experience gained with the use of dimethacrylate monomers in composite restorative materials. As a further development adducts of $POCl_3$ with hydroxy groups in methacrylate monomers are used in the hope that phosphonate groups, formed by hydrolysis, would favor ionic bonding to the hydroxyapatite component of dentin (24). A quite different approach is to use a hydrophilic combination of glutaraldehyde and hydroxyethyl methacrylate (25). Possibly the result is an interpenetrating network between a synthetic polymer and a more densely crosslinked collagen. These and other dentin bonding agents (26-28) have been reported to prevent marginal leakage in extracted teeth and to give a tensile adhesive strength as high as 10 MPa. For comparison, the two materials which are bonded together have tensile strengths as follows: dentin, 40-50 MPa; dental composite, 30-60 MPa.

In dental applications where crosslinked polymers do not need to be prepared in the mouth, higher degrees of crosslinking can be achieved merely by heating to higher temperatures. A high degree of crosslinking is desirable in crown and bridge prostheses where a high modulus of elasticity is sought in order to replace metals, such as gold alloys. The major problem is to increase the rigidity of polymeric materials yet without incurring the esthetically unacceptable appearance attending the use of most rigid filler particles. The best approach has been to use a variety of hydrophobic dimethacrylate copolymers which yield stiff polymer chains and high degrees of crosslinking (29,30). However, in pushing in this direction account must be taken of the onset of embrittlement (31). An interesting possibility for amelioration is the development of polymer blends (32).

Current ways in which many scientists are addressing problems related to the above applications can be judged by reference to abstracts of the International Association of Dental Research (33). For the remainder of this article, a few lines of investigation in just one laboratory will be outlined in the belief that they include points of general interest to workers on other aspects of highly crosslinked polymers.

Dynamic Mechanical Analysis (DMA)

Dynamic testing of proprietary dental composites has been used to determine Young's modulus, damping (5), and to assign a value of Tg from a maximal value of tan δ (34). However for present purposes, attention will be concentrated on unfilled networks made from mixtures of known components (35). The following mixture was polymerized by exposure to light of wavelength > 400 nm: BIS-GMA (75 wt%)+ triethylene glycol dimethacrylate (25%) containing dl-camphoroquinone (0.2%) and N,N-dimethylaminoethyl methacrylate (0.1%). Only about 50% of the double bonds reacted, as determined by calorimetry (36).

The BIS-GMA copolymer was tested at 11 Hz on an Autovibron apparatus. The transition from the glassy to the rubbery state occurs over a wide temperature range of more than 100°C (Fig. 1). Following a convention used in DMA, a value of Tg = 48°C may be assigned by reference to the maximal value of mechanical loss, i.e. of tan δ. However, an alternative assignment may be made at the temperature where the modulus of elasticity (E') first begins to decrease, i.e. at Tg = -25°C. The higher value, from tan δ, is inconsistent with the observation that the polymeric product is flexible at room temperature. Also the lower value from E' is closer to the value of Tg assigned by reference to changes in the coefficient of thermal expansion. As this question of assignment is of general importance it will be illustrated for the case of a more highly crosslinked polymer (c.f. ref. 37) made by - irradiation of BIS-GMA (Fig. 2). It will be seen that simultaneous measurements of specimen length versus temperature (Fig. 3) gave a value of Tg = 145°C which is closer to the value from E' (138°C) than to that from tan δ (195°C). In summary of this and other related work on epoxy systems (38), it appears that though, generally, it is more difficult to assign a precise value of Tg by reference to E', nevertheless this value is in closer agreement with the classical method of assignment by reference to the coefficient of thermal expansion.

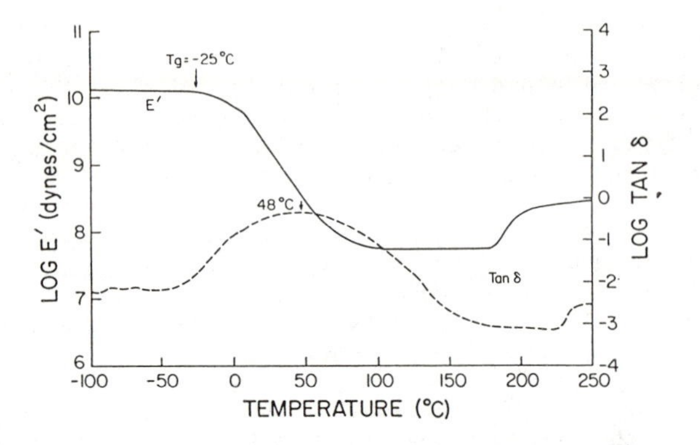

Fig. 1 Dynamic mechanical properties of a BIS-GMA/TGDM copolymer prepared by photopolymerization.

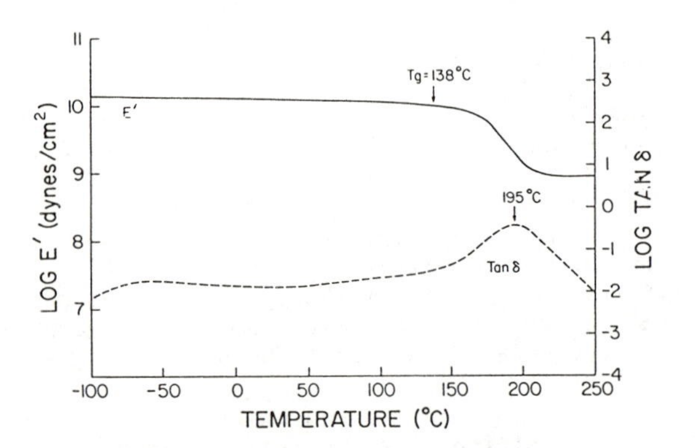

Fig. 2 Dynamic mechanical properties of a polymer prepared by γ-irradiation of BIS-GMA:Dose=75 Mrad.

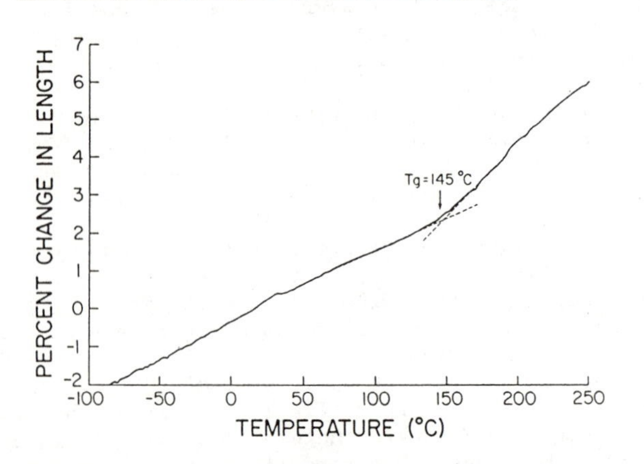

Fig. 3 Percent change in length of the polymer prepared by γ-irradiation of BIS-GMA.

From a practical point of view it is important to know whether similar results are obtained for the proprietary composites currently used to restore teeth. In the examples studied so far this has been found to be the case (35), in agreement with earlier results (34). Thus, there is agreement about results but a difference in how they should be interpreted. The view that there is a low value of Tg, within the range of temperatures (0-50°C) encountered in service at the surface of a restoration, would not seem to bode well for wear resistance. It would seem to be important to be able to control the polymerization reaction, for example by means of light intensity (39), in order to impose any desired value of Tg and then to evaluate its influence on wear.

Fillers

The influence of fillers has been studied mostly at high volume fractions (40-42). However, in addition, it is instructive to study low volume fractions in order to test conformity with theoretical predictions that certain mechanical properties should increase monotonically as the volume fraction of filler is increased (43). For example, Einstein's treatment of fluids predicts a linear increase in viscosity with an increasing volume fraction of rigid spheres. For glassy materials related comparisons can be made by reference to properties which depend mainly on plastic deformation, such as yield stress or, more conveniently, indentation hardness. Measurements of Vickers hardness number were made after photopolymerization of the BIS-GMA recipe, detailed above, containing varying amounts of a silanted silicate filler with particles of tens of microns. Contrary to expectation, a minimum value was obtained (44,45), for a volume fraction of 0.03-0.05 (Fig. 4). Subsequently, similar results (46) were obtained with all 5 other fillers tested (Table 1).

Table 1. Influence of fillers on Vickers Hardness Number

Filler	Volume fraction at minimum	% Decrease in VHN*
Silicate (45) (surface treated)	0.03-0.05	10
Silicate (surface treated)	0.05	34
Silicate (ashed)	0.05	7
Silica (surface treated)	0.03	16
Silica (ashed)	0.03	18
Tribasic calcium phosphate	0.03	13
Hydroxyapatite	0.03	5

*100 [VHN (no filler) - VHN (at minimum)]/VHN (no filler)

The particles varied in size from colloidal dimensions up to tens of microns and differed in surface characteristics, including treatment with silane coupling agents. A general explanation was based on the knowledge that rigid inclusions result in a highly localized stress concentration on application of an external force (47). It was suggested that this may result in a localized plastic deformation and hence in a reduction of macroscopic properties which depend on yielding, such as yield stress and indentation hardness. With higher volume fractions of filler, effects due to isolated particles become unimportant and there is an eventual increase in property values, as predicted theoretically.

Two lines of evidence are consistent with the above suggestion. First, fracture surfaces obtained with the silanted silicate have linear features which originate from particles and which extend away from the direction in which the crack tip advanced (Fig. 5). In fact such features had been predicted on the grounds that the particles would serve to rotate the tensile stress field ahead of the crack tip. Rotation of the stress field was expected to generate intersecting microcracks, for reasons discussed previously (48). From the present point of view the most pertinent point is that, at higher magnification, the linear features can be seen to have a rounded cross-section. In contrast, microcrack intersections in very brittle materials give features with sharp edges. Therefore, the rounded cross-section is an important observation which has been interpreted as evidence of localized plastic deformation.

A second line of evidence is provided by an experiment in which the degree of crosslinking of photopolymerized specimens was progressively increased by exposure to γ-rays. The expectation was that crosslinking should reduce plastic deformation by preventing macromolecules from slipping past one another. Therefore if the softening is, indeed, due to localized plastic deformation then the difference between unfilled specimens and ones with a low volume fraction of filler should be reduced by radiation crosslinking. This was found to be the case (Fig. 6). In fact after the highest dose, the specimen containing filler was as hard as the unfilled specimen.

A practical consequence of this work is that an isolated filler particle can act as a site of mechanical weakness. In lightly crosslinked materials this can result in plastic deformation. Presumably in highly cross-linked materials this might result in brittle fracture, especially in fatigue. Isolated filler particle situations can be envisaged in a variety of service applications. In the field of dental materials, this might occur at an interface between a dentin bonding agent and a composite filling and thereby constitute a zone of mechanical weakness.

Water Sorption

Water sorption of dental materials can result in undesirable changes in dimensions and to a deterioration in mechanical properties. Studies have been made of BIS-GMA copolymers of the kind mentioned above (49,50) and also of polymers of potential use

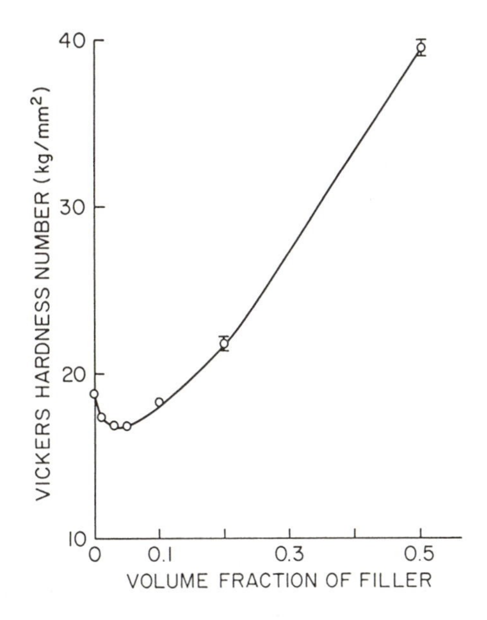

Fig. 4 Influence of volume fraction of rigid filler particles on Vickers hardness number of a BIS-GMA copolymer.

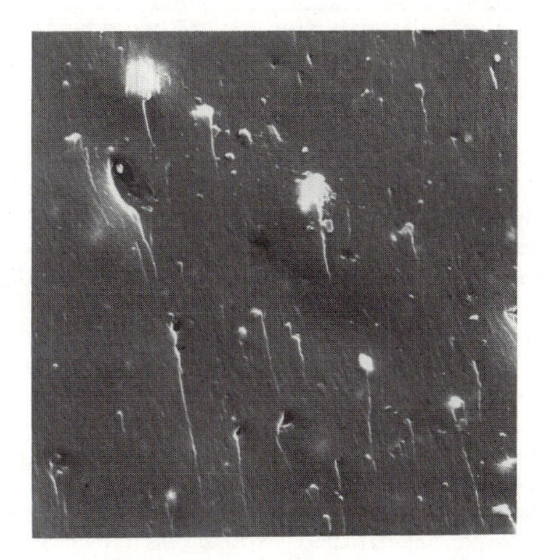

Fig. 5 Generation of linear features by particles of filler (X1200). Volume function of filler = 0.05. Crack propagated from top to bottom.

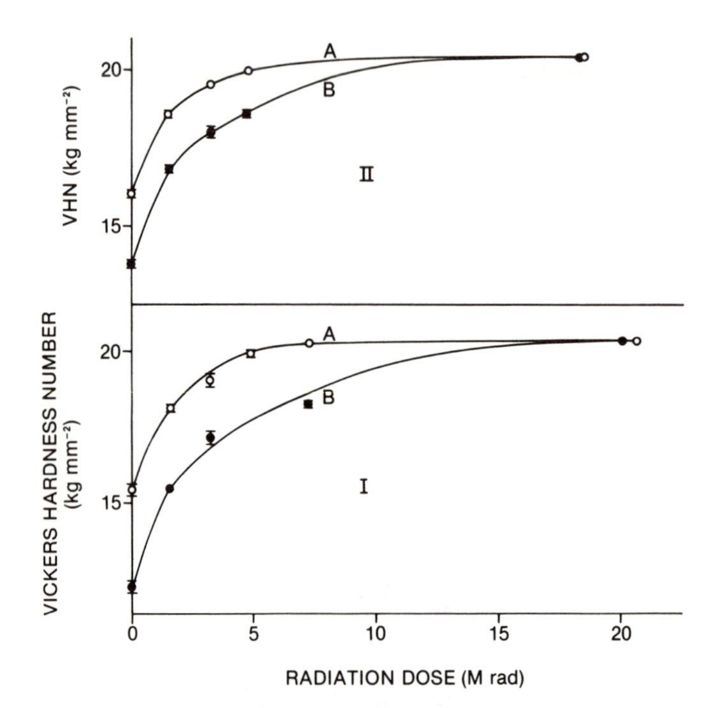

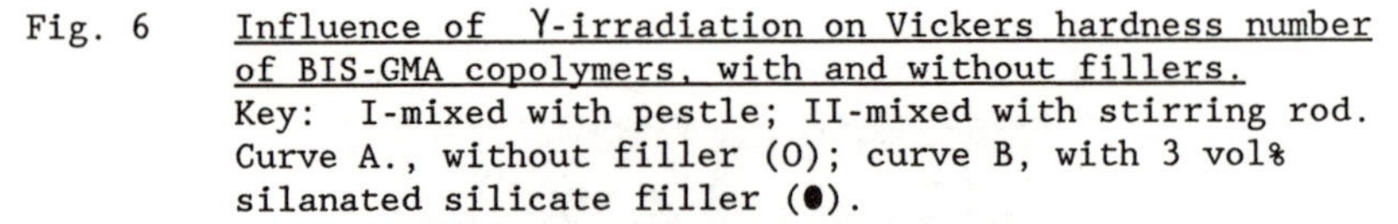
Fig. 6 Influence of γ-irradiation on Vickers hardness number of BIS-GMA copolymers, with and without fillers. Key: I-mixed with pestle; II-mixed with stirring rod. Curve A., without filler (O); curve B, with 3 vol% silanated silicate filler (●).

in crown and bridge work (29,30,51). Here attention will be confined to more systematic results obtained with networks made by copolymerization of methyl methacrylate (MMA) and triethylene glycol dimethacrylate (TGDM), using a redox initiator (52).

The diffusion coefficient for water was found to be rather insensitive to the proportion of the dimethacrylate crosslinker, especially in sorption (Fig. 7). This insensitivity might be rationalized in a number of ways. One of these is consistent with the view that these systems have a lightly crosslinked matrix which does little to impede the diffusion of water molecules.

The considerable increase in water uptake, as much as two-fold (Fig. 8), caused by replacing MMA with TGDM is surprising. It does not seem to be due to incomplete polymerization because similar results were calculated from data previously reported (29) for specimens made at the much higher temperature of 120°C. Neither can it be attributed to the more hydrophilic nature of TGDM, because similar results were obtained for copolymers of MMA and ethylene glycol dimethacrylate. An alternative hypothesis was explored that crosslinking results in less efficient macromolecular packing, and hence to increased accommodation of water in microvoids. However, this hypothesis was tested by

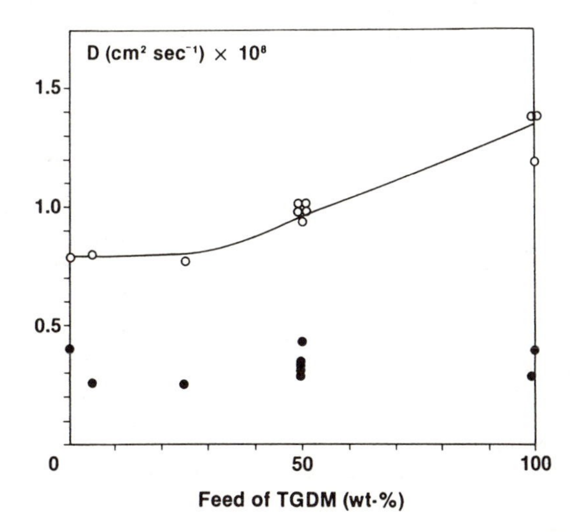

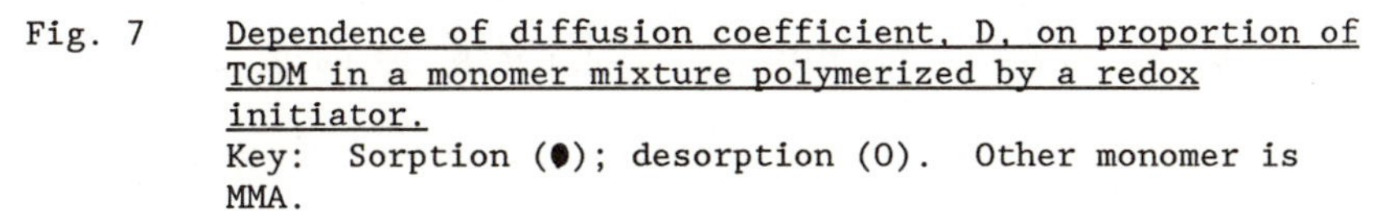

Fig. 7 Dependence of diffusion coefficient, D, on proportion of TGDM in a monomer mixture polymerized by a redox initiator.
Key: Sorption (●); desorption (O). Other monomer is MMA.

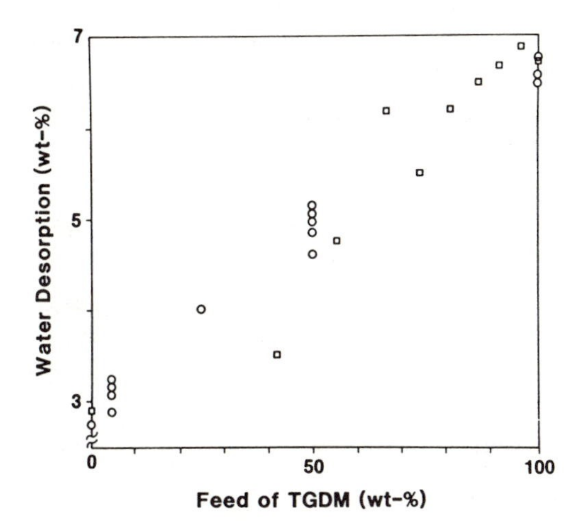

Fig. 8 Dependence of uptake of water on proportion of TGDM in a monomer mixture with MMA.
Key: present work (O); results calculated from ref. 29 (◘).

monitoring changes in density (53) and judged to be inadequate (54). Thus the influence of crosslinking in increasing water uptake remains unexplained.

Concluding Remarks

It appears that networks formed from dimethacrylates are not uniformly crosslinked, as was often assumed in pioneering studies (9,10). Instead they have some resemblance to the "porridge" microstructure first attributed by Houwink to Bakelite (55) and subsequently adopted to account for microstructural observations on other networks prepared by stepwise polymerization reactions (56). As far as is known, microgel particles have not been observed in networks formed by chain polymerization reactions. However, it seems necessary to invoke their formation in order to account for turbidimetric observations (57), the onset of gelation, and gel partition (39). In the present work a case has been made for invoking something like a "porridge" microstructure in order to account for some mechanical properties.

Acknowledgments

This work was supported by NIH grants DE-02668, DE 06201 and RR 0533.

Literature Cited

1. Phillips, R.W., "Skinner's science of dental materials", 9th ed., Saunders, Philadelphia, 1982.
2. Craig, R.G. ed. "Restorative dental materials", 6th ed., C.V.Mosby Company, St. Louis, 1980.
3. Greener, E.H., Harcourt, J.K., Lautenschlager, E.P., "Materials Science in Dentistry", Williams and Wilkins Company, Baltimore, 1972.
4. Taylor, D.F., Leinfelder, K.F., eds. "Posterior Composites", Intern. Symp., Chapel Hill, NC 1982.
5. Braem, M. "An In-Vitro Investigations into the Physical Durability of Dental Composites", Doctoral thesis, Leuven (Belgium), 1985.
6. American Dental Society Council on Dental Materials, "Posterior Composite Resins", JADA 1986, 112 707.
7. Roulet, J.F. "Degradation of Dental Polymers", Karger, Basel, 1987.
8. Ruyter, I.E., Oysaed, H. J. Biomed. Mater. Res. 1987, 21, 11.
9. Loshaek, S., Fox, T.G. J. Amer. Chem. Soc. 1953, 75, 3544.
10. Hwa, J.C.H. J. Polym. Sci., 1962, 58, 715.
11. Horie, K., Otawaga, A., Muraoka, M., Mita, I., J. Polym, Sci. (Chem. Ed), 1975, 13, 445.
12. Ruyter, I.E., Svendsen, S.A. Acta Odont. Scand. 1978, 36, 75.
13. Asmussen, E. Acta Odont. Scand. 1975, 33, 337.
14. Ferracane, J.L., Greener, E.H. J. Dental Res. 1984, 63, 1093.
15. Going, R.E. JADA 1972, 84, 1349.
16. Bausch, J.R., deLange, K., Davidson, C.L., Peters, A., deGee, A.J. J. Prosthetic Dent. 1982, 48, 59.
17. Asmussen, E. Acta Odont. Scand. 1975, 33, 129.

18. Ruyter, I.E., Sjovik, I.J. Acta Odont. Scand. 1981, 39, 133.
19. Bowen, R.L. Dental filling material comprising vinyl silane treated fused silica and a binder consisting of the reaction product of bisphenol A and glycidyl acrylate. USP 3066112 (1962) et seq.
20. Thompson, V.P., Williams, E.F, Bailey, W.J. J. Dental Res. 1979, 58, 1522.
21. Bailey, W.J., Amone, M.J., Polymer Reprints, 1987, 28(1), 45.
22. Fan, P.L. JADA 1984, 108, 240.
23. Brannstrom, M., Vojinovic, O. J. Dent. Child. 1976, 43, 83.
24. Eliades, G.C., Caputo, A.A., Vougiouklakis, G.J. Dental Mater. 1985, 1(5), 170.
25. Munksgaard, E.C., Asmussen, E. J. Dental Res. 1984, 63, 1087.
26. Nakabayashi, N. CRC Critical Reviews in Biocompatability, 1981, 1, 23.
27. Bowen, R.L., Cobb, E.N., Rapson, J.E., J. Dental Res. 1982, 61, 1070.
28. Lundeen, T.F., Turner, D.T. J. Biomed. Mater. Res. 1983, 17, 679.
29. Atsuta, M., Hirasawa, T., Masuhara, E. J. Japan Soc. Dent. Apparatus and Materials (in Japanese), 1969, 10, 52.
30. Suzuki, S., Nakabayashi, N., Masuhara, E. J. Biomed. Mater. Res. 1982, 16, 275.
31. Atsuta, M., Turner, D.T. Polym. Eng. and Sci., 1982, 22, 438 and 1199.
32. Atsuta, M., Turner, D.T. J. Mater. Sci. 1983, 18, 675.
33. Abstracts of papers: 65th General Session of IADR. J. Dent. Res. 1987, 66, Special Issue.
34. Greener, E.H., Bakir, N. Abst. No. 450 (AADR), J. Dent. Res. 1986, 65, 219.
35. Wilson, T.W., Turner, D.T. Characterization of poly(dimethacrylates) and their composites by dynamic mechanical analysis, J. Dental Res., 1987, in press.
36. Antonucci, J.M., Toth, E.E. J. Dental Res. 1983, 62, 121.
37. Thompson, D., Song, J.H., Wilkes, G.L. Polymeric Materials Science and Eng., Preprints (Amer. Chem. Soc.), 1987, 56, 754.
38. Wilson, W.T. "Effect of Radiation on the Dynamic Mechanical Properties of Epoxy Resins and Graphite Fiber/Epoxy Composites", 1986, Ph.D. Thesis, North Carolina State University, Raleigh, NC.
39. Kloosterboer, J.G. and Lijten, G.F.C.M. in "Biological and Synthetic Networks", Ed. O. Kramer, Elsevier Appl. Science, London, 1987.
40. St. Germain, H., Swartz, M.L., Phillips, R.W., Moore, B.K., Roberts, T.A. J. Dental Res., 1985, 64, 155.
41. Atsuta, M, Turner, D.T. Polymer Composites 1982, 3, 83.
42. Atsuta, M., Nagata, K., Turner, D.T. J. Biomed. Mater. Res. 1983, 17, 679.
43. Nielsen, L.E. J. Compos. Mater. 1967, 1, 100.
44. Kalnin, M., Turner, D.T. J. Mater. Sci. Lett. 1985, 4, 1479.
45. Kalnin, M., Turner, D.T. Polymer Composites, 1986, 7, 9.
46. Haque, Z.U., Turner, D.T. J. Mater. Sci., in press.
47. Goodier, J.N. J. Appl. Mech. 1933, 1, 39.

48. Turner, D.T. in ACS Symp. on "Characterization of highly crosslinked polymers" (Eds. S.S. Labana and R.A. Dickie), 1984, Symp. No. 243, 185
49. Soderholm, K.J., J. Biomed. Mater. Res. 1984, 18, 271.
50. Kalachandra, S., Turner, D.T. J. Biomed. Mater. Res. 1987, 21, 329.
51. Cowperthwaite, G.F., Foy, J.J., Malloy, M.A. in "Biomedical and dental applications of polymers" (Eds C.G. Gebglein and F.K. Koblitz), Plenum Press, New York, 1981, p. 379.
52. Turner, D.T., Abell, A.K. Polymer 1987, 28, 297.
53. Turner, D.T. Polymer 1982, 23, 197.
54. Haque, Z.U., Turner, D.T. unpublished work.
55. Houwink, R. "Elasticity, plasticity and structure of matter", 2nd ed., Dover, New York, 1958.
56. Morgan, R.J., O'Neal, J.E. J. Mater. Sci., 1977, 12, 1966.
57. Roschupkin, V.P., Ozerkovskii, B.V., Kalmykov, Y.B., Korolev, G.V., Vysokomol soed. 1977, A19, 699.

RECEIVED October 7, 1987

Chapter 30

Monomer Interaction

Effects on Polymer Cure

José A. Ors[1], Ivan M. Nunez[2], and L. A. Falanga[2]

[1]Engineering Research Center, AT&T, Box 900, Princeton, NJ 08540
[2]AT&T Bell Laboratories, Whippany, NJ 07981

The interactions between the components that make up a photopolymer are extremely important in arriving at a working formulation. Here we show that inclusion of pyrrolidone derivatives like NVP or NMP in acrylate systems enhances the ambient cure of a film. From the reactivity parameters of some simple systems we have derived an empirical scheme for the formulation of fully and/or partially reactive systems based on the molar equivalent ratios of the acrylate to pyrrolidone components. The data support the presence of a synergistic effect between NVP and the acrylate components.

Two of the main considerations in the development of totally reactive liquid photopolymer systems are the resin(s) and the reactive diluents (monomers). The resins play a major role in determining the end properties and therefore the applications of the cured polymer. The reactive diluents are used to provide a fully reactive system with the appropriate reactivity, viscosity, coatability before cure and the desired crosslink density, chemical resistance and dielectric character once it is cured. The photoreactive monomers most commonly used are acrylate based derivatives because of the properties they impart, and their high reactivity and wide solubility range.

We focus here on a different type of monomer, N-vinyl pyrrolidone (NVP). This monomer is extensively used in the coating industry to add strength, dye receptivity, hardness, hydrophylicity and improved adhesion to copolymers of acrylate systems. Further note has been made of NVP use because of its low viscosity and its ability to enhance curing. **(1-2)**

0097–6156/88/0367–0439$06.00/0

This ambient cure enhancement effect agrees with an earlier report (3) where addition of either NVP or NMP (N-methylpyrrolidone), the non-reactive N-methyl analog, showed an increase in the photopolymerization rate of acrylates used in laser disc fabrication. The non-reactivity of the NMP solely refers to its lack of olefinic group hence incorporation into the polymer network via a vinyl moiety. The reduced oxygen effect and high cure rate suggest the possible involvement of charge-transfer assisted polymerization in these neat acrylate monomers. (4) Further observations showed that when the cure was effected under nitrogen only a small rate enhancement remained. This suggests that enhancement under ambient conditions is a result of enhanced oxygen consumption by the NVP. (5)

The marked contrast of these properties prompted us to investigate the behavior of these pyrrolidone derivatives and to find the correct blend of monomers that will yield the optimal reactivity, cured properties and morphology under the cure process conditions of a liquid negative acting resist. The contrast between NVP and NMP allows us to determine the type of monomer contribution to the cure of the mixtures and differentiate between synergistic reactivity and other contributions such as reduced oxygen inhibition. The cure process entails the coating of a film ranging from 25 to 100 μm (wet thickness) followed by a brief off-contact exposure to collimated *uv* light and an image development step and a subsequent final cure with a high dose of *uv* light.

EXPERIMENTAL

All the materials in this study are commercially available and were used as received. The compositions of the mixtures are given in Table I. The components can be described as follows: *Resin* is a proprietary blend of acrylated epoxy resins with an number average molecular weight (Mn) of ~4800, based on acrylated diglycidyl bisphenol A, *DGEBAcr* is the diacrylate derivative of diglycidyl bisphenol A, *IBOA* is isobornyl acrylate, *TMPTA* is trimethyloltriacrylate and *DMPA* is 2,2-dimethoxy-2-phenyl acetophenone.

The infrared data were obtained using a Nicolet FT-IR Spectrometer Model 7199 in a single beam mode. Each data point was an average of two sets of 32 scans. Film samples were irradiated at room temperature through NaCl plates using a filtered Xe/Hg arc lamp source giving an effective wavelength range of 334 ± 20 nm and an intensity of 0.55 mW/cm^2. All values for the reaction efficiencies are reported relative to the rate constant for neat IBOA, which serves as a standard. The rate comparison is done by monitoring the following wavelengths: a) an *overlap band* at 1643-1566 cm^{-1} corresponding to the C=C stretch band of the olefinic moiety α to the carbonyl with contributions from the vinyl group of the NVP, b) an *NVP-band* at 840 cm^{-1}, and c) an *acrylate-band* at 810 cm^{-1}.

Table I. Mixture Compositions

Mixture	Components (Weight Percent)							Ξ†
	Resin	DGEBAcr	IBOA	TMPTA	NVP	NMP*	DMPA	
I-1			78.0		20.0		2.0	0.32
I-2			68.0		30.0		2.0	0.45
I-3			59.0		39.0		2.0	0.55
I-4			49.0		49.0		2.0	0.66
I-5			39.0		59.0		2.0	0.74
I-6			30.0		68.0		2.0	0.81
I-7			20.0		78.0		2.0	0.88
I-8			10.0		88.0		2.0	0.94
T-1				88.0	10.0		2.0	0.91
T-2				49.0	49.0		2.0	0.53
T-3				29.0	69.0		2.0	0.33
T-4				20.0	78.0		2.0	0.22
D-1		88.0			10.0		2.0	0.79
D-2		69.0			29.0		2.0	0.50
D-3		49.0			49.0		2.0	0.30
D-4		30.0			68.0		2.0	0.15
NV-1	57.3		38.0		3.1		1.6	0.93
NV-2	57.0		33.1		8.3		1.6	0.76
NV-3	57.1		23.1		18.2		1.6	0.54
NV-4	57.0		18.1		23.3		1.6	0.44
NV-5	57.0		28.2		13.2		1.6	0.64
NV-6‡	48.0		32.0		16.0		1.6	0.48
NM-1	59.4		34.1			4.9	1.6	0.82
NM-2	59.1		29.4			9.9	1.6	0.64
NM-3	59.4		19.4			19.6	1.6	0.40
NM-4	59.1		9.5			29.8	1.6	0.22
NM-5	59.0					39.4	1.6	0.09

* *A functionality of one is assumed for the non-reactive NMP.*
The molar equivalent (X) is calculated based on the degree of functionality divided by Σ (MW)
† *The parameter* $\Xi = \frac{X_{Acr}}{X_{Acr}+X_N}$ *is used to simplify the plotting of the data.*
of the acrylate components. Hence, $\Xi = 0.5$ *indicates an equimolar ratio of components.*
‡ *NV-6: the remaining 2.4% consists of additives like pigment, flow modifiers, etc.*

The sol fraction data were obtained from the methylene chloride extraction of films immediately after exposure at ambient conditions with an intensity of 1 mW/cm^2. (6) The degree of oxygen inhibition was obtained by measuring the thickness of uncured material on the surface (δ_{O_2}) of a film exposed in the presence of air with a Hg arc lamp with an intensity output (300-400 nm range) of $\approx 9\ mW/cm^2$. Light intensity variations were done using optical density filters. Following the initial exposure the uncured layer at the film surface was wiped with a cloth dampened with 1,1,1-trichloroethane (TCE). The still soft film was then fully cured with a dose of $\approx 2.8\ Joules/cm^2$. The thickness difference between the wiped and unwiped portions of the film was then measured with a Dektak surface profilometer.

RESULTS and DISCUSSION

Our preliminary formulation work using NVP agreed with reported observations of cure enhancement using this monomer under ambient conditions. To investigate the component interactions we decided to examine the behavior of the neat monomers followed by the interactions of monomer-NVP, resin-NVP, and finally more complex mixtures containing either NVP or NMP.

Monomers. Figure 1 shows the contrast in FTIR curing profile for the neat materials. The component reactivity were based on the first order rate constants of the initial slopes of the disappearance of the respective olefinic moieties. The results TMPTA (3.4) > IBOA (1.0) > DGEBAcr (0.7) > NVP (0.14) show that the neat monomers follow the expected cure profile vinyl < acrylate under similar conditions (e.g. photoinitiator concentration, light intensity, etc.). Even though a decrease in the extent of conversion with longer irradiation times is observed in the higher viscosity DGEBAcr resin, its initial rate is faster than the NVP. (7) The change in slope follows the two-regime cure profile observed in more complex systems. (8) The slow cure is in sharp contrast with the observed rate enhancement when this pyrrolidone derivative is mixed with acrylates. Monitoring the individual reactivity by using the distinct *ir* frequencies for the various components was used to quantify the controlling parameters and lead to improved formulation schemes.

Acrylate Monomer/NVP. To observe the interaction between monomers, we compared the behavior of two monomer mixtures, I and T. These mixtures contrast the interaction of NVP and a monofunctional acrylate (IBOA), which should yield to linear polymerization, with a trifunctional monomer (TMPTA) which results in a crosslinked network.

The FTIR cure profiles of some I mixtures using the overlap and the acrylate bands (Figure 2) show a fast initial rate followed by a slow down with extended irradiation times. The change in cure regime takes place at

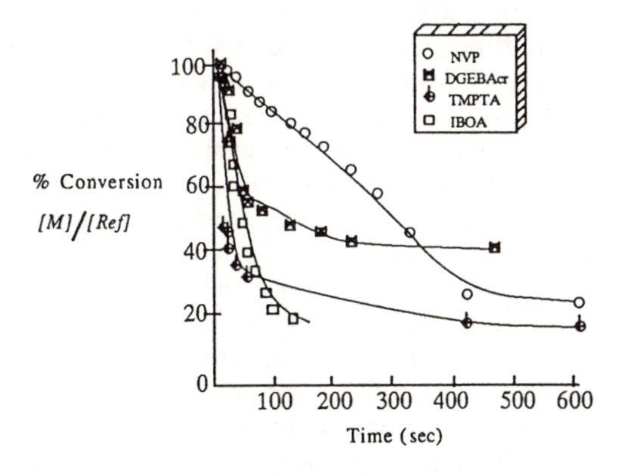

Figure 1
Comparison of Monomer Reactivity. Cure Study - FTIR Data

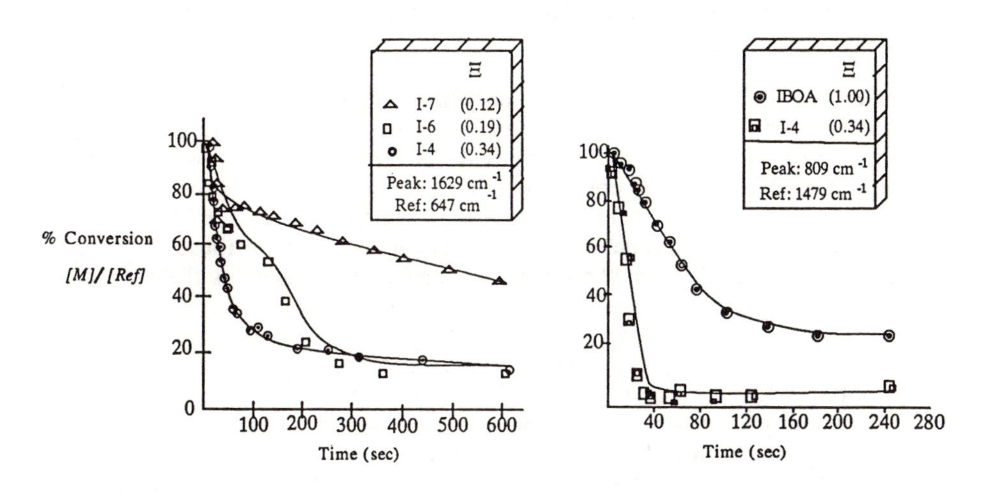

Figure 2
Cure Study I Mixtures - FTIR Profile. Overlap Band (left), Acrylate Band (right)

different conversion levels (I-4 > I-7 > I-8) which appear to vary inversely with the NVP concentration (I-4 < I-7 < I-8). The subsequent slope approximates the rate of the neat NVP. Monitoring the acrylate-band shows an enhancement in the initial IBOA rate that appears to be proportional to the NVP concentration. The slope change is also observed with extended irradiation times, but in the NVP rich mixtures a high acrylate conversion level (≥ 80%) before any slow-down is noticeable.

To interpret these results the initial rate constants (relative to IBOA), for both the overlap band and the acrylate moiety were plotted against a molar equivalent ratio parameter, Ξ, in the form of $\Xi = X_{Acr}/(X_{Acr} + X_N)$. Figures 3 and 4 compare the changes in reactivity with compositional variation of the overlap rate with the acrylate and NVP components respectively. The profiles show that the reactivity of each functional group increases as $\Xi \rightarrow 0.5$ ($X_{Acr}/X_N \rightarrow 1$), followed by a decrease in the composite rate down to the value of the neat NVP as the ratio decreases ($\Xi \rightarrow 0$). In contrast, the IBOA reactivity appears to reach a plateau at the higher NVP concentrations indicating a selective cure pattern arising from the reactivity difference of each component, leading to possible heterogeneity in the final film. **(9)** The heterogeneous cure can result in a mostly poly-NVP film that can either contain grafted segments of IBOA-NVP copolymer and/or poly-IBOA in its matrix.

A similar dual rate behavior is found for the trifunctional TMPTA in the T mixtures, again monitoring the overlap band shows an optimum rate as $\Xi \rightarrow 0.5$, Figure 5. The TMPTA emphasizes the equivalency dependence on the acrylate moiety and supports an earlier report with this monomer combination. **(10)**

Resin/NVP Mixtures. Since acrylates show similar cure profiles regardless of functionality, the next step was to investigate the curing behavior of the DGEBAcr resin with NVP. Here, series D, the information regarding the NVP component is difficult to obtain since the resin exhibits an absorption band at 834 cm^{-1} which interferes with the 840 cm^{-1} NVP band. Comparison of the curing profiles of these mixtures with the neat DGEBAcr show the expected enhancement in the reaction rate along with an increase in the extent of conversion at the higher NVP concentrations (Figure 6). The combination of lower viscosity and enhanced reactivity significantly improves the through cure of the film and supports earlier reports where the weight percent NVP ranged from 15 to 25% of the resin system still below the molar equivalency.

The sol fraction data in Figure 7 show a decrease in extractables as $\Xi \rightarrow 0.5$ followed by an increase in residual unreacted material at higher NVP concentration under the same exposure dose. Gas chromatographic analysis showed that the NVP becomes the main component of the soluble fraction, in the solvent extraction, particularly at the higher initial concentrations. This supports the earlier observation on heterogeneity of cure,

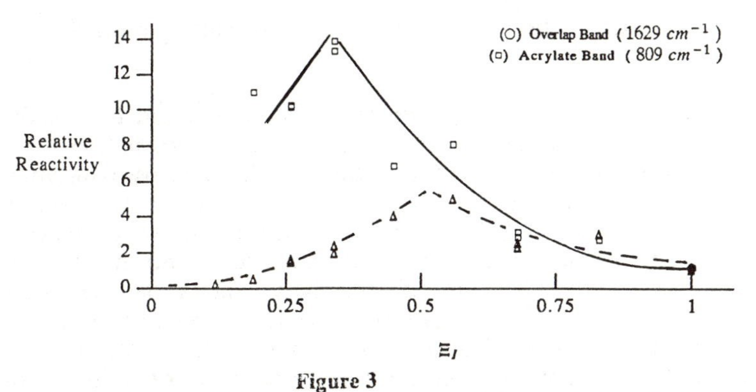

Figure 3
I Mixtures - Relative Reactivity versus Molar Equivalent Ratio.

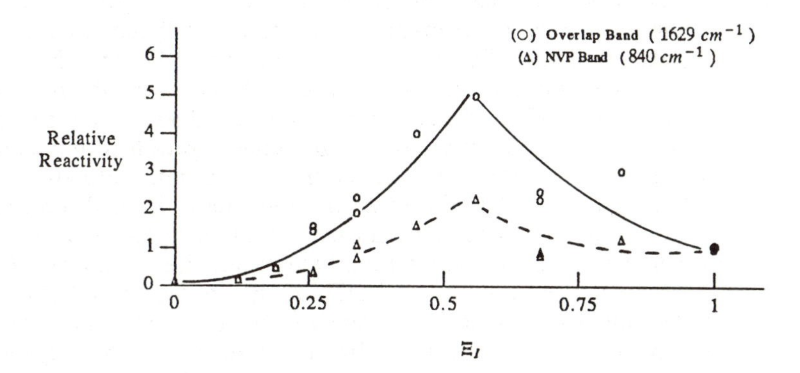

Figure 4
I Mixtures - Relative Reactivity versus Molar Equivalent Ratio.

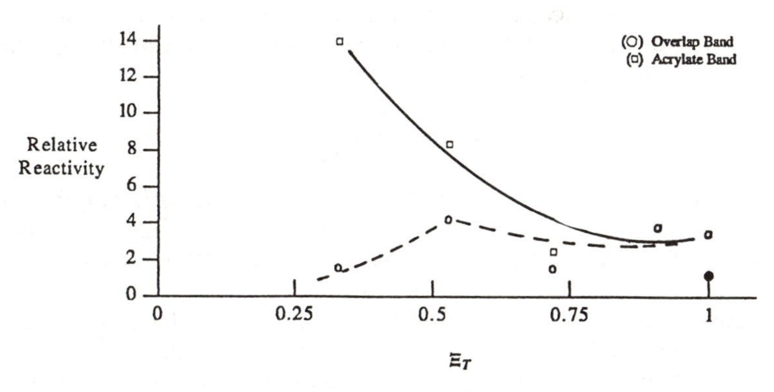

Figure 5
T Mixtures - Relative Reactivity versus Molar Equivalent Ratio.

where the acrylate portion cures preferentially leaving unreacted NVP in the film.

Complex Mixtures. In the ternary mixtures NM and NV (Table I) the contrast between the reactive versus non-reactive pyrrolidones should help clarify some of the specific interactions of these derivatives with the acrylates. Since the oxygen concentration is limited to the dissolved oxygen **(11-12)** , because of FTIR sample configuration, we presume that comparison of the rates yields the enhancement difference between NVP and NMP. Figures 8 and 9 show the reactivity contrast between the two sets of mixtures. While no significant change in relative reactivity is observed in the NM series, the NV mixtures show the previously noted enhancement trend. The slight increase in the acrylate reactivity of NM-5 could be attributed in part to changes in the polarity of the medium, causing a possible selectivity in the affinity of the acrylate groups and even an agglomeration of acrylate groups. Spectroscopic data *(uv-vis)* gave no evidence for ground state complex formation between the NVP and IBOA.

Solvent extraction data (Figure 10) from a series of NV films, irradiated under ambient conditions, show the expected reduction in the sol fraction with increasing NVP concentration even at values below molar equivalency ($\Xi = 0.44$). The effect of oxygen concentration is apparent when contrasting these results with the optimum reaction ratio of 0.5 obtained from the FTIR data of the overlap band, under limited oxygen concentration conditions. In contrast, the extraction data for NM films show a steady increase in the sol fraction. However, the values do not correlate directly with the amount of NMP in the initial formulations. This observation along with the lack of reactivity enhancement, under controlled environment, imply involvement in the reduction of the oxygen effect that lead to a higher extent of acrylate reaction. This effect should be more evident at the surface of the film, where the oxygen concentration is higher, and therefore lead to a reduction in the thickness of the inhibition layer (δ_{O_2}). The inhibited surface appears as an uncured (wet) layer on the surface of the partially cured film whose thickness is dependent on the incident light intensity (I_o), the exposure time (t_{uv}), and the photoinitiator concentration ($[PI]$) according to equation (1)

$$\delta_{O_2} = \frac{A + B/t_{uv}}{I_o\,[PI]} \qquad (1)$$

where A and B are empirically obtained constants, specific for the light source, photoinitiator and monomer/resin systems used. Figures 11 and 12 exemplify the δ_{O_2} dependence on the light intensity and the exposure time parameters for an NVP containing mixture.

Figure 13 shows the reduction in the thickness of the oxygen inhibition layer at two different irradiation doses (45 and 540 mJ/cm^2) for both series. As expected the effect is more pronounced in the NV series and

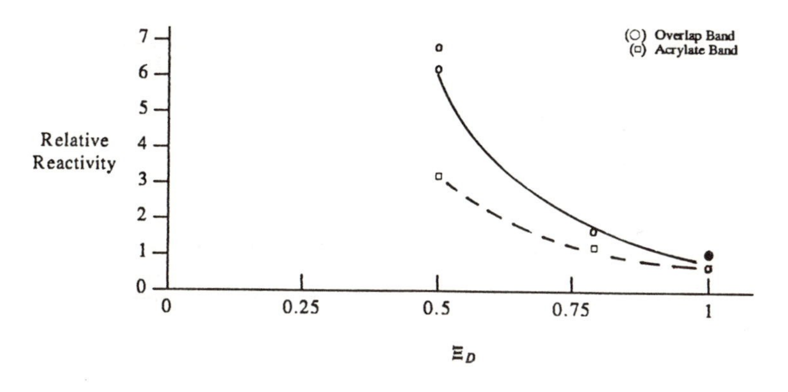

Figure 6
D Mixtures - Relative Reactivity versus Molar Equivalent Ratio.

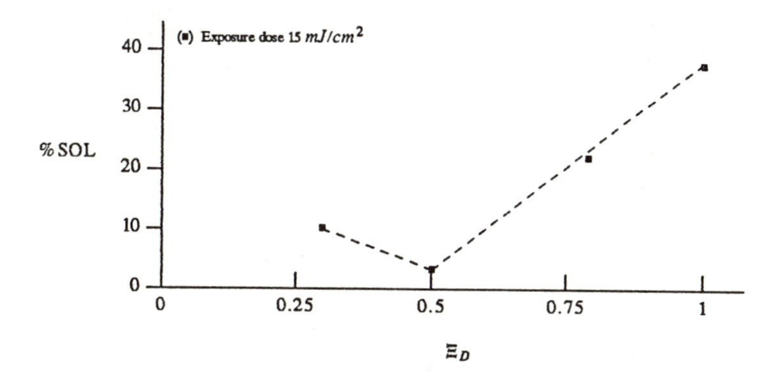

Figure 7
D Mixtures - Sol Fraction versus Molar Equivalent Ratio.

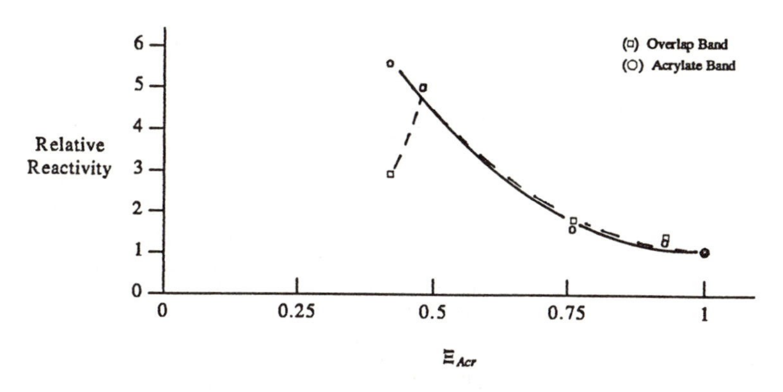

Figure 8
NV Mixtures - Relative Reactivity versus Molar Equivalent Ratio.

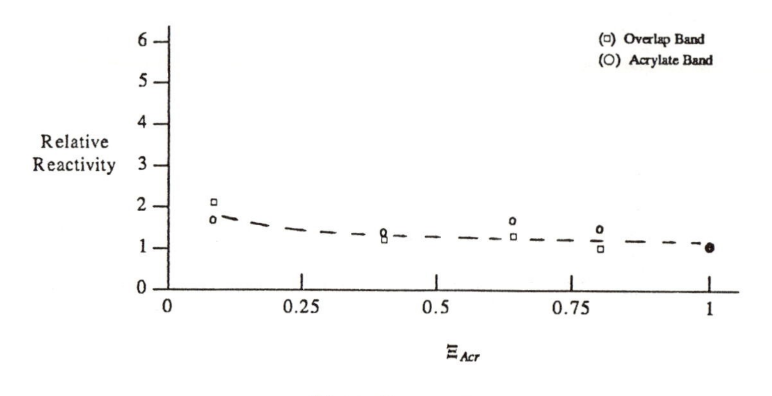

Figure 9
NM Mixtures - Relative Reactivity versus Molar Equivalent Ratio.

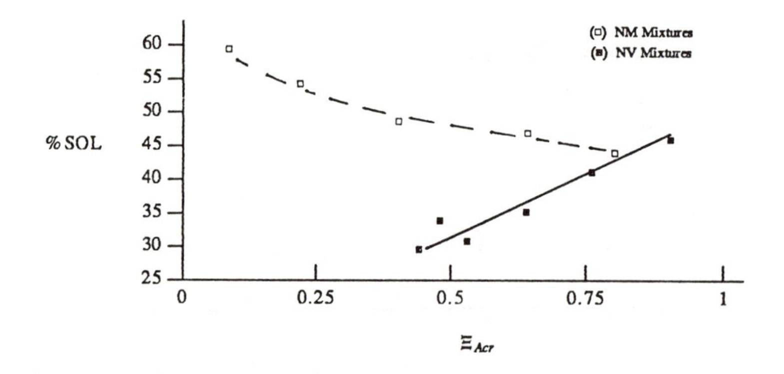

Figure 10
Sol Fraction versus Molar Equivalent Ratio. Exposure Dose: 60 mJ/cm^2.

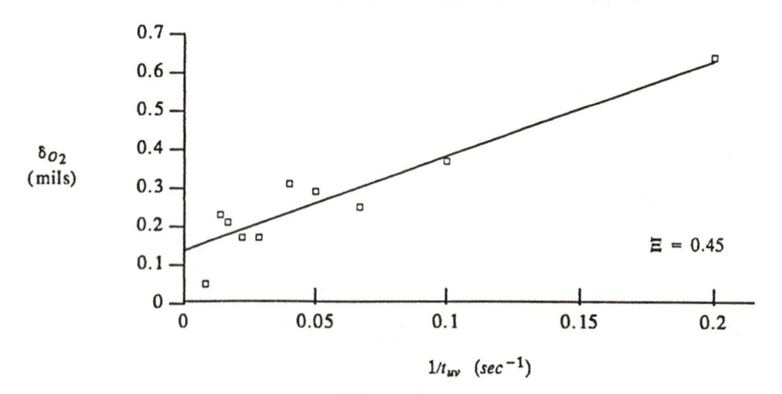

Figure 11
Dependence of Oxygen Inhibited Layer (δ_{O_2}) on Exposure Time (t_{uv}).

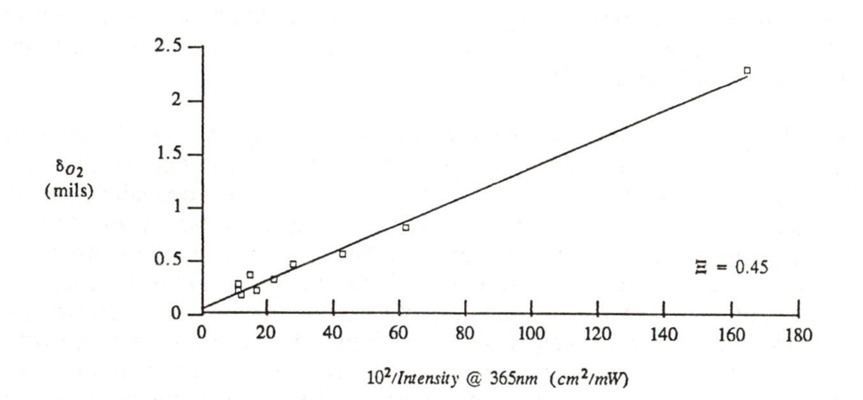

Figure 12
Dependence of Oxygen Inhibited Layer (δ_{O_2}) on Light Intensity (I).

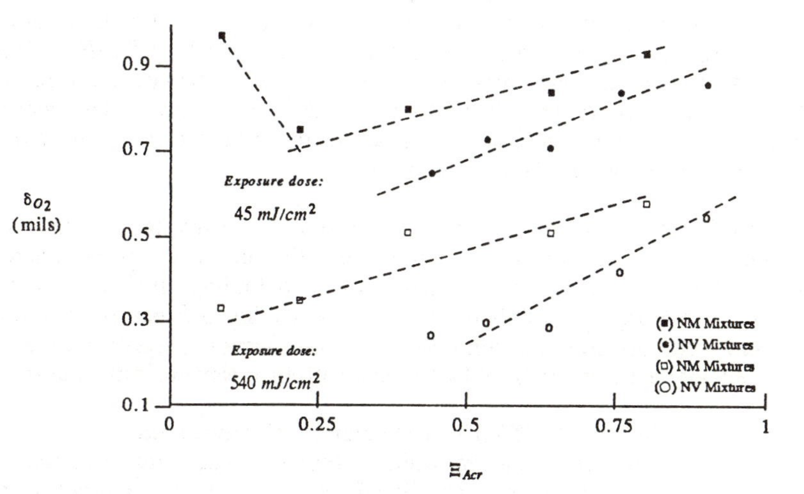

Figure 13
Variations in δ_{O_2} with Component Equivalent Ratio.

can to be attributed to a combination of oxygen scavenging with increased component reactivity. The NM mixtures exhibit a similar trend except for a marked increase in the thickness of the inhibited layer at $\Xi = 0.22$ with the low exposure dose. The latter may be attributed to a dilution effect that when coupled with the selective increase in acrylate reactivity indicates heterogeneous cure.

Reactivity Enhancement. The preceding data suggest that the cure of these acrylates is enhanced by the presence of the pyrrolidone derivatives. Two major pathways appear to be operative, which are dependent on the concentration of oxygen and the nature of the pyrrolidone derivative. First, in the absence or at low concentrations of oxygen, the contrast in initial rate enhancement of the NV versus the NM mixtures could be mostly attributed to the interaction of the acrylate and NVP, since no significant variation was noted with NMP (Figure 8). The rate increase with added NVP can, in turn, be related to the reactivity of the monomer and general solvent effects such as the polarity of the medium, the reduction in viscosity, improved miscibility, etc. Secondly, in the presence of higher concentrations of oxygen (eg. Ambient conditions) the rate enhancement can be a result of the reduction in the oxygen inhibition observed with increasing NVP or NMP concentration. The effect could involve a combination of solvent effects and/or a tertiary amine-type behavior by the lactam in scavenging the oxygen in the film as described by Roffey. **(13)** It appears that in the NV mixtures both of these pathways are operative. Furthermore, hydrogen abstraction needed to generate the relatively unreactive peroxide can lead to formation of active radicals that could be involved in the polymerization sequence.

Lactam-Benzoin Initiator Interaction. We have shown that whenever one of these pyrrolidone derivatives is present, the ambient cure is enhanced. A variety of reasons can be responsible, including the ones proposed above, however a possibility not discussed is the interaction between an excited state of the photoinitiator and the lactam. Spectroscopic data showed no changes in the DMPA absorption spectrum with addition of NVP.

This initiator (DMPA) is particularly effective because of its dual fragmentation nature, after the initial Norrish-I type photofragmentation **(14)** the generated dimethoxybenzyl radicals can further fragment to yield a methyl radical and methylbenzoate. However, the initial radicals can be assume come from the triplet (T_1) state of the excited initiator. **(15)** This triplet could form an exciplex or ion pair with the pyrrolidone leading to radical formation through a hydrogen abstraction of the latter. **(16)** This pathway can compete with oxygen inhibition thus showing an increase in acrylate reactivity when these lactams are added. However, since we have shown that the DMPA initiated polymerization of NVP is slower than of the acrylate components used in this study any complex formation between the DMPA and the NVP can not solely account for the enhanced rate.

SUMMARY

As discussed at the onset, this study was focused on determining the reactivity interaction of two lactam derivatives with acrylates in a variety of systems ranging from simple monomer to more complex mixtures. The definition of such interplay allows us to derive a simplistic scheme for the formulation of fully reactive blends containing NVP as a reactive monomer. The data show that:

1. An enhancement effect exists between NVP and the acrylate components in the cure of these systems. This behavior is interpreted as synergistic effect resulting from a variety of complex interactions between the components and not simply as a solvent effect.

2. Homogeneity of the cured films can be maintained in these systems as long as $\Xi \geq 0.5$ ($X_{Acr}/X_N \geq 1$). Of course, the optimum ratio will depend on specific components and their interactions.

3. At high concentrations of NVP, $\Xi << 0.5$ ($X_{Acr}/X_N << 1$), heterogeneity in the cure of the film will be observed.

4. Both NVP and NMP appear to reduce the effect of oxygen in ambient cures, with NVP showing the greater effect because of its dual role.

5. The added possibility that a complex formation between the lactam and an excited state of the photoinitiator exists and can lead to a more efficient reaction pathway, should be investigated.

LITERATURE CITED

1. "V-PYROL Vinyl Pyrrolidone. The versatile monomer", Technical Bulletin 9653-011, GAF Corporation, New York.

2. "V-PYROL/RC Vinyl Pyrrolidone Radiation-Curable Grade for Coating Systems", Technical Bulletin 2302-081R, GAF Corporation, New York, 1970.

3. Kloosterboer, J.G., Lippits, G.J.M., and Meinders, H.C., *Phillips Tech. Rev.,1982,* **40,298.**

4. Dowbenko, R., Friedlander, C., Gruber, G., Prucnal, P., and Wismer, M., *Progress in Organic Coatings*, 1993, **11**, 71.

5. Kloosterboer, J.G. and Lippits, G.J.M., 1984, *J. Rad. Cure*, 10.

6. Small, R.D., Ors, J.A. and Royce, B.S.H., 1984, *Polymers in Electronics, ACS Symposium Series No.* **242**, 325.

7. The viscosity values for the materials used, reported in cps at 25°C: TMPTA = 125 ; IBOA = 7.5 ; DGEBAcr ~ 10^6 ; NVP = 2 .

8. Scarlata, S.F. and Ors, J.A., *Polym. Comm.*, 1985, **27**, 41.

9. Al-Issa, M.A., Davis, T.P., Huglin, M.B. and Yip, D.C.F., *POLYMER*, 1985, **26**, 1869.

10. Tu, R.S., 1983, *J. of Rad. Cure*, 17.

11. Royce, B.S.H., Teng, Y.C. and Ors, J.A., *IEEE Meeting*, Chicago 1981.

12. Teng, Y.C. and Ors, J.A., *8th Annual FACSS*, Philadelphia 1981.

13. Roffey, C. G., *Photopolymerization of Surface Coatings*, John Wiley & Sons, New York, 1982, Chapter 3.

14. Hageman, H.J., *Progress in Organic Coatings*, 1985, **13**, 123.

15. Ledwidth, A., *Photoinitiation by Aromatic Carbonyl Compounds*,*J.O.C.C.A.*, 1976, **59**, 157.

16. Yates, S.F. and Schuster, G.B., *J. Org. Chem.*, 1984, **49**, 3349.

RECEIVED December 4, 1987

Chapter 31

Amide-Blocked Aldehyde-Functional Monomers

Synthesis and Chemistry

R. K. Pinschmidt, Jr., W. F. Burgoyne, D. D. Dixon, and J. E. Goldstein

Air Products and Chemicals, P.O. Box 538, Allentown, PA 18105

A new class of functional comonomers exemplified by acrylamidobutyraldehyde dialkyl acetals 1 and their interconvertible cyclic hemiamidal derivatives 2 were prepared and their chemistry was investigated for use in polymers requiring post-crosslinking capability. These monomers do not possess volatile or extractable aldehyde components and exhibit additional crosslinking modes not found with conventional amide/formaldehyde condensates, eg, loss of ROH to form enamides 9 or 10 and facile thermodynamically favored reaction with diols to form cyclic acetals.

Polymer post-crosslinking and substrate reaction play an important role in binders, adhesives and coatings in upgrading water and solvent resistance, strength, adhesion and block resistance. Although a number of crosslinking and substrate reactive chemistries are available for this purpose (1-5), aminoplast systems have proved the most fully and generally compatible with long shelf life water based delivery systems. These acid cured crosslinkers are formally hemiamidals derived from the condensation of an electron deficient amine and an aldehyde, usually formaldehyde. The reactions of one such material, N-methylolacrylamide (NMA) are shown in Scheme 1. NMA is copolymerized into free radical addition polymers and undergoes both efficient self-crosslinking and reaction with other active hydrogen containing groups in the polymer or the substrate to which it is applied. Unfortunately, even cured hemiamidals based on formaldehyde show at least a low level of reversibility to starting materials.

Replacement of formaldehyde in aminoplasts by substitution with higher aldehydes (or ketones) is even more problematic; the equilibrium of the condensation shifts strongly towards starting materials, except in cases where the formaldehyde replacement contains electron withdrawing substituents (e.g. with glyoxal

0097-6156/88/0367-0453$06.00/0

(8, 9) or glyoxylic acid derivatives (2)) to compensate for the increased steric bulk. In any case the reaction is still reversible and only the identity of the liberated carbonyl component is changed.

In an attempt to preserve the desirable features of aminoplast chemistry, while eliminating the release of aldehydic components, a related, but surprisingly different system, the amide/blocked aldehydes 1 and related cyclic hemiamidals 2 (R=Me, Et), was investigated. These linked amide/aldehydes avoid the loss of aldehyde by covalently attaching it to the amide. In cases where five or six membered rings can form, the equilibrium also shifts strongly to the cyclized hemiamidal side.

This paper reviews the synthesis and chemistry of this system. The companion paper (6) discusses some of the properties of derived copolymers.

Experimental

Product identification was by GC/MS, NMR, and IR. Fundamental crosslinking chemistry was explored using swell measurements on simple solution copolymers and swell and tensile measurements with vinyl acetate (VAc), vinyl acetate/butyl acrylate (VAc/BA) or vinyl acetate/ethylene (VAE) emulsion copolymers. Polymer synthesis is described in a subsequent paper (6). Homopolymer Tg was measured by DSC on a sample polymerized in isopropanol. Mechanistic studies were done in solution, usually at room temperature, with 1, 2 and the acetyl analogs 1', 2' (R'=CH_3).

Reproducible, if non-linear, analysis for most components employed capillary gas chromatography using, eg, OV-101 or DB-5 coated quartz columns (packed columns failed in our experiments).

Synthesis of Acrylamidobutyraldehyde Dimethyl Acetal (1, R=Me). To a rapidly stirred two phase mixture containing 100 parts of 4 (R=Me) in 200 parts of CH_2Cl_2, 72 parts of 50% NaOH and 94 parts of water were added 69 parts of flash distilled acryloyl chloride with efficient cooling or under reflux at reduced pressure to allow the reaction to be maintained at 30-35°C. The reaction was stirred for 30 min after acid chloride addition and the phases were separated. The organic phase was neutralized to pH 8 with HOAc (checked by pH measurements on small aliquots shaken with water). The organic phase was concentrated under reduced pressure in the presence of a small quantity of sodium carbonate and 200 ppm of phenothiazine to give a 95+% yield of about 95% pure 1 as a colorless to slightly yellow oil. The product was customarily used directly but may be vacuum distilled (bp 135°C, 0.7 torr) if adequately inhibited and buffered (eg, methylene blue plus sodium carbonate): NMR: (CD_2Cl_2): δ1.59 (m, 4, CH_2), 3.29 (s+m, 8, CH_3, NCH_2), 4.33 (~ t, 1, J~5.3 Hz, CH), 5.58 (dd, 1, J=3.1 Hz, J=8.4 Hz, vinyl), 6.11 (~dd, 1, J=16.9 Hz, J=8.4 Hz, vinyl), 6.18 (~ dd, J=3.1 Hz, J=16.9 Hz, vinyl) and 6.3 ppm (1, NH); IR: (film) 3295, 1660, 1625, 1550, 1135, 1060 cm^{-1}; Elemental: Calcd for $C_9H_{17}NO_3$: C, 57.73; H, 9.15; N, 7.47; Found: C, 57.68; H, 9.23, N, 7.34.

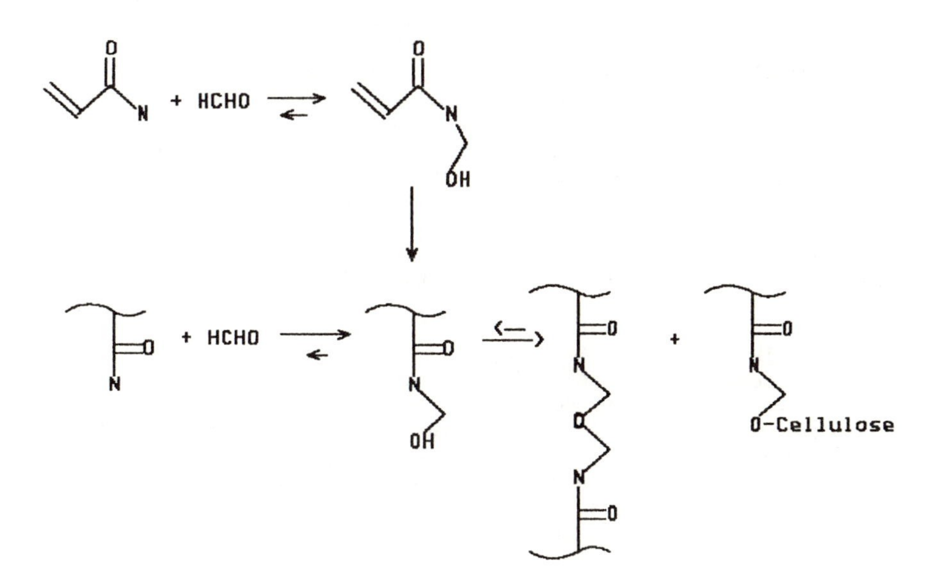

Scheme 1. N-Methylolacrylamide Chemistry

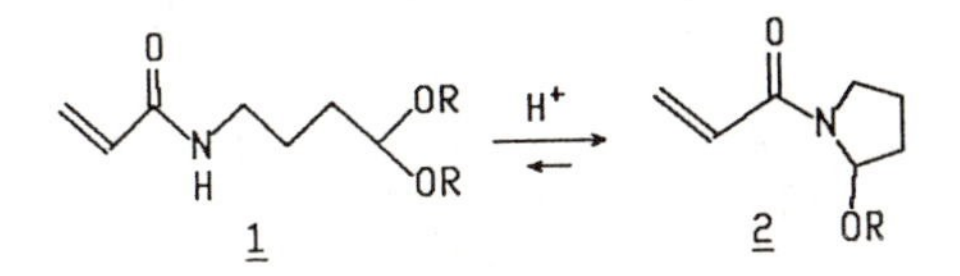

Synthesis of N-Acryloyl-2-methoxypyrroline (2, R=Me). A solution of 1 (R=Me) in a three-fold weight excess of 3:1 CH_2Cl_2 and methanol was treated with 2 g/100 g total solution of macroreticular strong acid ion exchange resin (eg, Rohm and Haas XN-1010) at room temperature with slow agitation. The reaction was followed by capillary gas chromatography (aliquots were neutralized with alcoholic KOH to avoid isomerization on the injection port). Cyclization to 2 was over 90% complete after 4 h and the reaction contained only minor amounts of N-acryloyl-2-pyrroline 9 and dimers. Overreaction increased 9 and dimers. (A 20% solution in dioxane also worked but required slight heating and gave slightly higher dimer formation.) The solution was decanted from the resin, neutralized to eliminate traces of base and concentrated at reduced pressure. This material was typically used as is, but the product could be futher purified by distillation. Kugelrohr distillation (90-96°C, 0.75 torr) of the 2-ethyl derivative (2, R=Et) gave a 92% yield of cyclized product: IR (film): 1645, 1610, 1445 cm^{-1}; ^1H NMR (D_2O): δ 6.45 (8 peak m, overlapping dd's, 1, J=10.4, J=16.8 Hz, vinyl), 6.12 (overlapping dd's, 1, J=16.8, J~1.6 Hz, vinyl), 5.70 (overlapping dd's, 1, J~10, J~1.6 Hz, vinyl) 5.33 (d, ~0.3, J=4.8 Hz, CH) + 5.21 (d, ~0.6, J=4 Hz, CH), 3.45 (m, 3, OCH_2, NCH_2), 3.19 (m, 1, NCH_2), 2.1-1.4 (m, 4, CH_2CH_2), 1.03 ppm (overlapping t's, 3, J=6.5 Hz, CH_3); M/S: 41, 55, 70, 86, 96, 112, 124, 125, 140 (CI (NH_3): M^+=169).

Use of water as the solvent gave N-acryloyl-2-hydroxypyrrolidine 2 (R = H) and, as a transient, the precursor aldehyde. The hydroxy derivative was typically used as the aqueous solution, but could be salted out and extracted into a volatile solvent. It readily dimerizes to 6 on heating.

Observation of 4-Acrylamidobutyraldehyde. Compound 1 (R=Me), 5 g, was stirred at ambient temperatures in 50 mL of H_2O at pH 2.2 (HCl). Aliquots were neutralized to pH 7 and analyzed by GLPC on a short (5 m) capillary column. After 4 h 1 (ret. time 6.2 min) was essentially consumed, producing 2 (R=Me) at 4.6 min and poorly formed overlapping peaks at approximately 4.4 (compound 2 (R=OH)) and 5.1 min. The sample was neutralized to pH 7, extracted with CH_2Cl_2 and concentrated at ambient temperature under vacuum to yield an oil showing a new ir band at 1720 cm-1 and a complex ^1H NMR spectrum in CD_2Cl_2 with new peaks at δ9.73 (t, 0.08 H, J=1.2 Hz, CHO) and 2.49 ppm (td, 0.16 H, J=1.2, J=7.2 Hz, CH_2). Reanalysis of this sample with time showed continued conversion to 2 (R=H).

Synthesis of 4-Acetamidobutyraldehyde Diethyl Acetal, 1' (R'=CH_3, R=Et). This reaction followed the procedure for 1, but using 50 g (0.31 mol) of 4, 24.3 g (0.31 mol) of acetyl chloride, 330 mL of CH_2Cl_2 and 50 mL of 14 N NaOH. Thirty min after the acid chloride addition was completed the reaction was adjusted to pH 7.7 with 30% H_2SO_4 and solid CO_2 was added to give a pH of 6.4. The separated aqueous layer was extracted with CH_2Cl_2 and the combined organic layers were washed with brine, dried with anhydrous $MgSO_4$ and concentrated to give 69 g of pale yellow liquid. This sample plus 1000 ppm of MEHQ and 0.6 g of copper bronze were distilled on a kugelrohr apparatus (120°C, 0.02 torr) to give 45 g of colorless oil (65% yield): IR: (film) 3290, 1650, 1555, 1440 cm^{-1}; ^{1}H NMR ($CDCl_3$): δ 6.40 (br s, 1, NH), 4.64 (t, impurity or CH rotamer), 4.47 (t, ~1, J=5 Hz, CH), 3.4-3.75 (m, 4, OCH_2), 3.23 (q, 2, J=6.5 Hz, NCH_2), 2.62 (m, 4, CH_2CH_2), 1.96 (s, 3, CH_3), 1.20 ppm (t, 6, J=7.2 Hz, CH_3).

Reactions of N-Acetyl-4-aminobutyraldehyde Diethyl Acetal, 1' (R'=Me, R=Et).

In MeOH. N-Acetyl-4-aminobutyraldehyde diethyl acetal, 1' (15 g) was added to 185 mL of 3:1 CH_2Cl_2/MeOH. Rohm and Haas XN-1010 strong acid macroreticular ion exchange resin (4 g) was added and the mixture was stirred slowly at ambient temperature. After 15 min, analysis by GLPC showed additional components at shorter retention times than 1'. At 1.5 h the XN-1010 was filtered off and the solution was concentrated to give 13.8 g of light yellow liquid analyzing as: 70% N-acetyl-2-methoxy-pyrrolidine, 2' (R=Me), retention time 5.66 min; 6.4% N-acetyl-2-ethoxy-pyrrolidine, 2' (R=Et), r.t. 5.99 min; 17.3% N-acetyl-4-aminobutyraldehyde dimethyl acetal, 1' (R=Me), r.t. 7.05 min; 3% N-acetyl-4-aminobutyraldehyde ethyl methyl acetal, 1' (R=Me + Et), r.t. 7.34 min; and 1.6% N-acetyl-2-pyrroline, 9, r.t. 4.74 min. Anal: 2' (R=Me): ^{1}H NMR ($CDCl_3$): δ 5.43 (d, 0.4, J=4.6 Hz, NCHO) + 4.96 (d, 0.6, J=4 Hz, NCHO), 3.48 + 3.39 (s's, 3, CH_3), 3.8-3.3 (m, NCH_2 + OCH_2CH_3 imp?), 3.30 (s, MeOH?), 2.16 + 2.09 (s's, 3, CH_3), 2.2-1.6 ppm (m, 4, CH_2CH_2). This sample contains signals attributable to the ethyl acetal and ethanol by GLPC. GC/MS: m/e 43, 70, 100, 113, 143 (confirmed by NH_3-CI). 2" (R=Et): GC/MS: m/e 43, 70, 85, 86, 113, 128, 142 (M+=157 by CI). 1' (R=Me): GC/MS: m/e 43, 70, 75, 85, 100, 128, 144, 160 (M+=175 by CI). 1' (R=Me+Et): GC/MS: m/e, 43, 61, 70, 85, 89, 100, 114, 144, 158, 174 (M+=189 by CI). 1' (R=Et): GC/MS: m/e 43, 47, 70, 75, 103, 114, 158, 174, 202 (M+=203 by CI). 9: GC/MS: m/e 43, 68, 69, 111 (M+=112 by CI).

In CH_2Cl_2: To 1 g of 1' in 12 mL of CH_2Cl_2 was added 0.27 g of XN-1010. The mixture was stirred at room temperature. Less than 1% 1' remained after 1.5 h, with 2" (R=Et) as the major product (88%) by GLPC. Minor amounts of N-acetyl-2-hydroxypyrrolidine, 2" (R=H) as a broad peak at r.t. ~5.6 min (1.9 area %), and N-acetyl-2-pyrroline, 9, r.t. ~4.74 min (9.6%) were also formed: 2" (R=H): GC/MS: m/e 43, 59, 68, 70, 72, 86, 101, 111, 114, 129 (NH_3 CI: 70, 77, 112, 129, 145).

Back Isomerization of N-Acetyl-2-methoxypyrrolidine, 2'. Six mL of the product of reaction in CH_2Cl_2 above was diluted with 2 mL of MeOH and catalyzed with 0.13 g of XN-1010. After 1 h at room temperature 2' decreased from 90.8 to 77.2% and 1' (acetamidobutyraldehyde dimethyl acetal) increased from less than 1% to 4.9%. The other major product was 9 (14.2%).

Reaction of N-Acetyl-4-aminobutyraldehyde Diethyl Acetal, 1', with 2,4-Pentandiol. One g of 1' and 0.51 g of 2,4-pentandiol (epimeric mix) were heated in 1 g of H_2O with p-toluenesulfonic acid (10 mg at 50°C for 1 h, then an additional 10 mg and heat at 70°C for 2.5 h). An aliquot was neutralized with KOH/EtOH, extracted with H_2O and CH_2Cl_2 (2X each), back extracted with brine and concentrated. GLPC analysis showed two major product peaks (8.34 and 8.71 min). These were separated by preparatory GLPC to give the epimeric cyclic 2,4-pentandiol acetals 11: ^1H NMR ($CDCl_3$): δ 5.83 (br s, 0.9, NH), 4.53 (t, 1, CHO_2), 3.72 (m, 2, OCH), 3.24 (qm, 2, NCH_2), 1.94 (s, 3, CH_3), 1.1-1.7 (m, ~6, CH_2), 1.19 ppm (d, ~6, CH_3); δ 5.80 (br s, 1, NH), 4.85 (~t, 1, CH), 4.26 (~quintet, 1, CH), 3.92 (m, 1, CH), 3.24 (m, 2, CH_2N), 1.94 (s, 3, CH_3), 2.0-1.5 (m's, ~6, CH_2), 1.3 (d, 3, CH_3), 1.2 ppm (d, 3, CH_3). The latter sample was contaminated with material from the first. GC/MS (8 34 min): m/e 43, 69, 86, 111, 115, 130, 146, 172, 214, 215; (8.71 min): m/e 43, 69, 86, 111, 115, 130, 146, 172, 214, 215.

Reaction of N-Acetyl-2-methoxypyrrolidine, 2' (R=Me), with 2,4-Pentandiol. One g of N-acetyl-2-methoxypyrrolidine (ca 80% of 2' plus lesser an amounts of 1' and other 2' ethers), 0.9 g of 2,4-pentandiol and 240 mg of N-methyl-2-pyrrolidone internal standard were heated at 50-55°C in 1 mL of H_2O. A separate sample was heated in 1 mL of EtOAc. Results in both samples were similar: rapid initial formation of a small peak at 8.02 min (cyclic hemiamidal of the diol?), followed by growth of major product peaks at 8.35 and 8.73 min, the epimeric cyclic acetals 11. Small quantities of dehydrated product 9 (r.t. 4.75 min), and, in the aqueous sample, N-acetyl-2-hydroxypyrrolidine, 2" (R=H), (at ~5.6 min) were also produced. One of the diols appeared to react faster and to a greater extent than the other. Measurements at 3-4 h were close to the overnight values: ~87% conversion of 1', 60-70% of 2' (R=Me) and essentially quantitative conversion of 1' (R=Et) and 2' (R=Et).

Reaction of 9 with Pentanediol, Water and Methanol. To a solution of 493 mg of toluene, N-acetylpyrroline, (9, 62 mg), N-methylpyrrolidone (internal standard) and 2,4-pentanediol (153 mg) was added 53 mg of XN-1010 resin. The magnetically stirred sample was heated to 80°C under an N_2 blanket and monitored by GLPC. Over 24 h 9 was 90% consumed, the 2 diols decreased by 60 and 70% and new peaks corresponding to the cyclic acetals 11 formed. Separate experiments without the diol in water or methanol showed no formation of 2' (R=Me, R'=Me) or 2" (R=H, R'=Me)

Amine Incorporation By Aminoplast Polymers. Three vinyl acetate/ethylene (VAE) emulsion copolymers (Tg ~0°C) were prepared under similar conditions at 45-55% solids containing a) no crosslinker, b) 5% NMA, and c) 6% ABDA (1, R=Et) respectively. Samples of each were catalyzed with NH_4OAc, p-toluenesulfonic acid (PTSA) or both (wgt./wgt. polymer solids) and cast as films on polyester. The air dried films were then heated for 3 min at 150°C and analyzed in duplicate for %N by combustion. The cured films were also weighted, soaked in dimethylformamide (DMF) for one h, dried briefly with a tissue to remove surface solvent, and reweighed. The ratio of the DMF swollen weight to the initial weight is the Swell Index. The results are shown on Table II.

Results and Discussion

The most convenient synthetic approach to compound 1, acrylamidobutyraldehyde dialkyl acetal (ABDA), is shown in Scheme 2.

Compounds 3 (when R=Me) and 5 are available in commercial quantities. Compound 4 may also be purchased or prepared in high yield via hydrogenation of 3 over a promoted Raney nickel catalyst (C. G. Coe, unpublished results). The condensation of 4 with distilled acryloyl chloride proceeds in over 95% yield.

The major by-products are small amounts of a dimer and higher oligomers 7 formed from base catalyzed Michael's addition of the monomer during synthesis (Scheme 3).

Numerous other derivatives could be prepared from 4 or its 2-carbon shorter homolog and commercially available acid chlorides, or via direct addition of the amines to appropriate isocyanates. Lower cost synthetic routes leading to the methacrylamide derivative or to 1 from acrylate esters have also been demonstrated (A. F. Nordquist, unpublished results).

Conversion to cyclic hemiamidal 2 and further products (Scheme 4) occurs readily at room temperature in the presence of acids (HCl, strong acid ion exchange resin, etc.).

Monomer Selection. In practice the amide/blocked aldehyde precursor 1 (ABDA) proved more readily accessible than 2. The two forms were completely interconvertible and equally useful as self- and substrate reactive crosslinkers (6). In our addition polymer systems, the acrylamide derivative 1 ($R=CH_3$) provided a good blend of accessibility, physical properties, and ready copolymerizability with most commercially important monomers. Structure/property relationships for other related monomers will be reported elsewhere.

Monomer Properties (Table I). The amide/acetal 1 (R=Me or Et) and its amine precursor are Ames negative. Compound 1 shows a very high LD_{50} and low skin or eye irritation. A low vapor pressure, its liquid form and infinite miscibility with water and common solvents are additional pluses.

Reactivity is typical of an acrylamide. For example, compound 1 shows essentially 1:1 copolymerizability with butyl acrylate. Copolymerizability has also been demonstrated with styrene, other acrylates and methacrylates, vinyl acetate (VAc), VAc/ethylene and vinyl chloride/ethylene. High molecular weight polymers and copolymers remain soluble, indicating any chain transfer to polymer, e.g. through abstraction of the acetal hydrogen, is minor.

Table I. ABDA Monomer Properties

Physicals:	MW=187; bp: 135°C (0.7 torr); mp: 16-19°C; Flash point (closed cup): >230°F; d(25°): 1.0323
Miscibility:	water, VAc, ether, CH_2Cl_2, toluene; limited miscibility: alkanes
Stability:	Indefinite at room temperature (O_2/MEHQ, etc.); no hydrolysis in water at pH >7
Toxicity Data:	Ames negative (non-mutagenic): Et or Me acetals Skin irritation index: 0.00 (rabbit) Eye irritation: slight to mild (rabbit) Oral LD_{50}: 4.8 g/kg (rat)
Homopolymer Tg:	~-68°C

Acid Catalyzed Reactions

The surprisingly rich chemistry of this system is summarized in Schemes 4, 5 and 6.

Exchange Reactions in Hydroxylic Media. Compounds 1' and 2' (Scheme 4) interconvert readily at room temperature under acid catalysis. The equilibrium favors the latter. Only 4.8% of 1' (R'=Me) forms from 2' in excess MeOH. Unblocked aldehyde (Scheme 4) is observable (GC, NMR) under certain conditions as an unstable intermediate in the aqueous hydrolysis of 1' to 2' (R=H). It is not detectable in the IR or NMR spectrum of 2'. Although kinetically accessible, the aldehyde is thermodynamically disfavored. As a result, the degradative chain transfer and rapid air oxidation observed with unblocked aldehyde containing monomers and polymers (10) is avoided.

Implicit in the interconversion of 1' and 2' is the comparably fast exchange of ether groups with other alcohols or water (e.g. Scheme 5, 2'⇄2"). In the reactions of 2', structure 8 is the assumed, but well precedented intermediate.

Crosslinking Reactions in Hydroxylic Media. Crosslinking proceeds via several routes depending on conditions. From intermediate 8, the crosslinking condensates 6 (11) originally anticipated by analogy to NMA chemistry (Scheme 1) are observed in aqueous media under acidic conditions (e.g. in the model acetyl system by GC/MS). But their formation is reversible. They are accompanied, particularly under more vigorous conditions or in the absence of active hydrogen species, by dimer 10 and its double bond isomers. These arise from loss of a proton from 8 to give 9 and condensation of 9 with additional 8 (12). Their formation is essentially irreversible. (The reaction in fact has yet another layer of complexity: GC/MS indicates that model compounds 10 readily redox disproportionate to derivatives of 10 with 2H more or less, but this is unimportant to the crosslinking reactions at hand.)

Under appropriate conditions (acid catalyzed isomerization in dilute CH_2Cl_2, NH_4Cl catalysis at room temperature in water), 9 is the predominate product and can be isolated as an

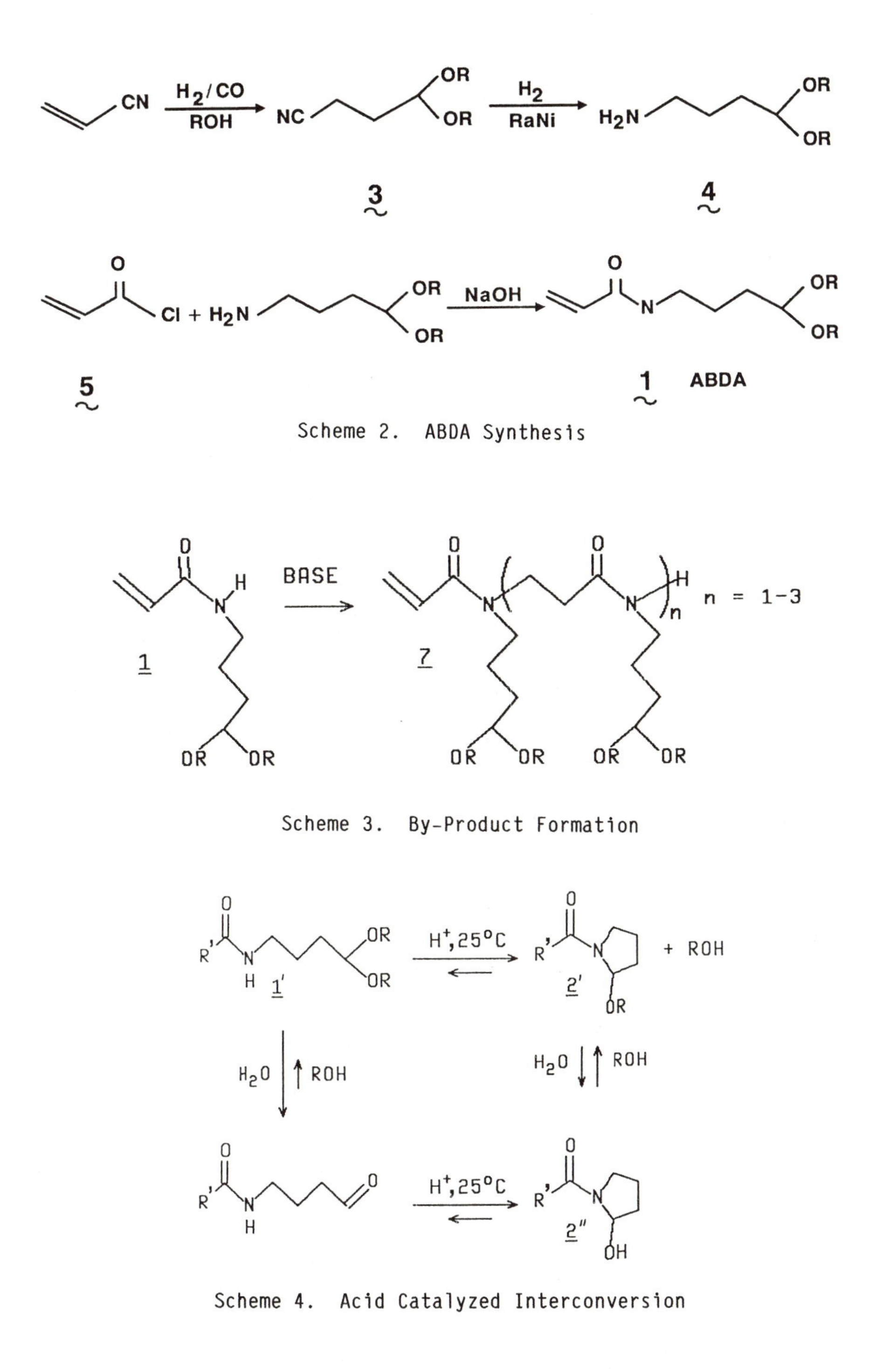

Scheme 2. ABDA Synthesis

Scheme 3. By-Product Formation

Scheme 4. Acid Catalyzed Interconversion

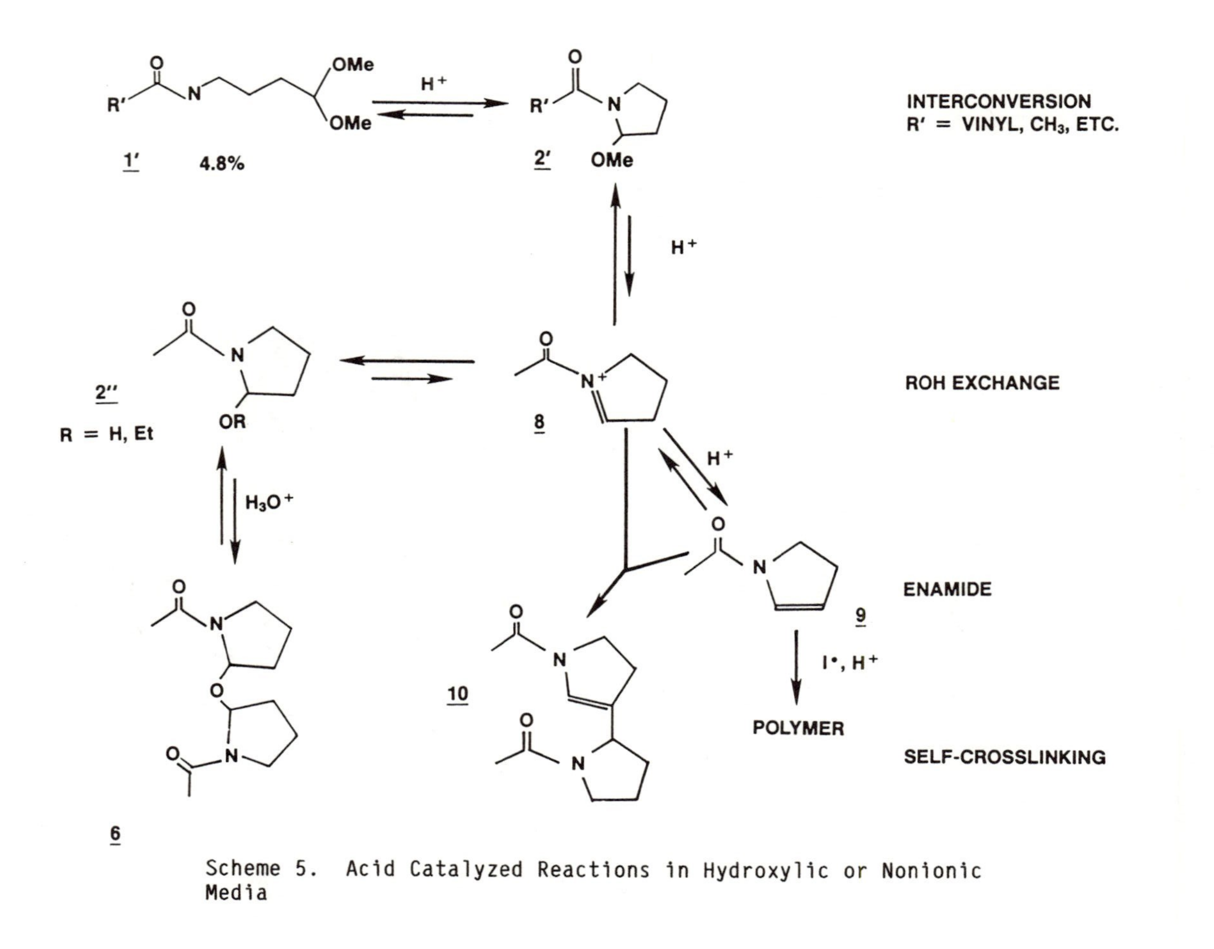

Scheme 5. Acid Catalyzed Reactions in Hydroxylic or Nonionic Media

unstable oil. It reacts further under free radical catalysis or heating in acid. This reactivity may offer options for introducing polymerizable double bond functionality into a finished addition polymer.

Two Component Crosslinking

In the presence of amines or diols the chemistry changes again (Schemes 6 and 7).

Reaction With Diols - With pentan-2,4-diol, both 1' and 2' (R'=Me) are converted with high selectivity to open chain cyclic acetals 11. In excess methanol or water with one equivalent of added diol, acetals 11 form in 80% yield at room temperature. (Pentandiol is a mixture of d,l and meso isomers and gives rise to two major acetal products, but at different rates.)

Compound 9, which is not detectably rehydrated by water or attacked by simple alcohols, is also converted in high yield to cyclic acetals 11 at 70°C, pointing again to the significant thermodynamic driving force of cyclic acetal formation. This strongly favored acetal reaction is one explanation for the excellent adhesion performance of this system on cellulosics and glass.

Reaction With Amine Acid Salts. A reaction which appears to be less well-appreciated is the ready condensation of some aminoplasts with simple amines to selectively generate secondary and tertiary amine linked condensates (Scheme 7). Even in the presence of excess ammonia or primary amine, crosslinked products 13 (plus, conceivably, a trisubstituted product) which contain more highly substituted secondary (and tertiary) amines predominate over 12 (13). These particularly stable, high crosslink density products form with simple amine or diamine acid salts under unexpectedly mild, even room temperature conditions (6).

This mechanism could be demonstrated via nitrogen elemental analysis of polymers and copolymers treated with amine acid salts and thermally cured (Table II and Experimental). In a control experiment, ammonium acetate was added in excess to a vinyl acetate/ethylene emulsion copolymer without aminoplast crosslinker to confirm that essentially all of the ammonia volatilized from the unfunctionalized polymer during cure (much poorer volatilization was observed if NH_4Cl was used in place of NH_4OAc).

Analysis of otherwise nearly identical emulsion copolymers functionalized with N-methylolacrylamide or ABDA and cured with a nitrogen-free strong acid catalyst, p-toluenesulfonic acid, likewise gave the expected % N based on the calculated nitrogen content of the copolymerized crosslinkers. In marked contrast, the same copolymers were then cured with added volatile NH_4OAc and demonstrated a significantly larger % N. The amounts cannot be accounted for by the comonomer itself or by possible formation of a nonvolatile ammonium p-toluenesulfonate salt. The simplest interpretation is that the amine became incorporated into the polymer, both in the ABDA case and with N-methylolacrylamide. The

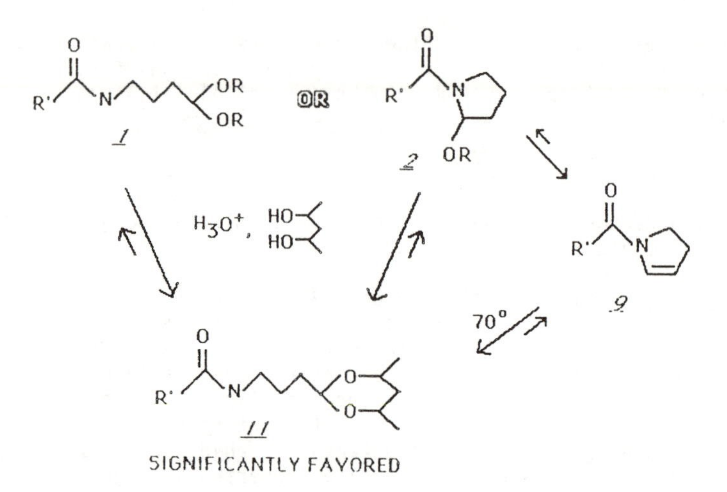

Scheme 6. Reactions with Diols

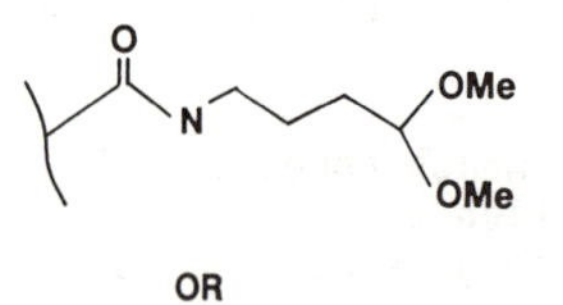

OR

OMe

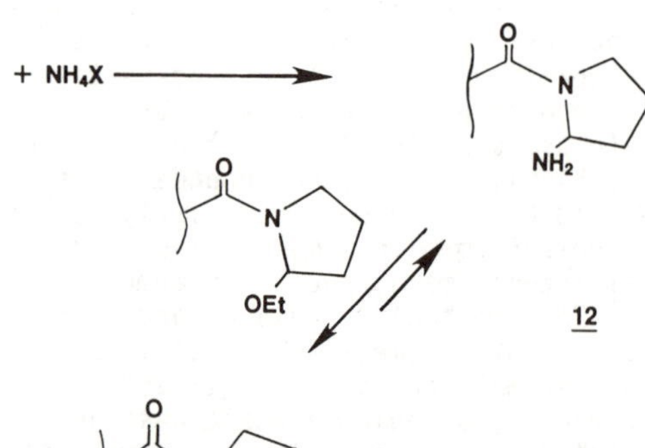

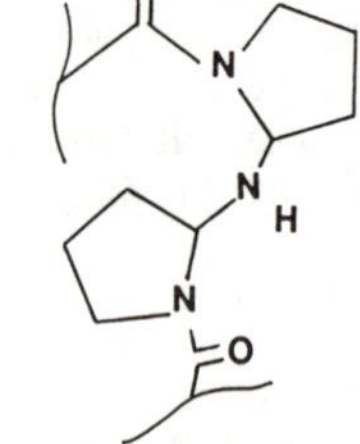

HIGH CROSSLINK DENSITY
LOW TEMPERATURE, "NEUTRAL" CURE

Scheme 7. Amine Acid Salt Mediated Crosslinking

increases in % N agree well with the amount calculated for a crosslinking mechanism involving two crosslinker units linked by a nitrogen (Scheme 7). Inefficient formation of more sterically congested structures containing three crosslinker units per nitrogen is also consistent with the data.

Table II. Amine Incorporation By Aminoplast Polymers

Emulsion Polymer	Additive/ Catalyst	%N Added	Cured Polymer % N Observed*	Theoretical	Swell Index
VAE	5% NH_4OAc	1.1	0.09	--	(Dissolved)
VAE/5% NMA	1% PTSA	--	0.73 ± .01	0.69	2.3
	5% NH_4OAc, 1% PTSA	1.1	1.07	1.04	O N 2 N
VAE/6% ABDA	1% PTSA	--	0.38 ± .01	0.39	4.5
	2% NH_4OAc, 1% PTSA	0.36	0.51	0.52 (n=3) 0.59 (n=2)	3.4 O N n N

* Duplicate analysis

More significantly, the ABDA copolymer films with added ammonia gave better swell values (tighter crosslinking) than identical samples without the ammonium acetate.

The companion paper (6) discusses the practical consequences of this reaction and other unique features of the chemistry of 1 and 2 on copolymer crosslinking (efficiency, rate and pot life) and substrate adhesion.

Literature Cited

1. Rosthauer, J. W.; Williams; J. L., Proc. Amer. Chem., Soc. Div. Polym. Mat. Sci. Engr., 1984, 50, 344.
2. Ley, D. A., ibid, 1984, 50, 353.
3. Fravel, J. G., Jr.; Cranley, P. E., Adhesives Age, 1984, 18, October
4. Temin, S. E.; J. Macromol. Sci., Macromol. Chem. Phys., 1982, C-22, 131.
5. Grave, C. H. G.; Bufkin, B. G., J. Coat. Technol., 1978, 50 (641), 41; (643), 67; (647), 73.
6. See companion paper, Pinschmidt, R. K., Jr.; Davidowich, G. E.; Burgoyne, W. F.; Dixon, D. D.; Goldstein, J. E.
7. Breslow, D. S.; Hulse, G. E.; Matlock, A. S., J. Amer. Chem. Soc., 1957, 79, 3760.
8. Frick, J. G., Jr.; Harper, R. J., Jr., Textile Res. J., 1983, 53, 758.
9. Falgiatore, D. R.; Emmons, W. D., U.S. Pat. 4 199 643 (1980).

10. Zabransky, J.; Houska, M., Plichta, Z.; Kalal J., Makromol. Chem., 1985, 186, 23.
11. Cue, B. W., Jr.; Chamberlain, N., Org. Prep. Proc. Int., 1979, 11, (6) 285.
12. Hubert, J. L.; Wijnberg, J. B. P.; Speckamp, W. N., Tetrahedron, 1975, 31, 1437.
13. Kosugi, K.; Hamagadi, H.; Nagasaka, T.; Ozawa, N.; Ohki, S., Heterocycles, 1980, 14, 9.

RECEIVED December 15, 1987

Chapter 32

Amide-Blocked Aldehyde-Functional Monomers

Cross-Linkable Substrate-Reactive Copolymers

R. K. Pinschmidt, Jr., G. E. Davidowich, W. F. Burgoyne, D. D. Dixon, and J. E. Goldstein

Air Products and Chemicals, P.O. Box 538, Allentown, PA 18105

Vinyl substituted cyclic hemiamidals 2 and their interconvertible acetal precursors (eg. acrylamido-butyraldehyde dimethyl acetal 1) were incorporated as latent crosslinkers and substrate reactive functional comonomers in solution and emulsion copolymers. Some use and applications data for copolymers prepared with these new monomers are presented. They show low energy cure potential, long shelf life and high catalyzed pot stability in solvent and aqueous media, good substrate reactivity and adhesion, and good product water and solvent resistance. They lack volatile or extractable aldehyde (eg. formaldehyde) components and show enhanced reactivity and hydrolytic stability with amines and diol functional substrates.

Post-crosslinkable and substrate reactive polymers are widely used to improve water and solvent resistance, strength, substrate adhesion and block resistance in binders, adhesives and coatings. The surprisingly rich chemistry of a new class of functional monomers (eg. 1 and 2) related to standard amide/aldehyde (aminoplast) condensates, but which eliminate aldehyde emissions, was elucidated by monomeric model and mechanistic studies and discussed in the preceeding paper (1). Results with these monomers in copolymer systems are reported here.

Experimental

Acrylamidobutyraldehyde dialkyl acetal (1, ABDA, R=Me, Et) and N-acryloyl-2-ethoxypyrrolidine (2, AEP) were synthesized as described in the previous paper (1) and used directly.

Acrylamidoacetaldehyde Dimethyl Acetal (AADMA, 14) was prepared via a route different than the published procedure (2). To 125 g (1.19 mol) of aminoacetaldehyde dimethyl acetal in a rapidly

0097-6156/88/0367-0467$06.00/0

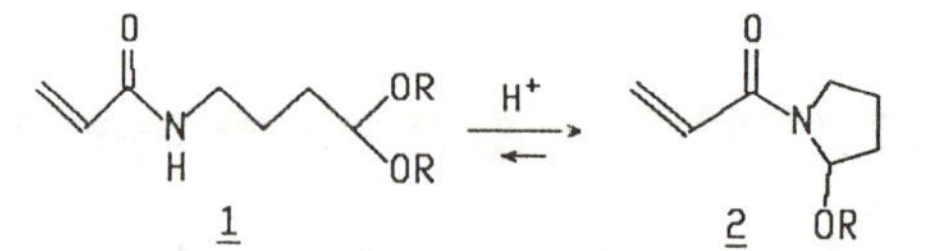

stirred two phase mixture of 390 mL of methyl t-butyl ether (MTBE) and 125 mL of 14 N aqueous NaOH at 18°C was added 107.5 g (1.19 mol) of acryloyl chloride over 1 h (NaCl precipitate). The reaction temperature was maintained below 30°C. After 15 min the pH was adjusted to 7.1-7.8 (dil. NaOH) and the layers were separated. The product was extracted into H_2O (hexane added to the MTBE to drive the equilibrium). GC analysis showed 181 g of AADMA in 700 mL of aqueous solution, plus 10 g AADMA in the brine layer (100% yield). The product could be isolated from the MTBE layer by concentration after brine extraction and distilled with severe losses to give an oil which resinified after several days at 0°C: IR (film): 3290 (br), 1660, 1624, 1610 (sh), 1545 cm^{-1}; ^{1}H NMR ($CDCl_3$): δ 6.3 (dd, 1, J=16.5, J=1.9 Hz, vinyl), 6.2 (dd, 1, J=16.5, J=6.9 Hz, vinyl + buried NH), 5.65 (dd, 1, J=9.6, J=1.9 Hz, vinyl), 4.43 (br t, 1, CH) 3.48 (apparent t, 2, NCH_2), 3.4 ppm (s, 6, CH_3).

Synthesis of 4-(Aminoethyl)butyraldehyde Diethyl Acetal. Acetamidobutyraldehyde diethyl acetal (25 g, 0.123 mol, 1'(R'=Me, R=Et)) was slowly added to 4.7 g (0.123 mol) of lithium tetrahydridoaluminum in 150 mL of dry THF at reflux. After 1 h the excess $LiAlH_4$ was destroyed by adding EtOAc and the reaction was treated with 1-2 mL of saturated aqueous Na_2SO_4. The green gelatinous mixture was vacuum filtered and the organic phase was concentrated on a rotary evaporator. The inorganic phase was extracted with EtOAc and again with THF and the combined concentrated organic phases were distilled on a kugelrohr apparatus (105-130°C at 0.2 torr) to yield 9.26 g of yellow liquid. Capillary GC analysis revealed the mixture to be 75% 4-(aminoethyl)butyraldehyde diethyl acetal (30% yield): IR: (film) 1655 (w), 1450, 1126, 1063 cm^{-1}; ^{1}H NMR ($CDCl_3$): 5.51 (tt, ~1, J=5 Hz, CH), 3.4-3.75 (m, 4, OCH_2), 2.7-2.3 (m, ~4, CH_2NCH_2), 1.61 (m, ~4, CH_2CH_2), 1.21 (t, J=3.5 Hz), 1.11 (t, J=3.5 Hz), 1.02 ppm (t, J=3.5 Hz) last three in 62:17:20 ratio, rotamer mix.

Synthesis of N-Ethylacrylamidobutyraldehyde Diethyl Acetal (Et-ABDA, 13) 4-(Aminoethyl)butyraldehyde diethyl acetal (8.5 g, 0.045 mol) in 56 mL of CH_2Cl_2 was rapidly stirred with 8 mL of 14 N NaOH at 20°C. Acryloyl chloride (3 g, 0.045 mol) was added slowly to maintain the temperature below 30°C. GC analysis at 30 min showed no remaining starting material and a new product at 8.65 min. The mixture was neutralized with 30% H_2SO_4 and buffered with solid CO_2. The organic phase was concentrated and distilled (kugelrohr) to give 6.5 g of yellow liquid (100-125°C, 0.15 torr). GC analysis showed 82% of the 8.65 min peak (53% yield) plus 4 minor components: ^{1}H NMR ($CDCl_3$): δ 6.56 (dd, <1, J=10.3 Hz, J=16.2 Hz, vinyl), 6.32 (dd, <1, J=16.2 Hz, J=2.0 Hz, vinyl), 5.65 (dd, <1, J=10.3 Hz, J=2.0 Hz, vinyl), 4.47 (br t, 1, methine), 3.8-3.2 (m, ~8, NCH_2, OCH_2), 1.65 (m, 4, $(CH_2)_2$), 1.2 ppm (t, 9, J-6.7 Hz, CH_3).

Solketal Acrylate, 17 has been reported previously (3) and was synthesized in this case via standard (4) ester exchange (aluminum isopropoxide catalyzed) from methyl acrylate and the commercially available acetone ketal of glycerine (solketal, Aldrich Chemical Co.).

Pentaerythritol Diacetonide (PEDA). Pentaerythritol, 27 g (0.2 mol), and acetone, 75 g, were heated overnight with 1 g of Rohm and Haas XN-1010 strong acid ion exchange resin in a Soxhlet extraction apparatus containing 3A zeolite. The mixture was diluted with additional acetone and unreacted pentaerythritol was filtered off. The solution was concentrated under reduced pressure and the product was partitioned between water and ether. The organic layer was dried with $MgSO_4$ and concentrated. The solid product was recrystallized from petroleum ether to yield 10 g (23 % yield) of slightly yellowish platelets, mp 115-116°C (lit. (5): 117-117.5°C).

Butyl Acrylate (BA) Solution Copolymers. Polymerizations were run with 0.30 mmol of comonomer per calculated g of polymer solids premixed with BA: Reactor Charge: X g comonomer, (50-X) g of butyl acrylate, 120 g of dry toluene, 0.15 g of 2,2'-azobisisobutyronitrile (AIBN).

The first three components were mixed at room temperature and heated at 65°C in a 250 mL round bottom flask equipped with a magnetic stirring bar, reflux condenser and N_2 blanket. The AIBN was then added to start the reaction. Monitoring of the unreacted free monomers (butyl acrylate and comonomer) was done by GLPC. Additional 10 mg amounts of AIBN were added as needed. The temperature was also increased to 75°C to finish the polymerization. Most of the reactions took longer than 24 h to reduce the free monomer below 0.7%. A control BA homopolymer and copolymers with butoxymethylacrylamide (BNMA), 2 (R = Et; AEP), acrylamidoacetaldehyde dimethyl acetal (AADMA), and N-ethyl-1 (Et-ABDA, 13) were prepared in this way.

High Functionality Butyl Acrylate/Methyl Methacrylate (BA/MMA) Solution Copolymers, approximately 1400 MW between crosslinks (Mn ~ 20-30,000), were prepared at 50% solids in butanol/toluene using a chain transfer agent and delayed monomer feeds. A 3 L jacketed polymerization flask equipped with a reflux condenser, double blade stirrer, nitrogen blanket, thermometer and delay feed was charged with 20 g of n-butanol and 30 g of toluene. A delay feed was prepared by mixing 128 g of 1 (R = Me), 1.25 g of acrylic acid, 225 g of BA, 137 g of MMA, 180 g of butanol, 270 g of toluene, 5 g of dodecylmercaptan and 11.4 g of a 75 % actives solution of t-butylperoxypivalate. The reactor was purged with nitrogen for 20 min and heated to 60°C. The monomer delay was then added over 3 h. The reaction was stirred at temperature for an additional 7 h and at room temperature for 24 h. The product analyzed as 53% solids, $\bar{M}n$ = 24,200, $\bar{M}w$ = 212,000 (GPC, vs polystyrene). A control BNMA terpolymer (107 g of BNMA, 235 g of BA and 143 g of MMA) was prepared similarly and showed Mn = 39,400, $\bar{M}w$ = 777,000.

Emulsion Polymerizations, eg. vinyl acetate [VAc]/ABDA, VAc/ethylene [VAE]/ABDA, butyl acrylate [BA]/ABDA, were done under nitrogen using mixed anionic/nonionic or nonionic surfactant systems with a redox initiator, eg. t-butyl hydroperoxide plus sodium formaldehyde sulfoxylate. Base monomer addition was batch or batch plus delay; comonomer additions were delay.

Polymer Evaluation.

Swell Index and Percent Solubles Measurements. Polymer films were cast on polyester film at 25% solids with various post additives. The films were air dried (16-48 h) then cured for 3 and 10 min at 150°C in a convected oven. Small samples (50-100 mg) of film were weighed, soaked in dimethylformamide (DMF) for 1 h, briefly pat-dried and reweighed in an Al weighing pan. The samples were then redried at 150°C (20 torr) or 170°C, 1 atm for 30 min.

$$\text{Swell index} = \frac{\text{sample weight swollen in DMF}}{\text{original sample weight}}$$

$$\%\ \text{Solubles} = 100\ [1-(\frac{\text{weight of dry film after DMF swell}}{\text{original dry film weight}})]$$

Percent solubles were between 5 and 15% for crosslinked polymers.

Tensile Tests. Polymers were padded on filter paper at 10% add-on, cured at 150°C for 3 or 10 min, and tested as in Tappi Useful Method #UM656. Tensile measurements on padded samples were done with 1" strips on an Instron Tester after conditioning in a constant temperature and humidity room. Wet, methyl ethyl ketone (MEK) and perchloroethylene (PCE) tensiles were run after brushing the sample with 1% aqueous Aerosol OT or solvent.

Catalysts. Catalyst addition was weight percent on polymer solids. Solution polymers employed, for example, 2-hydroxycyclohexane-p-toluene sulfonate or Nacure 155 (King Industries, dialkylnaphthalene disulfonic acid) as organic soluble acids. A variety of catalysts were tested with emulsion polymers, the best choice varied with base polymer hydrophobicity and functional monomer distribution. Some of these included p-toluenesulfonic acid (PTSA), ammonium chloride, $AlCl_3 \cdot 6H_2O$, phosphoric acid, and hexamethylenediamine diacid salts (HMDA•2HOAc, HMDA•2HCl).

Catalyzed Pot Stability. A sample of the above BA/MMA/25% ABDA copolymer in butanol/toluene was catalyzed with 1% Nacure (S/S). The viscosity was observed qualitatively and an initial aliquot was drawn down as a 2 mil film on glass and cured (150°C/3 min or room temperature for 35-45 d). After 24 h at ambient temperature one equivalent of methanol based on ABDA was detected by GLPC in the solution. After 5 d at ambient and 24 h at 40°C, two equivalents of MeOH were present, indicating complete alcohol exchange. The solution polymer showed no thickening. Films were also drawn down with these samples and evaluated (Table VI).

Results and Discussion

Scheme 1 summarizes the reactions of copolymers containing compounds 1 and 2 as suggested by model studies discussed in the companion paper (1).

Table I shows a comparison of free film swell index results as a function of crosslinker at a constant comonomer level (0.3 moles/kg of polymer). ABDA and AEP gave crosslinking performance identical to a conventional crosslinker, BNMA. However, derivatives which cannot cyclize, either because the amide has an additional substituent, as in 13, or the chain connecting the amide to the blocked aldehyde is too short, as in 14, did not exhibit efficient crosslinking. They also showed significant discoloration, presumably due to increased aldol condensation relative to 1, 2, or BNMA.

The poor performance with 13 and 14 could be substantially improved by adding a blocked tetraol, pentaerythritol diacetonide (pentaerythritol itself is insolubile in toluene), as a co-curing agent. This option is discussed in more detail below.

Table I. Self-Crosslinkable Solution Polymer Performance*

Co-Monomer		Swell Index (60 min. in DMF)	
		3 min. Cure	10 min. Cure
None		Dissolved	Dissolved
BNMA	 NHCH$_2$OBu	2.5	2.1
ABDA 1	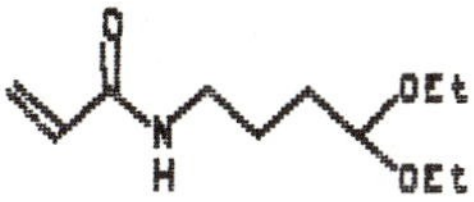 OEt, OEt, N H	2.7	1.8
AEP 2	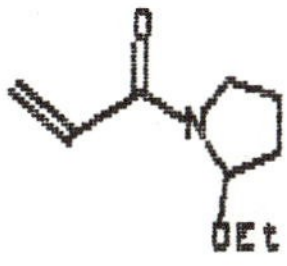 N, OEt	2.5	2.3
AADMA 14	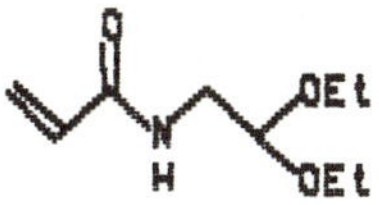 OEt, OEt, N H	Disintegr.	4.0
AADMA + "Tetraol" 15		3.0	2.7
Et-ABDA 13	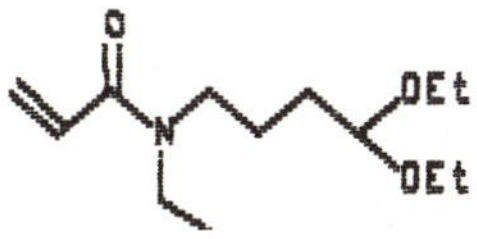 OEt, OEt, N	Dissolved	4.2
Et-ABDA + "Tetraol" 15		3.5	3.5

*Butyl acrylate/0.3 moles/kg crosslinker copolymers in toluene. See Experimental for a definition of swell index.

Amine Salt Catalyzed Cure. In general, catalyst selection with amide/blocked aldehyde copolymers was much more critical than with conventional aminoplasts. Usually stronger acids or, preferably, an amine acid salt rather than H_3PO_4 gave the best performance.

Experiments supporting the actual participation of amine acid salts in the crosslinking of cyclic hemiamidals were described in the companion paper (1). Amine acid salt catalysts are milder than mineral acids, causing less degradation of the substrate; the resulting secondary amine appears to be a less reversible crosslink than the ether formed with mineral acids; and amine crosslink centers have the potential to involve up to three functional monomer units in the crosslink, leading to a tighter network (Scheme 2).

Diamine salts, even those with extremely low acidity such as hexamethylenediamine·2HOAc, formally offer higher crosslink densities and give excellent swell indices (implying high crosslink densities) under very mild, even ambient cure conditions (Table II). Unfortunately, on cellulose the aliphatic diamines gave poor performance, possibly indicating substrate degradation by the aliphatic diamine (Table III).

Table II. Cure Catalyst Effects

Polymer	Catalyst	Cure (°C/min)	Swell Index
VAE/5% ABDA	H_3PO_4	150°/3	11.3
Emulsion	1% PTSA	150°/3	8.9
Polymer	1% NH_4Cl	150°/3	6.2
	1% PTSA/2% NH_4OAc	150°/3	4.5
	1.6% HMDA•2HOAc	110°/1	5.6
	1.6% HMDA•2HOAc	110°/1 + 27 d @ RT	3.7

Table III. LTC† and Amine Salt Cured Systems

Copolymer	Catalyst	Wet Tensile (pli)† std	LTC
VAE/6% ABDA	NH_4Cl	7.4	3.9
	H_3PO_4	5.9	5.6
	HMDA•2HOAc	4.4	3.3
E-1129*	Na sesqui-carbonate	6.8	6.5

* E-1179, a Rohm and Haas LTC acrylate polymer.

† std: 150°C/3 min; Low Temperature Cure: 90 sec at 150°C, 30 d at ambient

Surprisingly, for low temperature cure on cellulose, H_3PO_4, which performed poorly in normal cure modes gave the best result. It may be postulated that the relative importance and formation rates of 6 vs 10 and 12 is changing as the cure temperature changes.

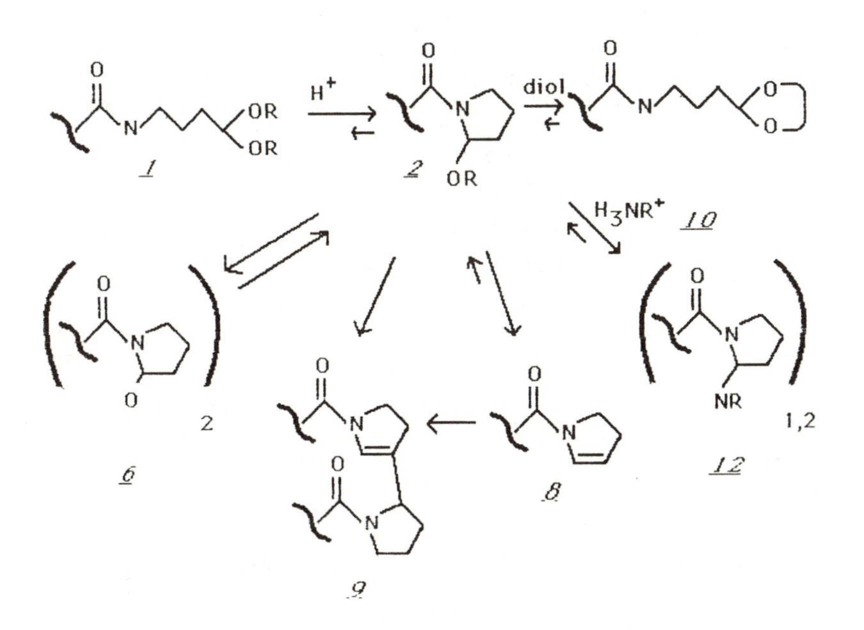

Scheme 1. Amide/Blocked Aldehyde Reaction Summary

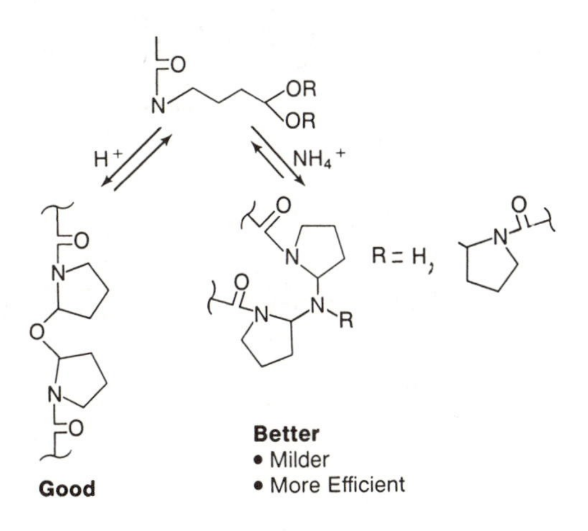

Scheme 2. Amine Salt Crosslinking

Performance vs Crosslinker Level. Table IV presents tensile and swell index results on a series of approximately 0°C T_g VAE emulsion copolymers. ABDA containing copolymers had fair swell index values which deteriorated rapidly as ABDA levels declined, but showed slow declines in wet tensile performance with reduced levels. Possible reasons for this are discussed below.

Table IV. VAE Nonwoven Binder Performance

Crosslinker	Crosslinker Moles per kg Polymer	Tensiles (pli) Wet	Dry	PCE	Swell Index
6.0% ABDA	0.28	7.5	18.5	9.7	3.4-4.5
3.0% ABDA	0.14	6.5	16.8		6.7-11
2.0% ABDA	0.1	6.0	15.3	6.0	high solubles
1.0% ABDA	0.05	5.1	8.2		dissolved

Diol Two Component Cure. Amide/blocked aldehyde systems show high reactivity with 1,2 or 1,3 diols and give thermodynamically favored cyclic acetals. As mentioned above, even monomers which self-crosslink poorly because they cannot cyclize to the hemiamidal (i.e., 13, 14) will react efficiently to form crosslinked bis-cycloacetals 16 in the presence of a tetraol (or, to impart organic solubility, a blocked tetraol, 15, Scheme 3).

Incorporation into a co- or terpolymer of a diol containing monomer such as solketal acrylate (SA, 17, a comonomer readily hydrolyzed by aqueous acid to a diol) likewise increases crosslinking (Scheme 4). Such polymers exhibit improved dry, wet and solvent resistance (Table V). Alternatively, half of the ABDA comonomer can be substituted with SA, without significantly degrading properties vs the initial ABDA only sample.

This reactivity with diols provides a unique aspect to the linked amide/blocked aldehydes and offers a basis for explaining their good performance at low molar levels of incorporation on cellulose (Table IV).

Table V. VAc Copolymer Films

Comonomer	Tensile (pli) Wet	Dry	MEK+	PCE*	Swell Index
3% ABDA	5.5	17.1	3.6	8.4	6.0
1.5% ABDA/1.3% SA	5.1	17.9	3.7	5.5	6.6
3% ABDA/3% SA	5.1	16.3	5.5	8.2	3.4

+ Methyl ethyl ketone
* Perchloroethylene

Adhesion to Polyester and Glass. Figure 1 shows peel adhesion of three closely matched VAE emulsion polymers: a control, one with a low level of acrylic acid and the third a low level of ABDA. These polymers are not represented as exceptional adhesives for polyester, but do show an adhesion promoting effect of the ABDA monomer over acrylic acid. Excellent resistance to delamination in boiling water was also observed for both ABDA and AEP containing polymers on glass (Table VI). Similar effects were, however, not observed on aluminum, steel or polyolefins.

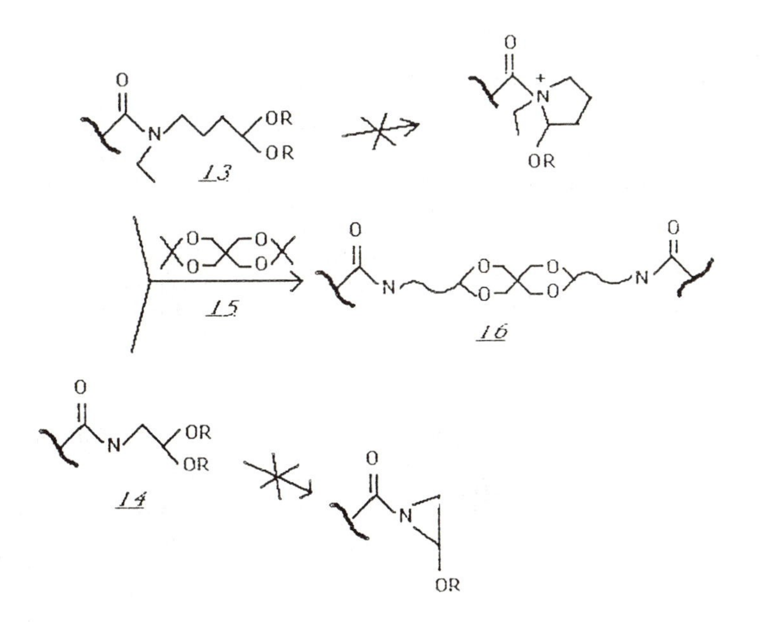

Scheme 3. Tetraol (Blocked Tetraol) Crosslinking

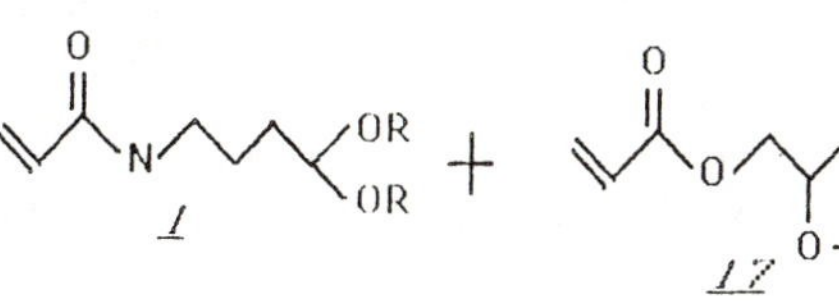

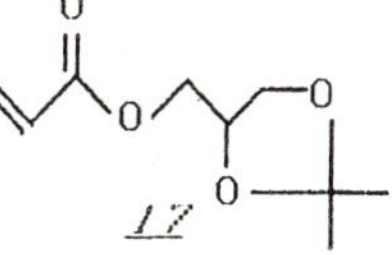

solketal acrylate

I°

H+

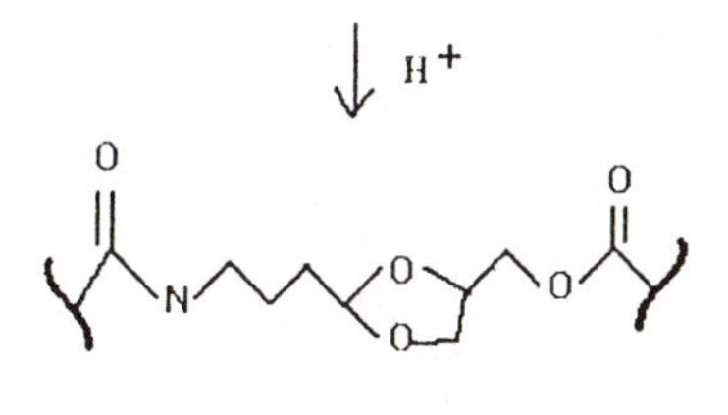

18

Scheme 4. Diol/Aldehyde Terpolymers

Catalyzed Pot Life. Under appropriate conditions ABDA containing copolymers showed good catalyzed pot stability. As detailed in Experimental, a BA/MMA/25% ABDA copolymer in butanol/toluene was actively catalyzed by 1% Nacure 155 aralkyldisulfonic acid. The polymer released one equivalent of methanol on ABDA to the solution [copoly-1 (R=Me) → copoly-2 (R=Me) + MeOH] after 24 h at room temperature, and a second equivalent [copoly-2 (R=Me) + BuOH ⇄ copoly-2 (R=Bu) + MeOH] at 40°C or longer time. Nonetheless, as demonstrated on Table VI, the catalyzed solution polymer did not gain significantly in viscosity due to crosslinking. Film drawdowns on glass showed equivalent performance (efficient crosslinking) across the test period. Even long term cure at ambient temperature gave good crosslink development.

Table VI. Catalyzed Pot Life vs. Cured Film Properties†

Catalyst	Catalyzed Solution Viscosity	Cure†	Pencil Hardness	Acetone Double Rub	Boiling H_2O Blush
2% Nacure 155	fluid	150	H-3H	60	None
		RT	H	45	
1% Nacure/ 24 h @ RT	same	150	3H	55	None
		RT	HB	30	
1% Nacure/5 d @ RT/24 h @ 40°C	same	150	3H	52	None
		RT	2H	35	

† Cure 150°C/3 min. or room temperature for 35-45 d, 2 mil films of BA/MMA/25% ABDA solution copolymer on glass.

Alternate Crosslinking Modes. In addition to the crosslinking modes previously described, (co)polymers containing 1 and 2 may be cured by other means. For example, under appropriate acidic conditions with limited availability of active hydrogen species cyclic hemiamidals 2 will lose ROH to form the enamide 9 (Scheme 5). This has been demonstrated on model systems, e.g., 2 where vinyl is replaced by methyl (1). The product, N-acetylpyrroline, has in turn been converted to nonvolatile products (oligomers) under free radical catalysis. These systems may thus be considered for application in the UV/EB or catalyzed free radical cure field.

Alternatively, 1 or 2 could be incorporated into a polymer not by radical copolymerization but rather by reaction with, e.g. diol functionality. The pendant acrylamide portion of the structure would then provide a site for light or initiator catalyzed free radical cure to give a hydrolysis resistant crosslink. These are areas of ongoing research and the results will be reported in detail at a later date.

Conclusion

Copolymers of amide/blocked aldehyde-cyclic hemiamidal monomers show a rich and broad chemistry and offer a number of desirable features as specialty copolymers. They provide high shelf and

Figure 1. ABDA Adhesion to Polyester

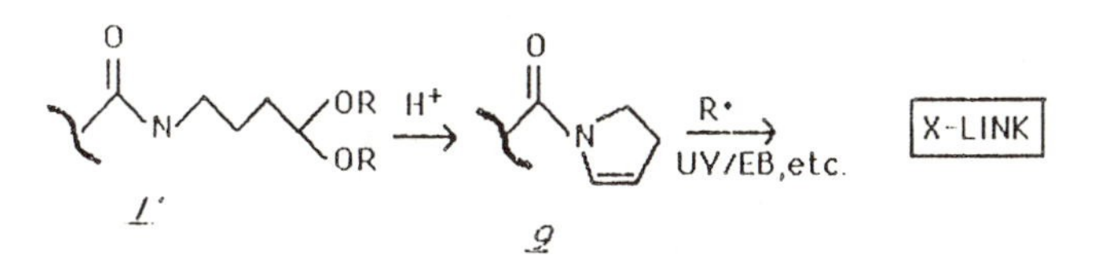

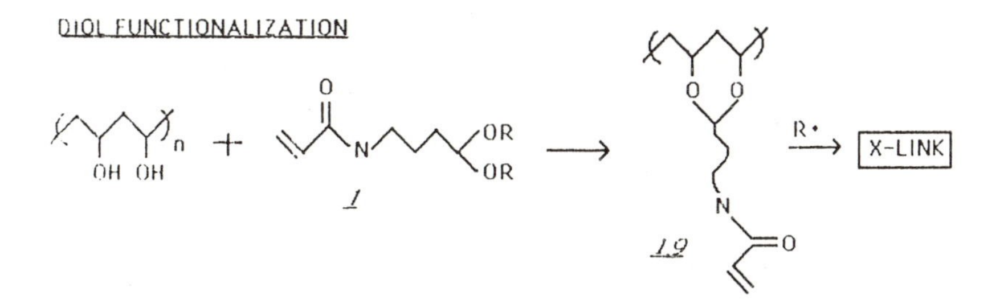

Scheme 5. Free Radical Crosslinking

catalyzed pot stability in water or solvent media, efficient self and two component crosslinking and enhanced adhesion to a number of substrates.

Literature Cited

1. Pinschmidt, R. K., Jr.; Burgoyne, W. F.; Dixon, D. D.; Goldstein, J. E.; preceeding paper.
2. Epton, R.; McLaren, J. V.; Thomas, T. H., Polymer, 1974, 15, 564.
3. D'Alelio, G. F.; Caiola, R. J. J. Polym. Sci., 1967, A-1, 5, 287.
4. Rehberg, C. E. In Organic Synthesis; Horning, E. C., Ed.; John Wiley and Sons: New York, 1955; Collective Vol. 3, pp 146-148.
5. Orthner, L.; Chem. Ber., 1928, 61B, 116; C.A., 1928, 22, 1327.

RECEIVED October 14, 1987

Author Index

Affiliation Index

Subject Index

E

F

P

T

U

V

W

Y

Production by Meg Marshall
Indexing by Deborah H. Steiner
Jacket design by Carla L. Clemens

Elements typeset by Hot Type Ltd., Washington, DC
Printed and bound by Maple Press, York, PA